TOPOLOGICAL RINGS

NORTH-HOLLAND MATHEMATICS STUDIES 178
(Continuation of the Notas de Matemática)

Editor: Leopoldo NACHBIN †

Centro Brasileiro de Pesquisas Físicas
Rio de Janeiro, Brazil
and
University of Rochester
New York, U.S.A.

NORTH-HOLLAND – AMSTERDAM • LONDON • NEW YORK • TOKYO

TOPOLOGICAL RINGS

Seth WARNER

Department of Mathematics
Duke University
Durham, NC, U.S.A.

1993

NORTH-HOLLAND – AMSTERDAM • LONDON • NEW YORK • TOKYO

ELSEVIER SCIENCE PUBLISHERS B.V.
Sara Burgerhartstraat 25
P.O. Box 211, 1000 AE Amsterdam, The Netherlands

ISBN: 0 444 89446 2

© 1993 ELSEVIER SCIENCE PUBLISHERS B.V. All rights reserved.

No part of this publication may be reproduced, stored in a retrieval system or transmitted in any form or by any means, electronic, mechanical, photocopying, recording or otherwise, without the prior written permission of the publisher, Elsevier Science Publishers B.V., Copyright & Permissions Department, P.O. Box 521, 1000 AM Amsterdam, The Netherlands.

Special regulations for readers in the U.S.A. – This publication has been registered with the Copyright Clearance Center Inc. (CCC), Salem, Massachusetts. Information can be obtained from the CCC about conditions under which photocopies of parts of this publication may be made in the U.S.A. All other copyright questions, including photocopying outside of the U.S.A., should be referred to the publisher.

No responsibility is assumed by the publisher for any injury and/or damage to persons or property as a matter of products liability, negligence or otherwise, or from any use or operation of any methods, products, instructions or ideas contained in the material herein.

Transferred to digital printing 2005

To Susan, Sarah, Michael, and Lawrence

PREFACE

This text brings to the frontiers of much current research in topological rings a reader having an acquaintance with some very basic point-set topology and algebra, which is normally presented in semester courses at the beginning graduate level or even at the advanced undergraduate level. Many results not in the text and many illustrations by example of theorems in the text are included among the exercises, sufficient hints for the solution of which with references to the pertinent literature are offered so that solving them does not become a major research effort for the reader. Within mentioned constraints, a bibliography intended to be complete is given. Expectations of a reader include some familiarity with Hausdorff, metric, compact and locally compact spaces and basic properties of continuous functions, also with groups, rings, fields, vector spaces and modules, and with Zorn's Lemma.

In view of the readers for whom the book is written, the exposition is more detailed than would be necessary for readers who are mature mathematicians. In addition, quite a bit of algebra, both commutative and noncommutative, is included, since many of those readers will need additional background in algebra to understand parts of the text. Obviously, there is considerable overlap with my earlier text, *Topological Fields*, in this series (North-Holland Mathematics Studies 157, Notas de Matemática (126)), since both require a common core of knowledge, but in some instances the presentation here of such material (*e.g.*, the completion of a commutative Hausdorff group) is quite different from that in *Topological Fields*. I deeply regret the omission of all applications of categorical concepts to topological rings. To have included the requisite background for those for whom the book is written would have greatly lengthened an already long book and overbalanced any introduction to the use of categorical concepts in the theory of topological rings that could reasonably be presented.

This seems a natural place to record significant errors thus far discovered in *Topological Fields*, and an Errata correcting such errors is included.

The book is typeset by $\mathcal{A}_{\mathcal{M}}\mathcal{S}$-TEX, with the exception of the indices, which are typeset by Latex. I am deeply grateful to Dr. Yun-Liang Yu, systems programmer of the Duke Mathematics Department, who has patiently guided me through the intricacies of $\mathcal{A}\mathcal{M}\mathcal{S}$-TeX and Duke's computer system. When I began the task of typesetting this volume, I remarked to Dr. Yu, a recent arrival from China, that I felt like "an immigrant who has just gotten off the boat and doesn't know a word of English." Thanks to him, I now have a rudimentary grasp of the language.

Seth Warner
Mathematics Department, Duke University
Durham, North Carolina
15 March 1993

CONTENTS

Preface   vii

Chapter I. Topological Rings and Modules   1
  1 Examples of Topological Rings   1
  2 Topological Modules, Vector Spaces, and Algebras   11
  3 Neighborhoods of Zero   18
  4 Subrings, Ideals, Submodules, and Subgroups   23
  5 Quotients and Projective Limits of Rings and Modules   32

Chapter II. Metrizability and Completeness   44
  6 Metrizable Groups   44
  7 Completions of Commutative Hausdorff Groups   55
  8 Completions of Topological Rings and Modules   64
  9 Baire Spaces   68
  10 Summability   71
  11 Continuity of Inversion and Adversion   80

Chapter III. Local Boundedness   88
  12 Locally Bounded Modules and Rings   88
  13 Locally Retrobounded Division Rings   94
  14 Norms and Absolute Values   102
  15 Finite-dimensional Vector Spaces   112
  16 Topological Division Rings   120

Chapter IV. Real Valuations   134
  17 Real Valuations and Valuation Rings   134
  18 Discrete Valuations   141
  19 Extensions of Real Valuations   152

Chapter V. Complete Local Rings   166
  20 Noetherian Modules and Rings   166
  21 Cohen Subrings of Complete Local Rings   173
  22 Complete Discretely Valued Fields   186
  23 Complete Local Noetherian Rings   192
  24 Complete Semilocal Noetherian Rings   197

Chapter VI. Primitive and Semisimple Rings — 206
25 Primitive Rings — 206
26 The Radical of a Ring — 216
27 Artinian Modules and Rings — 226

Chapter VII. Linear Compactness and Semisimplicity — 232
28 Linearly Compact Rings and Modules — 232
29 Linearly Compact Semisimple Rings — 241
30 Strongly Linearly Compact Modules — 251
31 Locally Linearly Compact Semisimple Rings — 256
32 Locally Compact Semisimple Rings — 267

Chapter VIII. Linear Compactness in Rings with Radical — 283
33 Linear Compactness in Rings with Radical — 283
34 Lifting Idempotents — 295
35 Locally Compact Rings — 317
36 The Radical Topology — 330

Chapter IX. Complete Local Noetherian Rings — 359
37 The Principal Ideal Theorem — 359
38 Krull Dimension and Regular Local Rings — 368
39 Complete Regular Local Rings — 377
40 The Japanese Property — 391

Chapter X. Locally Centrally Linearly Compact Rings — 398
41 Complete Discretely Valued Fields and Division Rings — 398
42 Finite-dimensional Algebras — 406
43 Locally Centrally Linearly Compact Rings — 414

Chapter XI. Historical Notes — 424
44 Topologies on Commutative Rings — 424
45 Locally and Linearly Compact Rings — 428
46 Category, Duality, and Existence Theorems — 435

Bibliography — 440

Errata — 487

Index of Names — 489

Index of Symbols and Definitions — 492

CHAPTER I

TOPOLOGICAL RINGS AND MODULES

In this introductory chapter we shall define and give examples of topological rings, modules, and groups, show how they may be introduced by specifying the neighborhoods of zero, and present a few basic constructions.

1 Examples of Topological Rings

By a *ring* is meant an associative ring, not necessarily one having a multiplicative identity. A *ring with identity* is a ring possessing a multiplicative identity 1 such that $1 \neq 0$. Thus a *zero ring*, one having only one element, is not a ring with identity. A ring A is *trivial* if $xy = 0$ for all $x, y \in A$. Any commutative group is thus the additive group of a trivial ring. A zero ring is a particularly trivial ring.

We shall denote by $\mathbb{N}$, $\mathbb{Z}$, $\mathbb{Q}$, $\mathbb{R}$, $\mathbb{C}$, and $\mathbb{H}$ the set of natural numbers (including zero), integers, rationals, real numbers, complex numbers, and quaterions respectively. The set of real numbers greater than zero is denoted by $\mathbb{R}_{>0}$, and those greater than or equal to zero by $\mathbb{R}_{\geq 0}$.

If A is a ring, A^* denotes the set of its nonzero elements, and if A is a ring with identity, $A^\times$ denotes the multiplicative group of its invertible elements.

If X and Y are sets, $X \setminus Y$ denotes the relative complement of Y in X, that is, $X \setminus Y = \{x \in X : x \notin Y\}$, and Y^X denotes the set of all functions from X to Y. The cardinality of a set X is denoted by card(X).

A topological ring is simply a ring furnished with a topology for which its algebraic operations are continuous:

1.1 Definition. *A topology $\mathcal{T}$ on a ring A is a **ring topology** and A, furnished with $\mathcal{T}$, is a **topological ring** if the following conditions hold:*

(TR 1) $(x, y) \to x + y$ is continuous from $A \times A$ to A

(TR 2) $x \to -x$ is continuous from A to A

(TR 3) $(x, y) \to xy$ is continuous from $A \times A$ to A

where A is given topology $\mathcal{T}$ and $A \times A$ the cartesian product topology determined by $\mathcal{T}$.

A ring topology on a ring A clearly induces a ring topology on any subring of A, and unless the contrary is indicated, we shall assume that a subring of a topological ring is furnished with its induced topology.

Norms furnish examples of topological rings:

1.2 Definition. *A function N from a ring A to $\mathbb{R}_{\geq 0}$ is a **norm** if the following conditions hold for all $x, y \in A$:*

(N 1) $\qquad\qquad N(0) = 0$
(N 2) $\qquad\qquad N(x+y) \leq N(x) + N(y)$
(N 3) $\qquad\qquad N(-x) = N(x)$
(N 4) $\qquad\qquad N(xy) \leq N(x)N(y)$
(N 5) $\qquad\qquad N(x) = 0$ *only if* $x = 0$.

If N is a norm on a ring A, then d, defined by $d(x, y) = N(x - y)$ for all $x, y \in A$, is a metric. Indeed, (N 1) and (N 5) imply that $d(x, y) = 0$ if and only if $x = y$, (N 3) implies that $d(x, y) = d(y, x)$, and (N 2) yields the triangle inequality, since

$$d(x, z) = N(x - z) = N((x - y) + (y - z))$$
$$\leq N(x - y) + N(y - z) = d(x, y) + d(y, z).$$

If d is a complete metric, we say that N is a *complete norm*.

Often symbols similar to $\|..\|$ are used to denote norms.

1.3 Theorem. *Let N be a norm on a ring A. The topology given by the metric d defined by N is a ring topology.*

Proof. Let $a, b \in A$. For all $x, y \in A$,

$$d(x + y, a + b) = N((x + y) - (a + b)) = N((x - a) + (y - b))$$
$$\leq N(x - a) + N(y - b) = d(x, a) + d(y, b).$$

Hence (TR 1) holds. For all $x \in A$, $d(-x, -a) = N(-x + a) = N(x - a) = d(x, a)$ by (N 3). Hence (TR 2) holds. Finally, for all $x, y \in A$,

$$d(xy, ab) = N((x - a)(y - b) + a(y - b) + (x - a)b)$$
$$\leq N(x - a)N(y - b) + N(a)N(y - b) + N(x - a)N(b).$$

Hence (TR 3) holds. ●

1 EXAMPLES OF TOPOLOGICAL RINGS

1.4 Theorem. *Let N be a norm on a ring A. For all $x, y \in A$,*

$$|N(x) - N(y)| \leq N(x - y),$$

and hence N is a uniformly continuous function from A (for the metric defined by N) to $\mathbb{R}_{\geq 0}$.

Proof. $N(x) = N((x-y) + y) \leq N(x-y) + N(y)$, so $N(x) - N(y) \leq N(x-y)$. Hence also $N(y) - N(x) \leq N(y-x) = N(x-y)$. Therefore $|N(x) - N(y)| \leq N(x-y)$. •

In view of 1.3, we shall say that a topological ring is *normable* if its topology is defined by a norm, and in §14 we shall give criteria for a topological ring to be normable. A *normed ring* is simply a ring furnished with a norm and hence with the topology defined by that norm.

Norms on rings play a substantial role in analysis:

Example 1. Let X be a set, $\mathcal{B}(X)$ the ring of all bounded real-valued (or complex-valued) functions on X (a function f is bounded if $N(f) < +\infty$, where $N(f) = \sup\{|f(x)| : x \in X\}$). The function N just defined is a complete norm on $\mathcal{B}(X)$, so $\mathcal{B}(X)$ and each of its subrings is a topological ring for the topology defined by N. Special cases: (a) The ring of all bounded continuous functions on a topological space X. (b) The ring of all continuous functions f on a locally compact space X which "vanish at infinity," that is, such that for every $\epsilon > 0$ there is a compact subset K (depending on f) of X such that $|f(x)| \leq \epsilon$ for all $x \in X \setminus K$. (c) The ring of all continuous functions on a compact space X. (A topological space X is *compact* if it is Hausdorff and if every collection of open subsets of X whose union is X contains finitely many members whose union is X, and X is *locally compact* if it is Hausdorff and each point of X has a compact neighborhood.)

Example 2. Let A be the ring of all analytic functions on a connected open subset D of $\mathbb{C}$, and let K be an infinite compact subset of D. Then N, defined by $N(f) = \sup\{|f(z)| : z \in K\}$, is an incomplete norm on A (Exercise 1.2).

Example 3. Let D be a bounded connected open subset of $\mathbb{C}$, and let A be the ring of all continuous complex-valued functions on $\overline{D}$ whose restrictions to D are analytic functions. Then N, defined by

$$N(f) = \sup\{|f(z)| : z \in \overline{D} \setminus D\},$$

is a complete norm on A.

Example 4. Let A be the ring of all continuous real-valued functions f on a closed bounded interval $[a, b]$ such that f has a continuous derivative f'

on (a,b), and $\lim_{x\to a+} f'(x)$ and $\lim_{x\to b-} f'(x)$ both exist. Then N, defined by $N(f) = \sup\{|f(x)| : a \leq x \leq b\} + \sup\{|f'(x)| : a < x < b\}$, is a complete norm on A.

Example 5. Let $L^1(\mathbb{N})$ be the set of all sequences $(a_i)_{i\geq 0}$ of real numbers such that $\sum_{i=0}^{\infty} |a_i| < +\infty$, and let N be defined on $L^1(\mathbb{N})$ by

$$N((a_i)_{i\geq 0}) = \sum_{i=0}^{\infty} |a_i|.$$

Addition on $L^1(\mathbb{N})$ is defined by $(a_i)_{i\geq 0} + (b_i)_{i\geq 0} = (a_i+b_i)_{i\geq 0}$. Under either of the following two multiplications $L^1(\mathbb{N})$ is a ring and N is a complete norm on $L^1(\mathbb{N})$: (a) pointwise multiplication, i.e., $(a_i)_{i\geq 0}(b_i)_{i\geq 0} = (a_i b_i)_{i\geq 0}$; (b) convolution, i.e.,

$$(a_i)_{i\geq 0} * (b_i)_{i\geq 0} = \left(\sum_{j=0}^{i} a_j b_{i-j}\right)_{i\geq 0}.$$

For an example of a nonmetrizable (in particular, a nonnormable) topological ring, it suffices to take the cartesian product of uncountably many nonzero topological rings, in view of the following theorem:

1.5 Theorem. *The cartesian product of a family $(A_\lambda)_{\lambda\in L}$ of topological rings is a topological ring.*

We shall prove a more general theorem:

1.6 Theorem. *Let $(A_\lambda)_{\lambda\in L}$ be a family of topological rings, let A be a ring, and let $(f_\lambda)_{\lambda\in L}$ be a family of functions such that for each $\lambda \in L$, f_λ is a homomorphism from A to A_λ. The weakest topology on A for which each f_λ is continuous is then a ring topology.*

Proof. That topology has as a basis of open sets all finite intersections of sets of the form $f_\lambda^{-1}(O_\lambda)$ where $\lambda \in L$ and O_λ is open in A_λ. It follows at once that a function g from a topological space B to A is continuous for this topology if and only if $f_\lambda \circ g$ is continuous from B to A_λ for each $\lambda \in L$. In particular, let $B = A \times A$, and let g be either addition or multiplication on A, g_λ the corresponding composition on A_λ. By the preceding, to show that g is continuous, it suffices to show that $f_\lambda \circ g$ is continuous from $A \times A$ to A_λ for all $\lambda \in L$. But $f_\lambda \circ g = g_\lambda \circ (f_\lambda \times f_\lambda)$, where $f_\lambda \times f_\lambda$ is the function $(x,y) \to (f_\lambda(x), f_\lambda(y))$ from $A \times A$ to $A_\lambda \times A_\lambda$. Since f_λ and g_λ are continuous, so is $g_\lambda \circ (f_\lambda \times f_\lambda)$. Thus g is continuous, and hence the topology is a ring topology. •

Theorem 1.5 thus follows by applying 1.6 to the case where

$$A = \prod_{\mu \in L} A_\mu$$

and for each $\lambda \in L$, $f_\lambda = pr_\lambda$, the canonical projection from $\prod_{\mu \in L} A_\mu$ to A_λ (defined by $pr_\lambda((x_\mu)_{\mu \in L}) = x_\lambda$).

1.7 Corollary. *If $(\mathcal{T}_\lambda)_{\lambda \in L}$ is a family of ring topologies on a ring A, then $\sup_{\lambda \in L} \mathcal{T}_\lambda$ is a ring topology.*

Proof. That topology is the weakest on A such that for each λ, the identity mapping from A to A, furnished with topology $\mathcal{T}_\lambda$, is continuous. •

If $\mathcal{T}_1 \ldots \mathcal{T}_p$ are topologies on a ring A defined by norms $N_1, \ldots N_p$, it is easy to see that $\sup_{1 \leq i \leq p} N_i$ is a norm defining the topology $\sup_{1 \leq i \leq p} \mathcal{T}_i$. This permits us to construct some unusual norms, for example, on the field $\mathbb{C}$ of complex numbers. For this, we first observe that the only continuous automorphisms of $\mathbb{C}$ are the identity automorphism and the conjugation automorphism $z \to \bar{z}$. Indeed, if σ is a continuous automorphism of $\mathbb{C}$, then $\sigma(x) = x$ for all $x \in \mathbb{Q}$, the prime subfield of $\mathbb{C}$, so as σ and the identity function must agree on a closed set, $\sigma(x) = x$ for all $x \in \mathbb{R}$. On the other hand, as $\sigma(i)^2 = \sigma(i^2) = \sigma(-1) = -1$, $\sigma(i)$ must be either i or $-i$. It readily follows that σ is the identity automorphism in the former case, the conjugation automorphism in the latter.

By the general theory of algebraically closed fields, however, there are nondenumerably many automorphisms of $\mathbb{C}$, so there exists a noncontinuous automorphism σ. We may further assume, by replacing σ with its composite with the conjugation isomorphism, if necessary, that $\sigma(i) = i$. Let $N(z) = \sup\{|z|, |\sigma(z)|\}$. Then N is a norm inducing the usual absolute value on the subfield $\mathbb{Q}(i)$ of C, but, as we shall see later (Corollary 13.13), the completion of $\mathbb{C}$ for the metric defined by N may be identified with the ring $\mathbb{C} \times \mathbb{C}$ and hence contains proper zero-divisors (i.e., nonzero zero-divisors).

1.8 Definition. *Let K be a division ring. An **absolute value** on K is a norm A such that $A(xy) = A(x)A(y)$ for all $x, y \in K$.*

It follows that $A(1) = 1$ since $A(1) = A(1)A(1)$ and $A(1) \neq 0$; more generally, if z is a root of unity, (i.e., if $z^n = 1$ for some $n \geq 1$), then $A(z) = 1$.

The most familiar absolute values, of course, are the usual absolute values on $\mathbb{R}, \mathbb{C}, \mathbb{H}$, and their subfields.

If A is an absolute value on a division ring K, the elements x of A satisfying $A(x) < 1$ may be characterized topologically as those elements x

such that $\lim_{n\to\infty} x^n = 0$; in any topological ring, such an element is called a *topological nilpotent*.

For any division ring K, the function A_d, defined by $A_d(0) = 0$ and $A_d(x) = 1$ for all $x \in K^*$, is called the *improper absolute value* since the topology it defines is the discrete topology. Moreover, it is the only absolute value on K defining the discrete topology. Indeed, if A is an absolute value other than A_d, then $A(x) \neq 1$ for some $x \in K^*$, so either x or x^{-1} is a topological nilpotent, and therefore the topology defined by A is not the discrete topology. In particular, the only absolute value on a finite field is the improper absolute value. An absolute value on a division ring is *proper* if it is not the improper absolute value.

1.9 Definition. *Absolute values on a division ring are* **equivalent** *if they define the same topology.*

1.10 Theorem. *Let A_1 and A_2 be proper absolute values on a division ring K. The following statements are equivalent:*

1° A_1 and A_2 are equivalent.
2° The topology defined by A_2 is weaker than that defined by A_1.
3° For all $x \in K$, if $A_1(x) < 1$, then $A_2(x) < 1$.
4° $A_2 = A_1^r$ for some $r > 0$.

Proof. If 2° holds and if $A_1(x) < 1$, then x is a topological nilpotent for the topology defined by A_1 and a fortiori for the weaker topology defined by A_2, so $A_2(x) < 1$. Assume 3°. As A_1 is proper, there exists $x_0 \in K$ such that $A_1(x_0) > 1$. Then $A_1(x_0^{-1}) < 1$, so $A_2(x_0^{-1}) < 1$, and therefore $A_2(x_0) > 1$. Let

$$r = \log A_2(x_0) / \log A_1(x_0).$$

Let $x \in K^*$, and let $s \in \mathbb{R}$ be such that $A_1(x) = A_1(x_0)^s$. Let $m, n \in \mathbb{Z}$, $n > 0$. If $m/n > s$, then $A_1(x) < A_1(x_0)^{m/n}$, so $A_1(x^n x_0^{-m}) < 1$, thus $A_2(x^n x_0^{-m}) < 1$, and therefore $A_2(x) < A_2(x_0)^{m/n}$. Similarly, if $m/n < s$, then $A_2(x) > A_2(x_0)^{m/n}$. Hence $A_2(x) = A_2(x_0)^s$, so

$$\log A_2(x) = s \log A_2(x_0) = sr \log A_1(x_0) = \log A_1(x_0)^{sr} = \log A_1(x)^r,$$

and therefore $A_2(x) = A_1(x)^r$. •

1.11 Theorem. *Let A be an absolute value on division ring K. The set J of numbers $r > 0$ such that A^r is an absolute value is an interval of $\mathbb{R}_{>0}$ containing $(0, 1]$. Moreover, the following statements are equivalent (where, for any $n \in \mathbb{N}$, $n.1 = 1 + \cdots + 1$ (n terms)):*

1° $J = \mathbb{R}_{>0}$.
2° For all $n \in \mathbb{N}$, $A(n.1) \leq 1$.
3° For all $x, y \in K$, $A(x+y) \leq \sup\{A(x), A(y)\}$.

Proof. Let $0 < r \leq 1$. For any $c \in (0,1)$, $0 < 1-c < 1$, so $c^r \geq c$ and $(1-c)^r \geq 1-c$, and therefore $c^r + (1-c)^r \geq 1$. Applying this inequality to $c = A(x)/(A(x) + A(y))$ where $x, y \in K^*$, we obtain

$$A(x)^r + A(y)^r \geq (A(x) + A(y))^r \geq A(x+y)^r.$$

Thus $r \in J$. Consequently, if $s \in J$ and $0 < t < s$, then $A^t = (A^s)^{(1/s)t}$, so A^t is an absolute value as $0 < t/s < 1$.

For any absolute value $|..|$, $|n.1| \leq n$ for all $n \in \mathbb{N}$ by induction. Hence if 1° holds, then for all $r > 0$, $A(n.1)^r \leq n$ and hence $A(n.1) \leq n^{1/r}$, so $A(n.1) \leq 1$. Clearly 3° implies 1°.

Assume 2°. As $A(y+z) \leq A(y) + A(z) \leq 2\sup\{A(y), A(z)\}$ for all $y, z \in K$, an inductive argument establishes that for any sequence $(y_i)_{1 \leq i \leq 2^r}$ of 2^r terms,

$$A(y_1 + \cdots + y_{2^r}) \leq 2^r \sup\{A(y_i) : 1 \leq i \leq 2^r\}.$$

Let $x \in K$. Then for any $r \in \mathbb{N}$, if $n = 2^r - 1$,

$$A(1+x)^n = A((1+x)^n) \leq 2^r \sup\{A(\binom{n}{k}x^k) : 0 \leq k \leq n\}$$
$$\leq 2^r \sup\{A(x^k) : 0 \leq k \leq n\} = (n+1)\sup\{1, A(x)^n\},$$

so $A(1+x) \leq (n+1)^{1/n} \sup\{1, A(x)\}$. Hence $A(1+x) \leq \sup\{1, A(x)\}$. Thus, for any $x, y \in K^*$,

$$A(x+y) = A(x)A(1 + A(x^{-1}y))$$
$$\leq A(x)\sup\{1, A(x^{-1}y)\} = \sup\{A(x), A(y)\}. \bullet$$

1.12 Definition. An absolute value A on a division ring K is **nonarchimedean** if $A(x+y) \leq \sup\{A(x), A(y)\}$ for all $x, y \in K$, **archimedean** if it is not nonarchimedean.

By 1.11, an absolute value A on a division ring K is archimedean if and only if $A(n.1) > 1$ for some $n \in \mathbb{N}$. Consequently, as a finite field admits only the improper absolute value, a field admitting an archimedean absolute value has characteristic zero.

Some important examples of nonarchimedean absolute values are defined as follows: Let K be the quotient field of a principal ideal domain D, and let P be a representative system of primes in D. As D is a unique factorization domain, for each $x \in K^*$ there exist a unique unit u of D and a unique family $(v_p(x))_{p \in P}$ of integers such that $v_p(x) = 0$ for all but finitely many $p \in P$ and
$$x = u \prod_{p \in P} p^{v_p(x)}.$$

For each $p \in P$, we also define $v_p(0) = +\infty$. The function v_p from K to $\mathbb{Z} \cup \{+\infty\}$ clearly satisfies

$$v_p(x) = +\infty \text{ if and only if } x = 0,$$
$$v_p(xy) = v_p(x) + v_p(y),$$
$$v_p(x+y) \geq \inf\{v_p(x), v_p(y)\}$$

for all $x, y \in K$; v_p is called the *p-adic valuation* on K. If $c > 1$, then $x \to c^{-v_p(x)}$ (with the convention $c^{-\infty} = 0$) is a nonarchimedean absolute value, denoted by $|..|_{p,c}$ and called the *p-adic absolute value to base c*. If $c > 1$ and $d > 1$ and if $r = \log_c d$, then $|x|_{p,d} = |x|_{p,c}^r$ for all $x \in K$, so p-adic absolute values to different bases are equivalent. The *p-adic topology* on K is the topology defined by the p-adic absolute values. For a sequence $(x_n)_{n \geq 1}$ of nonzero elements to converge to zero for the p-adic topology, it is necessary and sufficient that for each $r \in \mathbb{N}^*$ there exists $m \geq 1$ such that for all $n \geq m$, x_n can be expressed as a fraction whose numerator is an element of D divisible by p^r and whose denominator is an element of D relatively prime to p. For the special case where $K = \mathbb{Q}$ and $D = \mathbb{Z}$, for any prime integer p the p-adic absolute value $|..|_p$ on $\mathbb{Q}$ is the one to base p; thus
$$|x|_p = p^{-v_p(x)}$$
for all $x \in \mathbb{Q}$.

1.13 Theorem. *Let K be the quotient field of a principal ideal domain D, and let P be a representative system of primes in D. The proper absolute values A on K such that $A(x) \leq 1$ for all $x \in D$ are precisely the p-adic absolute values.*

Proof. If $x \in D$, then $v_p(x) \geq 0$, so $|x|_{p,c} \leq 1$. Conversely, let A be a proper absolute value on K such that $A(x) \leq 1$ for all $x \in D$. Let $V = \{x \in K : A(x) \leq 1\}$, $M = \{x \in K : A(x) < 1\}$. As A is nonarchimedean, V is a subring of K containing D by 1.12, $V \setminus M$ is the set of all invertible elements of V, and hence M is the only maximal ideal of V. In particular,

M is a prime ideal of V, so $M \cap D$ is a prime ideal of D. If $M \cap D = (0)$, then $A(x) = 1$ for all $x \in D^*$, so A would be the improper absolute value, a contradiction. Therefore $M \cap D$ is a nonzero prime ideal of D and hence is Dp for some $p \in P$. Let $c = A(p)^{-1}$. For any $x \in K^*$, let $x = ap^n/b$ where $n = v_p(x)$ and a and b are elements of D^* relatively prime to p; then $a, b \in D \setminus Dp$, so $A(a) = A(b) = 1$, and consequently

$$A(x) = A(a)A(b)^{-1}A(p^n) = A(p)^n = c^{-n} = |x|_{p,c}.$$

Thus A is $|..|_{p,c}$. ∎

1.14 Corollary. *If A is a proper nonarchimedean absolute value on $\mathbb{Q}$, there exist a prime p and $s > 0$ such that $A(x) = |x|_p^s$ for all $x \in \mathbb{Q}$.*

Proof. Clearly $A(n) \leq 1$ for all $n \in \mathbb{Z}$. By 1.13 A is $|..|_{p,c}$ for some prime p and some $c > 1$; we need only let $s = \log_p c > 0$. ∎

Often, the usual archimedean absolute value on $\mathbb{Q}$ is denoted by $|..|_\infty$; thus $|n|_\infty = |-n|_\infty = n$ for all $n \in \mathbb{N}$. The following theorem completes the identification of all proper absolute values on $\mathbb{Q}$:

1.15 Theorem. *If A is an archimedean absolute value on $\mathbb{Q}$, then there exists $s \in (0, 1]$ such that $A(x) = |x|_\infty^s$ for all $x \in \mathbb{Q}$.*

Proof. We shall first show that for any integers $m > 1$ and $n > 1$,

$$(1) \qquad \frac{\log A(n)}{\log n} = \frac{\log A(m)}{\log m}.$$

Indeed, expanding m to base n, we obtain integers $(a_k)_{0 \leq k \leq r}$ in $[0, n-1]$ such that

$$m = a_0 + a_1 n + \ldots + a_r n^r$$

and $a_r \neq 0$. Thus

$$A(m) \leq A(a_0) + A(a_1)A(n) + \cdots + A(a_r)A(n)^r,$$

and since $0 \leq A(a_i) \leq a_i < n$ for all $i \in [0, r]$, we conclude that

$$A(m) < n(1 + A(n) + \cdots + A(n)^r) \leq n(r+1)\sup\{1, A(n)\}^r.$$

Since $m \geq n^r$, $r \leq (\log m)/(\log n)$, and therefore

$$A(m) < n[\frac{\log m}{\log n} + 1]\sup\{1, A(n)\}^{(\log m)/(\log n)}.$$

Replacing m by m^s for any positive integer s, we have

$$A(m)^s = A(m^s) < n[\frac{s \log m}{\log n} + 1] \sup\{1, A(n)\}^{(s \log m)/(\log n)}.$$

Taking sth roots, we obtain

$$A(m) < n^{1/s} [\frac{s \log m}{\log n} + 1]^{1/s} \sup\{1, A(n)\}^{(\log m)/(\log n)}.$$

Since $\lim_{s \to \infty} (as + b)^{1/s} = 1$ for any positive real numbers a and b, we therefore conclude

(2) $$A(m) \leq \sup\{1, A(n)\}^{(\log m)/(\log n)}.$$

Since A is archimedean, $A(q) > 1$ for some integer $q > 1$ by 1.11. Replacing m by q in (2), we obtain

$$1 < \sup\{1, A(n)\}^{(\log q)/(\log n)},$$

whence $A(n) > 1$ as $(\log q)/(\log n) > 0$. Thus $A(n) > 1$ for all $n > 1$, so (2) becomes

$$A(m) \leq A(n)^{(\log m)/(\log n)}.$$

Taking logarithms we conclude that

(3) $$\frac{\log A(m)}{\log m} \leq \frac{\log A(n)}{\log n}.$$

Interchanging m and n in (3), we obtain (1). Let s be the common value of $(\log A(n))/(\log n)$ for all integers $n > 1$. Then for all such integers, $\log A(n) = s \log n = \log n^s$, so $A(n) = n^s$. It readily follows that $A(x) = |x|_\infty^s$ for all $x \in \mathbb{Q}$. Since $2^s = A(2) \leq 2$, $s \in (0, 1]$. ●

Exercises

1.1 Show directly or by citing theorems of analysis that the norm of Example 1 is complete, and that the subrings defined in (a) and (b) are closed and hence also complete.

1.2 Let N be the function of Example 2. (a) What theorem of complex analysis implies the validity of (N 5)? (b) Show that N is incomplete. [Show first that there exists $a \in D$ such that $|a| > \sup\{|z| : z \in K\}$, and then consider the functions $(f_n)_{n \geq 0}$ where

$$f_n(z) = \sum_{k=0}^{n} \left(\frac{z}{a}\right)^k$$

for all $z \in D$].

1.3 Let N be the function of Example 3. What theorem of complex analysis implies (a) the validity of (N 5)? (b) that N is complete?

1.4 Show directly or by citing theorems of analysis that the function N of Example 4 is a complete norm.

1.5 Show that the function N of Example 5 is a complete norm.

1.6 (a) If N is a function from a ring A to $\mathbb{R}_{\geq 0}$ satisfying (N 2)–(N 4) but not (N 1), then $N(x) \geq \max\{1, \frac{1}{2}N(0)\}$ for all $x \in A$. (b) Let Q be a norm on a ring A, let $c \geq 1$, and let $N(x) = Q(x) + c$ for all $x \in A$. Then N satisfies (N 2)–(N 4) but not (N 1).

2 Topological Modules, Vector Spaces, and Algebras

By an *A-module* is meant a left module over a ring A, not necessarily one possessing a multiplicative identity. An A-module E is *unitary* if A possesses a multiplicative identity 1 and $1.x = x$ for all $x \in E$. An A-module E is *trivial* if $\lambda.x = 0$ for all $\lambda \in A$, $x \in E$. If E is an A-module where A is a ring with identity, then E contains a largest unitary submodule M, namely, $\{x \in E : 1.x = x\}$, and a largest trivial submodule T, namely, $\{x \in E : 1.x = 0\}$, and E is the direct sum of M and T since for any $x \in E$, $1.x \in M$ and $x - 1.x \in T$.

If G is a commutative group, denoted additively, there is a unique scalar multiplication making G a unitary $\mathbb{Z}$-module, namely, that satisfying $n.x = x + x + \ldots x$ (n terms) for all $x \in G$, $n \in \mathbb{N}$. Whenever x belongs to a commutative group G and $n \in \mathbb{Z}$, $n.x$ refers to this scalar multiplication.

A topological module is simply a module over a topological ring furnished with a topology for which its algebraic operations are continuous:

2.1 Definition. *Let A be a topological ring, E an A-module. A topology $\mathcal{T}$ on E is an **A-module topology** (or simply a **module topology** if no confusion results) and E, furnished with $\mathcal{T}$, is a **topological A-module** (or simply a **topological module**) if the following conditions hold:*

(TM 1) $(x,y) \to x+y$ *is continuous from $E \times E$ to E*

(TM 2) $x \to -x$ *is continuous from E to E*

(TM 3) $(\lambda, x) \to \lambda x$ *is continuous from $A \times E$ to E*

*where E is given topology $\mathcal{T}$, $E \times E$ the cartesian product topology determined by $\mathcal{T}$, $A \times E$ the cartesian product topology determined by the topology of A and $\mathcal{T}$. If K is a division ring furnished with a ring topology and if E is a K-vector space, a topology $\mathcal{T}$ on E is a **K-vector topology** (or simply a **vector topology** if no confusion results) and E, furnished with $\mathcal{T}$,* is a **topological K-vector space** (*or simply a* **topological vector space**) *if T is a K-module topology.*

For example, any topological ring A may be regarded as a topological A-module, where scalar multiplication is the given multiplication.

A module topology on an A-module E clearly induces a module topology on any submodule of E, and unless the contrary is indicated, we shall assume that a submodule of a topological module is furnished with its induced topology.

If E is a topological A-module and if B is a subring of A, the B-module E obtained by restricting scalar multiplication to $B \times E$ is clearly a topological module. Also, if E is a topological A-module, E, with its given topology, is still a topological module over the ring A furnished with a stronger ring topology.

If A is a commutative ring with identity, an A-**algebra** (or simply an **algebra**) E is a unitary A-module furnished with a multiplicative composition that makes E into a ring and satisfies

$$\lambda(xy) = (\lambda x)y = x(\lambda y)$$

for all $\lambda \in A$ and all $x, y \in E$.

2.2 Definition. *Let A be a commutative topological ring with identity and E an A-algebra. A topology T on E is an A-**algebra topology** (or simply an **algebra topology** if no confusion results) and E, furnished with T is a **topological A-algebra** (or simply a **topological algebra**) if T is both a ring and an A-module topology.*

Norms furnish examples of topological vector spaces:

2.3 Definition. *Let K be a division ring furnished with an absolute value $|..|$, and let E be a K-vector space. A function N from E to $\mathbb{R}_{\geq 0}$ is a **norm** on E (relative to $|..|$) if (N 1)–(N 3) and (N 5) of Definition 1.2 hold and also*

$$N(\lambda x) = |\lambda| N(x)$$

*for all $\lambda \in K$ and all $x \in E$. If K is a field and E a K-algebra, an **algebra norm** (or simply a **norm**) on E is a function which is a norm on both the underlying ring and K-vector space.*

A proof similar to that of 1.3 yields:

2.4 Theorem. *The topology defined by a norm on a vector space [algebra] over a division ring [field] is a vector [algebra] topology.*

Thus a **normed space** [**normed algebra**] is simply a vector space [algebra] furnished with a norm relative to a given absolute value on its division ring [field] of scalars and hence with the topology defined by that norm. For example, the rings of Examples 1–5 of §1 may be viewed as algebras over either $\mathbb{R}$ or $\mathbb{C}$, and each of the norms defined is an algebra norm.

A topological group is simply a group furnished with a topology for which its algebraic operations are continuous:

2.5 Definition. *A topology $\mathcal{T}$ on a group G, denoted multiplicatively, is a* **group topology** *and G, furnished with $\mathcal{T}$, is a* **topological group** *if the following conditions hold:*

(TG 1) $(x, y) \to xy$ *is continuous from $G \times G$ to G*

(TG 2) $x \to x^{-1}$ *is continuous from G to G*

where G is given topology $\mathcal{T}$ and $G \times G$ the cartesian product topology determined by $\mathcal{T}$.

For example, the additive group of a topological ring or module is a commutative topological group, since (TR 1)–(TR 2) and (TM 1)–(TM 2) become (TG 1)–(TG 2) in additive notation.

A group topology on a group G clearly induces a group topology on any subgroup of G, and unless the contrary is indicated, we shall assume that a subgroup of a topological group is furnished with its induced topology.

Topologies on noncommutative groups do arise naturally in the study of topological rings. For example, if A is a topological ring with identity, the topology of A induces a topology on the group $A^\times$ that satisfies (TG 1) but may not satisfy (TG 2), and for certain questions it is important to know whether $A^\times$ is, indeed, a topological group.

2.6 Definition. *A ring topology $\mathcal{T}$ on a field [division ring] K is a* **field [division ring] topology** *and K, furnished with $\mathcal{T}$, is a* **topological field [topological division ring]** *if multiplicative inversion is continuous on K^*.*

The material presented here, however, will be needed only for discussions of topologies on the additive group of a ring or module. Consequently, we shall use additive notation throughout, even when commutativity is not used in a given discussion, and sometimes we shall include commutativity among the hypotheses of a theorem about topological groups even though a noncommutative generalization is available.

A composition $*$ on a set E induces in a natural way a composition, again denoted by $*$, on the set of all subsets of E, given by

$$X * Y = \{x * y : x \in X, \, y \in Y\}$$

for all subsets X, Y of E. It is also customary to denote $\{a\} * X$ by $a * X$ and $X * \{a\}$ by $X * a$ for any $a \in E$. We shall frequently employ this notation for subsets of a ring or group and its additive or multiplicative compositions. Similarly, if $+$ is a group composition on E, we define

$$-X = \{-x \in E : x \in X\}$$

for any subset X of E; we shall say X is *symmetric* if $X = -X$. Clearly the largest symmetric subset contained in a subset X of a group E is $X \cap (-X)$.

For the next two theorems about topologies on a group G we introduce the following notation: Let j be inversion, defined by

$$j(x) = -x$$

for all $x \in G$. Let s and t be addition and subtraction from $G \times G$ to G, defined by

$$s(x, y) = x + y,$$
$$t(x, y) = x - y$$

for all $(x, y) \in G \times G$, and let k be the function from $G \times G$ into $G \times G$ defined by

$$k(x, y) = (x, -y)$$

for all $(x, y) \in G \times G$. Finally, for each $a \in G$, let i_a be the function from G to $G \times G$ defined by

$$i_a(x) = (a, x)$$

for all $x \in G$. Clearly i_a is continuous for any topology on G and the cartesian product topology it defines on $G \times G$.

2.7 Theorem. *Let G be a topological group, and let $a \in G$. The functions $x \to -x$, $x \to a + x$, and $x \to x + a$ are homeomorphisms from G to G. Consequently, for any subset X of G, $\overline{-X} = -\overline{X}$, $\overline{a + X} = a + \overline{X}$, $\overline{X + a} = \overline{X} + a$, and for any open [closed] subset P of G, $-P$ and $a + P$ are open [closed].*

Proof. Since $j^{-1} = j$, j is a homeomorphism. The function

$$s \circ i_a : x \to a + x$$

is continuous as s and i_a are, and its inverse $s \circ i_{-a}$ is similarly continuous. Thus $x \to a + x$ is a homeomorphism, and similarly $x \to x + a$ is a homeomorphism. •

2 TOPOLOGICAL MODULES, VECTOR SPACES, AND ALGEBRAS

2.8 Theorem. *A topology $\mathcal{T}$ on a group G is a group topology if and only if*
$$(x,y) \to x - y$$
is continuous from $G \times G$ to G, where G is furnished with $\mathcal{T}$ and $G \times G$ the cartesian product topology determined by $\mathcal{T}$.

Proof. Necessity: By hypothesis, s and j are continuous. Hence k is continuous, and as $t = s \circ k$, t is also continuous. Sufficiency: By hypothesis t is continuous. Hence as $j = t \circ i_0$, j is also continuous. Consequently, k is continuous, so as $s = t \circ k$, s is also continuous. •

A *neighborhood* of a point [subset] of a topological space T is any subset of T that contains an open subset containing that point [subset]. Thus a subset of T is open if and only if it is a neighborhood of each of its points.

2.9 Theorem. *Let G be a topological group.*

(1) *If V is a neighborhood of zero, there is a neighborhood U of zero such that $U + U \subseteq V$.*

(2) *If V is is a neighborhood of zero, so is $-V$.*

(3) *Every [open] neighborhood U of zero contains a symmetric [open] neighborhood of zero, namely, $U \cap (-U)$.*

Proof. (1) follows from the continuity of addition at $(0,0)$, (2) follows from 2.7, and (3) follows from (2). •

2.10 Corollary. *Let G be a topological group. If U is a neighborhood of zero and $n \geq 1$, there is a symmetric neighborhood V of zero such that $V + \cdots + V$ (n terms) $\subseteq U$.*

Proof. The assertion follows by induction from 2.9.•

2.11 Theorem. *Let A be a topological ring.*

(1) *For each $b \in A$, the functions $x \to xb$ and $x \to bx$ are continuous from A to A; if b is invertible, they are homeomorphisms.*

(2) *$(x,y) \to yx$ is continuous from $A \times A$ to A.*

(3) *If f and g are functions from a topological space T to A that are continuous at $t \in T$, then $f + g$, $-f$, and fg are continuous at t.*

(4) *If A is a commutative ring with identity and if $h \in A[X_1, \ldots, X_n]$, the ring of polynomials in n indeterminates over A, then the polynomial function $(x_1, \ldots, x_n) \to h(x_1, \ldots, x_n)$ from $A \times \cdots \times A$ (n terms) to A is continuous.*

Proof. The proof of (1) is similar to that of 2.7. (2) The function m from $A \times A$ to A, defined by $m(x,y) = xy$, and the function q from $A \times A$ to $A \times A$, defined by $q(x,y) = (y,x)$, are continuous, so $m \circ q$ is also continuous. (3)

Let $f \times g$ be the function from T to $A \times A$ defined by $(f \times g)(s) = (f(s), g(s))$. As f and g are continuous at t, so is $f \times g$; $f + g$ and fg are simply the composites of that function with addition and multiplication. (4) follows by induction from (3). ●

2.12 Theorem. *Let E be a topological A-module.*

(1) For each $b \in E$, $\lambda \to \lambda b$ is continuous from A to E, and for each $\beta \in A$, $x \to \beta x$ is continuous from E to E; if β is invertible, $x \to \beta x$ is a homeomorphism.

(2) If f is a function from a topological space T to E that is continuous at $t \in T$, then for each $\lambda \in A$, λf is continuous at t.

The proof is similar to that of 2.11.

2.13 Definition. *Let G_1, G_2, and G be commutative groups. A function f from $G_1 \times G_2$ to G is $\mathbb{Z}$-bilinear if for each $a \in G_1$, the function $y \to f(a, y)$ is a homomorphism from G_2 to G, and for each $b \in G_2$, the function $x \to f(x, b)$ is a homomorphism from G_1 to G.*

For example, multiplication of a ring and scalar multiplication of a module are $\mathbb{Z}$-bilinear functions on the underlying additive groups.

If f is $\mathbb{Z}$-bilinear from $G_1 \times G_2$ to G, clearly

$$f(x, 0) = 0 = f(0, y),$$
$$f(-x, y) = -f(x, y) = f(x, -y),$$
$$f(-x, -y) = f(x, y)$$

for all $x \in G_1$, $y \in G_2$.

2.14 Theorem. *Let G_1, G_2, and G be commutative topological groups, and let f be a $\mathbb{Z}$-bilinear function from $G_1 \times G_2$ to G. If, for each $a \in G_1$, the function $y \to f(a, y)$ is continuous at zero from G_2 to G, if, for each $b \in G_2$, the function $x \to f(x, b)$ is continuous at zero from G_1 to G, and if f is continuous at $(0, 0)$, then f is continuous from $G_1 \times G_2$ to G.*

Proof. Let $(a_1, a_2) \in G_1 \times G_2$, and let Y be a neighborhood of $f(a_1, a_2)$ in G. We are to show that there exist neighborhoods W_1 of a_1 and W_2 of a_2 such that $f(x_1, x_2) \in Y$ for all $(x_1, x_2) \in W_1 \times W_2$. By 2.7 there is a neighborhood T of zero in G such that $f(a_1, a_2) + T = Y$, and by 2.10 there is a neighborhood W of zero in G such that $W + W + W \subseteq T$. By hypothesis there exist neighborhoods U_1 and V_1 of zero in G_1 and neighborhoods U_2 and V_2 of zero in G_2 such that $f(a_1, u_2) \in W$ for all $u_2 \in U_2$, $f(u_1, a_2) \in W$ for all $u_1 \in U_1$, and $f(v_1, v_2) \in W$ for all $(v_1, v_2) \in V_1 \times V_2$. Let $W_1 = a_1 + (U_1 \cap V_1)$, $W_2 = a_2 + (U_2 \cap V_2)$. By 2.7 W_1 and W_2 are neighborhoods of a_1 and a_2 respectively. Let $x_1 \in W_1$, $x_2 \in W_2$. Then $x_1 - a_1 \in U_1 \cap V_1$ and $x_2 - a_2 \in U_2 \cap V_2$, so

$$f(x_1, x_2) = f(a_1, a_2) + f(x_1 - a_1, a_2) + f(x_1 - a_1, x_2 - a_2) + \\ + f(a_1, x_2 - a_2) \in f(a_1, a_2) + W + W + W \\ \subseteq f(a_1, a_2) + T = Y. \bullet$$

2.15 Theorem. *If a topology T on a ring A satisfies (TR 1) and (TR 2) of Definition 1.1, then T satisfies (TR 3) if and only if it satisfies the following two conditions:*

(TR 4) $(x, y) \to xy$ *is continuous at* $(0, 0)$

(TR 5)   *For each $b \in A$, $x \to bx$ and $x \to xb$ are continuous at zero.*

The condition is necessary by (1) of 2.11 and sufficient by 2.14.

2.16 Theorem. *Let A be a topological ring, E an A-module. If a topology T on E satisfies (TM 1) and (TM 2) of Definition 2.1, then T satisfies (TM 3) if and only if it satisfies the following three conditions:*

(TM 4) $(\lambda, x) \to \lambda x$ *from $A \times E$ to E is continuous at* $(0, 0)$

(TM 5)   *For each $b \in E$, $\lambda \to \lambda b$ from A to E is continuous at zero*

(TM 6)   *For each $\beta \in A$, $x \to \beta x$ from E to E is continuous at zero.*

The condition is necessary by (1) of 2.12 and sufficient by 2.14.
Analogues of 1.5–1.7 hold for modules:

2.17 Theorem. *Let A be a topological ring, let E be an A-module, let $(E_\lambda)_{\lambda \in L}$ be a family of topological A-modules, and let $(f_\lambda)_{\lambda \in L}$ be a family of functions such that for each $\lambda \in L$, f_λ is a homomorphism from E to E_λ. The weakest topology on E for which each f_λ is continuous is an A-module topology.*

The proof is similar to that of 1.6.

2.18 Corollary. *The cartesian product of a family of topological A-modules is a topological A-module.*

2.19 Corollary. *If $(T_\lambda)_{\lambda \in L}$ is a family of A-module topologies on an A-module E, then $\sup\{T_\lambda : \lambda \in L\}$ is an A-module topology.*

Exercises

2.1 (a) If T is a topology on a ring A with identity such that (TR 1) and (TR 3) hold, then (TR 2) holds. (b) If A is a topological ring with identity and if T is a topology on a unitary A-module E such that (TM 1) and (TM 3) hold, then (TM 2) holds.

2.2 Let T be the set of all subsets P of $\mathbb{Z}$ such that for each $a \in P$, there exists $q \geq 1$ such that $a + \mathbb{N}q \subseteq P$. Show that T is a topology on the trivial ring whose additive group is $\mathbb{Z}$ that satisfies (TR 1) and (TR 3) but not (TR 2).

2.3 Let T be the set of all subsets P of $\mathbb{R}$ such that for each $a \in P$ there exists a nonzero integer q such that $a + \mathbb{Z}q \subseteq P$. (a) T is a topology on $\mathbb{R}$ satisfying (TR 1), (TR 2), and (TR 4), but not (TR 5). (b) The topology induced on $\mathbb{Q}$ by T is a ring topology, but multiplicative inversion on $\mathbb{Q}^*$ is not continuous at 1.

2.4 If N is a norm on a ring A with identity, then $A^\times$ is a topological group, i.e., multiplicative inversion is continuous on $A^\times$.

2.5 (a) An additive group topology on a trivial ring [module] is a ring [module] topology. (b) If A is a discrete topological ring, (i.e., its topology is the discrete topology) and if E is an A-module, an additive group topology on E satisfying (TM 6) is an A-module topology.

2.6 If K is a nondiscrete topological field and if E is a nonzero K-vector space, then the discrete topology on E is an additive group topology satisfying (TM 4) and (TM 6) but not (TM 5).

3 Neighborhoods of Zero

We recall that a set $\mathcal{F}$ of subsets of a set E is a *filter* on E if $E \in \mathcal{F}, \emptyset \notin \mathcal{F}$, the intersection of any two members of $\mathcal{F}$ again belongs to $\mathcal{F}$, and any subset of E containing a member of $\mathcal{F}$ also belongs to $\mathcal{F}$. For example, in a topological space T, the set of all neighborhoods of a point [subset] of T is a filter.

A set $\mathcal{B}$ of subsets of E is a *filter base* on E if the set of all subsets F of E for which there exists $B \in \mathcal{B}$ such that $B \subseteq F$ is a filter, called the filter *generated* by $\mathcal{B}$. Thus $\mathcal{B}$ is a filter base if and only if $\mathcal{B} \neq \emptyset, \emptyset \notin \mathcal{B}$, and the intersection of two members of $\mathcal{B}$ contains a member of $\mathcal{B}$. Consequently, a filter base on E is also a filter base on any set containing E. In a topological space T, a *fundamental system of neighborhoods* of a point [subset] of T is any filter base generating the filter of neighborhoods of that point [subset]. For example, the open neighborhoods of a point in a topological space form a fundamental system of neighborhoods of that point.

If $\mathcal{V}$ is the filter of neighborhoods of zero for a group topology on a group G, then by 2.7, for each $a \in G$, $a + \mathcal{V}$ is the filter of neighborhoods of a. Consequently, a group topology is completely determined by the filter of neighborhoods of zero; that is, distinct group topologies have distinct filters of neighborhoods of zero. For commutative groups, the following theorem gives necessary and sufficient conditions for a filter to be the filter of neighborhoods of zero for a group topology on G:

3.1 Theorem. *If $\mathcal{V}$ is the filter of neighborhoods of zero for a group topology on a group G, then*

(TGN 1) For each $V \in \mathcal{V}$, there exists $U \in \mathcal{V}$ such that $U + U \subseteq V$
(TGN 2) If $V \in \mathcal{V}$, then $-V \in \mathcal{V}$.

Conversely, if $\mathcal{V}$ is a filter on a commutative group G satisfying (TGN 1) *and* (TGN 2), *then there is a unique group topology on G for which $\mathcal{V}$ is the filter of neighborhoods of zero.*

Proof. The first statement is part of 2.9. Conversely, let $\mathcal{V}$ be a filter on a commutative group G satisfying (TGN 1) and (TGN 2). We have just seen that there is only one candidate for a group topology having $\mathcal{V}$ as its filter of neighborhoods of zero; since a set is open if and only if it is a neighborhood of each of its points, that candidate is the set $\mathcal{T}$ of all subsets O of G satisfying

(*) For each $a \in O$ there exists $V \in \mathcal{V}$ such that $a + V \subseteq O$.

Clearly $\emptyset$ and G satisfy (*), and the union of a family of subsets satisfying (*) again satisfies (*). Let $O_1, O_2 \in \mathcal{T}$, and let $a \in O_1 \cap O_2$. There exist $V_1, V_2 \in \mathcal{V}$ such that $a + V_1 \subseteq O_1$ and $a + V_2 \subseteq O_2$; then $V_1 \cap V_2 \in \mathcal{V}$, and $a + (V_1 \cap V_2) \subseteq O_1 \cap O_2$. Thus $O_1 \cap O_2$ satisfies (*). Hence $\mathcal{T}$ is indeed a topology.

Next we shall show that for each $V \in \mathcal{V}$, $0 \in V$. By (TGN 1) there exists $U \in \mathcal{V}$ such that $U + U \subseteq V$, and by (TGN 2), $U \cap (-U) \in \mathcal{V}$, so there exists $a \in U \cap (-U)$. Then a and $-a$ belong to U, so $0 = a + (-a) \in U + U \subseteq V$.

From this we may establish that for each $a \in G$ and each $V \in \mathcal{V}$, $a + V$ is a neighborhood of a for $\mathcal{T}$. Let

$$O = \{b \in G : \text{there exists } U \in \mathcal{V} \text{ such that } b + U \subseteq a + V\}.$$

Clearly $a \in O$. By the preceding paragraph, $O \subseteq a + V$. To show that O satisfies (*), let $b \in O$. By definition, there exists $U \in \mathcal{V}$ such that $b + U \subseteq a + V$. By (TGN 2) there exists $W \in \mathcal{V}$ such that $W + W \subseteq U$. Then $b + W \subseteq O$, for if $w \in W$, $b + w + W \subseteq b + W + W \subseteq b + U \subseteq a + V$. Thus $O \in \mathcal{T}$, and hence $a + V$ is a neighborhood of a for $\mathcal{T}$.

To show that $a + \mathcal{V}$ is the filter of neighborhoods of a for $\mathcal{T}$, therefore, we need only show that if W is a neighborhood of a for $\mathcal{T}$, there exists

$U \in \mathcal{V}$ such that $a + U = W$. As W is a neighborhood of a, there exists O satisfying (*) such that $a \in O \subseteq W$. By definition, there exists $V \in \mathcal{V}$ such that $a + V \subseteq O$. Let $U = -a + W$. Then $V \subseteq -a + O \subseteq U$, so $U \in \mathcal{V}$, and $a + U = W$.

Finally, to show that $\mathcal{T}$ is a group topology, let $a, b \in G$. By 2.8 it suffices to show that $(x, y) \to x - y$ is continuous at (a, b), or equivalently, that for any $V \in \mathcal{V}$, there exists $U \in \mathcal{V}$ such that if $x \in a + U$ and $y \in b + U$, then $x - y \in a - b + V$. By 2.9 there exists a symmetric $U \in \mathcal{V}$ such that $U + U \subseteq V$. If $x = a + u$ and $y = b + v$ where $u, v \in U$, then

$$x - y = (a + u) - (b + v) = (a - b) + (u - v)$$
$$\in a - b + U + U \subseteq a - b + V. \bullet$$

3.2 Corollary. *If $\mathcal{V}$ is a fundamental system of neighborhoods of zero for a group topology on a group G, then*

(TGB 1) *For each $V \in \mathcal{V}$ there exists $U \in \mathcal{V}$ such that $U + U \subseteq V$*

(TGB 2) *For each $V \in \mathcal{V}$ there exists $U \in \mathcal{V}$ such that $U \subseteq -V$.*

Conversely, if $\mathcal{V}$ is a filter base on a commutative group G such that (TGB 1) and (TGB 2) hold, then there is a unique group topology on G for which $\mathcal{V}$ is a fundamental system of neighborhoods of zero.

In the proof of 3.1, the hypothesis of commutativity was needed for the equality $(a + u) - (b + v) = (a - b) + (u - v)$. A generalization of 3.1 to noncommutative groups is given in Exercise 3.1.

In view of 3.2, to specify a group topology on a commutative group, we need only specify a filter base $\mathcal{V}$ satisfying (TGB 1) and (TGB 2). This method, in fact, is the most frequent way of defining a group topology. For example, if $\mathcal{V}$ is a filter base of subgroups of a commutative group G, then $\mathcal{V}$ is a fundamental system of neighborhoods of zero for a group topology on G.

3.3 Theorem. *Let G be a topological group, let $\mathcal{V}$ be a fundamental system of neighborhoods of zero, and let $A \subseteq G$.*

(1) The open symmetric neighborhoods of zero form a fundamental system of neighborhoods of zero.

(2) For any [open] neighborhood V of zero, $A + V$ is a neighborhood [an open neighborhood] of A.

(3) $\bar{A} = \cap \{A + V : V \in \mathcal{V}\}$; in particular, $\overline{\{0\}} = \cap \{V : V \in \mathcal{V}\}$.

(4) The closed symmetric neighborhoods of zero form a fundamental system of neighborhoods of zero.

Proof. (1) follows from (3) of 2.9, and (2) follows from 2.7. To prove (3), we may, without loss of generality, assume that each member of $\mathcal{V}$ is symmetric, in view of 2.9. First let $b \in \bar{A}$, and let V be a symmetric neighborhood of zero. Then $A \cap (b+V) \neq \emptyset$, so for some $v \in V$, $b+v \in A$, whence $b \in A+V$. Conversely, let $b \in \cap \{A+V : V \in \mathcal{V}\}$, and let W be a neighborhood of b. By 2.7 there exists $V \in \mathcal{V}$ such that $b+V \subseteq W$ and there exists $a \in A$ such that $b \in a+V$, so as V is symmetric, $a \in A \cap (b+V) \subseteq A \cap W$. Thus $b \in \bar{A}$.

(4) If V is a neighborhood of zero, there is a neighborhood U of zero such that $U+U \subseteq V$ by 2.9, and by (3) $\bar{U} \subseteq U+U \subseteq V$. Thus every neighborhood of zero contains a closed neighborhood of zero. If U is a closed neighborhood of zero, $U \cap (-U)$ is a closed symmetric neighborhood of zero contained in U by 2.7. •

A topological space T is *regular* if T is Hausdorff and for each $a \in T$ the closed neighborhoods of a form a fundamental system of neighborhoods of a.

3.4 Theorem. *Let G be a topological group. The following statements are equivalent:*

1° $\{0\}$ *is closed.*

2° $\{0\}$ *is the intersection of the neighborhoods of zero.*

3° G *is Hausdorff.*

4° G *is regular.*

Proof. 1° and 2° are equivalent by (3) of 3.3, and 3° and 4° are equivalent by (4) of 3.3. We therefore need only show that 2° implies 3°. Let $x, y \in G$, $x \neq y$. Then $x - y \neq 0$, so by 2° there exists a neighborhood V of zero such that $x - y \notin V$. By 2.9 there is a symmetric neighborhood U of zero such that $U + U \subseteq V$. Then $x + U$ and $y + U$ are disjoint neighborhoods of x and y respectively, for if $z \in (x+U) \cap (y+U)$, then

$$x - y = -(z-x) + (z-y) \in U + U \subseteq V,$$

a contradiction. •

3.5 Theorem. *Let A be a ring. If $\mathcal{V}$ is a fundamental system of neighborhoods of zero for a ring topology on A, then $\mathcal{V}$ satisfies* (TGB 1), (TGB 2) *and the following conditions:*

(TRN 1) *For each $V \in \mathcal{V}$ there exists $U \in \mathcal{V}$ such that $UU \subseteq V$*

(TRN 2) *For each $V \in \mathcal{V}$ and each $b \in A$ there exists $U \in \mathcal{V}$ such that $bU \subseteq V$ and $Ub \subseteq V$.*

Conversely, if $\mathcal{V}$ is a filter base on A satisfying (TGB 1), (TGB 2), (TRN 1), and (TRN 2), then there is a unique ring topology on A for which $\mathcal{V}$ is a fundamental system of neighborhoods of zero.

Proof. Conditions (TRN 1) and (TRN 2) restate (TR 4) and (TR 5) of 2.15. Hence the theorem follows from 3.2 and 2.15. •

The most frequent way of defining a ring topology on a ring A is to specify a filter base satisfying the conditions of 3.5. Those conditions are satisfied, for example, by a filter base of ideals of A. Any ring topology having a fundamental system of neighborhoods of zero consisting of ideals is called an *ideal topology*.

3.6 Theorem. *Let A be a topological ring, E and A-module. If $\mathcal{V}$ is a fundamental system of neighborhoods of zero for an A-module topology on E, then $\mathcal{V}$ satisfies* (TGB 1), (TGB 2), *and the following conditions:*

(TMN 1)   For each $V \in \mathcal{V}$ there exist a neighborhood T of zero in A and $U \in \mathcal{V}$ such that $TU \subseteq V$

(TMN 2)   For each $V \in \mathcal{V}$ and each $b \in E$ there exists a neighborhood T of zero in A such that $Tb \subseteq V$

(TMN 3)   For each $V \in \mathcal{V}$ and each $\beta \subset A$ there exists $U \in \mathcal{V}$ such that $\beta U \subseteq V$.

Conversely, if $\mathcal{V}$ is a filter base on E satisfying (TGB 1), (TGB 2), (TMN 1), (TMN 2), and (TMN 3), then there is a unique A-module topology on E for which $\mathcal{V}$ is a fundamental system of neighborhoods of zero.

Proof. Conditions (TMN 1)–(TMN 3) restate (TM 4)–(TM 6) of 2.16. Hence the theorem follows from 3.2 and 2.16. •

Exercises

3.1 Let G be a group, denoted multiplicatively, and let e be its neutral element. Modify the proof of 3.1 to establish the following: (a) If $\mathcal{V}$ is the filter of neighborhoods of e for a group topology on G, then

(TGN 1)   For each $V \in \mathcal{V}$ there exists $U \in \mathcal{V}$ such that $UU \subseteq V$

(TGN 2)   If $V \in \mathcal{V}$, then $V^{-1} \in \mathcal{V}$

(TGN 3)   If $V \in \mathcal{V}$, then for each $b \in G$, $bVb^{-1} \in \mathcal{V}$.

(b) Conversely, if $\mathcal{V}$ is a filter base on G satisfying (TGN 1)–(TGN 3), then there is a unique group topology on G for which $\mathcal{V}$ is the filter of neighborhoods of e.

3.2 There is a unique topology on the additive group $\mathbb{R}$ for which $x \to a+x$ is continuous for each $a \in \mathbb{R}$ and the sets V_n, defined by $V_n = \{x \in \mathbb{R} :$

$|x| < 2^{-n}$ and $x \neq \pm 2^k$ for all $k \in \mathbb{Z}\}$ for each $n \geq 1$, form a fundamental system of neighborhoods of zero. For this topology, $x \to -x$ is continuous, but $(x, y) \to x + y$ is not continuous at $(0, 0)$.

3.3 Let $\mathcal{T}$ be an additive group topology on a ring A. The subset B of A, consisting of all $b \in A$ such that $x \to bx$ and $x \to xb$ are continuous at zero, is a subring of A; furthermore, if A has an identity, $B^\times = B \cap A^\times$.

3.4 If $\mathcal{T}$ is a ring topology on a finite ring A, there is an ideal J of A such that the neighborhoods of zero for $\mathcal{T}$ are precisely the subsets of A containing J.

3.5 Let K be a nondiscrete topological field and let E be a K-vector space. If $\mathcal{V}$ is a filter base of subspaces of E whose intersection is $\{0\}$, then $\mathcal{V}$ is a fundamental system of neighborhoods of zero for a Hausdorff additive group topology on E and satisfies (TMN 1) and (TMN 3) but not (TMN 2).

3.6 Let p be a prime, let $\mathbb{Q}$ be furnished with the p-adic absolute value, let S be the unit ball of $\mathbb{Q}$, let E be the subspace of the $\mathbb{Q}$-vector space $\mathbb{Q}^\mathbb{N}$ generated by $S^\mathbb{N}$, and let $\mathcal{V}$ be the filter of neighborhoods of zero in $S^\mathbb{N}$ for the cartesian product topology. Then $\mathcal{V}$ is a fundamental system of neighborhoods of zero for a Hausdorff additive group topology on E and satisfies (TMN 1) and (TMN 2) but not (TMN 3). [Show that (TMN 3) fails if $\beta = 1/p$.]

3.7 Let E be the $\mathbb{R}$-vector space $\mathbb{R}^\mathbb{N}$ of all sequences of real numbers indexed by $\mathbb{N}$. For each $r \in \mathbb{R}_{>0}$, let

$$V_r = \{(x_n)_{n \geq 0} \in E : |x_n| < r \text{ for all } n \in \mathbb{N}\}.$$

Then $\{V_r : r > 0\}$ is a fundamental system of neighborhoods of zero for a Hausdorff additive group topology on E and satisfies (TMN 1) and (TMN 3) but not (TMN 2).

3.8 Let K be a nondiscrete Hausdorff topological field, let L be a proper extension field of K, let L_d be L furnished with the discrete topology, and regard L as a one-dimensional vector space over L_d. The filter of all neighborhoods of zero in K is a fundamental system of neighborhoods of zero for a Hausdorff additive group topology on L and satisfies (TMN 1) and (TMN 2) but not (TMN 3).

4 Subrings, Ideals, Submodules, and Subgroups

The closure of a subring, ideal, submodule, or subgroup is again one:

4.1 Theorem. *If H is a subgroup of a topological group G, then $\overline{H}$ is a subgroup.*

Proof. The continuous function $(x,y) \to x - y$ from $G \times G$ to G takes $H \times H$ into H and hence takes the closure $\overline{H} \times \overline{H}$ of $H \times H$ into the closure $\overline{H}$ of H. Thus $\overline{H}$ is a subgroup. •

4.2 Theorem. *If B is a subring [ideal, left ideal, right ideal] of a topological ring A, so is $\overline{B}$. If A is a dense subring of a topological ring A' and if J is an ideal [left ideal, right ideal] of A, then the closure of J in A' is an ideal [left ideal, right ideal].*

The proof is similar to that of 4.1.

4.3 Theorem. *If M is a submodule of a topological module E, so is $\overline{M}$.*

The proof is similar to that of 4.1.

We shall call a Hausdorff topological group, ring, or module simply a Hausdorff group, ring, or module.

4.4 Theorem. *Let A be a Hausdorff ring, and let B be a subring of A.*

(1) *If B is commutative, so is $\overline{B}$.*
(2) *A multiplicative identity of B is also a multiplicative identity of $\overline{B}$.*
(3) *The center of A is closed.*

Proof. As A is Hausdorff, continuous functions from a topological space T to A agree on a closed subset of T. (1) As $(x,y) \to yx$ is continuous from $A \times A$ to A by (2) of 2.11, it agrees with $(x,y) \to xy$ on a closed subset of $A \times A$ containing $B \times B$ and hence on $\overline{B} \times \overline{B}$; thus $\overline{B}$ is commutative. (2) Assume that B has a multiplicative identity e. As the continuous functions $x \to xe$, $x \to ex$, and the identity function agree on a closed subset of A containing B, they agree on $\overline{B}$; thus e is the identity of $\overline{B}$. (3) Let C be the center of A. As the functions of (1) agree on a closed subset of $A \times A$ containing $A \times C$, they agree on its closure $A \times \overline{C}$; thus $\overline{C} \subseteq C$, so C is closed.

The conclusions of 4.4 need not hold in a non-Hausdorff topological ring. Indeed, they do not hold in a noncommutative ring furnished with the trivial topology, for in such a topological ring A, the closure of every nonempty subset is A.

4.5 Theorem. *The connected component C of zero in a topological ring A is a closed ideal, and $a+C$ is the connected component of a for each $a \in A$.*

Proof. The second assertion follows from 2.7. Hence if $a \in C$, then $C \cap (a+C) \neq \emptyset$, so $C \cup (a+C)$ is a connected set containing zero, and thus $a + C \subseteq C$. Therefore $C + C \subseteq C$. Also $-C$ is a connected subset of A containing zero by 2.7, so $-C \subseteq C$. Hence C is an additive subgroup. For each $b \in A$, bC and Cb are connected sets containing zero by (1) of 2.11, so $bC \subseteq C$ and $Cb \subseteq C$. Thus C is a (closed) ideal. •

A topological space T is *totally disconnected* if for each $t \in T$, $\{t\}$ is the connected component of t. By 4.5, a topological ring is totally disconnected if and only if $\{0\}$ is the connected component of zero. As connected components are closed, a totally disconnected ring is Hausdorff by 3.4.

A *proper* subset of a set X is any subset of X other than X itself. Thus, for example, $\emptyset$ is a proper subset of every nonempty set.

4.6 Corollary. *If A is a topological ring having no proper nonzero closed ideals, then the topology of A is either Hausdorff and connected, or Hausdorff and totally disconnected, or the trivial topology.*

Proof. If A is not Hausdorff, then $\overline{\{0\}}$ is A by 4.2 and hypothesis, so for each $a \in A$
$$\overline{\{a\}} = \overline{a + \{0\}} = a + \overline{\{0\}} = A$$
by 2.7; therefore A is the only nonempty closed subset, so the topology of A is $\{A, \emptyset\}$. By 4.5, the topology of A is either connected or totally disconnected. •

A topological ring having no nonzero proper ideals satisfies the hypothesis of 4.6. In particular:

4.7 Corollary. *A ring topology on a division ring is either Hausdorff and connected, or Hausdorff and totally disconnected, or the trivial topology.*

The ring of all linear operators on a finite-dimensional vector space over a division ring also has no proper nonzero ideals, but if the dimension of the vector space exceeds one, it does have proper nonzero left and right ideals.

4.8 Theorem. *An open subgroup H of a topological group G is closed.*

Proof. Each left coset of H is open by 2.7; as H is the complement of the union of all left cosets of H other than H itself, H is closed. •

4.9 Theorem. *If a subgroup H of a topological group G has an interior point, then H is open.*

Proof. By 2.7 there exist $a \in H$ and an open neighborhood U of zero such that $a + U \subseteq H$. The subgroup H is a neighborhood of each of its points h, as
$$h \in h + U = (h - a) + (a + U) \subseteq H.$$
Thus H is open. •

4.10 Corollary. *The subgroup of a topological group G generated by a neighborhood of zero is both open and closed.*

4.11 Theorem. *Let H be a subgroup of a topological group G. If for some $a \in H$ there is a neighborhood V of a such that $V \cap H$ is closed in the topological space V, then H is closed.*

Proof. By 2.7 and 2.9 there is a symmetric open neighborhood U of zero such that $a + U \subseteq V$. Clearly $(a + U) \cap H$ is a closed subset of $a + U$. As

$$(a + U) \cap H = (a + U) \cap (a + H) = a + (U \cap H)$$

and as $x \to a + x$ is a homeomorphism from G to G, $U \cap H$ is a closed subset of U. Let $x \in \overline{H}$. Then there exists $h \in H \cap (x + U)$, so $x \in h + U$ as U is symmetric. As $h + U$ is open and as $h \in H$,

$$(h + U) \cap \overline{H} \subseteq \overline{(h + U) \cap H} = \overline{(h + U) \cap (h + H)}$$
$$= \overline{h + (U \cap H)} = h + \overline{U \cap H}$$

by 2.7. Thus

$$x \in (h + U) \cap \overline{H} \subseteq (h + \overline{U \cap H}) \cap (h + U)$$
$$= h + (\overline{U \cap H} \cap U) = h + (U \cap H) \subseteq H. \bullet$$

4.12 Corollary. *A locally compact subgroup H of a Hausdorff group G is closed.*

Proof. If V is a neighborhood of zero such that $V \cap H$ is compact, then $V \cap H$ is closed in G and hence in V. $\bullet$

4.13 Corollary. *If a subgroup H of a topological group G has an isolated point, then H is discrete. If G is Hausdorff and if H is a discrete subgroup, then H is closed.*

Proof. Let $a \in H$ be such that $\{a\} = (a + U) \cap H$ for some neighborhood U of zero. Then for each $h \in H$, $(h - a) + H = H$, so

$$(h + U) \cap H = [(h - a) + (a + U)] \cap [(h - a) + H]$$
$$= (h - a) + [(a + U) \cap H] = (h - a) + \{a\} = \{h\}.$$

Thus H is discrete. The second assertion follows from 4.12. $\bullet$

In contrast, if G is a topological group whose topology is not Hausdorff, then $\{0\}$ is a compact, discrete subgroup of G that is not closed.

4 SUBRINGS, IDEALS, SUBMODULES, AND SUBGROUPS

4.14 Theorem. *Let G be a topological group, and let K be a compact subset, F a closed subset of G.*

(1) Every neighborhood of K contains a closed neighborhood of K; if G is locally compact, every neighborhood of K contains a compact neighborhood of K.

(2) For any neighborhood U of K there is a neighborhood W of zero such that $K + W \subseteq U$ and $W + K \subseteq U$.

(3) If $K \cap F = \emptyset$, there is a neighborhood V of zero such that

$$(K+V) \cap (F+V) = \emptyset = (V+K) \cap (V+F).$$

Proof. (1) Let U be a neighborhood of K. By 2.7 and (4) of 3.3, for each $x \in K$ there is a closed neighborhood V_x of zero such that $x + V_x \subseteq U$. Since $\{x + V_x^\circ : x \in K\}$ is an open cover of K (where V_x° denotes the interior of V_x), there exist $x_1, \ldots, x_n \in K$ such that if

$$W = \bigcup_{i=1}^{n} (x_i + V_{x_i}),$$

then $K \subseteq W$. Thus W is a closed neighborhood of K contained in U. If G is locally compact, we may assume that each V_x is compact, in which case W is also.

(2) For each $x \in K$, let V_x be a neighborhood of zero such that $x + V_x \subseteq U$, and let W_x be an open neighborhood of zero such that $W_x + W_x \subseteq V_x$. Then $\{x + W_x : x \in K\}$ is an open cover of K, so there exist $x_1, \ldots, x_n \in K$ such that

$$\bigcup_{i=1}^{n} (x_i + W_{x_i}) \supseteq K.$$

Let

$$W_1 = \bigcap_{i=1}^{n} W_{x_i}.$$

If $x \in K$ and $y \in W_1$, then $x = x_i + w$ for some $i \in [1, n]$ and some $w \in W_{x_i}$, so

$$x + y = x_i + w + y \in x_i + W_{x_i} + W_{x_i} \subseteq x_i + V_{x_i} \subseteq U.$$

Thus $K + W_1 \subseteq U$. Similarly, there exists a neighborhood W_2 of zero such that $W_2 + K \subseteq U$. Finally, let $W = W_1 \cap W_2$.

(3) By (2) applied to the neighborhood $G \setminus F$ of K, there is a neighborhood W of zero such that $(K + W) \cap F = \emptyset = (W + K) \cap F$. Clearly

$$(K+V) \cap (F+V) = \emptyset = (V+K) \cap (V+F)$$

where V is any symmetric neighborhood of zero such that $V + V \subseteq W$. •

If P is a topological space, the connected component C_x of $x \in P$ is contained in every subset of P that is both open and closed and contains x, but, in general, C_x is not the intersection of all such subsets. If, however, P is compact, C_x is the intersection of all open and closed subsets containing x, a fact we prove under the additional assumption that P is a subspace of a topological group:

4.15 Theorem. *If P is a compact subset of a topological group G, for each $x \in P$ the connected component of x in the topological space P is the intersection of all open and closed subsets of P that contain x.*

Proof. For each symmetric neighborhood V of zero, we define $(A_{x,V,k})_{k \geq 0}$ recursively by

$$A_{x,V,0} = \{x\},$$
$$A_{x,V,k+1} = (A_{x,V,k} + V) \cap P,$$

and we define

$$A_{x,V} = \bigcup_{k=0}^{\infty} A_{x,V,k}.$$

First, $A_{x,V}$ is open in P, for if $y \in A_{x,V,k}$, the neighborhood $(y+V) \cap P$ of y in P is contained in $A_{x,V,k+1}$. Second, $A_{x,V}$ is closed in P, for if $y \in P \backslash A_{x,V}$, then

$$((y + V) \cap P) \cap A_{x,V} = \emptyset;$$

otherwise, there would exist $v \in V$ such that $y + v \in P \cap A_{x,V,k}$ for some $k \geq 0$, whence

$$y \in (A_{x,V,k} + V) \cap P = A_{x,V,k+1} \subseteq A_{x,V},$$

a contradiction.

Let $A_x = \cap \{A_{x,V} : V$ is a symmetric neighborhood of zero$\}$; it suffices to prove that A_x is connected. In the contrary case, $A_x = B \cup C$ where B and C are nonempty closed subsets of P such that $x \in B$ and $B \cap C = \emptyset$. As B is closed and hence compact, there is a neighborhood U of zero such that $(B + U) \cap (C + U) = \emptyset$ by (3) of 4.14. Let W be an open symmetric neighborhood of zero such that $W + W \subseteq U$, and let

$$H = P \backslash ((B + W) \cup (C + W)).$$

Then H is a closed and hence compact subset of P by 2.7. We shall show that if V is any symmetric neighborhood of zero such that $V \subseteq W$, then

$$H \cap A_{x,V} \neq \emptyset.$$

Indeed, as

$$A_{x,V} \supseteq A_x = B \cup C$$

and as

$$C \subseteq P \setminus (B + W),$$

there is a largest integer m such that $A_{x,V,m} \subseteq B + W$. Thus there exists

$$y \in A_{x,V,m+1} \setminus (B + W) \subseteq P \setminus (B + W).$$

Also

$$\begin{aligned} A_{x,V,m+1} &= (A_{x,V,m} + V) \cap P \subseteq (B + W + V) \cap P \\ &\subseteq (B + U) \cap P \subseteq P \setminus (C + U) \subseteq P \setminus (C + W). \end{aligned}$$

Thus $y \in H \cap A_{x,V}$. Consequently, as $\{A_{x,V}: V$ is a symmetric neighborhood of zero contained in $W\}$ is a filter base of closed subsets of compact P, $A_x \cap H \neq \emptyset$, a contradiction of the identity $A_x = B \cup C$. Thus A_x is connected. •

4.16 Theorem. *Let G be a locally compact group. If the connected component C of zero is compact, then the compact open subgroups of G form a fundamental system of neighborhoods of C.*

Proof. We shall first prove that if Q is a compact neighborhood of C, there is a neighborhood U of C contained in Q that is both open and closed in G. Indeed, let $B = Q \setminus Q^\circ$ (where Q° is the interior of Q), the boundary of Q, a compact set. Let $\mathcal{L}$ be the set of all subsets of Q that contain zero and are both open and closed in the topological space Q. By 4.15, $C = \cap \mathcal{L}$. If $L \cap B \neq \emptyset$ for all $L \in \mathcal{L}$, then by compactness,

$$\emptyset \neq \bigcap_{L \in \mathcal{L}} L \cap B = C \cap B \subseteq Q^\circ \cap (Q \setminus Q^\circ) = \emptyset,$$

a contradiction. Hence there exists $U \in \mathcal{L}$ such that $U \cap B = \emptyset$, whence $U \subseteq Q^\circ$. As U is closed in compact Q, U is closed in G; as U is an open subset of topological space Q that is contained in Q°, U is open in Q° and hence in G.

To prove the theorem, let P be a neighborhood of C. By (1) of 4.14 there is a compact neighborhood Q of C contained in P, and by the preceding there is an open and closed subset U that contains C and is contained in Q. Since U is compact and open, by (2) of 4.14 there is a a neighborhood V of zero such that $U + V \subseteq U$. Let W be a symmetric neighborhood of zero such that $W \subseteq U \cap V$. Then

$$W + W \subseteq U + V \subseteq U,$$

and an inductive argument establishes that for all $n \geq 1$,

$$W + W + \cdots + W \ (n \text{ terms}) \subseteq U.$$

Thus the subgroup H of G generated by W is contained in U and hence in P. By 4.10, H is both open and closed and hence compact. •

4.17 Corollary. *If G is a totally disconnected locally compact group, the compact open subgroups of G form a fundamental system of neighborhoods of zero.*

4.18 Theorem. *If E is a topological module over a topological ring A and if K is a compact subset of E, then for each neighborhood V of zero in E there is a neighborhood U of zero in A such that $UK \subseteq V$.*

Proof. For each $c \in E$, $(\lambda, x) \to \lambda x$ is continuous at $(0, c)$, so there exist an open neighborhood P_c of c and an open neighborhood U_c of zero in A such that $U_c P_c \subseteq V$. Since $\{P_c : c \in K\}$ is an open cover of K, there is a finite subset M of K such that $\cup_{c \in M} P_c \supseteq K$. Let $U = \cap_{c \in M} U_c$, an open neighborhood of zero in A. Then $UK \subseteq V$. •

4.19 Corollary. *If K is a compact subset of a topological ring A, for any neighborhood V of zero there is a neighborhood U of zero such that $UK \subseteq V$ and $KU \subseteq V$.*

Proof. A is clearly a topological left and right module over itself, so by 4.18 there exist open neighborhoods U_1 and U_2 of zero such that $U_1 K \subseteq V$ and $KU_2 \subseteq V$; let $U = U_1 \cap U_2$. •

4.20 Theorem. *If A is a compact totally disconnected ring, the open ideals of A form a fundamental system of neighborhoods of zero, that is, the topology of A is an ideal topology.*

Proof. By 4.17 the compact open additive subgroups form a fundamental system of neighborhoods of zero. Let H be a compact open additive subgroup. By 4.19 there is an open neighborhood U of zero such that $AU \subseteq H$ and $U \subseteq H$, and there is an open symmetric neighborhood L of zero such that $L \subseteq U$ and $LA \subseteq U$. Then L, AL, LA, and ALA are all subsets of H, so the ideal J of A generated by L, which is simply the additive group generated by $L \cup AL \cup LA \cup ALA$, is contained in H, and J is open by 4.10. •

4.21 Theorem. *If A is a locally compact totally disconnected ring, the compact open subrings of A form a fundamental system of neighborhoods of zero.*

Proof. By 4.17 we need only show that a compact open additive subgroup H contains an open subring. By 4.19 there is an open neighborhood U of zero such that $U \subseteq H$ and $UH \subseteq H$. Then $UU \subseteq UH \subseteq H$, and an inductive argument establishes that for each $n \geq 1$,

$$UU \ldots U \ (n \text{ terms}) \subseteq H.$$

Consequently, the subring B generated by U is contained in H, and B is open and closed (and hence compact) by 4.10. •

We conclude with a useful theorem relating the neighborhoods of zero in a topological group to those in a dense subgroup.

4.22 Theorem. *If G is a dense subgroup of a Hausdorff group G_1, the closures in G_1 of a fundamental system of neighborhoods of zero in G form a fundamental system of neighborhoods of zero in G_1.*

Proof. Let V be a neighborhood of zero in G. Then there is an open neighborhood U of zero in G_1 such that $U \cap G \subseteq V$. Hence

$$U = U \cap \overline{G} \subseteq \overline{U \cap G} \subseteq \overline{V},$$

so $\overline{V}$ is a neighborhood of zero in G_1. Conversely, any neighborhood of zero in G_1 contains a closed neighborhood W by 3.4, and W contains the closure $\overline{W \cap G}$ of the neighborhood $W \cap G$ of zero in G. •

Exercises

4.1 A closed discrete subset of a connected locally compact group is countable. [Use 4.10.]

4.2 Let S be a subset of a Hausdorff topological ring A. (a) The centralizer of S, consisting of all $x \in A$ such that $xs = sx$ for all $s \in S$, is closed. (b) The left [right] annihilator of S, consisting of all $x \in A$ such that $xs = 0$ [$sx = 0$] for all $s \in S$, is closed.

4.3 Let J be a left ideal of a topological ring A. For each $x \in J$, the left annihilator of x in A (Exercise 4.2) is open if either (a) J is discrete for its induced topology, or (b) A is locally connected and J totally disconnected for its induced topology.

4.4 If A is a nondiscrete topological ring having no nonzero zero-divisors and if the center of A is open, then A is commutative.

4.5 If A is a closed subring of a topological ring B and if

$$A_0 = \{x \in B : xA \subseteq A \text{ and } Ax \subseteq A\},$$

then A_0 is closed and is the largest subring of B of which A is an ideal.

4.6 Let K be a commutative topological ring with identity, and let A be a topological K-algebra. Let A_1 be the K-algebra obtained by adjoining an identity to A; thus $A_1 = K \times A$, with addition, multiplication, and scalar multiplication defined by

$$(\lambda, x) + (\mu, y) = (\lambda + \mu, x + y),$$
$$(\lambda, x)(\mu, y) = (\lambda\mu, \lambda y + \mu x + xy),$$
$$\alpha(\lambda, x) = (\alpha\lambda, \alpha x).$$

Show that A_1, furnished with the cartesian product topology, is a topological K-algebra.

4.7 (Correl [1958]) (a) Let K be a commutative topological ring with identity, A a topological K-algebra. The open K-submodules of A form a fundamental system of neighborhoods of zero for a weaker K-algebra topology on A. (b) In particular, if A is a topological ring, the open additive subgroups of A form a fundamental system of neighborhoods of zero for a weaker ring topology on A.

5 Quotients and Projective Limits of Rings and Modules

Let f be a function from S to T. We shall say that f is *injective* or an *injection* if for all $x, y \in S$, $f(x) = f(y)$ implies that $x = y$, that f is *surjective* or a *surjection* if the range $f(S)$ of f is T, and that f is *bijective* or a *bijection* if f is both injective and surjective.

If f is a function from one group [ring, A-module, A-algebra] to another, f is a *monomorphism* [*epimorphism*, *isomorphism*] is f is an injective [surjective, bijective] homomorphism.

Let f be a function from a topological space S to a set T. Of all the topologies on T for which f is continuous, there is a strongest, namely, $\{O \subseteq T : f^{-1}(O) \text{ is open in } S\}$, for that collection of subsets of T is easily seen to be a topology on T.

Let H be a subgroup of a group G. The *canonical surjection* from G to G/H is the surjection ϕ_H defined by $\phi_H(x) = x + H$ for all $x \in H$; if H is a normal subgroup, ϕ_H is an epimorphism, called the *canonical epimorphism* from G to G/H. Similarly, if J is an ideal of a ring or algebra A, ϕ_J is called the *canonical epimorphism* from A to A/J, and if M is a submodule of a module E, ϕ_M is called the *canonical epimorphism* from E to E/M.

5.1 Definition. *Let J be an ideal of a topological ring or algebra A. The* **quotient topology** *of A/J is the strongest topology on A/J for which the canonical epimorphism ϕ_J from A to A/J is continuous.*

We similarly define the *quotient topology* of E/M where M is a submodule [subgroup] of a topological module [group] E.

The following theorems, stated for quotient rings determined by ideals of topological rings, are also valid (with essentially the same proof) for quotient modules [groups] determined by submodules [normal subgroups] of topological modules [groups].

A function f from a topological space S to a topological space T is *open* if for every open subset O of S, $f(O)$ is open in T.

5.2 Theorem. *If J is an ideal of a topological ring A, the canonical epimorphism ϕ_J from A to A/J is continuous and open.*

Proof. By 5.1, ϕ_J is continuous. If O is an open subset of A, $\phi_J^{-1}(\phi_J(O)) = O + J$, an open subset of A by (2) of 3.3, so $\phi_J(O)$ is open in A/J. •

We shall use the following theorem in proving that the quotient topology of a quotient ring of a topological ring is a ring topology:

5.3 Theorem. *Let R, S, and T be topological spaces, let h be a continuous open surjection from R to S, and let q be a function from S to T. If $q \circ h$ is continuous [open], then q is continuous [open].*

Proof. If O is an open subset of T and if $q \circ h$ is continuous, then

$$q^{-1}(O) = h(h^{-1}(q^{-1}(O))) = h((q \circ h)^{-1}(O)),$$

an open subset of S. If O is an open subset of S and if $q \circ h$ is open, then

$$q(O) = q(h(h^{-1}(O))) = (q \circ h)(h^{-1}(O)),$$

an open subset of T. •

5.4 Theorem. *If J is an ideal of a topological ring A, the quotient topology of A/J is a ring topology.*

Proof. Let $\phi_J \times \phi_J$ be the function from $A \times A$ to $(A/J) \times (A/J)$ defined by $(\phi_J \times \phi_J)(x,y) = (\phi_J(x), \phi_J(y))$ for all $(x,y) \in A \times A$. As ϕ_J is a continuous open surjection by 5.2, so is $\phi_J \times \phi_J$. If q is either subtraction or multiplication from $A \times A$ to A, and if q_J is the corresponding function from $(A/J) \times (A/J)$ to A/J, then

$$q_J \circ (\phi_J \times \phi_J) = \phi_J \circ q.$$

As q is continuous, so is $\phi_J \circ q$; hence q_J is continuous by 5.3. •

5.5 Theorem. *If $\mathcal{V}$ is a fundamental system of neighborhoods of zero in a topological ring A and if J is an ideal of A, then $\phi_J(\mathcal{V})$ is a fundamental system of neighborhoods of zero for the quotient topology of A/J.*

Proof. As ϕ_J is open, $\phi_J(V)$ is a neighborhood of zero in A/J for each $V \in \mathcal{V}$. Conversely, if U is a neighborhood of zero in A/J, then as ϕ_J is continuous, $\phi_J^{-1}(U)$ is a neighborhood of zero in A, so there exists $V \in \mathcal{V}$ such that $V \subseteq \phi_J^{-1}(U)$, whence

$$\phi_J(V) \subseteq \phi_J(\phi_J^{-1}(U)) = U. \bullet$$

5.6 Corollary. *If the topology of a topological ring A is an ideal topology, then for any ideal J of A, the quotient topology of A/J is an ideal topology.*

5.7 Theorem. *Let J be an ideal of a topological ring A.*

(1) *A/J is Hausdorff if and only if J is closed.*
(2) *A/J is discrete if and only if J is open.*

Proof. (1) follows from 3.4, 5.1, and the identity

$$\phi_J^{-1}((A/J) \setminus \{J\}) = A \setminus J.$$

(2) follows from 2.7, 5.1, and the identity

$$\phi_J^{-1}(\{J\}) = J. \bullet$$

If B is a subring and J an ideal of a ring A such that $J \subseteq B$, then the quotient ring B/J is actually a subring of A/J. Happily, if A is a topological ring, the quotient topology of B/J is identical with the topology induced on the subring B/J of A/J by the quotient topology of A/J:

5.8 Theorem. *Let B be a subring and J an ideal of a topological ring A such that $J \subseteq B$. The quotient topology of B/J is identical with the topology induced on the subring B/J of A/J by the quotient topology of A/J.*

Proof. Let $\phi_{B,J}$ and $\phi_{A,J}$ be the canonical epimorphism from B to B/J and from A to A/J respectively. First, let O be open for the quotient topology of B/J. Then $\phi_{B,J}^{-1}(O)$ is open in B, so $\phi_{B,J}^{-1}(O) = B \cap Q$ for some open subset Q of A. To show that O is open for the topology induced on B/J by the quotient topology of A/J, it suffices to show that

$$O = (B/J) \cap \phi_{A,J}(Q)$$

as $\phi_{A,J}$ is open. Clearly

$$O \subseteq (B/J) \cap \phi_{A,J}(Q).$$

Conversely, let $\beta \in (B/J) \cap \phi_{A,J}(Q)$. Then $\beta = b + J$ for some $b \in B$ and $\beta = q + J$ for some $q \in Q$. Hence $q - b \in J$, so $q \in J + B = B$. Consequently,

$$q \in B \cap Q = \phi_{B,J}^{-1}(O),$$

so $\beta = q + J \in O$.

Second, let O be open in B/J for the topology on B/J induced by the quotient topology of A/J. Then $O = (B/J) \cap P$ for some open subset P of A/J. Clearly

$$\phi_{B,J}^{-1}(O) = B \cap \phi_{A,J}^{-1}(P),$$

an open subset of B, so O is open for the quotient topology of B/J. •

5.9 Corollary. *Let B be a subring and J an ideal of a topological ring A. The quotient topology of $(B+J)/J$ is identical with the topology induced on it by the quotient topology of A/J.*

5.10 Definition. *Let f be a function from a topological ring [module, group] A to another B. The function f is a **topological isomorphism** if f is both an isomorphism and a homeomorphism; f is a **topological homomorphism** is f is a continuous homomorphism and is also an open mapping from A onto its range $f(A)$; f is a **topological epimorphism** [**monomorphism**] if f is a surjective [injective] topological homomorphism.*

Thus f is a topological epimorphism if and only if f is a continuous open epimorphism. If J is an ideal of a topological ring A, ϕ_J is a topological epimorphism from A to A/J by 5.2. If f is a homomorphism from A to B and if f_1 is the epimorphism obtained from f by restricting its codomain to its range, then clearly f is a topological homomorphism if and only if f_1 is a topological epimorphism.

5.11 Theorem. *Let f be a homomorphism from a topological ring A to a topological ring B, and let J be an ideal contained in the kernel K of f. The homomorphism g from A/J to B satisfying $g \circ \phi_J = f$ is continuous [open, a topological homomorphism] if and only if f is. In particular, if $J = K$, g is a topological isomorphism [monomorphism] if and only if f is a topological epimorphism [homomorphism].*

The assertion follows from 5.2 and 5.3.

5.12 Corollary. *If H and J are ideals of a topological ring A such that $J \subseteq H$, then the canonical epimorphism $f : x + J \to x + H$ from A/J to A/H is a topological epimorphism.*

The assertion follows from 5.11 applied to ϕ_H.

5.13 Corollary. *If H and J are ideals of a topological ring such that $J \subseteq H$, the canonical isomorphism g from $(A/J)/(H/J)$ to A/H is a topological isomorphism.*

The assertion follows by applying 5.11 to the epimorphism of 5.12.

5.14 Theorem. *Let A be a dense subring of a topological ring B, and let J be a closed ideal of A, $\bar{J}$ its closure in B. Then $g : x + J \to x + \bar{J}$ is a topological isomorphism from A/J to the dense subring $(A + \bar{J})/\bar{J}$ of $B/\bar{J}$. If J is an open ideal of A, then $\bar{J}$ is an open ideal of B, and g is an isomorphism from A/J to $B/\bar{J}$.*

Proof. $\bar{J}$ is indeed an ideal of B by 4.2. The kernel of the restriction to A of $\phi_{\bar{J}}$ is $\bar{J} \cap A = J$, so g is a continuous isomorphism from A/J to $(A + \bar{J})/\bar{J}$ by 5.11. As A is dense in B and as $\phi_{\bar{J}}$ is continuous, $\phi_{\bar{J}}(A)$ (which is $(A + \bar{J})/\bar{J}$) is dense in $B/\bar{J}$.

To show that g is open, let O be an open subset of A/J and let $P = \phi_J^{-1}(O)$. Then $g(O) = \phi_{\bar{J}}(P)$, and $P + J = P$. As P is open in A, $P = U \cap A$ for some open subset U of B. We shall show that $(U + \bar{J}) \cap A = P$. Indeed, let $u + h \in A$ where $u \in U$ and $h \in \bar{J}$. As U is a neighborhood of u, there exists a symmetric neighborhood V of zero such that $u + V \subseteq U$. As $h \in \bar{J}$, $(V + h) \cap J \neq \emptyset$, so there exists $z \in V$ such that $z + h \in J$. Consequently,

$$u + h = (u - z) + (z + h) \in (u + V) + J \subseteq U + J.$$

Thus for some $j \in J$,
$$u + h - j \in U \cap A = P$$

since $u + h \in A$, so $u + h \in P + J = P$. Therefore

$$g(O) = \phi_{\bar{J}}(P) = \phi_{\bar{J}}(U) \cap ((A + \bar{J})/\bar{J}),$$

an open subset of $(A + \bar{J})/\bar{J}$, for if $x + \bar{J} \in \phi_{\bar{J}}(U)$ where $x \in A$, then

$$x \in \phi_{\bar{J}}^{-1}(\phi_{\bar{J}}(U)) \cap A = (U + \bar{J}) \cap A = P,$$

whence $x + \bar{J} \in \phi_{\bar{J}}(P) = g(O)$. The final assertion follows from 4.22, 4.9, and (2) of 5.7. ∎

5.15 Corollary. *If A is a subring of a topological ring B, if J is a closed ideal of A, and if $\bar{A}$ and $\bar{J}$ are the closures of A and J respectively in B, then $g : x + J \to x + \bar{J}$ is a topological isomorphism from A/J to the dense subring $(A + \bar{J})/\bar{J}$ of $\bar{A}/\bar{J}$.*

Proof. We need only let $B = \bar{A}$ in 5.14. •

5.16 Theorem. *If C is the connected component of zero in a topological ring A, then A/C is totally disconnected.*

Proof. It suffices to show that if D is a closed subset of A/C such that $\phi_C^{-1}(D)$ is disconnected, then D is disconnected; for then, if the connected component K of zero in A/C contained more than one point, $\phi_C^{-1}(K)$ would properly contain C and hence be disconnected, so K would also be disconnected, a contradiction. Let X and Y be nonempty closed subsets of $\phi_C^{-1}(D)$ (and hence of A) such that $X \cup Y = \phi_C^{-1}(D)$ and $X \cap Y = \emptyset$. For each $x \in X$, $x + C$ is a connected subset of $\phi_C^{-1}(D)$ and hence is contained in X; thus

$$X = X + C = \phi_C^{-1}(\phi_C(X)),$$

and similarly

$$Y = \phi_C^{-1}(\phi_C(Y)).$$

Therefore

$$\phi_C(X) \cap \phi_C(Y) = \phi_C(X \cap Y) = \emptyset,$$

and

$$(A/C) \setminus \phi_C(X) = \phi_C(A \setminus X),$$

an open set by 5.2; thus $\phi_C(X)$ is closed in A/C, and similarly $\phi_C(Y)$ is also closed. As

$$\phi_C(X) \cup \phi_C(Y) = \phi_C(\phi_C^{-1}(D)) = D,$$

we conclude that D is disconnected. •

5.17 Theorem. *Let J be a closed ideal of a locally compact ring A, and let C be the connected component of zero in A.*

(1) A/J is a locally compact ring.
(2) C is the intersection of all open subrings of A.
(3) A/J is totally disconnected if and only if $J \supseteq C$.
(4) A is connected if and only if the additive subring generated by each neighborhood of zero is A.

Proof. (1) follows from 5.2, 5.4, and (1) of 5.7. In particular, A/C is a totally disconnected locally compact ring by 5.16, so $\{C\}$ is the intersection of all open subrings of A/C by 4.21. But if L is an open subring of A/C, $\phi_C^{-1}(L)$ is an open subring of A. Therefore the intersection of all open subrings of A is contained in and thus, by 4.8, identical with $\phi_C^{-1}(\{C\})$, which is C.

(3) If J does not contain C, $\phi_J(C)$ is a connected subset of A/J containing more than one point. To prove the converse, assume that $J \supseteq C$. By 5.13, A/J is topologically isomorphic to $(A/C)/(J/C)$; replacing A and J respectively by A/C and J/C, we may by 5.16 further assume that A is totally disconnected. But then by 4.21, 5.2 and 5.5, the open subrings of A/J form a fundamental system of neighborhoods of zero, so A/J is totally disconnected. (4) follows from (2), 4.9, and 4.8. •

5.18 Theorem. *Let f be a homomorphism from a topological group G to a topological group H, and let $\mathcal{V}$ be a fundamental system of neighborhoods of zero in G.*

(1) f is continuous if and only if f is continuous at zero.

(2) f is open if and only if for every $V \in \mathcal{V}$, $f(V)$ is a neighborhood of zero in H.

Proof. (1) Assume f is continuous at zero. Let $a \in G$, and let U be a neighborhood of $f(a)$ in H. Then $U = f(a) + V$ for some neighborhood V of zero in H, and $f^{-1}(V)$ is a neighborhood of zero in G by hypothesis. Consequently, $a + f^{-1}(V)$ is a neighborhood of a, and as f is a homomorphism,

$$f(a + f^{-1}(V)) = f(a) + f(f^{-1}(V)) \subseteq f(a) + V = U.$$

Therefore f is continuous at a.

(2) The condition is clearly necessary: Sufficiency: To show that f is open, let O be an open subset of G. For each $x \in O$ there exists $V_x \in \mathcal{V}$ such that $x + V_x \subseteq O$. As f is a homomorphism, for each $x \in O$,

$$f(x) + f(V_x) = f(x + V_x) \subseteq f(O),$$

and by hypothesis, $f(x) + f(V_x)$ is a neighborhood of $f(x)$. Thus $f(O)$ is a neighborhood of each of its points, and so $f(O)$ is open in H. •

A *direction* $\leq$ on a set L is a reflexive, transitive, cofinal relation. Thus for all $\lambda \in L$, $\lambda \leq \lambda$; for all $\lambda, \mu, \nu \in L$, if $\lambda \leq \mu$ and $\mu \leq \nu$, then $\lambda \leq \nu$; and for all $\lambda, \mu \in L$ there exists $\gamma \in L$ such that $\lambda \leq \gamma$ and $\mu \leq \gamma$. A *directed set* is a set furnished with a direction.

5.19 Definition. *Let $(E_\lambda)_{\lambda \in L}$ be a family of nonempty sets indexed by a directed set L, and for each pair (λ, μ) of elements of L such that*

$\lambda \leq \mu$, let $f_{\lambda\mu}$ be a function from E_μ to E_λ. We shall say that $(E_\lambda)_{\lambda \in L}$ is a **projective family of sets relative to** $(f_{\lambda\mu})$ if

(PF) $$f_{\lambda\mu} \circ f_{\mu\nu} = f_{\lambda\nu}$$

for all $\lambda, \mu, \nu \in L$ such that $\lambda \leq \mu \leq \nu$. If (PF) holds, the **projective limit** of $(E_\lambda)_{\lambda \in L}$ relative to $(f_{\lambda\mu})$, denoted by $\varprojlim_{\lambda \in L}(E_\lambda, f_{\lambda\mu})$, or simply $\varprojlim_{\lambda \in L} E_\lambda$ if no confusion results, is the set of all $x \in \prod_{\lambda \in L} E_\lambda$ such that

$$f_{\lambda\mu}(pr_\mu(x)) = pr_\lambda(x)$$

for all $\lambda, \mu \in L$ satisfying $\lambda \leq \mu$ (where for each $\alpha \in L$, pr_α is the canonical projection from $\prod_{\lambda \in L} E_\lambda$ to E_α).

5.20 Theorem. *If $(E_\lambda)_{\lambda \in L}$ is a projective family of Hausdorff topological spaces relative to continuous functions $(f_{\lambda\mu})$, $\varprojlim_{\lambda \in L} E_\lambda$ is a closed subset of $\prod_{\lambda \in L} E_\lambda$.*

Proof. Let $E = \prod_{\lambda \in L} E_\lambda$. For all $\lambda, \mu \in L$ such that $\lambda \leq \mu$, the set $A_{\lambda\mu}$ of all $x \in E$ such that $f_{\lambda\mu}(pr_\mu(x)) = pr_\lambda(x)$ is closed since $f_{\lambda\mu} \circ pr_\mu$ and pr_λ are continuous functions from E to E_λ. By definition,

$$\varprojlim_{\lambda \in L} E_\lambda = \bigcap_{\lambda \leq \mu} A_{\lambda\mu}$$

and hence is closed. •

If $(E_\lambda)_{\lambda \in L}$ is a projective family of rings [A-modules, groups] relative to homomorphisms $(f_{\lambda\mu})$, it is easy to see that $\varprojlim_{\lambda \in L} E_\lambda$ is a subring [A-submodule, subgroup] of the ring [A-module, group] $\prod_{\lambda \in L} E_\lambda$.

As before, the following theorems are stated only for topological rings, but their analogues for A-modules or groups are also valid, with essentially the same proof.

5.21 Theorem. *Let A be a Hausdorff ring, and let $(J_\lambda)_{\lambda \in L}$ be a family of closed ideals of A indexed by a directed set L such that for all $\lambda, \mu \in L$, if $\lambda \leq \mu$, then $J_\lambda \supseteq J_\mu$. Let*

$$g : A \to \prod_{\lambda \in L} (A/J_\lambda)$$

be the function defined by

$$g(x) = (x + J_\lambda)_{\lambda \in L},$$

and for all $\lambda, \mu \in L$ such that $\lambda \leq \mu$, let $f_{\lambda\mu}$ be the canonical epimorphism from A/J_μ to A/J_λ defined by

$$f_{\lambda\mu}(x + J_\mu) = x + J_\lambda$$

for all $x \in A$. For each $\lambda \in L$ let A/J_λ be furnished with a ring topology $\mathcal{T}_\lambda$ such that the canonical epimorphism ϕ_λ from A to A/J_λ is continuous, and let $\prod_{\lambda \in L}(A/J_\lambda)$ be topologized with the cartesian product topology determined by $(\mathcal{T}_\lambda)_{\lambda \in L}$. Under the following conditions, g is a topological isomorphism from A to a dense subring A_0 of $\varprojlim_{\lambda \in L}(A/J_\lambda)$:

(1) For all $\lambda, \mu \in L$ such that $\lambda \leq \mu$, $f_{\lambda\mu}$ is continuous.
(2) For every neighborhood U of zero in A, there exists $\lambda \in L$ such that $J_\lambda \subseteq U$.
(3) For every neighborhood U of zero in A, there exists $\beta \in L$ such that $\phi_\beta(U)$ is a neighborhood of zero for $\mathcal{T}_\beta$.

Proof. For each $x \in A$, clearly

$$f_{\lambda\mu}(pr_\mu(g(x))) = pr_\lambda(g(x))$$

whenever $\lambda \leq \mu$, so $g(x) \in \varprojlim_{\lambda \in L}(A/J_\lambda)$. The kernel of g is $\bigcap_{\lambda \in L} J_\lambda$, so by (2) g is a monomorphism. As ϕ_λ is continuous from A to A/J_λ for each $\lambda \in L$, g is also continuous. To show that g is an open mapping from A to A_0, it suffices by 5.18 to show that if U is a neighborhood of zero in A, $g(U)$ is a neighborhood of zero in A_0. By (4) of 3.3 there is a closed symmetric neighborhood V of zero such that $V + V \subseteq U$. By (2) and (3) there exist $\lambda, \beta \in L$ such that $J_\lambda \subseteq V$ and $\phi_\beta(V)$ is a neighborhood for zero for $\mathcal{T}_\beta$. As L is directed, there exists $\mu \in L$ such that $\lambda \leq \mu$ and $\beta \leq \mu$. As $f_{\beta\mu}$ is continuous by (1) and as $\phi_\mu(V) = f_{\beta\mu}^{-1}(\phi_\beta(V))$, $\phi_\mu(V)$ is a neighborhood of zero in A/J_μ for $\mathcal{T}_\mu$. Therefore $A_0 \cap pr_\mu^{-1}(\phi_\mu(V))$ is a neighborhood of zero in A_0. But

$$A_0 \cap pr_\mu^{-1}(\phi_\mu(V)) \subseteq g(U),$$

for if $g(x) \in pr_\mu^{-1}(\phi_\mu(V))$ where $x \in A$, then

$$\phi_\mu(x) = x + J_\mu = pr_\mu(g(x)) \in \phi_\mu(V),$$

so

$$x \in \phi_\mu^{-1}(\phi_\mu(V)) = V + J_\mu \subseteq V + J_\lambda \subseteq V + V \subseteq U,$$

whence $g(x) \in g(U)$.

To show, finally, that A_0 is dense in $\varprojlim_{\lambda \in L}(A/J_\lambda)$, let U be a nonempty open subset of $\varprojlim_{\lambda \in L}(A/J_\lambda)$, and let $z \in U$. Then there is a family $(U_\lambda)_{\lambda \in L}$

5 QUOTIENTS AND PROJECTIVE LIMITS OF RINGS AND MODULES

of sets and a finite subset K of L such that $U_\lambda = A/J_\lambda$ for all $\lambda \in L \setminus K$, U_λ is an open subset of A/J_λ for all $\lambda \in K$, and, if $V = \prod_{\lambda \in L} U_\lambda$, then

$$z \in V \cap \varprojlim_{\lambda \in L}(A/J_\lambda) \subseteq U.$$

As L is directed, there exists $\beta \in L$ such that $\lambda \le \beta$ for all $\lambda \in K$. Let $a \in A$ be such that $a + J_\beta = pr_\beta(z)$. For each $\lambda \in K$,

$$pr_\lambda(g(a)) = a + J_\lambda = f_{\lambda\beta}(a + J_\beta) = f_{\lambda\beta}(pr_\beta(z)) = pr_\lambda(z) \in U_\lambda.$$

Hence $g(a) \in V \cap A_0 \subseteq U \cap A_0$. •

A common example of a projective limit is that arising from the special case of 5.21 where $\mathcal{J}$ is a filter base of closed ideals indexed by itself with direction $\le$ defined to be $\supseteq$, where A/J is given its quotient topology for all $J \in \mathcal{J}$, and where $f_{J,K}$ is the canonical homomorphism from A/K to A/J whenever $J \supseteq K$. Unless otherwise indicated, these are the underlying assumptions in any discussion of $\varprojlim_{J \in \mathcal{J}}(A/J)$ whenever $\mathcal{J}$ is a filter of closed ideals of a topological ring A. Thus $\varprojlim_{J \in \mathcal{J}}(A/J)$ is the subring of $\prod_{J \in \mathcal{J}}(A/J)$ consisting of all $(x_J + J)_{J \in \mathcal{J}}$ such that for all $J, K \in \mathcal{J}$ satisfying $J \supseteq K$, $x_J + K = x_K + K$. In this case, the mapping $g: x \to (x + J)_{J \in \mathcal{J}}$ from A to $\varprojlim_{J \in \mathcal{J}}(A/J)$ is called the *canonical homomorphism*.

5.22 Corollary. *If A is a Hausdorff topological ring and $\mathcal{J}$ a filter base of closed ideals of A such that every neighborhood of zero contains a member of $\mathcal{J}$, then the canonical homomorphism g from A to $\varprojlim_{J \in \mathcal{J}}(A/J)$ is a topological isomorphism from A to a dense subring A_0 of $\varprojlim_{J \in \mathcal{J}}(A/J)$.*

5.23 Theorem. *Let A be a compact, totally disconnected ring. There is a fundamental system of neighborhoods of zero $\mathcal{J}$ consisting of open ideals; for any such $\mathcal{J}$, A is topologically isomorphic to $\varprojlim_{J \in \mathcal{J}}(A/J)$.*

Proof. The first assertion is a restatement of 4.20. For any such $\mathcal{J}$, the range of the canonical homomorphism g from A to $\varprojlim_{J \in \mathcal{J}}(A)$ is compact and hence closed as g is continuous. Therefore by 5.22, g is a topological isomorphism. •

5.24 Corollary. *A topological ring A is compact and totally disconnected if and only if it is topologically isomorphic to the projective limit of a projective family of discrete finite rings.*

Proof. Necessity: If J is an open ideal of A, A/J is discrete by (2) of 5.7, but A/J is also compact as it is the continuous image of compact A; hence A/J is finite. The assertion therefore follows from 5.23. Sufficiency: By 5.20 the projective limit A of a family of finite, discrete rings is a closed subset of their cartesian product, a compact ring by Tikhonov's theorem that is totally disconnected. Therefore A is also a compact, totally disconnected ring.

Corollary 5.24 may be used to prove half of a theorem illustrating the power, in the context of topological rings, of the assumption that a ring has an identity element: A compact ring is totally disconnected if and only if it is a topological subring of a compact ring with identity.

5.25. Theorem. *If A is a compact, totally disconnected ring, A is a topological subring of a compact ring with identity.*

Proof. A finite ring B of m elements is a subring of a ring with identity having m^2 elements; indeed, B may be regarded as an algebra over $\mathbb{Z}/(m)$ isomorphic to a subalgebra of $(\mathbb{Z}/(m)) \times B$, where addition is defined componentwise and multiplication by

$$(\lambda, x)(\mu, y) = (\lambda\mu, \lambda y + \mu x + xy).$$

By 5.24 A is topologically isomorphic to a subring of the cartesian product of a family of discrete finite rings, and hence to a subring of the cartesian product of a family of discrete finite rings with identity.

The converse of 5.25 will be proved in §32.

Exercises

5.1 If X and Y are connected [compact] subsets of a topological [Hausdorff] ring, then $X + Y$ and XY are connected [compact].

5.2 If J is an ideal of a topological ring A and if J and A/J are both Hausdorff, then A is Hausdorff.

5.3 If C is an ideal of a topological ring A, then C is the connected component of zero if and only if C is connected and A/C is totally disconnected.

5.4 Let C be the connected component of zero in a topological ring A. (a) If J is an ideal of A contained in C, then C/J is the connected component of zero in A/J. [Use Exercise 5.3.] (b) C is the smallest of the ideals J of A such that A/J is totally disconnected.

5.5 If J is an ideal of a topological ring A and if J and A/J are both connected, then A is connected. [Use Exercise 5.4.]

5.6 If J is a closed ideal of a locally compact ring A, then A is compact if and only if J and A/J are compact.

5.7 If C is the connected component of zero in a locally compact ring A and if J is a closed ideal of A, then $(C + J)/J$ is the connected component of zero in A/J. [Use 5.17 and Exercise 5.3.]

5.8 (a) If $r \in \mathbb{R}_{>0}$, the topological group $\mathbb{R}/r\mathbb{Z}$ is compact. [Show that it is the continuous image of a compact subset of $\mathbb{R}$.] (b) Exhibit a topological isomorphism from the compact additive group $\mathbb{R}/2\pi\mathbb{Z}$ to the compact multiplicative group $\{z \in \mathbb{C} : |z| = 1\}$.

5.9 If $(A_\lambda)_{\lambda \in L}$ is a family of topological rings and if J_λ is an ideal of A_λ for each $\lambda \in L$, exhibit a topological isomorphism

$$f : (\prod_{\lambda \in L} A_\lambda)/(\prod_{\lambda \in L} J_\lambda) \to \prod_{\lambda \in L} (A_\lambda/J_\lambda).$$

5.10 Let $\mathbb{Q}$ be furnished with the usual topology it inherits from $\mathbb{R}$, and let E be the projective limit of the additive groups $(\mathbb{Q}/n\mathbb{Z})_{n \geq 1}$. The canonical mapping $g : \mathbb{Q} \to E$, defined by

$$g(x) = (x + n\mathbb{Z})_{n \geq 1},$$

is a continuous monomorphism from the additive topological group $\mathbb{Q}$ to E. Let $\mathcal{T}$ be the topology on $\mathbb{Q}$ making g a topological isomorphism from $\mathbb{Q}$ to $g(\mathbb{Q})$. Then $\mathcal{T}$ is an additive group topology on the one-dimensional $\mathbb{Q}$-vector space $\mathbb{Q}$ satisfying (TM 5) and (TM 6) of 2.16 but not (TM 4).

CHAPTER II

METRIZABILITY AND COMPLETENESS

Our first main result is that the First Axiom of Countability is not only necessary but also sufficient for a Hausdorff group topology on a group G to be metrizable, in which case the topology may be defined by a metric d satisfying $d(a+x, a+y) = d(x,y)$ for all $a, x, y \in G$. Such a metric on a commutative group defines a group topology, and the definition of a Cauchy sequence depends only on that topology. This enables us to define completeness for arbitrary Hausdorff commutative groups and to show that each such group is a dense subgroup of an essentially unique complete Hausdorff commutative group. To establish this, we assume familiarity with the theorem that each metric space is a dense subspace of an essentially unique complete metric space in considering first the case of metrizable commutative groups. These results may easily be applied to show that every Hausdorff ring [module] is a dense subring [submodule] of an essentially unique complete Hausdorff ring [module]. We conclude by discussing conditions for and consequences of the continuity of inversion in a topological ring with identity.

6 Metrizable Groups

A metric space satisfies the First Axiom of Countability, that is, each point has a countable fundamental system of neighborhoods. Happily, the converse holds for Hausdorff group topologies: If one point (and hence each point) in a Hausdorff group G has a countable fundamental system of neighborhoods, then the topology is not only metrizable, but there exists a metric d defining the topology that satisfies

$$d(a+x, a+y) = d(x, y)$$

for all $a, x, y \in G$. To establish this and other results, we need the following theorem:

6.1 Theorem. *Let G be a group, denoted additively, and let $(U_n)_{n \in \mathbb{Z}}$ be a family of symmetric subsets of G such that*

$$G = \bigcup_{n \in \mathbb{Z}} U_n$$

and for all $k \in \mathbb{Z}$,

$$0 \in U_k$$
$$U_{k+1} + U_{k+1} + U_{k+1} \subseteq U_k.$$

Let $g : G \to \mathbb{R}$ be defined by

$$g(x) = 0 \text{ if } x \in \bigcap_{n \in \mathbb{Z}} U_n$$
$$g(x) = 2^{-k} \text{ if } x \in U_k \setminus U_{k+1}.$$

For all $x, y \in G$,
  (1) $g(x) \geq 0$, and $g(x) = 0$ if and only if $x \in \bigcap_{n \in \mathbb{Z}} U_n$
  (2) $g(-x) = g(x)$
  (3) $U_k = g^{-1}([0, 2^{-k}])$ for all $k \in \mathbb{Z}$
and, if each U_n is a subgroup,
  (4) $g(x + y) \leq \sup\{g(x), g(y)\}$.

Let $f : G \to \mathbb{R}$ be defined by

$$f(x) = \inf\{\sum_{i=1}^{p} g(z_i) : z_1, z_2, \ldots, z_p \in G \text{ and } z_1 + z_2 + \cdots + z_p = x\}.$$

For all $x, y \in G$,
  (5) $f(x) \geq 0$, and $f(x) = 0$ if and only if $x \in \bigcap_{n \in \mathbb{Z}} U_n$
  (6) $f(-x) = f(x)$
  (7) $f(x + y) \leq f(x) + f(y)$
  (8) $|f(x) - f(y)| \leq f(x - y)$
  (9) $U_k \subseteq f^{-1}([0, 2^{-k}]) \subseteq U_{k-1}$ for all $k \in \mathbb{Z}$
and, if each U_n is a subgroup,
  (10) $f(x) = g(x)$, whence $f(x + y) \leq \sup\{f(x), f(y)\}$.

Proof. The assertions concerning g are evident. We shall first prove by induction that for any sequence $(z_i)_{1 \leq i \leq p}$ of elements of G,

(*) $$\tfrac{1}{2} g(z_1 + z_2 + \cdots + z_p) \leq \sum_{i=1}^{p} g(z_i).$$

The assertion clearly holds if $p = 1$ or if $\sum_{i=1}^{p} g(z_i) = 0$; indeed, in the latter case,

$$z_i \in \bigcap_{n \in \mathbb{Z}} U_n$$

for all $i \in [1,p]$, whence for every $k \in \mathbb{Z}$,
$$z_1 + z_q + \cdots + z_p \in U_{k+1} + U_{k+2} + \cdots + U_{k+p} \subseteq U_k.$$

Assume that (*) holds for any sequence of p terms whenever $p < q$, and let $z_1, \ldots, z_q \in G$ be such that $a > 0$ where $a = \sum_{i=1}^{q} g(z_i)$. Let h be the smallest of the integers k such that
$$\sum_{i=1}^{k} g(z_i) > \frac{a}{2}.$$

Then
$$\sum_{i=1}^{h-1} g(z_i) \leq \frac{a}{2},$$

and
$$\sum_{i=h+1}^{q} g(z_i) = a - \sum_{i=1}^{h} g(z_i) < \frac{a}{2},$$

so by our inductive hypothesis,
$$g(z_1 + z_2 + \cdots + z_{h-1}) \leq a,$$
$$g(z_{h+1} + \cdots + z_q) < a,$$

and, of course,
$$g(z_h) \leq a.$$

Let k be the smallest integer such that $2^{-k} \leq a$. Thus
$$z_1 + z_2 + \cdots + z_{h-1} \in U_k,$$
$$z_h \in U_k,$$
$$z_{h+1} + \cdots + z_q \in U_k,$$

so
$$z_1 + z_2 + \cdots + z_q \in U_k + U_k + U_k \subseteq U_{k-1},$$

whence
$$\tfrac{1}{2} g(z_1 + z_2 + \cdots + z_q) \leq 2^{-(k-1)-1} = 2^{-k} \leq a = \sum_{i=1}^{q} g(z_i).$$

Thus (*) holds for any $p \geq 1$.

Clearly $f(x) \geq 0$ and $f(x) = 0$ if $x \in \bigcap_{n \in \mathbb{Z}} U_n$. Conversely, suppose that $f(x) = 0$, and let $k \in \mathbb{Z}$. Then there exist $z_1, \ldots, z_p \in G$ such that $z_1 + z_2 + \cdots + z_p = x$ and
$$\sum_{i=1}^{p} g(z_i) \leq 2^{-(k+1)},$$
so by (*)
$$g(x) \leq 2 \cdot 2^{-(k+1)} = 2^{-k},$$
and therefore $x \in U_k$. Thus (5) holds. Also, (6) follows from (2), and (7) from the definition of f. By (7),
$$f(x) = f((x-y) + y) \leq f(x-y) + f(y),$$
and similarly by (6),
$$f(y) = f((y-x) + x) \leq f(y-x) + f(x) = f(x-y) + f(x),$$
so (8) follows.

To establish (9), we first note that if $x \in U_k$,
$$f(x) \leq g(x) \leq 2^{-k}.$$
Assume $f(x) \leq 2^{-k}$. Then there exist $z_1, \ldots, z_p \in G$ such that $x = z_1 + z_2 + \cdots + z_p$ and
$$\sum_{i=1}^{p} g(z_i) < 2^{-k+1}.$$
By (*),
$$\tfrac{1}{2} g(x) < 2^{-k+1},$$
so $g(x) < 2^{-k+2}$ and therefore $g(x) \leq 2^{-k+1}$. Consequently, $x \in U_{k-1}$ by (3).

Finally, assume that each U_k is a subgroup. We have already seen that $f(x) \leq g(x)$ for all $x \in G$ and $f(x) = g(x) = 0$ for all $x \in \bigcap_{n \in \mathbb{Z}} U_n$. Assume that $x \in U_n \setminus U_{n+1}$. If $(z_i)_{1 \leq i \leq p}$ is any sequence such that $z_1 + z_2 + \cdots + z_p = x$, then not all z_i can belong to U_{n+1}, so there exists $j \in [1, p]$ such that $g(z_j) > 2^{-(n+1)}$ and hence $g(z_j) \geq 2^{-n}$, and consequently
$$\sum_{i=1}^{p} g(z_i) \geq 2^{-n} = g(x)$$
by (3). Thus $f(x) \geq g(x)$. ∎

6.2 Theorem. *If F is a closed subset of a topological group G and if $a \in G \setminus F$, there is a continuous function h from G to $[0,1]$ such that $h(a) = 0$ and $h(x) = 1$ for all $x \in F$. In particular, the topology of a Hausdorff group is completely regular.*

Proof. If h_1 has the desired properties for zero and $F + (-a)$, then $h : x \to h_1(x - a)$ has the desired properties for a and F. Therefore we may assume that $a = 0$. By 2.10 there is a decreasing family $(U_n)_{n \in \mathbb{Z}}$ of symmetric neighborhoods of zero such that

$$U_n = G \text{ if } n < 0,$$
$$U_0 \subseteq G \setminus F,$$

and, for all $n \geq 0$,

$$U_{n+1} + U_{n+1} + U_{n+1} \subseteq U_n.$$

Let f be the function associated to $(U_n)_{n \in \mathbb{Z}}$ by 6.1. By (8) and (9) of that theorem, if $x - y \in U_k$, then $|f(x) - f(y)| \leq 2^{-k}$, so f is continuous from G to $\mathbb{R}$. If $x \in F$, then $x \in G \setminus U_0$, so by (9), $f(x) > \frac{1}{2}$. Consequently, h, defined by

$$h(x) = \inf\{2f(x), 1\},$$

has the desired properties. •

6.3 Definition. *A metric d on a group G is* **left invariant** *if*

$$d(a + x, a + y) = d(x, y)$$

for all $a, x, y \in G$. Similarly, d is **right invariant** *if*

$$d(x + a, y + a) = d(x, y)$$

for all $a, x, y \in G$, and d is an **invariant** *metric if d is both left and right invariant. A metric d on a set E is an* **ultrametric** *if*

$$d(x, z) \leq \sup\{d(x, y), d(y, z)\}$$

for all $x, y, z \in E$.

6.4 Theorem. *Let G be a Hausdorff group. If there is a countable fundamental system of neighborhoods of zero, there is a left [right] invariant metric on G defining its topology. If there is a countable family of open subgroups that is a fundamental system of neighborhoods of zero, there is a left [right] invariant ultrametric on G defining its topology.*

Proof. By 2.10 there is a fundamental sequence $(U_n)_{n \geq 1}$ of symmetric neighborhoods of zero such that $U_{n+1} + U_{n+1} + U_{n+1} \subseteq U_n$ for all $n \geq 1$.

Let $U_n = G$ for all $n \leq 0$. Let f be the function associated to $(U_n)_{n \in \mathbb{Z}}$ by 6.2. By that theorem, the functions d_1 and d_2, defined by

$$d_1(x,y) = f(-x+y)$$
$$d_2(x,y) = f(x-y)$$

are easily seen to be the desired left and right invariant metrics defining the topology of G. If, in addition, each U_n is a subgroup, then by (10) of 6.1, d_1 and d_2 are ultrametrics. •

6.5 Definition. *A function N from a commutative group G to $\mathbb{R}_{\geq 0}$ is a **norm** if N satisfies (N 1)–(N 3) and (N 5) of Definition 1.2, and N is an **ultranorm** if, in addition,*

(N 6) $$N(x+y) \leq \sup\{N(x), N(y)\}$$

for all $x, y \in G$.

Clearly (N 6) implies (N 2).

An ultranorm on a ring is a norm that is an ultranorm on the underlying additive group.

6.6 Theorem. *Let G be a commutative group. An invariant [ultra-]metric d on G defines a[n] [ultra]norm N_d on G by*

$$N_d(x) = d(x, 0),$$

and an [ultra]norm N on G defines an invariant [ultra]metric d_N by

$$d_N(x, y) = N(x - y).$$

Thus $d \to N_d$ is a bijection from the set of all invariant [ultra]metrics on G to the set of all [ultra]norms on G, and its inverse is $N \to d_N$. Any invariant metric on G defines a group topology.

Proof. The proof of the first statement is easy. The second follows from the identities

$$d_{N_d}(x, y) = N_d(x - y) = d(x - y, 0) = d(x, y)$$

and

$$N_{d_N}(x) = d_N(x, 0) = N(x).$$

The proof of the third is contained in the proof of 1.3. •

Consequently, the *topology defined by a norm* on a commutative group G is the topology defined by its associated invariant metric. From 6.4 and 6.6 we obtain:

6.7 Theorem. *Let G be a commutative Hausdorff group. The following statements are equivalent:*

$1°$ *There is a countable fundamental system of neighborhoods of zero [consisting of subgroups].*

$2°$ *The topology of G is given by a[n] [ultra]metric.*

$3°$ *The topology of G is given by an invariant [ultra]metric.*

$4°$ *The topology of G is given by a[n] [ultra]norm.*

In contrast with the situation in topology, where two metrics on a set may define the same topology but yield different Cauchy sequences, any two invariant metrics on a commutative group that define the same topology yield the same Cauchy sequences, which may be identified solely in terms of the topology they define:

6.8 Theorem. *Let d be an invariant metric on a commutative group G. A sequence $(x_n)_{n \geq 1}$ in G is a Cauchy sequence for d if and only if for each neighborhood U of zero there exists $p \geq 1$ such that for all $m, n \geq p$, $x_m - x_n \in U$, and $(x_n)_{n \geq 1}$ converges to $a \in G$ if and only if for each neighborhood U of zero there exists $p \geq 1$ such that for all $m \geq p$, $x_m - a \in U$.*

The proof follows readily from the identity $d(x,y) = d(x-y, 0)$.

Consequently, we may make the following definition:

6.9 Definition. *A commutative metrizable group is* **complete** *if every Cauchy sequence for an invariant metric on G defining its topology converges. A topological group $\widehat{G}$ is a* **completion** *of G if $\widehat{G}$ is a complete metrizable group of which G is a dense subgroup.*

To show that every metrizable commutative group has a completion, we shall use the following facts from the theory of metric spaces. (1) Every metric space has an essentially unique completion: that is, if ρ is a metric on T, there exist a set $\widehat{T}$ containing T and a complete metric $\widehat{\rho}$ on $\widehat{T}$ extending ρ such that T is a dense subset of $\widehat{T}$; and if σ is a complete metric on a set S containing T that extends ρ and if T is dense in S, then there is an isometry f from S to $\widehat{T}$ such that $f(t) = t$ for all $t \in T$. (2) A uniformly continuous function from a dense subset D of a metric space S to a complete metric space T is the restriction to D of a unique uniformly continuous function from S to T. (3) Let ρ be a metric on T. The function $\rho_\times$ from $(T \times T) \times (T \times T)$ to $\mathbb{R}_{\geq 0}$, defined by

$$\rho_\times((x,y),(u,v)) = \rho(x,u) + \rho(y,v),$$

is a metric on $T \times T$ yielding the cartesian product topology defined by the topology given by ρ. Consequently, if $\widehat{T}$ with metric $\widehat{\rho}$ is the completion of T with metric ρ, $\widehat{T} \times \widehat{T}$ with metric $\widehat{\rho_\times}$ is the completion of $T \times T$ with metric $\rho_\times$. Furthermore, ρ is uniformly continuous from $T \times T$ to $\mathbb{R}$, since

$$|\rho(x,y) - \rho(u,v)| \leq |\rho(x,y) - \rho(u,y)| + |\rho(u,y) - \rho(u,v)|$$
$$\leq \rho(x,u) + \rho(y,v) = \rho_\times((x,y),(u,v)).$$

6.10 Theorem. *If d is an invariant [ultra]metric on a commutative topological group G defining its topology, then G has a completion $\widehat{G}$ whose topology is defined by a unique invariant [ultra]metric $\widehat{d}$ that extends d.*

Proof. Let $\widehat{G}$ with metric $\widehat{d}$ be the completion of the metric space G with metric d. By statement (3), $\widehat{G} \times \widehat{G}$ with metric $\widehat{d_\times}$ is the completion of $G \times G$ for metric $d_\times$. Let s be the function from $G \times G$ to G defined by

$$s(x,y) = x + y.$$

Then s is uniformly continuous for the metrics $d_\times$ and d, for

$$d((s(x,y), s(u,v)) = d(x+y, u+v) = d(x+y-u-v, 0)$$
$$= d(x-u+y-v, 0) = d(x-u, v-y)$$
$$\leq d(x-u, 0) + d(0, v-y) = d(x,u) + d(y,v)$$
$$= d_\times((x,y),(u,v)).$$

Consequently s has a unique continuous extension $\widehat{s}$ from $\widehat{G} \times \widehat{G}$ to $\widehat{G}$. We define addition on $\widehat{G}$ by

$$x + y = \widehat{s}(x,y)$$

for all $x, y \in \widehat{G}$.

The functions f and g from $\widehat{G} \times \widehat{G} \times \widehat{G}$ to $\widehat{G} \times \widehat{G}$, defined by

$$f(x,y,z) = (x, \widehat{s}(y,z))$$

and

$$g(x,y,z) = (\widehat{s}(x,y), z)$$

are both continuous, so $\widehat{s} \circ f$ and $\widehat{s} \circ g$ are also continuous. As addition on G is associative, they agree on the dense subset $G \times G \times G$ of $\widehat{G} \times \widehat{G} \times \widehat{G}$; hence they agree on $\widehat{G} \times \widehat{G} \times \widehat{G}$, that is, addition on $\widehat{G}$ is associative. A similar argument establishes that addition is commutative on $\widehat{G}$ and that the zero element of G is the zero element for addition on $\widehat{G}$.

The function $j : x \to -x$ is uniformly continuous from G to G, since
$$d(-x,-y) = d(-x+x+y, -y+x+y) = d(y,x).$$
Hence j has a unique continuous extension $\widehat{j}$ from $\widehat{G}$ to $\widehat{G}$. Consequently, the function $x \to \widehat{s}(x, \widehat{j}(x))$ is continuous from $\widehat{G}$ to $\widehat{G}$; as it and the constant zero function agree on G, they agree on $\widehat{G}$, that is, $\widehat{j}(x)$ is the additive inverse of x for each $x \in \widehat{G}$. Therefore $\widehat{G}$ is a comutative topological group.

Let $a \in G$. As $L_a : (x,y) \to (a+x, a+y)$ is continuous from $\widehat{G} \times \widehat{G}$ to $\widehat{G} \times \widehat{G}$, $\widehat{d} \circ L_a$ is continuous from $\widehat{G} \times \widehat{G}$ to $\mathbb{R}$. As $\widehat{d} \circ L_a$ and $\widehat{d}$ agree on $G \times G$, they agree on $\widehat{G} \times \widehat{G}$, so
$$\widehat{d}(a+x, a+y) = \widehat{d}(x,y)$$
for all $x, y \in \widehat{G}$. For any $x, y \in \widehat{G}$, the function $z \to \widehat{d}(z+x, z+y)$ is continuous on $\widehat{G}$; we have just seen that it agrees with the constant function defined by the number $d(x,y)$ on G, so
$$\widehat{d}(z+x, z+y) = \widehat{d}(x,y)$$
for all $z \in \widehat{G}$. Thus $\widehat{d}$ is an invariant metric. Finally,
$$h : (x,y,z) \to \sup\{\widehat{d}(x,y), \widehat{d}(y,z)\} - \widehat{d}(x,z)$$
is continuous from $\widehat{G} \times \widehat{G} \times \widehat{G}$ to $\mathbb{R}$; so $h^{-1}(\mathbb{R}_{\geq 0})$ is closed. If d is an ultrametric, that set contains $G \times G \times G$ and hence is all of $\widehat{G} \times \widehat{G} \times \widehat{G}$, so $\widehat{d}$ is an ultrametric. •

6.11 Corollary. *If the topology of a commutative topological group G is given by a[n] [ultra]norm N, the topology of $\widehat{G}$ is given by a unique [ultra]norm $\widehat{N}$ that extends N.*

6.12 Theorem. *Let G be a commutative metrizable topological group, H a closed subgroup. Then G/H is a metrizable group. If G is complete, so is G/H.*

Proof. Let $(V_n)_{n \geq 1}$ be a fundamental system of symmetric neighborhoods of zero such that $V_{n+1} + V_{n+1} \subseteq V_n$ for all $n \geq 1$. Then $(\phi_H(V_n))_{n \geq 1}$ is a fundamental system of neighborhoods of the zero element H of G/H by the group analogue of 5.5, so G/H is metrizable by 6.4. Assume that G is complete, and let $(\alpha_n)_{n \geq 1}$ be a Cauchy sequence in G/H. Extracting a subsequence if necessary, we may assume that $\alpha_{n+1} - \alpha_n \in \phi_H(V_n)$ for all $n \geq 1$. We shall inductively obtain a sequence $(x_n)_{n\geq 1}$ in G such that $x_n \in \alpha_n$ and $x_{n+1} - x_n \in V_n$ for all $n \geq 1$. Indeed, assume that $x_1, \ldots, x_m$ satisfy $x_n \in \alpha_n$ for all $n \in [1, m-1]$ and $x_{n+1} - x_n \in V_n$ for all $n \in [1, m-1]$. Let $y \in G$ be such that $\alpha_{m+1} = y + H$. As $\alpha_{m+1} - \alpha_m \in \phi_H(V_m)$,

$$y - x_m \in \phi_H^{-1}(\phi_H(V_m)) = V_m + H,$$

so

$$y - x_m = v + h$$

for some $v \in V_m$ and some $h \in H$. Let $x_{m+1} = y - h$. Then $x_{m+1} \in \alpha_{m+1}$ and

$$x_{m+1} - x_m = v \in V_m.$$

Thus a sequence with the desired properties exists. For any $n \geq 1$, $p \geq 1$,

$$x_{n+p} \in x_{n+p-1} + V_{n+p-1} \subseteq x_{n+p-2} + V_{n+p-2} + V_{n+p-1} \subseteq \cdots$$
$$\subseteq x_n + V_n + V_{n+1} + \cdots + V_{n+p-1} \subseteq x_n + V_{n-1}.$$

Thus by 6.8, $(x_n)_{n\geq 1}$ is a Cauchy sequence in G and hence converges to some $c \in G$, so $(\alpha_n)_{n\geq 1}$ converges to $c + H \in G/H$. •

We have seen from 6.7, in particular, that if a metric on a group defines a group topology, that topology is also defined by a left [right] invariant metric. A much deeper theorem is that if a complete metric on a group defines a group topology (or even, merely, a topology for which translations are continuous), then the topology it defines is also given by a complete left [right] invariant metric:

6.13 Theorem. *If T is a topology on a group G defined by a complete metric such that for each $a \in G$, the functions $x \to a + x$ and $x \to x + a$ are continuous, then T is a group topology and is defined by a complete left [right] invariant metric.*

A proof is given in §7 of *Topological Fields*.

Another celebrated theorem concerning metrizable groups is the following "closed graph" theorem:

6.14 Theorem. *If g is an epimorphism from a complete metrizable group G to a complete separable metrizable group H whose graph is a closed subset of $G \times H$, then g is continuous.*

For a proof, see, for example, Theorem 8.8 of *Topological Fields*.

Exercises

6.1 If d is a left invariant metric on a group G and if H is a closed normal subgroup, the function d_H from $(G/H) \times (G/H)$ to $\mathbb{R}_{\geq 0}$, defined by

$$d_H(\alpha, \beta) = \inf\{d(a,b) : a \in \alpha,\ b \in \beta\},$$

is a left invariant metric on G/H defining its quotient topology.

6.2 (Freudenthal [1935]) Let G be a metrizable group. (a) If f is a topological epimorphism from G to a Hausdorff group H, then H is metrizable; if $(y_n)_{n \geq 1}$ is a sequence of points in H converging to $b \in H$, then for any $a \in G$ such that $f(a) = b$ there is a sequence $(x_k)_{k \geq 1}$ in G converging to a such that $(f(x_k))_{k \geq 1}$ is a subsequence of $(y_n)_{n \geq 1}$. (b) If K is a compact subgroup of G such that G/K is compact, then G is compact.

6.3 (Ng and Warner [1972]) Let H be a complete metrizable commutative group, and let s be a continuous function from H into H such that $s(0) = 0$. If f is a homomorphism from H into the additive group $\mathbb{R}$ such that for some $K > 0$,

$$f(x)^2 \leq K f(s(x))$$

for all $x \in H$, then f is continuous. [Suppose that $(a_k)_{k \geq 0}$ is a sequence such that $\lim_{k \to \infty} a_k = 0$ but $f(a_k) \geq e > 0$ for all $k \geq 0$. Let $m \in \mathbb{N}$ be such that $m \geq K/e$, and define $g : H \times K \to H$ by

$$g(x_1, x_2) = x_1 + m.s(x_2).$$

Define $(g_k)_{k \geq 0}$ recursively by $g_0(x) = x$ for all $x \in H$, and, if g_{k-1} is defined from H^k to H, g_k is defined from H^{k+1} to H by

$$g_k(x_1, \ldots, x_{k+1}) = g(x_1, g_{k-1}(x_2, \ldots, x_{k+1})).$$

Show that

$$g_k(x_1, \ldots, x_k, 0) = g_{k-1}(x_1, \ldots, x_k)$$

for all $k \geq 1$. Let $(V_n)_{n \geq 1}$ be a decreasing fundamental sequence of neighborhoods of zero such that $V_{n+1} + V_{n+1} \subseteq V_n$ for all $n \geq 1$. Show that there is a subsequence $(b_n)_{n \geq 0}$ of $(a_k)_{k \geq 0}$ such that

$$g_{n-k+1}(b_k, \ldots, b_n, b_{n+1}) - g_{n-k+1}(b_k, \ldots, b_n, 0) \in V_{n+1}$$

for all $k \in [0, n]$. Show that $(g_{p-k}(b_k, \ldots, b_p))_{p \geq k}$ has a limit c_k for each $k \geq 0$ and that

$$f(c_k) \geq e + e^{-1} f(c_{k+1})^2.$$

Infer that for any $r \geq 1$,

$$f(c_0) \geq re + e^{1 - 2^r}.]$$

7 Completions of Commutative Hausdorff Groups

To extend the definition of completeness to all Hausdorff commutative groups, we need some additional terminology.

Let E be a topological space, $\mathcal{B}$ a filter base on E. The filter base $\mathcal{B}$ *converges to* $c \in E$ if the filter generated by $\mathcal{B}$ contains the filter of neighborhoods of c, or equivalently, if every neighborhood of c contains a member of $\mathcal{B}$. If E is Hausdorff, $\mathcal{B}$ converges to at most one point of E, for if U and V are disjoint neighborhoods of two points of E, the filter generated by $\mathcal{B}$ cannot contain both U and V since then it would contain the empty set $U \cap V$.

If $(x_n)_{n \geq 1}$ is a sequence of points of E, the filter base *associated to* $(x_n)_{n \geq 1}$ is the filter base $\{F_n : n \geq 1\}$, where $F_n = \{x_m : m \geq n\}$ for each $n \geq 1$. If E is a topological space, a sequence in E clearly converges to a point of E if and only if the associated filter base does.

A point $c \in E$ is *adherent* to $\mathcal{B}$ (or a *cluster point* of $\mathcal{B}$) if c belongs to the closure of each member of $\mathcal{B}$; the *adherence* of $\mathcal{B}$ is the set of all points adherent to $\mathcal{B}$, that is, the intersection of the closures of the members of $\mathcal{B}$. If $\mathcal{B}$ converges to c, then c is adherent to $\mathcal{B}$, for if $B \in \mathcal{B}$ and if U is a neighborhood of c, then $U \cap B \neq \emptyset$ since $U \cap B$ belongs to the filter generated by $\mathcal{B}$.

If $(x_n)_{n \geq 1}$ is a sequence of points in a topological space E, then c is a *cluster point* of $(x_n)_{n \geq 1}$ if c is a cluster point of the filter base associated to $(x_n)_{n \geq 1}$, or equivalently, if for every neighborhood U of c and every $n \geq 1$ there exists $m \geq n$ such that $x_m \in U$.

The image of $\mathcal{B}$ under any function f from E to F is a filter base on F. If F is also a topological space, if $\mathcal{B}$ converges to c, and if f is continuous at c, then $f(\mathcal{B})$ converges to $f(c)$, for if V is any neighborhood of $f(c)$, the neighborhood $f^{-1}(V)$ of c contains a member B of $\mathcal{B}$, so $f(B) \subseteq V$.

7.1 Definition. *Let G be a commutative topological group. If V is a neighborhood of zero, a subset F of G is V-**small*** if $F + (-F) \subseteq V$, that is, if $x - y \in V$ for all $x, y \in F$. A filter [base] on G is a **Cauchy filter** [base] if it contains a V-small set for every neighborhood V of zero. A filter [base] $\mathcal{B}$ on a subset E of G is a **Cauchy filter** [base] on E if the filter it generates on G is a Cauchy filter.*

If d is an invariant metric on a commutative group G, then by 6.8 a sequence in G is a Cauchy sequence for d if and only if its associated filter base is a Cauchy filter base.

7.2 Theorem. *Let $\mathcal{B}$ be a filter base on a commutative topological group G, and let $c \in G$. Then $\mathcal{B}$ converges to c if and only if c is adherent to $\mathcal{B}$ and $\mathcal{B}$ is a Cauchy filter base.*

Proof. Necessity: Let V be a neighborhood of zero, and let W be a symmetric neighborhood of zero such that $W + W \subseteq V$. By hypothesis, there exists $B \in \mathcal{B}$ such that $B \subseteq c + W$. Consequently, B is V-small, for

$$B + (-B) \subseteq (c + W) + [-(c + W)] = W + (-W) = W + W \subseteq V.$$

Therefore $\mathcal{B}$ is a Cauchy filter base. Also, c is adherent to $\mathcal{B}$, for if U is a neighborhood of c and if $B \in B$, then $U \cap B$ contains a member of $\mathcal{B}$ and hence $U \cap B \neq \emptyset$.

Sufficiency: Let $\mathcal{B}$ be a Cauchy filter base to which c is adherent. Let V be a neighborhood of zero; we shall show that $c + V$ contains a member of $\mathcal{B}$. Let W be a neighborhood of zero such that $W + W \subseteq V$, and let B be a W-small member of $\mathcal{B}$. As $c \in \overline{B}$, there exists $b \in B \cap (c + W)$. Consequently as $(-b) + B \subseteq W$,

$$B \subseteq b + W \subseteq c + W + W \subseteq c + V.$$

Thus $\mathcal{B}$ converges to c. •

7.3 Definition. *Let G be a commutative Hausdorff group. A subset E of G is **complete** if every Cauchy filter on E converges to a point of E. A Hausdorff group $\widehat{G}$ is a **completion** of G if G is a dense topological subgroup of $\widehat{G}$ and $\widehat{G}$ is complete.*

The following theorem establishes that Definition 7.3 is an extension of Definition 6.9 to arbitrary Hausdorff groups:

7.4 Theorem. *Let G be a commutative metrizable topological group, d an invariant metric defining its topology. Then G is complete if and only if d is a complete metric.*

Proof. The condition is necessary, for we have just seen that a sequence is a Cauchy sequence for d if and only if its associated filter base is Cauchy, and by 6.8 a sequence converges for d if and only if the associated filter base converges. Sufficiency: Let $\mathcal{F}$ be a Cauchy filter on G and let $(V_n)_{n \geq 1}$ be a fundamental decreasing sequence of neighborhoods of zero. For each $p \geq 1$ let $F_p \in \mathcal{F}$ be V_p-small, and let

$$x_p \in \bigcap_{k=1}^{p} F_k.$$

If $m \geq p$ and $n \geq p$, then both x_m and x_n belong to F_p, so $x_m - x_n \in V_p$. Thus $(x_n)_{n \geq 1}$ is a Cauchy sequence for d by 6.8 and hence converges to a point c. To show that $\mathcal{F}$ converges to c, let U be a neighborhood of zero, and let $p \geq 1$ be such that $V_p + V_p \subseteq U$. As $(x_n)_{n \geq 1}$ converges to c, there exists $m \geq p$ such that $x_n - c \in V_p$ for all $n \geq m$. Hence $F_m \subseteq c + U$, for if $x \in F_m$, then

$$x = (x - x_m) + (x_m - c) + c \in V_m + V_p + c$$
$$\subseteq V_p + V_p + c \subseteq U + c. \bullet$$

7.5 Theorem. *Let E be a subset of a Hausdorff commutative group G. (1) If E is complete, so is every closed subset of E. (2) If E is complete, then E is closed in G. (3) If E is compact, then E is complete.*

Proof. (1) If F is a closed subset of E and if $\mathcal{F}$ is a Cauchy filter on F, then by hypothesis $\mathcal{F}$ converges in the space E to a point c of E; as each member of $\mathcal{F}$ is a subset of F and as c is adherent to $\mathcal{F}$ by 7.2, $c \in \overline{F} = F$, and also $\mathcal{F}$ converges to c in the space F.

(2) Let $c \in \overline{E}$, and let $\mathcal{V} = \{V \cap E : V \text{is a neighborhood of } c \in G\}$. Then $\mathcal{V}$ is a filter on E converging to c in the space G, so by 7.2 $\mathcal{V}$ is a Cauchy filter on E and hence converges to a point of E, which must be c as G is Hausdorff.

(3) The assertion follows from 7.2, since a filter base on a compact space has an adherent point.$\bullet$

7.6 Theorem. *If a Hausdorff commutative group G has a complete neighborhood V of zero, then G is complete.*

Proof. By (1) of 7.5 and 3.3, we may assume that V is symmetric. Let $\mathcal{F}$ be a Cauchy filter on G. Then $\mathcal{F}$ contains a V-small set L. Let $a \in L$, and let

$$\mathcal{F}_V = \{F + (-a) : F \in \mathcal{F} \text{ and } F + (-a) \subseteq V\}.$$

Since $F + (-a) \subseteq V$ if $F \subseteq L$, $\mathcal{F}_V$ is a filter on V. Let U be a neighborhood of zero. If F is a U-small subset contained in L, then $F + (-a)$ is a U-small subset of V, for

$$(F + (-a)) + [-(F + (-a))] = F + (-F) \subseteq U.$$

Therefore $\mathcal{F}_V$ is a Cauchy filter on V and thus converges to some $c \in V$. But then, as $x \to x + a$ is continuous, $\mathcal{F}_V + a$ and hence also $\mathcal{F}$ converge to $c + a$. $\bullet$

7.7 Corollary. *A commutative locally compact group is complete. In particular, a discrete commutative group is complete.*

7.8 Theorem. *Let G be the cartesian product of a family $G_{\lambda \in L}$ of commutative topological groups. (1) If $\mathcal{F}$ is a filter on G, then $\mathcal{F}$ is a Cauchy filter if and only if for all $\lambda \in L$, $pr_\lambda(\mathcal{F})$ is a Cauchy filter on G_λ (where pr_λ is the canonical epimorphism from G to G_λ). (2) G is complete if and only if G_μ is complete for all $\mu \in L$.*

Proof. (1) Let $V = \prod_{\lambda \in L} V_\lambda$, where each V_λ is a neighborhood of zero in G_λ and $V_\lambda = G_\lambda$ for all but finitely many $\lambda \in L$. Clearly F is V-small if and only if $pr_\lambda(F)$ is V_λ-small for all $\lambda \in L$. (2) A filter $\mathcal{F}$ on G converges to $(c_\lambda)_{\lambda \in L}$ if and only if for all $\lambda \in L$, $pr_\lambda(\mathcal{F})$ converges to c_λ. Necessity: Let $\mathcal{F}_\mu$ be a Cauchy filter on G_μ. For each $F \in \mathcal{F}_\mu$, let $F' = \prod_{\lambda \in L} F_{\lambda \mu}$, where $F_{\lambda \mu} = \{0\}$ if $\lambda \neq \mu$ and $F_{\mu \mu} = F$, and let $\mathcal{F} = \{F' : F \in \mathcal{F}_\mu\}$. Clearly $\mathcal{F}$ is a Cauchy filter base on G, and $pr_\mu(\mathcal{F}) = \mathcal{F}_\mu$. Therefore as $\mathcal{F}$ converges, so does $\mathcal{F}_\mu$. Sufficiency: By (1), for each $\lambda \in L$ there exists $c_\lambda \in G_\lambda$ such that $pr_\lambda(\mathcal{F})$ converges to c_λ. Therefore $\mathcal{F}$ converges to $(c_\lambda)_{\lambda \in L}$. •

7.9 Theorem. *A commutative Hausdorff group G has a completion.*

Proof. By set-theoretic considerations, we need only show that G is topologically isomorphic to a dense subgroup of a complete Hausdorff commutative group. Let $\mathcal{U}$ be the set of all sequences $(U_n)_{n \geq 1}$ such that for all $n \geq 1$, U_n is a closed symmetric neighborhood of zero and $U_{n+1} + U_{n+1} \subseteq U_n$. For each $U \in \mathcal{U}$, let U_n be the nth term of U, so that $U = (U_n)_{n \geq 1}$. We introduce a direction $\leq$ on $\mathcal{U}$ by

$$U \leq V \text{ if and only if } U_n \supseteq V_n \text{ for all } n \geq 1.$$

Clearly $\leq$ is an ordering of $\mathcal{U}$; it is a direction since for any $U, V \in \mathcal{U}$, if $W = (U_n \cap V_n)_{n \geq 1}$, then $W \in \mathcal{U}$, $U \leq W$, and $V \leq W$.

For each $U \in \mathcal{U}$, let

$$H_U = \bigcap_{n=1}^{\infty} U_n.$$

Clearly H_U is a closed subgroup of G. Let ϕ_U be the canonical epimorphism from G to G/H_U. By 3.1, 3.4, 6.4, and the group analogue of 5.5, $(\phi_U(U_n))_{n \geq 1}$ is a fundamental system of neighborhoods of zero for a metrizable group topology $\mathcal{T}_U$ on G/H_U weaker than the quotient topology induced by that of G. Indeed, for all $n \geq 1$,

$$\phi_U(U_{n+1}) + \phi_U(U_{n+1}) = \phi_U(U_{n+1} + U_{n+1}) \subseteq \phi_U(U_n),$$

and

$$\phi_U^{-1}(\bigcap_{n=1}^{\infty} \phi_U(U_n)) = \bigcap_{n=1}^{\infty} \phi_U^{-1}(\phi_U(U_n)) = \bigcap_{n=1}^{\infty} (U_n + H_U)$$
$$= \bigcap_{n=1}^{\infty} U_n = H_U$$

by (3) of 3.3, and thus

$$\bigcap_{n=1}^{\infty} \phi_U(U_n) = \{H_U\}.$$

Therefore by 6.10, G/H_U has a completion $(\widehat{G/H_U})$ for $\mathcal{T}_U$.

The hypotheses of the group analogue of 5.21 are satisfied by $(H_U)_{U \in \mathcal{U}}$: Indeed, let $U, V \in \mathcal{U}$ satisfy $U \leq V$. Clearly $H_U \supseteq H_V$, and the canonical epimorphism $f_{U,V}$ from G/H_V to G/H_U is continuous, since for all $n \geq 1$,

$$\phi_V(V_n) \subseteq \phi_V(U_n) \subseteq f_{U,V}^{-1}(\phi_U(U_n)).$$

By (4) of 3.3 and 3.2, for each neighborhood V of zero there exists $U \in \mathcal{U}$ such that $U_1 \subseteq V$, whence $H_U \subseteq V$ and $\phi_U(V)$ is a neighborhood of zero for $\mathcal{T}_U$. Thus by the group analogue of 5.21, G is topologically isomorphic to a subgroup G_0 of

$$\varprojlim{}_{U \in \mathcal{U}}(G/H_U),$$

itself a subgroup of

$$\prod_{U \in \mathcal{U}} (\widehat{G/H_U}).$$

The closure of G_0 in the latter is thus a completion of G_0 by 7.8 and 7.5. •

The definition of uniform continuity in the context of metric spaces can be carried over to topological spaces that are subsets of commutative topological groups:

7.10 Definition. Let G and G' be commutative topological groups. A function f from a subset E of G to G' is **uniformly continuous** if for every neighborhood V of zero in G' there is a neighborhood U of zero in G such that for all $x, y \in E$, if $x - y \in U$, then $f(x) - f(y) \in V$.

For example, for any $a \in G$, the translation $x \to a + x$ is uniformly continuous from G to G.

7.11 Theorem. Let E and E' be subsets respectively of commutative topological groups G and G'. If f is uniformly continuous from E to E', then f is continuous, and the image $f(\mathcal{B})$ of any Cauchy filter base $\mathcal{B}$ on E is a Cauchy filter base on E'.

The proof is easy.

The principal example of a uniformly continuous function is a continuous homomorphism:

7.12 Theorem. *Let f be a homomorphism from a commutative topological group G to a commutative topological group G'. The following statements are equivalent:*

1° *f is continuous at zero.*
2° *f is continuous.*
3° *f is uniformly continuous.*

Proof. Assume 1°. Then for any neighborhood V of zero in G', there is a neighborhood U of zero in G such that for all $s \in U$, $f(s) \in V$. Consequently, for all $x, y \in G$, if $x - y \in U$, then

$$f(x) - f(y) = f(x - y) \in V.$$

Thus 3° holds.●

The proofs of the following three theorems are also easy:

7.13 Theorem. *Let G and G' be commutative Hausdorff groups, let E and E' be subsets of G and G' respectively, and let f be a bijection from E to E'. If both f and f^{-1} are uniformly continuous, then E is complete if and only if E' is complete.*

7.14 Theorem. *If G and G' are commutative topological groups and if f is a topological isomorphism from G to G', then a subset E of G is complete if and only if $f(E)$ is.*

7.15 Theorem. *Let G, H, and K be commutative topological groups, and let D, E, and F be subsets of G, H, and K respectively. If $f : D \to E$ and $g : E \to F$ are uniformly continuous functions, then $g \circ f$ is uniformly continuous.*

The main theorem concerning uniformly continuous functions is the following:

7.16 Theorem. *Let E be a subset of a commutative topological group G, and let f be a uniformly continuous function from E to a complete commutative Hausdorff group G'. There is a unique continuous function g from $\overline{E}$ to G' extending f, and moreover, g is uniformly continuous.*

Proof. Since G' is Hausdorff, there is at most one continuous extension of f to $\overline{E}$. For each $c \in \overline{E}$, $\{V \cap E : V \text{ is a neighborhood of } c\}$, which we denote by $\mathcal{V}(c)$, is a convergent filter base on G and hence is a Cauchy filter on E. By 7.11, $f(\mathcal{V})$ converges to a unique point of G', which we denote by $g(c)$. If $c \in E$, then $f(\mathcal{V}(c))$ converges to $f(c)$ as f is continuous, so $g(c) = f(c)$; thus g is an extension of f. Consequently, we need only show that g is uniformly continuous.

Let V' be a neighborhood of zero in G', and let U' be a symmetric neighborhood of zero in G' such that $U' + U' + U' \subseteq V'$. By hypothesis there is a neighborhood U of zero in G such that if $x, y \in E$ and if $x - y \in U$, then $f(x) - f(y) \in U'$. Let V be a symmetric neighborhood of zero such that $V + V + V \subseteq U$. We shall show that if $x, y \in \overline{E}$ and if $x - y \in V$, then $g(x) - g(y) \in V'$. Since $g(x)$ is adherent to $f(\mathcal{V}(x))$ by 7.2,

$$g(x) \in \overline{f((V+x) \cap E)} \subseteq f((V+x) \cap E) + U'$$

by (3) of 3.3. Hence there exist $v \in V$ and $u' \in U'$ such that $v + x \in E$ and $g(x) = f(v + x) + u'$. Similarly, there exist $w \in V$ and $z' \in U'$ such that $w + y \in E$ and $g(y) = f(w + y) + z'$. Then

$$(v + x) - (w + y) = v + (-w) + (x - y) \in V + V + V \subseteq U,$$

so

$$f(v + x) - f(w + y) \in U'.$$

Therefore

$$\begin{aligned} g(x) - g(y) &= f(v + x) + u' - (f(w + y) + z') \\ &= f(v + x) - f(w + y) + u' + (-z') \in U' + U' + U' \subseteq V'. \end{aligned}$$ •

7.17 Theorem. *Let H be a dense subgroup of a commutative topological group G, and let f be a continuous homomorphism from H to a complete commutative Hausdorff group G'. There is a unique continuous homomorphism g from G to G' extending f. Moreover, if G is Hausdorff and complete and if f is a topological isomorphism from H to a dense subgroup H' of G', then g is a topological isomorphism from G to G'.*

Proof. For the first statement, it suffices by 7.12 and 7.16 to show that the unique continuous extension g of f is a homomorphism from the closure G of H to G'. The functions $(x, y) \to g(x + y)$ and $(x, y) \to g(x) + g(y)$ from $G \times G$ to G' are continuous and agree on the dense subset $H \times H$ of $G \times G$. Hence as G' is Hausdorff, they agree on $G \times G$, so g is a homomorphism. Suppose further that G is complete and that f is a topological isomorphism from H to a dense subgroup H' of G'. By what we have just proved, there is a unique continuous homomorphism h from G' to G extending f^{-1}. Then $h \circ g$ is a continuous function from G to G agreeing with the identity function on dense subgroup H and hence on all of G. Similarly, $g \circ h$ is the identity function on G'. Thus g is a continuous isomorphism whose inverse h is continuous, and hence g is a topological isomorphism. •

7.18 Corollary. *If G is a dense subgroup of complete, commutative, Hausdorff groups G_1 and G_2, then there is a unique topological isomorphism f from G_1 to G_2 such that $f(x) = x$ for all $x \in G$.*

Consequently by 7.9, each commutative Hausdorff group G has an essentially unique completion, which we shall normally denote by $\widehat{G}$. If H is a subgroup of G, the closure $\overline{H}$ of H in $\widehat{G}$ is a completion of H by (1) of 7.5, so we customarily identify $\widehat{H}$ with $\overline{H}$. Similarly if $(G_\lambda)_{\lambda \in L}$ is a family of Hausdorff groups, we customarily identify the cartesian product of $(\widehat{G}_\lambda)_{\lambda \in L}$ with the completion of the cartesian product of $(G_\lambda)_{\lambda \in L}$, in view of 7.8. Finally, if $\mathcal{H}$ is a filter base of closed subgroups of G that converges to zero and if G/H is complete for each $H \in \mathcal{H}$, then by the group analogue of 5.22, 5.20, and 7.8, we may identify the $\widehat{G}$ with $\varprojlim_{H \in \mathcal{H}} G/H$.

7.19 Theorem. *Let G_1 and G_2 be Hausdorff groups, and let f be a continuous homomorphism from G_1 to G_2. There is a unique continuous homomorphism $\widehat{f}$ from $\widehat{G}_1$ to $\widehat{G}_2$ extending f. Moreover, if f is a topological isomorphism, so is $\widehat{f}$.*

The statement is a consequence of 7.17.

7.20 Theorem. *Let G_1 and G_2 be Hausdorff groups, and let f be a continuous homomorphism from G_1 to G_2. If there is a fundamental system $\mathcal{V}$ of neighborhoods of zero in G_1 such that $f(V)$ is closed in the topological subgroup $f(G_1)$ of G_2 for each $V \in \mathcal{V}$, then the kernel of the continuous extension $\widehat{f} : \widehat{G}_1 \to \widehat{G}_2$ of f is the closure in $\widehat{G}_1$ of the kernel K of f; in particular, if f is a continuous monomorphism, so is $\widehat{f}$.*

Proof. Replacing G_2 with $f(G_1)$ if necessary, we may assume that f is an epimorphism. If X is a subset of G_1, we shall denote its closure in G_1 by $\overline{X}$ and its closure in $\widehat{G}_1$ by $\widehat{X}$, and similarly for subsets Y of G_2. Thus, for example, $\widehat{Y} \cap G_2 = \overline{Y}$.

As the kernel of $\widehat{f}$ is closed, it clearly contains $\widehat{K}$. To show that $\widehat{K}$ contains the kernel of $\widehat{f}$, let $a \in \widehat{G}_1$ be such that $\widehat{f}(a) = 0$. To show that $a \in \widehat{K}$, it suffices by 4.22 and (3) of 3.3 to show that for any neighborhood V of zero in G_1, $a \in \widehat{K} + \widehat{V}$. By hypothesis there is a symmetric neighborhood W of zero in G_1 such that $W + W \subseteq V$ and $f(W)$ is closed in G_2. As $a + \widehat{W}$ is a neighborhood of a in $\widehat{G}_1$ by 4.22, there exists $x \in (a + \widehat{W}) \cap G_1$; let $w \in \widehat{W}$ be such that $x = a + w$. Then

$$f(x) = \widehat{f}(a + w) = \widehat{f}(a) + \widehat{f}(w) = \widehat{f}(w) \in \widehat{f}(\widehat{W}) \subseteq \widehat{f(W)}.$$

Thus

$$f(x) \in \widehat{f(W)} \cap G_2 = \overline{f(W)} = f(W),$$

so
$$x \in f^{-1}(f(W)) = K + W \subseteq \widehat{K} + \widehat{W}.$$

Therefore as $\widehat{W}$ is also symmetric,
$$a = x - w \in \widehat{K} + \widehat{W} + \widehat{W} \subseteq \widehat{K} + \widehat{W+W} \subseteq \widehat{K} + \widehat{V}. \bullet$$

7.21 Corollary. *If T_1 and T_2 are Hausdorff group topologies on a commutative group G such that $T_1 \supseteq T_2$ and there is a fundamental system of neighborhoods of zero for T_1 each of which is closed for T_2, then any subset of G that is complete for T_2 is also complete for T_1.*

Proof. For $i = 1, 2$, let G_i be G furnished with T_i, and for any subset X of G, let $\widehat{X}_i$ be its closure in $\widehat{G}_i$. The identity map f from G_1 to G_2 is continuous, so for any subset A of G,
$$\widehat{f}(\widehat{A}_1) \subseteq \widehat{f(A)_2} = \widehat{A}_2.$$

Hence if $A = \widehat{A}_2$, then $\widehat{A}_1 = A$ as $\widehat{f}$ is injective by 7.20 and $f(A) = A$. $\bullet$

Exercises

7.1 Let G be a Hausdorff commutative group. (a) If $\mathcal{F}$ is a Cauchy filter on G and if $\mathcal{V}$ is a fundamental system of symmetric neighborhoods of zero, then $\mathcal{F} + \mathcal{V}$ is a Cauchy filter on G; moreover, $\mathcal{F}$ converges to $a \in G$ if and only if $\mathcal{F} + \mathcal{V}$ converges to a. (b) If K is a closed subgroup of G and if both K and G/K are complete, then G is complete. [Use (a).]

7.2 Let $\mathcal{H}$ be a filter based of closed subgroups of a Hausdorff commutative group G that converges to zero. If G/H is compact for all $H \in \mathcal{H}$, then $\widehat{G}$ is compact. [Use the group analogue of 5.22.]

7.3 Let G be a dense subgroup of a Hausdorff commutative group G_1. If $H_1, \ldots, H_n$ are open subgroups of G, then in G_1,
$$\overline{H}_1 \cap \cdots \cap \overline{H}_n = \overline{H_1 \cap \cdots \cap H_n}.$$

7.4 Let f be the function defined by $f(x) = x^2$ from $\mathbb{Q}$ into $\mathbb{R}$. Then f is continuous, the image under f of every Cauchy filter base on $\mathbb{Q}$ is a Cauchy filter base on $\mathbb{R}$, and f has a continuous extension $\widehat{f}$ from $\widehat{\mathbb{Q}} = \mathbb{R}$ into $\mathbb{R}$, but f is not uniformly continuous.

7.5 Let $(T_\lambda)_{\lambda \in L}$ be a family of complete Hausdorff group topologies on a group G. If for all $\alpha, \beta \in L$ there exists $\gamma \in L$ such that $T_\alpha \subseteq T_\gamma$ and $T_\beta \subseteq T_\gamma$, then $\sup_{\lambda \in L} T_\lambda$ is complete.

8 Completions of Topological Rings and Modules

A topological ring or module is *complete* if its underlying additive group is.

8.1 Theorem. *Let A be a dense subring of a topological ring B, and let f be a continuous homomorphism from A to a complete Hausdorff ring B'. There is a unique continuous homomorphism g from B to B' extending f. Moreover, if B is Hausdorff and complete and if f is a topological isomorphism from A to a dense subring A' of B', then g is a topological isomorphism from B to B'.*

Proof. By 7.17 we need only show that the unique continuous extension g of f preserves multiplication. But $(x,y) \to g(xy)$ and $(x,y) \to g(x)g(y)$ from $B \times B$ to B' are continuous and agree on the dense subset $A \times A$ of $B \times B$; hence as B' is Hausdorff, they agree on $B \times B$. •

A topological ring B is a *completion* of a topological ring A if B is complete and if A is a dense topological subring of B. The existence of a completion of a Hausdorff ring results from the following theorem:

8.2 Theorem. *Let E, F, and G be complete Hausdorff abelian groups, and let A and B be dense subgroups of E and F respectively. If f is a continuous $\mathbb{Z}$-bilinear function from $A \times B$ to G, then there is a unique continuous $\mathbb{Z}$-bilinear function g from $E \times F$ to G extending f.*

Proof. For each $x_0 \in E$, let $\mathcal{U}(x_0)$ be the set of intersections with A of the neighborhoods of x_0; as A is dense in E, $\mathcal{U}(x_0)$ is a filter on A. Similarly, for each $y_0 \in F$, the set $\mathcal{V}(y_0)$ of intersections with B of the neighborhoods of y_0 is a filter on B. We shall first show that for any neighborhood T of zero in G and any $a \in A$, $b \in B$, there exist $U \in \mathcal{U}(x_0)$ and $V \in \mathcal{V}(y_0)$ such that for all $x, x' \in U$ and all $y, y' \in V$, $f(x'-x, y'-y) \in T$, $f(a, y'-y) \in T$, and $f(x'-x, b) \in T$. Indeed, as f is continuous at $(0,0)$, as $y \to f(a,y)$ is continuous at zero, and as $x \to f(x,b)$ is continuous at zero, there exist closed neighborhoods P and Q of zero in A and B respectively such that $f(P \times Q) \subseteq T$, $f(\{a\} \times Q) \subseteq T$, and $f(P \times \{b\}) \subseteq T$. By 4.22 the closure $\overline{P}$ of P in E is a neighborhood of zero in E, so there exists a symmetric neighborhood P_1 of zero in E such that $P_1 + P_1 \subseteq \overline{P}$; similarly there exists a symmetric neighborhood Q_1 of zero in F such that $Q_1 + Q_1 \subseteq \overline{Q}$. Let

$$U = (x_0 + P_1) \cap A \in \mathcal{U}(x_0),$$
$$V = (y_o + Q_1) \cap B \in \mathcal{V}(y_0).$$

If $x, x' \in U$, then

$$x' - x = (x' - x_0) - (x - x_0) \in P_1 + P_1 \subseteq \overline{P},$$

so $x' - x \in \overline{P} \cap A = P$; similarly, if $y, y' \in V$, then $y' - y \in Q$. Hence for all $x, x' \in U$ and all $y, y' \in V$, $f(x'-x, y'-y) \in T$, $f(a, y'-y) \in T$, and $f(x'-x, b) \in T$.

Next, we shall show that $f(\mathcal{U}(x_0) \times \mathcal{V}(y_0))$ is a Cauchy filter base on G. Indeed, let W be a neighborhood of zero in G, and let T be a symmetric neighborhood of zero such that $T+T+T+T \subseteq W$. By the preceding (with $a = b = 0$), there exist $U \in \mathcal{U}(x_0)$ and $V \in \mathcal{V}(x_0)$ such that $f(x'-x, y'-y) \in T$ for all $x, x' \in U$ and all $y, y' \in V$. Let $a \in U$, $b \in V$. Again, by the preceding, there exist $U' \in \mathcal{U}(x_0)$ and $V' \in \mathcal{V}(y_0)$ such that $U' \subseteq U$, $V' \subseteq V$, and for all $x, x' \in U'$ and all $y, y' \in V'$, $f(a, y'-y) \in T$ and $f(x'-x, b) \in T$. Also, as $U' \subseteq U$ and $V' \subseteq V$, $f(x'-x, y'-b) \in T$ and $f(x'-a, y'-y) \in T$. Hence

$$f(x', y') - f(x, y) = f(x'-x, b) + f(a, y'-y) + f(x'-x, y'-b) + \\ + f(x-a, y'-y) \in T+T+T+T \subseteq W.$$

We therefore define $g(x_0, y_0)$ to be the limit of $f(\mathcal{U}(x_0) \times \mathcal{V}(x_0))$ for all $(x_0, y_0) \in E \times F$. As f is continuous, g is an extension of f. To show that g is continuous at (x_0, y_0), let W be a closed neighborhood of $g(x_0, y_0)$. By the definition of $g(x_0, y_0)$, there exist open neighborhoods U of x_0 and V of y_0 such that

$$f((U \cap A) \times (V \cap B)) \subseteq W.$$

But then $g(U \times V)$ is contained in the closure of $f((U \cap A) \times (V \cap B))$ and hence in W, for if $u \in U$ and $v \in V$, then $g(u, v)$ is, by definition, the limit of and hence adherent to a filter base of which $f((U \cap A) \times (V \cap B))$ is a member. Thus g is continuous at (x_0, y_0).

The functions $(x, x', y) \to g(x + x', y)$ and $(x, x', y) \to g(x, y) + g(x', y)$ are continuous from $E \times E \times F$ to G and coincide on the dense subset $A \times A \times B$ of $E \times E \times F$. Hence they coincide on all of $E \times E \times F$, so

$$g(x + x', y) = g(x, y) + g(x', y)$$

for all $x, x' \in E$ and all $y \in F$. Similarly,

$$g(x, y + y') = g(x, y) + g(x, y')$$

for all $x \in E$ and all $y, y' \in F$. Thus g is $\mathbb{Z}$-bilinear. •

8.3 Theorem. *Let A be a Hausdorff ring. There is a complete Hausdorff ring $\hat{A}$ containing A as a dense subring. If A is commutative, so is $\hat{A}$. If 1 is the identity element for A, 1 is also the identity element of $\hat{A}$. If A is also a* dense subring of a complete Hausdorff ring B, there is a unique topological isomorphism h from $\widehat{A}$ to B satisfying $h(x) = x$ for all $x \in A$.

Proof. Let $\widehat{A}$ be the completion of the additive group A. We need only apply 8.2 to multiplication, viewed as a continuous $\mathbb{Z}$-bilinear function from $A \times A$ to $\widehat{A}$, to conclude that there is a continuous multiplication on $\widehat{A}$ that is distributive over addition and induces on A the given multiplication. Verifying the associativity of multiplication on $\widehat{A}$ and the remaining assertions about multiplication is similar to establishing the $\mathbb{Z}$-bilinearity of g in the proof of 8.2. The final assertion follows from 8.1. •

8.4 Theorem. *If f is a continuous homomorphism from a Hausdorff ring A_1 to a Hausdorff ring A_2, there is a unique continuous homomorphism $\widehat{f}$ from $\widehat{A}_1$ to $\widehat{A}_2$ extending f; moreover, if f is a topological isomorphism, so is $\widehat{f}$.*

The statement is a consequence of 8.1.

8.5 Theorem. *Let A be a topological ring, and let $\mathcal{J}$ be a filter base of closed ideals.*

(1) If A/J is complete for each $J \in \mathcal{J}$, then $\varprojlim_{J \in \mathcal{J}}(A/J)$ is complete.

(2) If A is Hausdorff, if $\mathcal{J}$ converges to zero, and if some $L \in \mathcal{J}$ is complete, then the canonical homomorphism g from A to $\varprojlim_{J \in \mathcal{J}}(A/J)$ is a topological isomorphism.

Proof. (1) follows from 7.8, 5.20, and 7.5. To prove (2), it suffices by 5.22 to prove that the range of g is $\varprojlim_{J \in \mathcal{J}}(A/J)$. Let $z \in \varprojlim_{J \in \mathcal{J}}(A/J)$. With the notation of 5.19, $pr_L(z) = a + L$ for some $a \in A$ and hence $pr_L(z)$ is complete by the remark following 7.10 and 7.13. Let
$$\mathcal{J}_L = \{J \in \mathcal{J} : J \subseteq L\}.$$
Then the set of all the subsets $pr_J(z)$ such that $J \in \mathcal{J}_L$ of A is a Cauchy filter base on $pr_L(z)$, for if V is a neighborhood of zero, there exists $J \in \mathcal{J}_L$ such that $J \subseteq V$, so the coset $pr_J(z)$ of J is V-small. Consequently, as each coset of each $J \in \mathcal{J}$ is closed, there exists
$$c \in \bigcap_{J \in \mathcal{J}_L} pr_J(z).$$
Thus for each $J \in \mathcal{J}_L$, c belongs to the coset $pr_J(z)$ of J, so
$$pr_J(g(c)) = c + J = pr_J(z),$$
and for any $K \in \mathcal{J}$, there exists $J \in \mathcal{J}_L$ such that $J \subseteq K$, so
$$pr_K(g(c)) = f_{K,J}(pr_J(g(c))) = f_{K,J}(pr_J(z)) = pr_K(z).$$
Thus $g(c) = z$, and the proof is complete. •

8.6 Theorem. Let E be a Hausdorff module over a Hausdorff ring A. There is a unique scalar multiplication from $\widehat{A} \times \widehat{E}$ to $\widehat{E}$ that makes $\widehat{E}$ into a topological $\widehat{A}$-module and extends the given scalar multiplication of the A-module E; moreover, if E is a unitary A-module, $\widehat{E}$ is a unitary $\widehat{A}$-module.

Proof. By 8.2 there is a continuous scalar multiplication from $\widehat{A} \times \widehat{E}$ to $\widehat{E}$ that extends the given scalar multiplication from $A \times E$ to E and satisfies

$$\lambda(x+y) = \lambda x + \lambda y,$$
$$(\lambda + \mu)x = \lambda x + \mu x$$

for all $x, y \in \widehat{E}$ and all $\lambda, \mu \in \widehat{A}$. A proof similar to that establishing the bilinearity of g in 8.2 establishes the identity

$$(\lambda\mu)x = \lambda(\mu x)$$

for all $\lambda, \mu \in \widehat{A}$ and all $x \in \widehat{E}$ and, if E is a unitary A-module, the identity $1x = x$ for all $x \in \widehat{E}$. •

Often we regard $\widehat{E}$ as an A-module by restricting scalar multiplication from $\widehat{A} \times \widehat{E}$ to $A \times \widehat{E}$. The analogues of 8.4 and 8.5 hold with essentially the same proofs:

8.7 Theorem. Let A be a Hausdorff ring, E_1 and E_2 Hausdorff A-modules. If u is a continuous homomorphism from E_1 to E_2, there is a unique continuous homomorphism $\widehat{u}$ from the $\widehat{A}$-module $\widehat{E}_1$ to the $\widehat{A}$-module $\widehat{E}_2$ extending u; moreover, if u is a topological isomorphism, so is $\widehat{u}$.

8.8 Theorem. Let E be a topological A-module, and let $\mathcal{M}$ be a filter base of closed submodules of E.
  (1) If E/M is complete for each $M \in \mathcal{M}$, then $\varprojlim_{M \in \mathcal{M}}(E/M)$ is complete.
  (2) If E is Hausdorff, if $\mathcal{M}$ converges to zero, and if some $L \in \mathcal{M}$ is complete, then the canonical homomorphism g from E to $\varprojlim_{M \in \mathcal{M}}(E/M)$ is a topological isomorphism.

Finally, the ring analogue of 6.11 holds:

8.9 Theorem. If the topology of a topological ring A is given by a[n] [ultra]norm N, the topology of $\widehat{A}$ is given by a[n] [ultra]norm $\widehat{N}$ that extends N.

Proof. By 6.11 there is a unique [ultra]norm $\widehat{N}$ on the additive group $\widehat{A}$ that extends N and defines the topology of $\widehat{A}$. Moreover, since $\widehat{N}$ is continuous by 1.4, the function

$$f : (x,y) \to \widehat{N}(x)\widehat{N}(y) - \widehat{N}(xy)$$

is continuous on $\widehat{A} \times \widehat{A}$, so $f^{-1}(\mathbb{R}_{\geq 0})$ is closed and contains $A \times A$ and hence is all of $\widehat{A} \times \widehat{A}$. Thus $\widehat{N}$ is a[n] [ultra]norm on the ring $\widehat{A}$. •

Exercises

8.1 If A is a complete metrizable ring, any homomorphism from A into the topological ring $\mathbb{R}$ is continuous. [Use Exercise 6.3.]

8.2 (a) The only complete separable metrizable ring topology on the field $\mathbb{R}$ is the usual topology. [Use 6.14 and Exercise 8.1.] (b) The only automorphism of the field $\mathbb{R}$ is the identity automorphism.

8.3 (Andrunakievich and Arnautov [1966]) Let A be a Hausdorff ring with identity 1 in which every nonzero left or right ideal is dense and in which there is a neighborhood V of zero such that for every neighborhood W of zero there exists $n \geq 1$ such that $V^m \subseteq W$ for all $m \geq n$. Let $a \in A^*$. (a) For any neighborhood U of zero there exists $x \in A$ such that $ax + 1 \in U$. (b) There exists $n \geq 1$ such that $V^{m+1} + V^m \subseteq V$ for all $m \geq n$. (c) There exists $y \in A$ such that $ay + 1 \in V^n$. [Use (a) and expand $(ax+1)^n$.] (d) $\sum_{k=1}^{r}(ay+1)^k \in V$ for all $r \geq 1$. [Use induction and (b).] (e) The sequence $(s_p)_{p \geq 1}$, defined by
$$s_p = \sum_{k=0}^{p} y(ay+1)^k,$$
is a Cauchy sequence. (f) Let
$$d = \lim_{p \to \infty} s_p \in \widehat{A}.$$
Then
$$yad = ya(d - s_p) + s_{p+1} - s_p - y.$$
[Use geometric series.] (g) $a(-d) = 1$. (h) Every nonzero element of A is invertible in $\widehat{A}$.

8.4 Let A be a commutative topological ring with identity whose topology is given by a norm. The following statements are equivalent:
1° Every nonzero ideal of A is dense.
2° There is a subfield F of $\widehat{A}$ containing A.
[Use Exercise 8.3.]

9 Baire Spaces

Here we shall use Baire category concepts to establish that a complete, metrizable additive group topology on a ring for which multiplication is separately continuous in each variable is actually a ring topology.

9 BAIRE SPACES

9.1 Definition. *Let E be a topological space. A subset X of E is* **rare** *(or* **nowhere dense**) *if the closure of X has empty interior (that is, if $\overline{X}^\circ = \emptyset$). A subset Y of E is* **meager** *(or a* **first Baire category** *subset of E) if Y is the union of countably many rare subsets.*

Clearly any subset of a rare [meager] subset of E is a rare [meager] subset, and the union of countably many meager subsets of E is meager.

9.2 Theorem. *The following properties of a topological space E are equivalent:*

$1°$ *The intersection of any countable family of dense open subsets of E is dense.*
$2°$ *No meager subset of E contains a nonempty open subset.*
$3°$ *Every nonempty open subset of E is nonmeager.*
$4°$ *The complement of any meager subset of E is dense.*

The proof follows readily from the fact that a subset of E is meager if and only if it is contained in the union of countably many closed sets, each having an empty interior.

9.3 Definition. *A topological space E is a* **Baire space** *if E satisfies the equivalent properties of Theorem 9.2.*

If d is a metric on E, the *diameter* of a nonempty subset X of E, denoted by $\operatorname{diam}(X)$, is defined to be $\sup\{d(x,y) : x, y \in X\}$.

9.4 Theorem. (1) *A locally compact space is a Baire space.* (2) *A topological space whose topology is given by a complete metric is a Baire space.*

Proof. Let E be either locally compact or a complete metric space, let $(U_n)_{n\geq 1}$ be a sequence of dense open subsets of E, and let P be a nonempty open subset. We shall show that

$$(\bigcap_{n=1}^{\infty} U_n) \cap P \neq \emptyset.$$

Since E is regular and since each U_n is dense, there is a decreasing sequence $(V_n)_{n\geq 1}$ of nonempty open sets such that $V_1 = P$ and

$$\overline{V}_{n+1} \subseteq P \cap V_n \cap U_n.$$

If E is locally compact, we may further assume that $\overline{V}_2$ is compact; then there exists

$$c \in \bigcap_{n=1}^{\infty} \overline{V}_n \subseteq (\bigcap_{n=1}^{\infty} U_n) \cap P.$$

If d is a complete metric defining the topology of E, we may further assume that $\operatorname{diam}(V_n) \leq 1/n$ for all $n \geq 2$; then if $c_n \in V_n$ for all $n \geq 1$, $(c_n)_{n \geq 1}$ is a Cauchy sequence for d, and if c is its limit,

$$c \in \bigcap_{n=2}^{\infty} \overline{V}_n \subseteq (\bigcap_{n=1}^{\infty} U_n) \cap P. \bullet$$

9.5 Theorem. *Let E, F, and G be commutative topological groups, and let f be a $\mathbb{Z}$-bilinear function from $E \times F$ into G such that for each $a \in E$, $y \to f(a, y)$ from F to G is continuous at zero, and for each $b \in F$, $x \to f(x, b)$ from E to G is continuous at zero. If E is metrizable and F a Baire space, then f is continuous.*

Proof. By 2.14, it suffices to show that f is continuous at $(0, 0)$. Let W be a neighborhood of zero in G, and let V be a closed neighborhood of zero in G such that $V + V \subseteq W$. Let $(U_n)_{n \geq 1}$ be a decreasing fundamental sequence of symmetric neighborhoods of zero in E. For each $n \geq 1$, let

$$T_n = \{y \in F : f(U_n \times \{y\}) \subseteq V\}.$$

Since $x \to f(x, y)$ is continuous for each $y \in F$,

$$F = \bigcup_{n=1}^{\infty} T_n.$$

Since V is closed and since $y \to f(x, y)$ is continuous for each $x \in E$, T_n is closed. Then for some $m \geq 1$, T_m has an interior point t as F is a Baire space. Let $T = T_m + (-T_m)$; then as $0 = t + (-t)$, zero is an interior point of T. As U_m is symmetric,

$$f(U_m \times T) \subseteq V + V \subseteq W.$$

Thus f is continuous at $(0, 0)$. $\bullet$

9.6 Theorem. *If $\mathcal{T}$ is a complete metrizable additive group topology on a ring A such that for each $a \in A$, $x \to ax$ and $x \to xa$ are continuous at zero, then $\mathcal{T}$ is a ring topology.*

The assertion follows from 9.4 and 9.5. Actually, a stronger result is available:

9.7 Theorem. *If $\mathcal{T}$ is a topology on a ring A defined by a complete metric such that for each $a \in A$, $x \to a + x$, $x \to ax$, and $x \to xa$ are continuous, then $\mathcal{T}$ is a ring topology.*

The assertion follows from 6.13 and 9.6.

9.8 Theorem. *Let A be a topological ring, and let E be an A-module furnished with an additive group topology such that for each $\alpha \in A$, $x \to \alpha x$ from E to E is continuous at zero and for each $c \in E$, $\lambda \to \lambda c$ from A to E is continuous at zero. If A is metrizable and E a Baire space, or if A is a Baire space and E metrizable, then E is a topological A-module.*

The assertion follows from 9.4 and 9.5.

Exercises

9.1 Let E be a topological space. (a) If A is an open subset of E, then A is a meager subset of E if and only if there is a sequence $(U_n)_{n\geq 1}$ of open dense subsets of A such that

$$\bigcap_{n=1}^{\infty} U_n = \emptyset.$$

(b) If A is a meager open subset of E and if B is an open set of which A is a dense subset, then B is meager.

9.2 A separable metrizable group G that is a nonmeager subset of itself is a Baire space. [If G contains a nonempty meager open set P, show that the union of a maximal family of mutually disjoint open meager subsets of G is a dense, meager subset of G, and apply Exercise 9.1(b).]

9.3 A subset A of a topological space E is a nonmeager subset of itself if and only if A is a nonmeager subset of $\bar{A}$.

9.4 The cartesian product E of a family $(E_\lambda)_{\lambda \in L}$ of complete metric spaces is a Baire space. [Argue as in the proof of 9.4 by letting V_n be the cartesian product of $(V_{n,\lambda})_{\lambda \in L}$ where, if $V_{n,\lambda} \neq E_\lambda$, then $\text{diam}(V_{n,\lambda}) \leq 1/n$.]

10 Summability

A *net* in a set E is a family of elements of E indexed by a directed set. Thus a net in E is simply a function from a directed set to E. Let $(z_\alpha)_{\alpha \in D}$ be a net in E, and let $\leq$ be the direction of D. For each $\beta \in D$ let $F_\beta = \{z_\gamma : \gamma \geq \beta\}$. Then $\{F_\beta : \beta \in D\}$ is a filter base on E, called the filter base generated by $(z_\alpha)_{\alpha \in D}$. If E is a topological space, the net $(x_\alpha)_{\alpha \in D}$ *converges* to $c \in E$ if the associated filter base does, that is, if for every neighborhood V of c there exists $\beta \in D$ such that $x_\gamma \in V$ for all $\gamma \geq \beta$. Similarly, c is *adherent* to $(z_\alpha)_{\alpha \in D}$ if c is adherent to the associated filter base, that is, if

$$c \in \bigcap_{\beta \in D} \overline{F}_\beta.$$

If E is a topological group, a net in E is a *Cauchy net* if the associated filter base is a Cauchy filter base.

Here, we shall primarily be concerned with the directed set $\mathcal{F}(A)$ of all finite subsets of a set A, directed by the relation $\subseteq$.

10.1 Definition. *Let G be a Hausdorff commutative group, $(x_\alpha)_{\alpha \in A}$ a family of elements of G. An element s of G is the* **sum** *of $(x_\alpha)_{\alpha \in A}$ if the net $(s_J)_{J \in \mathcal{F}(A)}$ converges to s, where for each $J \in \mathcal{F}(A)$,*

$$s_J = \sum_{\alpha \in J} x_\alpha.$$

The family $(x_\alpha)_{\alpha \in A}$ is **summable** *if it has a sum.*

Thus s is the sum of $(x_\alpha)_{\alpha \in A}$ if and only if for each neighborhood V of s there exists $J_V \in \mathcal{F}(A)$ such that

$$\sum_{\alpha \in J} x_\alpha \in V$$

for all $J \in \mathcal{F}(A)$ containing J_V.

The sum s of a summable family $(x_\alpha)_{\alpha \in A}$ of elements of G is usually denoted by

$$\sum_{\alpha \in A} x_\alpha.$$

10.2 Theorem. *If $(x_\alpha)_{\alpha \in A}$ is a family of elements of a Hausdorff commutative group G having a sum s, then for any permutation σ of A, s is also the sum of $(x_{\sigma(\alpha)})_{\alpha \in A}$.*

Proof. Let V be a neighborhood of s. If

$$\sum_{\alpha \in J} x_\alpha \in V$$

for all finite subsets J of A containing J_V, then

$$\sum_{\alpha \in K} x_{\sigma(\alpha)} \in V$$

for all finite subsets K of A containing $\sigma^{-1}(J_V)$. ∎

10.3 Definition. *A family $(x_\alpha)_{\alpha \in A}$ of elements of a Hausdorff commutative group G satisfies* **Cauchy's Condition** *if for every neighborhood V of zero there is a finite subset J_V of A such that*

$$\sum_{\alpha \in K} x_\alpha \in V$$

for every finite subset K of A disjoint from J_V.

10.4 Theorem. Let $(x_\alpha)_{\alpha \in A}$ be a family of elements of a Hausdorff commutative group G. If $(x_\alpha)_{\alpha \in A}$ is summable, then $(x_\alpha)_{\alpha \in A}$ satisfies Cauchy's Condition. If G is complete, then $(x_\alpha)_{\alpha \in A}$ is summable if and only if $(x_\alpha)_{\alpha \in A}$ satisfies Cauchy's Condition.

Proof. Cauchy's Condition is equivalent to the statement that $(s_J)_{J \in \mathcal{F}(A)}$ is a Cauchy net (where $s_J = \sum_{\alpha \in J} x_\alpha$ for all $J \in \mathcal{F}(A)$). Indeed, let V be a neighborhood of zero, and let W be a symmetric neighborhood of zero such that $W + W \subseteq V$. If $(x_\alpha)_{\alpha \in A}$ satisfies Cauchy's Condition, there exists $J_W \in \mathcal{F}(A)$ such that $s_K \in W$ for all $K \in \mathcal{F}(A)$ disjoint from J_W. Hence if J_1 and J_2 are any finite subsets of A containing J_W,

$$\begin{aligned} s_{J_1} - s_{J_2} &= (s_{J_1} - s_{J_W}) - (s_{J_2} - s_{J_W}) \\ &= s_{J_1 \setminus J_W} - s_{J_2 \setminus J_W} \in W + W \subseteq V. \end{aligned}$$

Conversely, if $(s_J)_{J \in \mathcal{F}(A)}$ is a Cauchy net, there exists $J_V \in \mathcal{F}(A)$ such that

$$s_{J_1} - s_{J_2} \in V$$

for all finite subsets J_1, J_2 of A containing J_V; hence for any finite subset K of A disjoint from J_V,

$$s_K = s_{K \cup J_V} - s_{J_V} \in V.$$

The assertions therefore follow from 7.2 and 7.3. •

10.5 Theorem. If $(x_\alpha)_{\alpha \in A}$ is a summable family of elements of a Hausdorff commutative group G, then for every neighborhood V of zero, $x_\alpha \in V$ for all but finitely many $\alpha \in A$. If G is complete and if the open subgroups of G form a fundamental system of neighborhoods of zero, then $(x_\alpha)_{\alpha \in A}$ is summable if and only if for every neighborhood V of zero, $x_\alpha \in V$ for all but finitely many $\alpha \in A$.

Proof. By 10.4 there is a finite subset K of A such that $x_\alpha \in V$ whenever $\{\alpha\} \cap K = \emptyset$, that is, whenever $\alpha \in A \setminus K$. Conversely, if U is an open subgroup and if K is a finite subset of A such that $x_\alpha \in U$ for all $\alpha \notin K$, then

$$\sum_{\alpha \in J} x_\alpha \in U$$

for all finite subsets J of A disjoint from K. •

10.6 Corollary. If G is a metrizable topological group and if $(x_\alpha)_{\alpha \in A}$ is a summable family of elements of G, then $x_\alpha = 0$ for all but countably many $\alpha \in A$.

10.7 Theorem. *If G is a complete Hausdorff commutative group and if $(x_\alpha)_{\alpha \in A}$ is a summable family of elements of G, then for any subset B of A, $(x_\alpha)_{\alpha \in B}$ is summable.*

Proof. If $(x_\alpha)_{\alpha \in A}$ satisfies Cauchy's Condition, then a fortiori $(x_\alpha)_{\alpha \in B}$ satisfies Cauchy's Condition, so the assertion follows from 10.4. •

10.8 Theorem. *If $(x_\alpha)_{\alpha \in A}$ is a summable family of elements of a Hausdorff commutative group G and if $(A_\lambda)_{\lambda \in L}$ is a partition of A such that $(x_\alpha)_{\alpha \in A_\lambda}$ is summable with sum s_λ for each $\lambda \in L$, then $(s_\lambda)_{\lambda \in L}$ is summable, and*

$$\sum_{\lambda \in L} s_\lambda = \sum_{\alpha \in A} x_\alpha.$$

Proof. Let

$$s = \sum_{\alpha \in A} x_\alpha,$$

and let V be a closed neighborhood of zero. For each finite subset J of A, let

$$s_J = \sum_{\alpha \in J} x_\alpha.$$

By hypothesis there is a finite subset J_V of A such that $s - s_J \in V$ for every finite subset J of A containing J_V. Let

$$K_V = \{\lambda \in L : A_\lambda \cap J_V \neq \emptyset\},$$

a finite subset of L. To show that

$$s - \sum_{\lambda \in K} s_\lambda \in V$$

for every finite subset K of L containing K_V, it suffices by (3) of 3.3 to show that for any neighborhood W of zero,

$$s - \sum_{\lambda \in K} s_\lambda \in V + W.$$

Let n be the number of elements in K. By 2.10 there is a symmetric neighborhood U of zero such that $U + U + \cdots + U$ (n terms) $\subseteq W$. By hypothesis, for each $\lambda \in K$ there is a finite subset J_λ of A_λ containing $J_V \cap A_\lambda$ such that for any finite subset I_λ of A_λ containing J_λ,

$$s_\lambda - \sum_{\alpha \in I_\lambda} x_\alpha \in U.$$

Let
$$J = \bigcup_{\lambda \in K} J_\lambda,$$
a finite subset of A. Then $J \supseteq J_V$, so as
$$\sum_{\alpha \in J} x_\alpha = \sum_{\lambda \in K}(\sum_{\alpha \in J_\lambda} x_\alpha),$$
we have
$$s - \sum_{\lambda \in K} s_\lambda = s - \sum_{\alpha \in J} x_\alpha - \sum_{\lambda \in K}(s_\lambda - \sum_{\alpha \in J_\lambda} x_\alpha) \in V + U + \cdots + U \subseteq V + W.$$
Thus as G is regular by 3.4,
$$s = \sum_{\lambda \in L} s_\lambda. \bullet$$

10.9 Theorem. Let $(x_\alpha)_{\alpha \in A}$ be a family of elements of a Hausdorff commutative group G. If $\{A_1, \ldots, A_n\}$ is a partition of A and if $(x_\alpha)_{\alpha \in A_k}$ is summable for each $k \in [1, n]$, then $(x_\alpha)_{\alpha \in A}$ is summable, and
$$\sum_{\alpha \in A} x_\alpha = \sum_{k=1}^n (\sum_{\alpha \in A_k} x_\alpha).$$

Proof. Let V be a neighborhood of zero. By 2.10 there is a symmetric neighborhood W of zero such that $W + W + \cdots + W$ (n terms) $\subseteq V$. For each $k \in [1, n]$ there is a finite subset J_k of A_k such that for any finite subset I_k of A_k containing J_k,
$$\sum_{\alpha \in A_k} x_\alpha - \sum_{\alpha \in I_k} x_\alpha \in W.$$
Let
$$J_V = \bigcup_{k=1}^n J_k.$$
If J is a finite subset of A containing J_V, then for each $k \in [1, n]$, $J \cap A_k \supseteq J_k$, so
$$\sum_{\alpha \in A_k} x_\alpha - \sum_{\alpha \in J \cap A_k} x_\alpha \in W,$$
and therefore
$$\sum_{k=1}^n (\sum_{\alpha \in A_k} x_\alpha) - \sum_{\alpha \in J} x_\alpha = \sum_{k=1}^n (\sum_{\alpha \in A_k} x_\alpha - \sum_{\alpha \in J \cap A_k} x_\alpha) \in W + W + \ldots W \subseteq V. \bullet$$

10.10 Theorem. *Let G be the cartesian product of a family $(G_\lambda)_{\lambda \in L}$ of Hausdorff commutative groups. Then s is the sum of a family $(x_\alpha)_{\alpha \in A}$ of elements of G if and only if $pr_\lambda(s)$ is the sum of $(pr_\lambda(x_\alpha))_{\alpha \in A}$ for each $\lambda \in L$.*

Proof. For each finite subset J of A and each $\lambda \in L$, let

$$s_J = \sum_{\alpha \in A} x_\alpha, \quad s_{\lambda, J} = \sum_{\alpha \in J} pr_\lambda(x_\alpha).$$

Then $pr_\lambda(s_J) = s_{\lambda, J}$. Therefore the net $(s_J)_{J \in \mathcal{F}(A)}$ converges to $(s_\lambda)_{\lambda \in L}$ if and only if for each $\lambda \in L$, the net $(s_{\lambda, J})_{J \in \mathcal{F}(A)}$ converges to s_λ. •

10.11 Theorem. *If f is a continuous homomorphism from a Hausdorff commutative group G to a Hausdorff commutative group G' and if $(x_\alpha)_{\alpha \in A}$ is a summable family of elements in G, then $(f(x_\alpha))_{\alpha \in A}$ is summable, and*

$$\sum_{\alpha \in A} f(x_\alpha) = f(\sum_{\alpha \in A} x_\alpha).$$

The proof is easy.

10.12 Corollary. *If $(x_\alpha)_{\alpha \in A}$ and $(y_\alpha)_{\alpha \in A}$ are summable families of elements of a Hausdorff commutative group G, then so are $(x_\alpha + y_\alpha)_{\alpha \in A}$, $(-x_\alpha)_{\alpha \in A}$, and $(m.x_\alpha)_{\alpha \in A}$ for any integer m, and moreover*

$$\sum_{\alpha \in A}(x_\alpha + y_\alpha) = \sum_{\alpha \in A} x_\alpha + \sum_{\alpha \in A} y_\alpha,$$

$$\sum_{\alpha \in A}(-x_\alpha) = -\sum_{\alpha \in A} x_\alpha$$

$$\sum_{\alpha \in A} m.x_\alpha = m. \sum_{\alpha \in A} x_\alpha.$$

Proof. The first equality is a consequence of 10.11 and the continuity of the homomorphism $(x, y) \to x + y$ from $G \times G$ to G. •

10.13 Theorem. *Let G be a complete commutative topological group whose topology is given by a norm N. If $(x_\alpha)_{\alpha \in A}$ is a family of elements of G such that $(N(x_\alpha))_{\alpha \in A}$ is a summable family of real numbers, then $(x_\alpha)_{\alpha \in A}$ is summable, and*

$$N(\sum_{\alpha \in A} x_\alpha) \le \sum_{\alpha \in A} N(x_\alpha).$$

Proof. Let
$$s = \sum_{\alpha \in A} N(x_\alpha).$$
If K is any finite subset of A, then
$$N(\sum_{\alpha \in K} x_\alpha) \leq \sum_{\alpha \in K} N(x_\alpha) \leq s.$$
Consequently, as Cauchy's Condition holds for $(N(x_\alpha))_{\alpha \in A}$ by 10.4, it holds also for $(x_\alpha)_{\alpha \in A}$, and therefore $(x_\alpha)_{\alpha \in A}$ is summable, and moreover,
$$N(\sum_{\alpha \in A} x_\alpha) \leq s. \bullet$$

10.14 Theorem. *Let G be a complete commutative topological group whose topology is given by an ultranorm N, and let $(x_\alpha)_{\alpha \in A}$ be a family of elements of G.*

(1) $(x_\alpha)_{\alpha \in A}$ is summable if and only if for every $e > 0$, $N(x_\alpha) \leq e$ for all but finitely many $\alpha \in A$.

(2) If $(x_\alpha)_{\alpha \in A}$ is summable, then
$$N(\sum_{\alpha \in A} x_\alpha) \leq \sup_{\alpha \in A} N(x_\alpha) < +\infty.$$

Proof. (1) follows from 10.5. (2) Let
$$s = \sum_{\alpha \in A} x_\alpha, \quad b = \sup_{\alpha \in A} N(x_\alpha).$$
By 10.4 there is a finite subset J of A such that $N(x_\alpha) \leq 1$ if $\alpha \in A \setminus J$. Consequently,
$$b \leq \sup\{1, \sup_{\alpha \in J} N(x_\alpha)\} < +\infty.$$
If $b = 0$, then $x_\alpha = 0$ for all $\alpha \in A$, so $s = 0$. If $b > 0$, there is a finite subset K of A such that
$$N(s - \sum_{\alpha \in K} x_\alpha) \leq b,$$
so
$$N(s) \leq \sup\{N(s - \sum_{\alpha \in K} x_\alpha), N(\sum_{\alpha \in K} x_\alpha)\} \leq \sup\{b, \sup_{\alpha \in K} N(x_\alpha)\} = b. \bullet$$

10.15 Theorem. Let E, F, and G be Hausdorff commutative groups, let f be a continuous $\mathbb{Z}$-bilinear function from $E \times F$ to G, and let $(x_\lambda)_{\lambda \in L}$ be a summable family of elements of E, $(y_\mu)_{\mu \in M}$ a summable family of elements of F.

(1) For each $a \in E$, $(f(a, y_\mu))_{\mu \in M}$ is summable, and

$$\sum_{\mu \in M} f(a, y_\mu) = f(a, \sum_{\mu \in M} y_\mu).$$

(2) For each $b \in F$, $(f(x_\lambda, b))_{\lambda \in L}$ is summable, and

$$\sum_{\lambda \in L} f(x_\lambda, b) = f(\sum_{\lambda \in L} x_\lambda, b).$$

(3) If $(f(x_\lambda, y_\mu))_{(\lambda, \mu) \in L \times M}$ is summable, then

$$\sum_{(\lambda, \mu) \in L \times M} f(x_\lambda, y_\mu) = f(\sum_{\lambda \in L} x_\lambda, \sum_{\mu \in M} y_\mu).$$

(4) If the open subgroups of G form a fundamental system of neighborhoods of zero, then $(f(x_\lambda, y_\mu))_{(\lambda, \mu) \in L \times M}$ is summable.

Proof. Since $y \to f(a, y)$ and $x \to f(x, b)$ are continuous homomorphisms, (1) and (2) follow from 10.11.

(3) Let

$$x = \sum_{\lambda \in L} x_\lambda, \quad y = \sum_{\mu \in M} y_\mu.$$

For each $\lambda \in L$, $(f(x_\lambda, y_\mu))_{\mu \in M}$ is summable and

$$\sum_{\mu \in M} f(x_\lambda, y_\mu) = f(x_\lambda, y)$$

by (1). Also, $(f(x_\lambda, y))_{\lambda \in L}$ is summable and

$$\sum_{\lambda \in L} f(x_\lambda, y) = f(x, y)$$

by (2). Thus by 10.8,

$$\sum_{(\lambda, \mu) \in L \times M} f(x_\lambda, y_\mu) = \sum_{\lambda \in L} (\sum_{\mu \in M} f(x_\lambda, y_\mu)) = \sum_{\lambda \in L} f(x_\lambda, y) = f(x, y).$$

(4) By (3) and 10.5 applied to $\widehat{G}$, it suffices to show that if U is a neighborhood of zero in G, then $f(x_\alpha, y_\beta) \in U$ for all but finitely many $(\alpha, \beta) \in L \times M$. As f is continuous, there exist neighborhoods V and W of zero in E and F respectively such that $f(V \times W) \subseteq U$. By 10.5 there exist finite subsets S of L and T of M such that $x_\alpha \in V$ for all $\alpha \in L \setminus S$ and $y_\beta \in W$ for all $\beta \in M \setminus T$. For each $\mu \in T$, $(f(x_\alpha, y_\mu))_{\alpha \in L}$ is summable by (2), so by 10.5 there is a finite subset S_μ of L such that $f(x_\alpha, y_\mu) \in U$ for all $\alpha \in L \setminus S_\mu$. Similarly, for each $\lambda \in S$, $(f(x_\lambda, y_\beta))_{\beta \in M}$ is summable by (1), so by 10.5 there is a finite subset T_λ of M such that $f(x_\lambda, y_\beta) \in U$ for all $\beta \in M \setminus T_\lambda$. Consequently, $f(x_\alpha, y_\beta) \in U$ for all

$$(\alpha, \beta) \notin [\bigcup_{\mu \in T} (S_\mu \times \{\mu\})] \cup [\bigcup_{\lambda \in S} (\{\lambda\} \times T_\lambda)],$$

a finite subset of $L \times M$. •

Theorem 10.15 applies, in particular, to scalar multiplication of a topological module. In particular, it applies to multiplication in a topological ring:

10.16 Corollary. *Let $(x_\lambda)_{\lambda \in L}$ and $(y_\mu)_{\mu \in M}$ be summable families of elements of a Hausdorff ring A. For any $c \in A$, $(cx_\lambda)_{\lambda \in L}$ and $(x_\lambda c)_{\lambda \in L}$ are summable, and*

$$\sum_{\lambda \in L} cx_\lambda = c \sum_{\lambda \in L} x_\lambda, \quad \sum_{\lambda \in L} x_\lambda c = \sum_{\lambda \in L} x_\lambda c.$$

If $(x_\lambda y_\mu)_{(\lambda, \mu) \in L \times M}$ is summable,

$$\sum_{(\lambda, \mu) \in L \times M} x_\lambda y_\mu = (\sum_{\lambda \in L} x_\lambda)(\sum_{\mu \in M} y_\mu).$$

If the open additive subgroups of A form a fundamental system of neighborhoods of zero, then $(x_\lambda y_\mu)_{(\lambda, \mu) \in L \times M}$ is summable.

Exercises

10.1 Let $(x_\alpha)_{\alpha \in A}$ be a family of real numbers. (a) If $x_\alpha \in \mathbb{R}_{\geq 0}$ for all $\alpha \in A$ and if

$$s = \sup_{J \in \mathcal{F}(A)} \sum_{\alpha \in J} x_\alpha,$$

then $(x_\alpha)_{\alpha \in A}$ is summable if and only if $s < +\infty$, in which case

$$\sum_{\alpha \in A} x_\alpha = s.$$

(b) $(x_\alpha)_{\alpha \in A}$ is summable if and only if $(|x_\alpha|)_{\alpha \in A}$ is summable.

10.2 A family $(z_\alpha)_{\alpha \in A}$ of complex numbers is summable if and only if $(|z_\alpha|)_{\alpha \in A}$ is summable.

10.3 If $(x_\lambda)_{\lambda \in L}$ and $(y_\mu)_{\mu \in M}$ are summable families of complex numbers, then $(x_\lambda y_\mu)_{(\lambda,\mu) \in L \times M}$ is summable. [Use Exercises 10.2 and 10.1.]

10.4 Let $\mathcal{B}(X)$ be the normed ring Example 1, §1, where X is an infinite set. Give an example of a summable family $(f_\alpha)_{\alpha \in X}$ of members of $\mathcal{B}(X)$ whose sum is the constant function 1 such that $(N(f_\alpha))_{\alpha \in X}$ is not summable.

10.5 Let A be a normed ring with norm N. If $(x_\lambda)_{\lambda \in L}$ and $(y_\mu)_{\mu \in M}$ are summable families of elements of A such that $(N(x_\lambda))_{\lambda \in L}$ and $(N(y_\mu))_{\mu \in M}$ are summable families of real numbers, then $(x_\lambda y_\mu)_{(\lambda,\mu) \in L \times M}$ is summable.

11 Continuity of Inversion and Adversion

The definition of a ring topology does not require that inversion on a topological ring A with identity (the function $x \to x^{-1}$ on $A^\times$) be continuous; if it is, we say that A is a topological ring *with continuous inversion*. It is easy to see that if A is a ring with identity and if $(\mathcal{T}_\lambda)_{\lambda \in L}$ is a family of ring topologies on A for which inversion is continuous, then inversion is continuous for $\sup_{\lambda \in L} \mathcal{T}_\lambda$. In particular, the supremum of a family of division ring topologies is a division ring topology.

To show that inversion is continuous on $A^\times$, it suffices to show that it is continuous at 1:

11.1 Theorem. *Let $\mathcal{T}$ be a topology on a group G, denoted multiplicatively, such that for each $c \in G$, the functions $x \to cx$ and $x \to xc$ are continuous from G to G. If inversion is continuous at 1, it is continuous everywhere.*

Proof. Let $c \in G$, and let V be a neighborhood of c^{-1}. Clearly $x \to cx$ is a homeomorphism, so cV is a neighborhood of 1. By hypothesis there is a neighborhood U of 1 such that $U^{-1} \subseteq cV$. Also, $x \to xc$ is a homeomorphism, so Uc is a neighborhood of c. Clearly

$$(Uc)^{-1} = c^{-1}U^{-1} \subseteq V. \bullet$$

Let A be a commutative ring with identity, and let T be the set of all cancellable elements of A (that is, the complement of the set of zero-divisors). A *total quotient ring* of A is a ring B containing A as a subring such that each $t \in T$ is invertible in B and $B = \{x/t : x \in A, t \in T\}$. The proof that each integral domain A is a subdomain of a field B may be carried over without essential alteration to show that each commutative ring A has a total quotient ring. Moreover, if B and B' are total quotient rings of A, there is a unique isomorphism f from B to B' such that $f(x) = x$ for all $x \in A$. Consequently, we may speak of *the quotient ring* $Q(A)$ of A.

A subset S of a commutative ring A with identity is *multiplicative* if $1 \in S$, $0 \notin S$, and $xy \in S$ whenever $x \in S$ and $y \in S$. For example, the subset T of cancellable elements is multiplicative. If S is a multiplicative subset of T, we denote by $S^{-1}A$ the subring of the total quotient ring $Q(A)$ of A consisting of all the elements x/s where $x \in A$ and $y \in S$; in particular, $Q(A) = T^{-1}A$. If A is an integral domain, then $T = A^*$, and $T^{-1}A$, or $Q(A)$, is the quotient field of A.

For any Hausdorff ring topology $\mathcal{T}$ on a field K there is a Hausdorff field topology $\mathcal{S}$ on K weaker than $\mathcal{T}$, a consequence of the following theorem:

11.2 Theorem. *Let $\mathcal{T}$ be a ring topology on a commutative ring A with identity, and let S be a multiplicative set of cancellable elements of A such that S is a neighborhood of 1 and, for each $s \in S$, $x \to sx$ is an open mapping from A to A. Of all the ring topologies on $S^{-1}A$ for which inversion is continuous and which induce on A a topology weaker than $\mathcal{T}$, there is a strongest $\mathcal{S}$. If $\mathcal{T}$ is Hausdorff, so is $\mathcal{S}$. If $\mathcal{V}$ is a fundamental system of symmetric neighborhoods of zero for $\mathcal{T}$ such that $1 + V \subseteq S$ for each $V \in \mathcal{V}$, then $\tilde{\mathcal{V}}$ is a fundamental system of symmetric neighborhoods of zero for $\mathcal{S}$, where*

$$\tilde{\mathcal{V}} = \{\tilde{V} : V \in \mathcal{V}\},$$

and for each $V \in \mathcal{V}$,

$$\tilde{V} = \{\frac{v}{1+w} : v, w \in V\}.$$

Proof. Clearly $\tilde{\mathcal{V}}$ is a filter base of symmetric subsets of $S^{-1}A$. If $V \in \mathcal{V}$, there exists $U \in \mathcal{V}$ such that $U + U + UU + UU \subseteq V$; easy calculations then establish that $\tilde{U} + \tilde{U} \subseteq \tilde{V}$ and $\tilde{U}\tilde{U} \subseteq \tilde{V}$. If $a \in A$, $s \in S$, and $V \in \mathcal{V}$, there exists $U \in \mathcal{V}$ such that $aU \subseteq V$ and $U \subseteq V$; as $x \to sx$ is open, there exists $W \in \mathcal{V}$ such that $W \subseteq U \cap sU$; therefore $as^{-1}\tilde{W} \subseteq a\tilde{U} \subseteq \tilde{V}$. Thus $\tilde{\mathcal{V}}$ is a fundamental system of neighborhoods of zero for a ring topology $\mathcal{S}$ on $S^{-1}A$.

If $\mathcal{T}$ is Hausdorff, so is $\mathcal{S}$. Indeed, let $a \in A^*$ and $s \in S$. Then there exists $U \in \mathcal{V}$ such that $a \notin U$. There exists $W \in \mathcal{V}$ such that $W + W \subseteq U$, and there exists $V \in \mathcal{V}$ such that $sV \subseteq W$ and $aV \subseteq W$. Then $s^{-1}a \notin \tilde{V}$.

To show that inversion is continuous at 1 for $\mathcal{S}$, let $V \in \mathcal{V}$. There exists $U \in \mathcal{V}$ such that $U + U \subseteq V$. Then $(1+\tilde{U})^{-1} \subseteq 1 + \tilde{V}$, for if $u, v \in U$, then

$$(1 + \frac{u}{1+v})^{-1} = \frac{1+v}{1+u+v} = 1 + \frac{-u}{1+u+v} \in 1 + \tilde{V}.$$

Thus inversion is continuous for S by 11.1.

Let S' be a ring topology on $S^{-1}A$ for which inversion is continuous and which induces on A a topology weaker than T, and let T be a neighborhood of zero for S'. As

$$(x, y) \to \frac{x}{1+y}$$

is continuous at $(0,0)$ for the cartesian product topology determined by S', there is a neighborhood W of zero for S' such that $W(1+W)^{-1} \subseteq T$. By assumption, there exists $V \in \mathcal{V}$ such that $V \subseteq W \cap A$. Hence $\tilde{V} \subseteq T$. Thus S is stronger than S'. •

11.3 Corollary. *If T is a Hausdorff ring topology on a field K, then of all the field topologies on K weaker than T there is a strongest S, and S is Hausdorff.*

Proof. We need only let $S = K^*$ in 11.2, for then $S^{-1}K = K$. •

11.4 Definition. **Circulation** or the **circle composition** on a ring A is the composition $\circ$ defined by

$$x \circ y = x + y - xy$$

for all $x, y \in A$. An element of A is [**left, right**] **advertible** if it is [left, right] invertible for $\circ$.

11.5 Theorem. *Let A be a ring. (1) Circulation on A is an associative composition with neutral element zero. (2) Circulation on A is commutative if and only if multiplication is commutative. (3) For any $a, b \in A$, if ab is left [right] advertible, so is ba.*

Proof. The proofs of (1) and (2) are easy. (3) If $y \circ ab = 0$, then $(bya - ba) \circ ba = 0$, and similarly if $ab \circ y = 0$, then $ba \circ (bya - ba) = 0$. •

By (1) of 11.5, if $x \in A$, there is at most one element $y \in A$ such that $x \circ y = 0 = y \circ x$. If such an element exists, it is called the *adverse* of x and denoted by x^a. We shall denote by A^a the group (under $\circ$) of all advertible elements of A, and call the function $x \to x^a$ from A^a to A^a (or any larger set) *adversion*. If A is a topological ring and adversion is continuous on A^a, we shall say that A is a ring *with continuous adversion*.

If A is a topological ring, circulation is clearly continuous from $A \times A$ to A. In particular, $x \to a \circ x$ and $x \to x \circ a$ are continuous functions from A to A for any $a \in A$, and if $c \in A^a$, then $x \to c \circ x$ and $x \to x \circ c$ are homeomorphisms from A to A.

In a ring with identity, circulation and adversion are essentially disguises of multiplication and inversion, as the following theorem shows. They are introduced since adversion is defined in any ring, whereas inversion is defined only in rings with identity.

11.6 Theorem. *Let A be a ring with identity. The function k from A to A, defined by*
$$k(x) = 1 - x$$
for all $x \in A$, is an isomorphism from the semigroup A under multiplication [circulation] to the semigroup A under circulation [multiplication].

Proof. It is easy to see that
$$(1-x) \circ (1-y) = 1 - xy$$
and that
$$1 - (x \circ y) = (1-x)(1-y)$$
for all $x, y \in A$. •

In particular, the restriction of k to the group $A^\times$ [the group A^a] is an isomorphism from $A^\times$ [A^a] to A^a [$A^\times$].

If A is a division ring, then $A^\times = A \setminus \{0\}$, so $A^a = A \setminus \{1\}$. For any ring, $0 \in A^a$, and no nonzero idempotent e belongs to A^a, for if
$$e + x - ex = 0,$$
then
$$0 = e(e + x - ex) = e^2 + ex - e^2 x = e.$$
If A is a trivial ring, then $A^a = A$ and, in fact, $x^a = -x$ for all $x \in A$.

11.7 Definition. *A topological ring A is **advertibly open** if A^a is an open subset of A.*

Thus if A is a topological ring with identity, A is advertibly open if and only if $A^\times$ is open. For example, a Hausdorff division ring is advertibly open. The cartesian product of infinitely many topological rings with identity is not advertibly open, however, as every neighborhood of zero contains a nonzero idempotent.

To show that a topological ring A is advertibly open, it suffices to show that A^a is a neighborhood of zero:

11.8 Theorem. *If A is a topological ring and if A^a contains an interior point, then A^a is open. If A is a topological ring with identity and if $A^\times$ contains an interior point, then $A^\times$ is open.*

Proof. Let U be a nonempty open subset of A contained in A^a, and let $c \in U$. For any $x \in A^a$, $x \circ c^a \circ U$ is an open set containing x and contained in A^a. •

Certain topological conditions imply the continuity of adversion:

11.9 Theorem. *If A is a complete, metrizable, advertibly open ring [with identity], then adversion on A^a [inversion on $A^\times$] is continuous.*

Proof. A theorem of topology (see, for example, Theorem 14.9 of *Topological Fields*) asserts that on any open subset of a complete metric space there is a complete metric defining its induced topology. The assertion therefore follows by applying 6.13 to A^a. •

A deep theorem asserts that local compactness may replace complete metrizability in 6.13 (for a proof, see §9 of *Topological Fields*):

11.10 Theorem. *If T is a locally compact topology on a group G, denoted multiplicatively, such that for all $c \in G$, $x \to cx$ and $x \to xc$ are continuous, then T is a group topology.*

Correspondingly, we obtain:

11.11 Theorem. *If A is an advertibly open, locally compact ring [with identity], then adversion on A^a [inversion on $A^\times$] is continuous.*

Proof. As A^a is open, it is locally compact for its induced topology, so we need only apply 11.10. •

Complete normed rings are advertibly open and have continuous adversion:

11.12 Theorem. *Let A be a ring [with identity] topologized by a complete norm N. Then A is an advertibly open ring with continuous adversion [inversion]. Specifically, if $N(x) < 1$, then x is advertible $[1-x$ is invertible$]$, $(x^n)_{n \geq 1}$ is summable, and*

$$x^a = -\sum_{n=1}^{\infty} x^n \qquad [(1-x)^{-1} = \sum_{n=0}^{\infty} x^n].$$

Proof. Let

$$s_m = \sum_{n=1}^{m} x^n$$

for all $m \geq 1$. If $m > p \geq 1$,

$$N(s_m - s_p) \leq \sum_{n=p+1}^{m} N(x)^n \leq N(x)^{p+1} \sum_{k=0}^{\infty} N(x)^k = N(x)^{p+1}[1-N(x)]^{-1}.$$

Consequently $(x^n)_{n \geq 1}$ is summable, and clearly

$$x \circ (-\sum_{n=1}^{\infty} x^n) = x - \sum_{n=1}^{\infty} x^n + \sum_{n=1}^{\infty} x^{n+1} = 0.$$

The continuity of adversion now follows from 11.9, but an elementary argument also establishes it. By the preceding,

$$N(x^a) \leq \sum_{n=1}^{\infty} N(x)^n = N(x)[1 - N(x)]^{-1},$$

so adversion is continuous at zero and hence everywhere on A^a by 11.1. •

To show that adversion is continuous, it suffices to show that its restriction to a dense subgroup of A^a is continuous:

11.13 Theorem. *Let T be a topology on a group G, denoted multiplicatively, such that $(x,y) \to xy$ is continuous from $G \times G$, furnished with the cartesian product topology defined by T, to G. If the restriction of inversion to a dense subgroup H of G is continuous, then inversion is continuous on G.*

Proof. By 11.1 it suffices to show that inversion is continuous at 1. Let W be a neighborhood of 1. By hypothesis there is a neighborhood V of 1 such that $VV \subseteq W$. Also by hypothesis there is a neighborhood U of 1 such that $(U \cap H)^{-1} \subseteq V \cap H$. Again, there exists by hypothesis a neighborhood T of 1 such that $TT \subseteq U$ and $T \subseteq V$. To show that $T^{-1} \subseteq W$, let $s \in T$. As $s \in \overline{H}$ and as Ts is a neighborhood of s in G, $Ts \cap H \neq \emptyset$. Thus there exists $t \in T$ such that $ts \in H$. Hence $ts \in U \cap H$, so

$$s^{-1}t^{-1} = (ts)^{-1} \in V,$$

whence

$$s^{-1} \in Vt \subseteq VT \subseteq VV \subseteq W. \bullet$$

11.14 Theorem. *If B is a Hausdorff ring [with identity] containing a dense advertibly open subring A with continuous adversion, then B is a ring with continuous adversion [inversion].*

Proof. Clearly A^a is a dense subgroup of its closure in B^a, which is $\overline{A^a} \cap B^a$. Consequently by 11.13, the restriction of adversion to $\overline{A^a} \cap B^a$ is continuous. By 4.22 and our hypothesis, $\overline{A^a}$ is a neighborhood of zero in B, so $\overline{A^a} \cap B^a$ is a neighborhood of zero in B^a. Consequently, adversion on B^a is continuous at zero and hence everywhere by 11.1. •

As we shall shortly see, the completion of a Hausdorff field need not be a field or even an advertibly open topological ring, but at least it has continuous inversion:

11.15 Corollary. *If K is a Hausdorff topological division ring, the completion $\hat{K}$ of K is a topological ring with continuous inversion; in particular, if $\hat{K}$ is algebraically a division ring, it is a topological division ring.*

11.16 Theorem. *If A is a complete, Hausdorff ring [with identity] whose open additive subgroups form a fundamental system of neighborhoods of zero, and if x is a topological nilpotent of A, then x is advertible [$1-x$ is invertible], $(x^n)_{n\geq 1}$ is summable, and*

$$x^a = -\sum_{n=1}^{\infty} x^n \qquad [(1-x)^{-1} = \sum_{n=0}^{\infty} x^n].$$

Proof. By 10.5, $(x^n)_{n\geq 1}$ is summable, and clearly

$$x - \sum_{n=1}^{\infty} x^n + x\sum_{n=1}^{\infty} x^n = 0. \bullet$$

Exercises

11.1 (a) The filter base of all nonzero ideals of $\mathbb{Z}$ is a fundamental system of neighborhoods of zero for a ring topology $\mathcal{T}$ on $\mathbb{Q}$ that is not a field topology. (b) For each integer $a > 0$, let

$$V_a = \{n/q : n \in \mathbb{Z}, q \in \mathbb{Z}^*, a\,|\,n, \text{ and } (a,q) = 1\}.$$

Show that $\{V_a : a \in \mathbb{Z}, a > 0\}$ is a fundamental system of neighborhoods of zero for the strongest field topology on $\mathbb{Q}$ weaker than $\mathcal{T}$.

11.2 With the terminology of 11.2, show that: (a) $A^\times$ is open for $\mathcal{T}$ if and only if A is open in $S^{-1}A$ for $\mathcal{S}$; (b) $A^\times$ is open for $\mathcal{T}$ and inversion is continuous on $A^\times$ if and only if every open subset of A for $\mathcal{T}$ is also open in $S^{-1}A$ for $\mathcal{S}$.

11.3 (Gould [1961]) Let A be a commutative topological ring with identity, S a multiplicative subset of A. Of all the ring topologies on A stronger than its given topology $\mathcal{T}$ such that for each $s \in S$, $x \to sx$ is an open mapping, there is a weakest $\mathcal{T}_S$. If $\mathcal{V}$ is a fundamental system of symmetric neighborhoods of zero for $\mathcal{T}$, then $\{sV : s \in S, V \in \mathcal{V}\}$ is a fundamental system of symmetric neighborhoods of zero for $\mathcal{T}_S$.

11.4 The supremum of a family of ring topologies on a ring A having continuous adversion is a ring topology having continuous adversion.

11.5 (Warner [1955]) A topological ring A is *advertibly complete* if every Cauchy filter $\mathcal{F}$ on A for which there exists $a \in A$ such that $\mathcal{F} \circ a$ and $a \circ \mathcal{F}$

converge to zero is convergent. (a) A complete topological ring is advertibly complete. (b) An advertibly open topological ring is advertibly complete. (c) A left or right ideal of an advertibly complete ring is advertibly complete.

11.6 (Warner [1955]) Let A be a ring topologized by a norm N. The following statements are equivalent:

1° For all $x \in A$, if $N(x) < 1$, then x is advertible.
2° For all $x \in A$, if $N(x) < 1$, then $(x^n)_{n \geq 1}$ is summable.
3° A is advertibly open.
4° A is advertibly complete

11.7 (a) The cartesian product of advertibly complete rings is advertibly complete. (b) Give an example of an advertibly complete ring that is neither complete nor advertibly open.

11.8 If X is a subset of a topological space T, let X' be its derived set, which consists of all $t \in T$ such that every neighborhood of t contains infinitely many elements of X. Let A be a topological ring. (a) If $0 \in (A^a)'$, then $A^a \subseteq (A^a)'$. (b) Either $A^a \cap (A^a)' = \emptyset$, or $A^a \subseteq (A^a)'$.

11.9 If A is a complete Hausdorff ring [with identity] whose open subrings form a fundamental system of neighborhoods of zero, and if the set of topological nilpotents is a neighborhood of zero, then A is an advertibly open ring with continuous adversion [inversion].

CHAPTER III

LOCAL BOUNDEDNESS

Normed rings and vector spaces are examples of locally bounded rings and modules, whose elementary properties are presented in §12. Straight division rings and locally retrobounded division rings, which include topological rings whose topology is given by a proper absolute value, are introduced in §13. A condition for a topological ring to be normable and relations between norms and absolute values on a field are discussed in §14. The simple and elegant theory of Hausdorff finite-dimensional vector spaces over complete straight division rings (in particular, over division rings whose topology is given by a complete absolute value) is presented in §15. Finally, in §16 we derive certain classical theorems concerning topological division rings: Pontriagin's theorem on connected locally compact division rings, the Extension Theorem for complete absolute values, the Gel'fand-Mazur Theorem on normed division algebras, and Ostrowski's Theorem identifying all archimedean absolute values.

12 Locally Bounded Modules and Rings

Bounded and locally bounded rings and modules constitute a central topic, which we introduce here.

12.1 Definition. *Let E be a topological module over a topological ring A. A subset B of E is* **bounded** *if for every neighborhood U of zero in E there is a neighborhood V of zero in A such that $V.B \subseteq U$.*

If S is a ring topology on A stronger than its given topology T, then E remains a topological module over A when A is retopologized with S, and every subset of E bounded when A is furnished with T remains bounded when A is furnished with S; in general, there may be additional bounded sets. For example, if S is the discrete topology on A, every subset of E is bounded since $(0).E \subseteq U$ for any neighborhood U of zero in E.

If T is a ring topology on a ring A, T is a module topology on the associated left and right A-modules A, whose scalar multiplications are simply the given multiplication of A.

12.2 Definition. *A subset B of a topological ring A is* **left [right] bounded** *if B is a bounded subset of the right [left] topological A-module A, and B is* **bounded** *if it is both left and right bounded.*

Thus B is left [right] bounded if and only if for every neighborhood U of zero there is a neighborhood V of zero such that $BU \subseteq V$ $[UB \subseteq V]$.

Any subset consisting of one element of a topological module [ring] is bounded by (TMN 2) of 3.6 [(TRN 2) of 3.5]. More generally:

12.3 Theorem. *A compact subset of a topological module or topological ring is bounded.*

The assertion is a restatement of 4.18 and 4.19.

Many operations are closed under the formation of bounded sets. For example, any subset of a bounded set is clearly bounded.

12.4 Theorem. *Let E be a topological module over a topological ring A, and let B_1 and B_2 be bounded subsets of E, C a right bounded subset of A. Then $\overline{B}_1$, $B_1 + B_2$, $B_1 \cup B_2$, and $C.B_1$ are bounded.*

Proof. Let U be a closed neighborhood of zero. There is a neighborhood V of zero in A such that $V.B_1 \subseteq U$, and as scalar multiplication is continuous,

$$V.\overline{B}_1 \subseteq \overline{V}.\overline{B}_1 \subseteq \overline{V.B_1} \subseteq \overline{U} = U.$$

Thus by (4) of 3.3, $\overline{B}_1$ is bounded.

Let W be a neighborhood of zero such that $W + W \subseteq U$, and let V_1, V_2 be neighborhoods of zero in A such that $V_1.B_1 \subseteq W$ and $V_2.B_2 \subseteq W$. Then

$$(V_1 \cap V_2).(B_1 \cup B_2) \subseteq V_1.B_1 \cup V_2.B_2 \subseteq W \subseteq U,$$

and

$$(V_1 \cap V_2).(B_1 + B_2) \subseteq V_1 B_1 + V_2 B_2 \subseteq W + W \subseteq U.$$

Finally, let T be a neighborhood of zero in A such that $TC \subseteq V$. Then

$$T.(C.B_1) = (TC).B_1 \subseteq V.B_1 \subseteq U. \bullet$$

Consequently, the union or sum of finitely many bounded subsets of a topological module is bounded.

12.5 Corollary. *If B and C are [left, right] bounded subsets of a topological ring, then so are $\overline{B}$, $B \cup C$, $B + C$, and CB.*

12.6 Theorem. *If u is a continuous homomorphism from a topological A-module E to a topological A-module F and if B is a bounded subset of E, then $u(B)$ is a bounded subst of F.*

Proof. Let U be a neighborhood of zero in F. Then $u^{-1}(U)$ is a neighborhod of zero in E, so there is a neighborhood V of zero in A such that $V.B \subseteq u^{-1}(U)$. Consequently,

$$V.u(B) = u(V.B) \subseteq u(u^{-1}(U)) \subseteq U. \bullet$$

12.7 Theorem. *If u is a topological epimorphism from a topological ring A to a topological ring A' and if B is a [left, right] bounded subset of A, then $u(B)$ is a [left, right] bounded subset of A'.*

Proof. We consider the left bounded case. Let U be a neighborhood of zero in A'. Then $u^{-1}(U)$ is a neighborhood of zero in A, so there is a neighborhood V of zero in A such that $BV \subseteq u^{-1}(U)$. But then $u(V)$ is a neighborhood of zero in A', and

$$u(B)u(V) = u(BV) \subseteq u(u^{-1}(U)) \subseteq U. \bullet$$

In general, the image of a bounded set under a continuous isomorphism need not be bounded. For example, a ring A furnished with the discrete topology is bounded, but A need not be bounded for a nondiscrete ring topology.

12.8 Theorem. *If E is the cartesian product of a family $(E_\mu)_{\mu \in M}$ of topological A-modules, then a subset B of E is bounded if and only if $pr_\lambda(B)$ is a bounded subset of E_λ for each $\lambda \in M$ (where pr_λ is the canonical projection from E to E_λ).*

Proof. The condition is necessary by 12.6. Sufficiency: Let U be the cartesian product of $(U_\mu)_{\mu \in M}$, where U_μ is a neighborhood of zero in E_μ for all $\mu \in M$ and for some finite subset Q of M, $U_\mu = E_\mu$ for all $\mu \in M \setminus Q$. By assumption, for each $\mu \in Q$ there is a neighborhood V_μ of zero in A such that $V_\mu.pr_\mu(B) \subseteq U_\mu$. Therefore

$$\left(\bigcap_{\mu \in Q} V_\mu\right).B \subseteq U. \bullet$$

12.9 Theorem. *If A is the cartesian product of a family $(A_\mu)_{\mu \in M}$ of topological rings, then a subset B of A is [left, right] bounded if and only if $pr_\lambda(B)$ is [left, right] bounded for all $\lambda \in M$.*

The proof is similar to that of 12.8.

12.10 Theorem. *If F is a submodule of a topological A-module E and if $B \subseteq F$, then B is a bounded subset of E if and only if it is a bounded subset of F.*

Proof. Clearly $V.B \subseteq U$ if and only if $V.B \subseteq U \cap F$. •

12.11 Theorem. *If B is a [left, right] bounded subset of a topological ring A and if A' is a subring of A, then $B \cap A'$ is a [left, right] bounded subset of A'.*

Proof. If $BV \subseteq U$, then $(B \cap A')(V \cap A') \subseteq U \cap A'$. •

In contrast, a bounded subset of a subring of A need not be a bounded subset of A (Exercise 12.7).

The condition given in the following theorem is the original definition of a bounded set in real topological vector spaces.

12.12 Theorem. *A necessary condition for a subset B of a topological A-module E to be bounded is that for every sequence $(x_n)_{n \geq 1}$ of elements of B and every sequence $(\lambda_n)_{n \geq 1}$ of scalars, if $\lim_{n \to \infty} \lambda_n = 0$, then $\lim_{n \to \infty} \lambda_n x_n = 0$. If A is metrizable, this condition is both necessary and sufficient for B to be bounded.*

Proof. If U is a neighborhood of zero in E, there is a neighborhood V of zero in A such that $V.B \subseteq U$; if $\lambda_n \in V$ for all $n \geq m$, then $\lambda_n x_n \in U$ for all $n \geq m$. Conversely, assume that A is metrizable, and let $(V_n)_{n \geq 1}$ be a fundamental decreasing sequence of neighborhoods of zero in A. Assume that B is not bounded. Then there is a neighborhood U of zero in E such that for each $n \geq 1$ there exist $\lambda_n \in V_n$ and $x_n \in B$ such that $\lambda_n x_n \notin U$. Then $\lim_{n \to \infty} \lambda_n = 0$, but $(\lambda_n x_n)_{n \geq 1}$ does not converge to zero. •

12.13 Definition. *A topological A-module E and its topology are called* **bounded** *if E is a bounded set; E and its topology are* **locally bounded** *if there is a bounded neighborhood of zero (and hence a fundamental system of bounded neighborhoods of zero). Similarly, a topological ring A is* **[left, right] bounded** *if A is a [left, right] bounded set, and A is* **locally [left, right] bounded** *if A has a [left, right] bounded neighborhood of zero (and hence a fundamental system of [left, right] bounded neighborhoods of zero).*

12.14 Theorem. *(1) If M is a submodule of a [locally] bounded module E, then both M and E/M are [locally] bounded.*

(2) If E is the cartesian product of a family $(E_\mu)_{\mu \in M}$ of topological A-modules, then E is bounded if and only if each E_μ is bounded, and E is locally bounded if and only if each E_μ is locally bounded and for all but finitely many $\mu \in M$, E_μ is bounded.

(3) If E is a Hausdorff [locally] bounded A-module, then $\hat{E}$ is [locally] bounded.

(4) A [locally] compact module is [locally] bounded.

Proof. (1) follows from 12.6, (2) from 12.8, (3) from 12.5 and 4.22, and (4) from 12.3. •

12.15 Theorem. (1) If J is an ideal of a [locally] bounded ring A, then A/J is [locally] bounded.

(2) If $(A_\mu)_{\mu \in M}$ is a family of topological rings and if A is their cartesian product, then A is bounded if and only if each A_μ is bounded, and A is locally bounded if and only if each A_μ is locally bounded and for all but finitely many $\mu \in M$, A_μ is bounded.

(3) If A is a Hausdorff [locally] bounded ring, then $\hat{A}$ is [locally] bounded.

(4) If A is a [locally] bounded ring, so is any subring. (5) A [locally] compact ring is [locally] bounded.

The proof is similar to that of 12.14.

12.16 Theorem. *Let A be a topological ring whose open additive subgroups form a fundamental system of neighborhoods of zero.*

(1) *If A is [left, right] bounded, the open [left, right] ideals of A form a fundamental system of neighborhoods of zero.*

(2) *If A is locally left [right] bounded, the open subrings of A form a fundamental system of neighborhoods of zero.*

A slight modification of the proofs of 4.20 and 4.21 yields (1) and (2) respectively.

12.17 Theorem. *Let A be a topological ring with identity possessing a subset C of invertible elements such that $0 \in \overline{C}$. If V is a bounded neighborhood of zero in a unitary topological A-module E, then $\{\lambda V : \lambda \in C\}$ is a fundamental system of neighborhoods of zero.*

Proof. If $\lambda \in A^\times$, then $x \to \lambda x$ is a homeomorphism from E to E, so λV is a neighborhood of zero. Let U be any neighborhood of zero in E. There exists a neighborhood W of zero in A such that $W.V \subseteq U$, and there exists $\lambda \in C \cap W$, so $\lambda V \subseteq U$. •

12.18 Theorem. *If A is a topological ring with identity and if zero is adherent to $A^\times$, the only Hausdorff bounded unitary A-module is the zero module.*

Proof. Let E be a Hausdorff bounded unitary A-module. By 12.17, $\{\lambda E : \lambda \in A^\times\}$ is a fundamental system of neighborhoods of zero. But $\lambda E = E$ for all $\lambda \in A^\times$, so as E is Hausdorff, $E = (0)$. •

12.19 Corollary. *If K is a division ring furnished with a ring topology and if E is a nonzero Hausdorff bounded K-vector space, then the topology of K is discrete.*

12.20 Corollary. *If A is a Hausdorff ring with identity and if zero is adherent to $A^\times$, then A is left and right unbounded. In particular, the only Hausdorff left or right bounded topology on a division ring is the discrete topology.*

By 12.3, we conclude:

12.21 Corollary. *A compact division ring is finite.*

Exercises

12.1 If $(a_n)_{n \geq 1}$ is a Cauchy sequence in a topological module E, then $\{a_n : n \geq 1\}$ is bounded.

12.2 Let A be a topological ring with identity such that $0 \in \overline{A^\times}$, and let E be a locally compact unitary A-module. A subset of E is relatively compact (that is, its closure is compact) if and only if it is bounded.

12.3 The topology defined in Exercise 11.1(b) on $\mathbb{Q}$ is a field topology that is not locally bounded.

12.4 Let A be a topological ring with identity, C a subset of $A^\times$ such that $0 \in \overline{C}$. (a) A subset B of A is left [right] bounded if and only if for every neighborhood U of zero there exists $\lambda \in C$ such that $B\lambda \subseteq U$ [$\lambda B \subseteq U$]. (b) A left [right] bounded subset of A remains left [right] bounded if A is furnished with a weaker ring topology.

12.5 A sequence $(B_n)_{n \geq 1}$ of bounded subsets of a topological module E is a *fundamental sequence of bounded subsets* if each bounded subset of E is contained in some B_n. (a) If a topological module E is a Baire space and has a fundamental sequence of bounded subsets, then E is locally bounded. (b) If A is a metrizable ring with identity such that $0 \in \overline{A^\times}$ and if E is a unitary locally bounded A-module, then E has a fundamental sequence of bounded subsets.

12.6 Let A be a topological ring with identity such that $0 \in \overline{A^\times}$, and let E be a Hausdorff unitary A-module. (a) No proper submodule of E is open. (b) If every neighborhood of zero in E contains a nonzero submodule, then E is not locally bounded. (c) In particular, if E is the cartesian product of infinitely many nonzero Hausdorff unitary A-modules, then the cartesian product topology on E is not locally bounded.

12.7 The subring $\mathbb{Z}$ of $\mathbb{Q}$, furnished with its usual discrete topology, is a bounded subset of itself but is not a bounded subset of $\mathbb{Q}$.

12.8 If A is a bounded ring, adversion is uniformly continuous on A^a. [First, assume that A has an identity.]

13 Locally Retrobounded Division Rings

An important class of Hausdorff division rings that includes those whose topology is defined by a proper absolute value is given in the following definition:

13.1 Definition. Let K be a Hausdorff topological division ring. A topological K-vector space E is **straight** if for every nonzero $c \in E$, $\lambda \to \lambda c$ is a homeomorphism from K to the one-dimensional subspace Kc of E. The Hausdorff topological division ring K is **straight** if every Hausdorff K-vector space is straight.

13.2 Theorem. If K is a division ring furnished with a Hausdorff ring topology $\mathcal{T}$ such that every isomorphism from the K-vector space K to a Hausdorff one-dimensional K-vector space is a homeomorphism, then $\mathcal{T}$ is minimal in the set of all Hausdorff ring topologies on K, ordered by inclusion.

Proof. Let $\mathcal{S}$ be a Hausdorff ring topology on K weaker than $\mathcal{T}$. Then K, furnished with $\mathcal{S}$, is a topological vector space over K, furnished with $\mathcal{T}$. By hypothesis, $\lambda \to \lambda.1 = \lambda$ is a homeomorphism from K, furnished with $\mathcal{T}$, to K, furnished with $\mathcal{S}$, so $\mathcal{S} = \mathcal{T}$. •

13.3 Theorem. If K is a straight division ring, its topology is minimal among the Hausdorff ring topologies on K, that is, there is no ring topology on K strictly weaker than its given topology.

If K is a field furnished with a minimal Hausdorff ring topology, that topology is necessarily a field topology by 11.3. Thus, for fields, the requirement in Definition 13.1 that the inversion be continuous for the given topology follows from the other requirements of the Definition.

13.4 Theorem. If K is a straight division ring, so are $\widehat{K}$ and any dense division subring of K.

Proof. If c is a nonzero element of a Hausdorff $\widehat{K}$-module E, then $u_c : \lambda \to \lambda c$ is a topological isomorphism from the $\widehat{K}$-module $\widehat{K}$ to the submodule $\widehat{K}c$ of E. Indeed, the restriction v of u_c to K is a topological isomorphism from K to Kc; as u_c is continuous, u_c is the unique continuous extension of v to $\widehat{K}$; by 8.7, therefore, u_c is a topological isomorphism from $\widehat{K}$ to $\widehat{K}c$. Thus to establish that $\widehat{K}$ is straight, it suffices by 11.15 to prove that $\widehat{K}$ is a division ring.

Let c be a nonzero element of $\widehat{K}$. We have just seen that $\widehat{K}c$ is topologically isomorphic to $\widehat{K}$ and hence is complete and therefore closed. Assume that $\widehat{K}c$ is a proper left ideal of $\widehat{K}$, let ϕ be the canonical epimorphism from the $\widehat{K}$-module $\widehat{K}$ to the Hausdorff $\widehat{K}$-module $\widehat{K}/\widehat{K}c$, and let

$e = 1 + \widehat{K}c \in \widehat{K}/\widehat{K}c$. Then for all $\lambda \in \widehat{K}$,

$$\phi(\lambda) = \lambda + \widehat{K}c = \lambda(1 + \widehat{K}c) = \lambda e = u_e(\lambda),$$

so by the preceding, ϕ is a topological isomorphism from $\widehat{K}$ to $\widehat{K}/\widehat{K}c$. Thus $\widehat{K}c = \phi^{-1}(0) = (0)$, so $c = 0$, a contradiction. Therefore $\widehat{K}c = \widehat{K}$ for all nonzero $c \in \widehat{K}$. Thus every nonzero element of $\widehat{K}$ has a left inverse, so $\widehat{K}$ is a division ring.

Let L be a dense division subring of K; then L is also a dense division subring of the straight division ring $\widehat{K}$. If c is a nonzero vector of a Hausdorff L-vector space E, then $\lambda \to \lambda c$ from L to Lc is simply the restriction to L of the topological isomorphism $\lambda \to \lambda c$ from $\widehat{K}$ to the subspace $\widehat{K}c$ of $\widehat{E}$, and so is a topological isomorphism. •

13.5 Definition. *Let K be a division ring furnished with a ring topology $\mathcal{T}$. A subset V of K that contains zero is* **retrobounded** *if $(K \setminus V)^{-1}$ is bounded. The topology $\mathcal{T}$ is* **locally retrobounded** *if $\mathcal{T}$ is Hausdorff and the retrobounded neighborhoods of zero form a fundamental system of neighborhoods of zero. A* **locally retrobounded division ring** *is a division ring furnished with a locally retrobounded topology.*

13.6 Theorem. *A division ring K topologized by an absolute value is locally retrobounded.*

Proof. If $r > 0$ and

$$V = \{x \in K : |x| \leq r\},$$

then

$$(K \setminus V)^{-1} = \{y \in K^* : |y| < r^{-1}\}. \bullet$$

13.7 Theorem. *If $\mathcal{T}$ is a locally retrobounded topology on a division ring K, then every neighborhood of zero is retrobounded, and $\mathcal{T}$ is a locally bounded division ring topology.*

Proof. Let V be a neighborhood of zero. By hypothesis there is a retrobounded neighborhood U of zero such that $U \subseteq V$ and $1 \notin UU$. Then

$$U \subseteq (K \setminus U)^{-1} \cup \{0\}$$

and hence is bounded, and V is retrobounded since

$$(K \setminus V)^{-1} \subseteq (K \setminus U)^{-1}.$$

In particular, $\mathcal{T}$ is a locally bounded topology.

Next, we shall show that if U is a neighborhood of zero, the restriction of inversion to $K \setminus U$ is uniformly continuous. Let V be a neighborhood of zero. As U is retrobounded, there is a neighborhood W of zero such that $W(K \setminus U)^{-1} \subseteq V$, and also there is a neighborhood T of zero such that $(K \setminus U)^{-1} T \subseteq W$. Thus

$$(K \setminus U)^{-1} T (K \setminus U)^{-1} \subseteq V.$$

Hence if $x, y \in K \setminus U$ and if $x - y \in T$, then

$$y^{-1} - x^{-1} = x^{-1}(x-y)y^{-1} \in (K \setminus U)^{-1} T (K \setminus U)^{-1} \subseteq V.$$

In particular, if U is a closed neighborhood of zero not containing 1, inversion is continuous on the open neighborhood $K \setminus U$ of 1, so inversion is continuous on K^* by 11.1. •

13.8 Theorem. *A nondiscrete locally retrobounded division ring K is straight. In particular, a division ring topologized by a proper absolute value is straight.*

Proof. By 13.7 a locally retrobounded division ring is a topological division ring. Let c be a nonzero vector in a Hausdorff vector space E over a locally retrobounded division ring K. Since $\lambda \to \lambda c$ is continuous, we need only show that if U is a neighborhood of zero in K, Uc is a neighborhood of zero in Kc. As E is Hausdorff, there is a neighborhood Y of zero in E such that $c \notin Y$. There exist neighborhoods W of zero in E and V of zero in K such that $VW \subseteq Y$ by 3.6. Since $(K \setminus U)^{-1}$ is bounded by 13.7 and since K is not discrete, there is a nonzero scalar λ such that $(K \setminus U)^{-1} \lambda \subseteq V$. By 2.12 λW is a neighborhood of zero in E; we shall show that $\lambda W \cap Kc \subseteq Uc$. Indeed, let $\mu c \in W$. If $\lambda \mu \notin U$, then

$$\mu^{-1} = (\mu^{-1} \lambda^{-1}) \lambda \in (K \setminus U)^{-1} \lambda \subseteq V,$$

whence

$$c = \mu^{-1}(\mu c) \in VW \subseteq Y,$$

a contradiction. Hence $\lambda \mu \in U$, so $\lambda \mu c \in Uc$. Thus $\lambda W \cap Kc \subseteq Uc$, and the proof is complete. •

13.9 Theorem. *The completion $\widehat{K}$ of a locally retrobounded division ring K is a locally retrobounded division ring.*

Proof. By 13.8 and 13.4, $\widehat{K}$ is a topological division ring, so by 4.22 we need only show that if V is a closed neighborhood of zero in K, its closure

$\overline{V}$ in $\widehat{K}$ is retrobounded. Let U be a closed neighborhood of zero in $\widehat{K}$. As V is retrobounded, there is a neighborhood W of zero in K such that

$$W(K \setminus V)^{-1} \subseteq U \cap K \text{ and } (K \setminus V)^{-1}W \subseteq U \cap K.$$

As $\overline{V} \cap (K \setminus V) = V \cap (K \setminus V) = \emptyset$, $0 \notin \overline{K \setminus V}$. Thus as $\widehat{K} \setminus \overline{V}$ is open,

$$\widehat{K} \setminus \overline{V} \subseteq \overline{(\widehat{K} \setminus \overline{V}) \cap K} = \overline{K \setminus V} \subseteq \widehat{K}^*.$$

Therefore as inversion is a topological automorphism of $\widehat{K}^*$ and as multiplication is continuous on $\widehat{K}$,

$$\overline{W}(\widehat{K} \setminus \overline{V})^{-1} \subseteq \overline{W}(\overline{K \setminus V})^{-1} = \overline{W}(\overline{(K \setminus V)^{-1} \cap \widehat{K}^*})$$
$$\subseteq \overline{\overline{W}(K \setminus V)^{-1}} \subseteq \overline{W(K \setminus V)^{-1}} \subseteq \overline{U \cap K} \subseteq U,$$

and similarly

$$(\widehat{K} \setminus \overline{V})^{-1}\overline{W} \subseteq U.$$

By 4.22 $\overline{W}$ is a neighborhood of zero in $\widehat{K}$. Thus $\overline{V}$ is a retrobounded subset of $\widehat{K}$. •

13.10 Theorem. *If the topology of a topological division ring K is given by an absolute value A, then $\widehat{K}$ is a topological division ring whose topology is given by a unique absolute value $\widehat{A}$ that extends A.*

Proof. By 13.6 and 13.9, $\widehat{K}$ is a division ring, and by 8.9 its topology is defined by a unique norm $\widehat{A}$ that extends A. Since $\widehat{A}$ is continuous by 1.4, the function

$$f : (x, y) \to \widehat{A}(x)\widehat{A}(y) - \widehat{A}(xy)$$

is continuous on $\widehat{K} \times \widehat{K}$. As $f(x,y) = 0$ for all $(x,y) \in A \times A$, therefore, f is the zero function on $\widehat{A} \times \widehat{A}$. Thus $\widehat{A}$ is an absolute value. •

13.11 Theorem. *Let K be a division ring [field] furnished with an absolute value A. If the topology of a K-vector space [K-algebra] E is given by a norm N relative to A, then the topology of $\widehat{E}$ is given by a unique norm $\widehat{N}$ relative to $\widehat{A}$ that extends N.*

Proof. By 6.11 [8.9], the topology of the additive group [ring] $\widehat{E}$ is given by a unique norm $\widehat{N}$ that extends N. The functions $(\lambda, x) \to \widehat{N}(\lambda x)$ and $(\lambda, x) \to \widehat{A}(\lambda)\widehat{N}(x)$ are continuous from $\widehat{K} \times \widehat{E}$ to $\widehat{E}$ and agree on the dense subset $K \times E$; hence they agree on $\widehat{K} \times \widehat{E}$. •

13.12 Theorem. *(Approximation Theorem) Let K be a division ring, let $\mathcal{T}_0, \mathcal{T}_1, \ldots, \mathcal{T}_n$ be distinct Hausdorff nondiscrete division ring topologies on K such that $\mathcal{T}_1, \ldots, \mathcal{T}_n$ are locally retrobounded and $\mathcal{T}_i \not\subseteq \mathcal{T}_0$ for all $i \in [1,n]$, let $\widehat{K}_0, \widehat{K}_1, \ldots, \widehat{K}_n$ be the completions of K for $\mathcal{T}_0, \mathcal{T}_1, \ldots, \mathcal{T}_n$ respectively, and let*

$$L = \prod_{i=0}^n \widehat{K}_i.$$

(1) If $U_0, U_1, \ldots, U_n$ are nonempty open subsets for $\mathcal{T}_0, \mathcal{T}_1, \ldots, \mathcal{T}_n$ respectively, then

$$\bigcap_{i=0}^n U_i \neq \emptyset.$$

(2) If K is furnished with $\sup_{0 \leq i \leq n} \mathcal{T}_i$, then the diagonal mapping Δ from K to L, defined by

$$\Delta(x) = (x, x, \ldots, x)$$

for all $x \in K$, is a topological isomorphism from K to the division subring $\Delta(K)$ of L, and $\Delta(K)$ is dense in L. (3) $\sup_{0 \leq i \leq n} \mathcal{T}_i$ is not the discrete topology.

Proof. Clearly Δ is a topological isomorphism from K, furnished with $\sup_{0 \leq i \leq n} \mathcal{T}_i$, to $\Delta(K)$. Therefore (2) follows from (1), and (3) follows from (2), for $\mathcal{T}_0$ is not discrete by hypothesis, and if $n > 0$, $\Delta(K) \neq \widehat{K}$. Thus it suffices to prove (1).

We shall prove (1) by induction on n. Clearly (1) is true if $n = 0$. Consequently, we shall prove (1) under the assumption that $n > 0$ and

$$\bigcap_{i=0}^{n-1} U'_i \neq \emptyset$$

whenever $U'_0, U'_1, \ldots, U'_{n-1}$ are nonempty open sets for distinct Hausdorff nondiscrete division ring topologies $\mathcal{T}'_0, \mathcal{T}'_1, \ldots, \mathcal{T}'_{n-1}$ on K such that $\mathcal{T}'_i \not\subseteq \mathcal{T}'_0$ for all $i \in [1, n-1]$ and $\mathcal{T}'_1, \ldots, \mathcal{T}'_{n-1}$ are locally retrobounded and $\mathcal{T}'_i \not\subseteq \mathcal{T}'_0$ for all $i \in [1, n-1]$.

Let $U_0, U_1, \ldots, U_n$ be nonempty open subsets of K for $\mathcal{T}_0, \mathcal{T}_1, \ldots, \mathcal{T}_n$ respectively. By 13.8 and 13.3, $\mathcal{T}_i \not\subseteq \mathcal{T}_1$ for all $i \in [2, n]$, so by our inductive hypothesis applied to $\mathcal{T}'_0 = \mathcal{T}_1, \mathcal{T}'_1 = \mathcal{T}_2, \ldots, \mathcal{T}'_{n-1} = \mathcal{T}_n$, $\bigcap_{i=1}^n U_i \neq \emptyset$. Let $\mathcal{T} = \sup_{1 \leq i \leq n} \mathcal{T}_i$. We therefore need only prove that $U_0 \cap U \neq \emptyset$ whenever U_0 and U are nonempty open subsets of K for $\mathcal{T}_0$ and $\mathcal{T}$ respectively. To do so, it suffices to prove (*): If V_0 is a neighborhood of 1 for $\mathcal{T}_0$ and W a neighborhood of zero for $\mathcal{T}$, then $V_0 \cap W \neq \emptyset$. Indeed, let $b \in U$; as $\mathcal{T}_0$ is not discrete, there is a nonzero $a \in b + U_0$; then $a^{-1}(-b + U_0)$ is an open neighborhood of 1 for $\mathcal{T}_0$ and $a^{-1}(-b+U)$ is an open neighborhood of zero for $\mathcal{T}$, so by (*),

$$a^{-1}(-b+U_0) \cap a^{-1}(-b+U) \neq \emptyset,$$

and therefore $U_0 \cap U \neq \emptyset$.

To prove (*), we shall first establish by induction that if $m \in [1, n]$ and if $B_1, \ldots B_m$ are subsets of K bounded for $\mathcal{T}_1, \ldots, \mathcal{T}_m$ respectively and if U is a neighborhood of zero for $\mathcal{T}_0$, then

$$U \not\subseteq \bigcup_{i=1}^{m} B_i.$$

Indeed, the statement is true if $m = 1$, for if $U \subseteq B_1$, then for any neighborhood W_1 of zero for $\mathcal{T}_1$ there would exist $a \in K^*$ such that $aB_1 \subseteq W_1$, whence $aU \subseteq W_1$; thus $\mathcal{T}_1 \subseteq \mathcal{T}_0$, a contradiction. Assume that the statement is true if $m < n$, and let $B_1, \ldots, B_{m+1}$ be subsets of K bounded for $\mathcal{T}_1, \ldots, \mathcal{T}_{m+1}$ respectively and U a neighborhood of zero for $\mathcal{T}_0$. Let V be a symmetric neighborhood of zero for $\mathcal{T}_0$ such that $V + V \subseteq U$. By our inductive hypothesis there exists

$$y \in V \setminus \bigcup_{i=1}^{m} C_i$$

where $C_1 = B_1 + B_1$, and $C_i = B_i$ for all $i \in [2, m]$; and there exists

$$z \in V \setminus \bigcup_{i=2}^{m+1} D_i$$

where $D_i = B_i \cup (y + (-B_i))$ for all $i \in [2, m+1]$. If $z \notin B_1$, then

$$z \in U \setminus \bigcup_{i=1}^{m+1} B_i.$$

Assume, therefore, that $z \in B_1$, and let $x = y - z \in V + V \subseteq U$. If $x \in B_1$, then $y = x + z \in C_1$, a contradiction. If $x \in B_i$ where $i \in [2, m+1]$, then $z = y - x \in y + (-B_i) \subseteq D_i$, a contradiction. Hence

$$x \in U \setminus \bigcup_{i=1}^{m+1} B_i.$$

To prove (*), let V_0 be a neighborhood of 1 for $\mathcal{T}_0$ and W a neighborhood of zero for $\mathcal{T}$. Then there exist retrobounded neighborhoods $U_1, \ldots, U_n$ of zero for $\mathcal{T}_1, \ldots, \mathcal{T}_n$ respectively such that $\cap_{i=1}^n U_i \subseteq W$, and there exists a neighborhood U_0 of zero for $\mathcal{T}_0$ such that $-1 \notin U_0$ and $(1+U_0)^{-1} \subseteq V_0$. For each $i \in [1,n]$ let $B_i = -1 + (K \setminus U_i)^{-1}$, a set bounded for $\mathcal{T}_i$ by hypothesis. By the preceding, there exists

$$x \in U_0 \setminus \bigcup_{i=1}^n B_i.$$

Then $1 + x \neq 0$, and

$$(1+x)^{-1} \in (\bigcap_{i=1}^n U_i) \cap V_0 \subseteq W \cap V_0. \bullet$$

Theorem 13.12 is called the Approximation Theorem because it implies that if $A_1, A_2, \ldots, A_n$ are pairwise inequivalent proper absolute values on a division ring K, if $c_1, c_2, \ldots, c_n$ are elements of K, and if $\epsilon > 0$, there exists $x \in K$ such that $A_i(x - c_i) < \epsilon$ for all $i \in [1,n]$.

13.13 Corollary. *Let K be a division ring, $(\mathcal{T}_\lambda)_{\lambda \in L}$ a family of distinct Hausdorff nondiscrete division ring topologies on K such that for some $\alpha \in L$, $\mathcal{T}_\lambda$ is locally retrobounded and $\mathcal{T}_\lambda \not\subseteq \mathcal{T}_\alpha$ for all $\lambda \in L \setminus \{\alpha\}$, and for each $\lambda \in L$, let $\widehat{K}_\lambda$ be the completion of K for $\mathcal{T}_\lambda$. Then the diagonal mapping*

$$\Delta : K \to \prod_{\lambda \in L} \widehat{K}_\lambda,$$

defined by $\Delta(x) = (x_\lambda)_{\lambda \in L}$, where $x_\lambda = x$ for all $\lambda \in L$, is a topological isomorphism from K, furnished with $\sup_{\lambda \in L} \mathcal{T}_\lambda$, to a dense division subring of $\prod_{\lambda \in L} \widehat{K}_\lambda$.

The assertion follows at once from 13.12 in view of the definition of the topology of a cartesian product of topological spaces.

13.14 Corollary. *If $(\mathcal{T}_\lambda)_{\lambda \in L}$ is a family of distinct topologies on a division ring K defined by proper absolute values, and if $\widehat{K}_\lambda$ is the completion of K for $\mathcal{T}_\lambda$ for each $\lambda \in L$, then $\widehat{\Delta}$ is a topological isomorphism from the completion $\widehat{K}$ of K for $\sup_{\lambda \in L} \mathcal{T}_\lambda$ to $\prod_{\lambda \in L} \widehat{K}_\lambda$.*

Exercises

13.1 A Hausdorff ring topology on a division ring K is *sequentially retrobounded* if for every sequence $(x_n)_{n \geq 1}$ in K^* that contains no bounded subsequence,
$$\lim_{n \to \infty} x_n^{-1} = 0.$$
(a) A metrizable ring topology on K is locally retrobounded if and only if it is sequentially retrobounded. (b) A ring topology on K is metrizable and locally retrobounded if and only if it is locally bounded, sequentially retrobounded, and there is a fundamental sequence of bounded subsets (Exercise 12.5).

13.2 Let K be a division ring furnished with a Hausdorff ring topology. (a) K is locally retrobounded if and only if for every filter $\mathcal{F}$ on K^*, if $K^* \setminus B \in \mathcal{F}$ for every bounded subset B of K, then $\mathcal{F}^{-1}$ converges to zero. (b) K is locally retrobounded if and only if for every subst B of K, if there is a neighborhood U of zero such that $1 \notin UB$, then B is bounded.

13.3 Give an example of a sequence of nonzero rationals converging to zero for the supremum of all the topologies on $\mathbb{Q}$ defined by proper absolute values.

13.4 Let $(\mathcal{T}_\lambda)_{\lambda \in L}$ be a family of distinct topologies on a division ring K, each defined by a proper absolute value, and let $\widehat{K}$ be the completion of K for $\sup_{\lambda \in L} \mathcal{T}_\lambda$. Then $\widehat{K}$ is a ring with continuous inversion; $\widehat{K}$ is a division ring if and only if L has only one element; and $\widehat{K}$ is advertibly open if and only if L is finite.

13.5 Let $(\mathcal{T}_\lambda)_{\lambda \in L}$ be a family of distinct locally retrobounded topologies on a division ring K, and let $M \subseteq L$. Let Δ_L and Δ_M be the diagonal mappings from K into $\prod_{\lambda \in L} \widehat{K}_\lambda$ and $\prod_{\lambda \in M} \widehat{K}_\lambda$ respectively. The identity map f of K is continuous from K, furnished with $\mathcal{T}_L$, the topology $\sup_{\lambda \in L} \mathcal{T}_\lambda$, to K, furnished with $\mathcal{T}_M$, the topology $\sup_{\lambda \in M} \mathcal{T}_\lambda$. Hence f has a continuous extension $\widehat{f}$ from the completion $\widehat{K}_L$ of K for $\mathcal{T}_L$ to the completion $\widehat{K}_M$ of K for $\mathcal{T}_M$. Describe explicitly the continuous homomorphism $\widehat{\Delta}_M \circ \widehat{f} \circ \widehat{\Delta}_L^{-1}$ from $\prod_{\lambda \in L} \widehat{K}_\lambda$ to $\prod_{\lambda \in M} \widehat{K}_\lambda$. What is its kernel?

13.6 Let $\leq$ be a total ordering on a field K such that for all $x, y, z \in K$, if $x \leq y$, then $x + z \leq y + z$, and, if $z > 0$, $xz \leq yz$. For each $a > 0$, let $V_a = \{x \in K : -a < x < a\}$. Then $\{V_a : a > 0\}$ is a fundamental system of neighborhoods of zero for a locally retrobounded topology on K.

13.7 (Baer and Hasse [1931]) If K is a totally disconnected locally retrobounded division ring, the open and closed subsets of K containing zero form a fundamental system of neighborhoods of zero. [If C is open and closed and $0 \notin C$, show that an open and closed subset of $\overline{C^{-1}}$ not containing zero is bounded, open and closed in K.]

14 Norms and Absolute Values

If N is a norm on a ring A, we shall say that a subset B of A is *norm-bounded* if there exists $r > 0$ such that $N(x) \leq r$ for all $x \in B$.

14.1 Theorem. *Let A be a topological ring whose topology is given by a norm N. Every norm-bounded subset of A is bounded; in particular, A is a locally bounded ring. If A is a ring with identity and if zero is adherent to $A^\times$, the [left, right] bounded subsets of A are precisely the norm-bounded subsets.*

Proof. For each $r > 0$, let $B_r = \{x \in A : N(x) \leq r\}$. Then B_r is bounded, for if $s > 0$, $B_{s/r} B_r \subseteq B_s$ and $B_r B_{s/r} \subseteq B_s$. Conversely, assume that $0 \in \overline{A^\times}$, and let B be left bounded. Then there exists $r > 0$ such that $BB_r \subseteq B_1$, and by hypothesis there exists $a \in B_r \cap A^\times$. Thus for each $x \in B$,
$$N(x) \leq N(xa)N(a^{-1}) \leq N(a^{-1}). \bullet$$

In contrast, any metrizable trivial ring A is a bounded normable ring by 6.7, but A itself need not be bounded in norm.

14.2 Definition. *The **core** of a norm N on a ring A is the set of all $h \in A^*$ such that*
$$N(hx) = N(h)N(x) = N(xh)$$
for all $x \in A$.

14.3 Theorem. *If the core H of a norm N on a ring A with identity is not empty, then $H \cap A^\times$ is a subgroup of $A^\times$, and for each $h \in H \cap A^\times$, $N(h^n) = N(h)^n$ for all $n \in \mathbb{Z}$.*

Proof. If H contains an element k, then $N(k) = N(k \cdot 1) = N(k)N(1)$, so $N(1) = 1$, and therefore $1 \in H$. Let $h \in H \cap A^\times$. For any $x \in A$,
$$N(h)N(x) = N(hx) = N(hxh^{-1}h) = N(h)N(xh^{-1})N(h),$$
so
$$N(x)N(h)^{-1} = N(xh^{-1}).$$
In particular, choosing $x = 1$, we obtain $N(h)^{-1} = N(h^{-1})$. Therefore
$$N(x)N(h^{-1}) = N(xh^{-1}).$$
Similarly, $N(h^{-1})N(x) = N(h^{-1}x)$. Thus $h^{-1} \in H \cap A^\times$. Clearly $H \cap A^\times$ is closed under multiplication, and an inductive argument establishes that $N(h^n) = N(h)^n$ for all $n \in \mathbb{Z}$. •

Here is a criterion for the topology of a topological ring to be given by a norm:

14.4 Theorem. *If A is a Hausdorff ring with identity that possesses a left or right bounded neighborhood V of zero [a left or right bounded open additive subgroup V] and an invertible topological nilpotent c such that $cV = Vc$, then the topology of A is given by a[n] [ultra]norm whose core contains an invertible topological nilpotent.*

Proof. Replacing V by $V \cap (-V)$ if necessary, we may assume that V is symmetric. Let
$$U = \{x \in A : Vx \subseteq V\}.$$
As V is a symmetric left bounded neighborhood of zero [left bounded open additive subgroup], U is a symmetric neighborhood of zero [an open additive subgroup]. Since $cVc^{-1} = V$, clearly $cUc^{-1} = U$; thus $cU = Uc$. For some $p \geq 1$, $c^p \in V$, so $c^p U \subseteq V$, whence $U \subseteq c^{-p}V$, a left bounded set. Thus U is left bounded, and clearly $1 \in U$ and $UU \subseteq U$. As $U + U + U$ is therefore left bounded, $(U + U + U)c^q \subseteq U$ for some $q \geq 1$; let $d = c^q$, an invertible topological nilpotent. Then $Ud = dU$, and for all $n \in \mathbb{Z}$,
$$Ud^{n+1} + Ud^{n+1} + Ud^{n+1} \subseteq Ud^n,$$
and in particular, $Ud^{n+1} \subseteq Ud^n$. By 12.17 $(Ud^n)_{n \in \mathbb{Z}}$ is a fundamental decreasing sequence of neighborhoods of zero. In particular, as A is Hausdorff,
$$\bigcap_{n \in \mathbb{Z}}^{\infty} Ud^n = (0).$$
Also,
$$\bigcup_{n \in \mathbb{Z}} Ud^n = A,$$
for if $x \in A$, then $\lim_{n \to \infty} xd^n = 0$, so $xd^r \in U$ for some $r \geq 1$, whence $x \in Ud^{-r}$.

Therefore we may apply 6.1 to $(Ud^n)_{n \in \mathbb{Z}}$; let g and f be the associated functions. For any nonzero $x, y \in A$,
$$g(xy) \leq g(x)g(y),$$
for if $g(x) = 2^{-i}$ and $g(y) = 2^{-j}$, then $x \in Ud^i$ and $y \in Ud^j$, whence
$$xy \in Ud^i Ud^j = UUd^i d^j \subseteq Ud^{i+j},$$
and therefore
$$g(xy) \leq 2^{-(i+j)} = g(x)g(y).$$

Consequently by 6.1, if V is a left bounded open additive subgroup, g is an ultranorm defining the topology of A. Now $d^{-1} \notin U$, for otherwise, as $UU \subseteq U$, $d^{-n} \in U$ for all $n \geq 1$, whence

$$1 = \lim_{n \to \infty} d^n d^{-n} = 0$$

by 12.12, a contradiction. Hence $1 \in U \setminus Ud$, so $d \in Ud \setminus Ud^2$, and therefore $g(d) = 1/2$. Moreover, $x \in Ud^n \setminus Ud^{n+1}$ if and only if $xd \in Ud^{n+1} \setminus Ud^{n+2}$, and also, if and only if

$$dx \in dUd^n \setminus dUd^{n+1} = Ud^{n+1} \setminus Ud^{n+2}.$$

Consequently,

$$g(xd) = (1/2)g(x) = g(d)g(x)$$

and

$$g(dx) = (1/2)g(x) = g(x)g(d).$$

In general, $f(xy) \leq f(x)f(y)$, for if

$$\sum_{i=1}^{p} x_i = x \text{ and } \sum_{j=1}^{q} y_j = y,$$

then

$$\sum_{j=1}^{q} \sum_{i=1}^{p} x_i y_j = xy,$$

so

$$f(xy) \leq \sum_{j=1}^{q} \sum_{i=1}^{p} g(x_i y_j) \leq \sum_{j=1}^{q} \sum_{i=1}^{p} g(x_i) g(y_j) = (\sum_{i=1}^{p} g(x_i))(\sum_{j=1}^{q} g(y_j)).$$

Consequently by 6.1, f is a norm defining the topology of A. In particular, if $x \in A$, then

$$f(xd) \leq f(x)f(d).$$

But if

$$\sum_{i=1}^{p} t_i = xd,$$

then

$$\sum_{i=1}^{p} t_i d^{-1} = x,$$

so
$$f(x)f(d) \le \sum_{i=1}^{p} g(t_i d^{-1})g(d) = \sum_{i=1}^{p} g(t_i).$$

Hence
$$f(x)f(d) \le f(xd).$$

Similarly, $f(dx) = f(d)f(x)$. Thus d belongs to the core of f. •

14.5 Corollary. *A Hausdorff ring topology T on a field is defined by a norm if and only T is locally bounded and there is a nonzero topological nilpotent for T.*

We begin our discussion of the relation between norms and absolute values on fields by defining spectral norms, which are intermediate between the two:

14.6 Definition. *A norm N on a ring A is a* **spectral norm** *if*

$$N(x^n) = N(x)^n$$

for all $x \in A$ and all $n \ge 1$.

The norms of Examples 1–3 of §1 are spectral norms, for example, whereas that of Example 4 is not.

To show that to every norm N on a field there is a largest spectral norm N_s smaller than N, we need the following theorem:

14.7 Theorem. *If $(x_n)_{n \ge 1}$ is a sequence in $\mathbb{R}_{>0}$ such that $x_{n+k} \le x_n x_k$ for all $n, k \ge 1$, then $\lim_{n \to \infty} x_n^{1/n}$ exists, and*

$$\lim_{n \to \infty} x_n^{1/n} = \inf_{n \ge 1} x_n^{1/n}.$$

Proof. By induction, $x_{qk} \le x_k^q$ for all $k, q \ge 1$. Let $x_0 = 1$, and for each $k \ge 1$ let
$$M_k = \sup_{0 \le r < k} x_r.$$
Let $k \ge 1$. For each $n \ge 1$, let $n = q_n k + r_n$ where $q_n, r_n \in \mathbb{N}$ and $0 \le r_n < k$. Then
$$x_n = x_{q_n k + r_n} \le x_k^{q_n} x_{r_n} \le M_k x_k^{q_n} = M_k x_k^{(1/k)(n-r_n)}.$$

Hence
$$x_n^{1/n} \le M_k^{1/n} x_k^{1/k} (x_k^{1/k})^{-r_n/n}.$$

Now
$$\lim_{n\to\infty} (x_k^{1/k})^{-r_n/n} = 1$$

as $\lim_{n\to\infty}(-r_n)/n = 0$, and also $\lim_{n\to\infty} M_k^{1/n} = 1$. Therefore

$$\limsup_{n\to\infty} x_n^{1/n} \leq x_k^{1/k}.$$

Consequently

$$\limsup_{n\to\infty} x_n^{1/n} \leq \inf_{k\geq 1} x_k^{1/k} \leq \liminf_{k\to\infty} x_k^{1/k}. \bullet$$

14.8 Theorem. *Let N be a norm on a field K. Of all the spectral norms M on K such that $M \leq N$, there is a largest, N_s, defined by*

$$N_s(x) = \lim_{n\to\infty} N(x^n)^{1/n}$$

for all $x \neq 0$ and $N_s(0) = 0$. Furthermore, for each $x \in K$, $N_s(x) = N(x)$ if and only if $N(x^n) = N(x)^n$ for all $n \geq 1$, every element of the core of N is also in the core of N_s, and if N is an ultranorm, so is N_s.

Proof. By 14.7 applied to the sequence $(N(x^n))_{n\geq 1}$, $N_s(x)$ is indeed defined, and if $x \neq 0$, $N_s(x) = \inf_{n\geq 1} N(x^n)^{1/n}$. Since $N(x^n) \leq N(x)^n$ for all $n \geq 1$, it follows that $N_s \leq N$ and for any spectral norm M such that $M \leq N$, $M \leq N_s$. Consequently, we need only show that N_s is a spectral norm.

To show (N 2) of Definition 1.2 holds for N_s, we first observe that for any $m \in \mathbb{N}$ and any $z \in K$, $N(m.z) \leq mN(z)$. Let $x, y \in K$, let $e > 0$, and let $m \geq 1$ be such that for all $n \geq m$,

$$N(x^n)^{1/n} \leq N_s(x) + e,$$

$$N(y^n)^{1/n} \leq N_s(y) + e.$$

Let $C > 1$ be such that

$$N(x^j)^{1/j} \leq C[N_s(x) + e],$$

$$N(y^j)^{1/j} \leq C[N_s(y) + e]$$

for all $j \in [1, m-1]$. Let $n > 2m$. For any $k \in [0, n]$, if $k < m$, then $n - k > m$, and if $n - k < m$, then $k > m$. Therefore

$$N((x+y)^n) = N(\sum_{k=0}^{n} \binom{n}{k} x^{n-k} y^k) \leq \sum_{k=0}^{n} N(\binom{n}{k} x^{n-k} y^k)$$

$$\leq \sum_{k=0}^{n} \binom{n}{k} N(x^{n-k}) N(y^k)$$

$$\leq \sum_{k=0}^{n} \binom{n}{k} C^m [N_s(x) + e]^{n-k} [N_s(y) + e]^k$$

$$= C^m [N_s(x) + N_s(y) + 2e]^n.$$

Thus

$$N_s(x+y) \leq C^{m/n} [N_s(x) + N_s(y) + 2e].$$

As $\lim_{n \to \infty} C^{m/n} = 1$, therefore,

$$N_s(x+y) \leq N_s(x) + N_s(y) + 2e.$$

Consequently,

$$N_s(x+y) \leq N_s(x) + N_s(y).$$

Assume further that N is an ultranorm and that $N_s(x) \leq N_s(y)$. Then $N(q.1) \leq 1$ for all $q \in \mathbb{Z}$. If $n \geq 2m$,

$$N((x+y)^n) = N(\sum_{k=0}^{n} \binom{n}{k} x^{n-k} y^k)$$

$$\leq \sup_{0 \leq k \leq n} N(x^{n-k} y^k) \leq \sup_{0 \leq k \leq n} N(x^{n-k}) N(y^k)$$

$$\leq \sup_{0 \leq k \leq n} C^m [N_s(x) + e]^{n-k} [N_s(y) + e]^k$$

$$\leq C^m [N_s(y) + e]^n.$$

Thus as before,

$$N_s(x+y) \leq N_s(y).$$

As $N(-t) = N(t)$ for all $t \in K$,

$$N_s(-x) = \lim_{n \to \infty} N((-x)^n)^{1/n} = \lim_{n \to \infty} N(x^n)^{1/n} = N_s(x)$$

for all $x \in K^*$. Hence (N 3) holds for N_s.

If $x, y \in K^*$, then

$$\begin{aligned} N_s(xy) &= \lim_{n\to\infty} N((xy)^n)^{1/n} = \lim_{n\to\infty} N(x^n y^n)^{1/n} \\ &\leq \lim_{n\to\infty} N(x^n)^{1/n} N(y^n)^{1/n} = N_s(x) N_s(y), \end{aligned}$$

so (N 4) holds for N_s.

From (N 2)–(N 4) we conclude that $\{x \in K : N_s(x) = 0\}$ is an ideal of K. But

$$N_s(1) = \lim_{n\to\infty} N(1)^{1/n} = 1.$$

Consequently, as K is a field, that ideal is the zero ideal, so (N 5) holds for N_s.

For each $x \in K$,

$$N_s(x) = \inf_{n\geq 1} N(x^n)^{1/n}$$

by 14.7, so $N_s(x) = N(x)$ if and only if $N(x^n) = N(x)^n$ for all $n \geq 1$. Finally, if x belongs to the core of N, then for each $y \in K$, $N(x^n y^n) = N(x^n) N(y^n)$ as $x^n \in H$ by 14.3, so $N_s(xy) = N_s(x) N_s(y)$, and consequently x belongs to the core of N_s. •

14.9 Theorem. *Let N be a spectral [ultra]norm on a field K, and let $c \in K^*$. There is a spectral [ultra]norm N_c on K such that:*

(1) $N_c \leq N$.

(2) *The core H of N is contained in that of N_c, and $N_c(x) = N(x)$ for all $x \in H$.*

(3) *c is in the core of N_c, and $N_c(c) = N(c)$.*

Proof. For each $x \in K$, the sequence $(N(xc^n) N(c)^{-n})_{n \geq 0}$ is clearly decreasing; we define N_c by

$$N_c(x) = \lim_{n\to\infty} N(xc^n) N(c)^{-n} = \inf_{n\geq 0} N(xc^n) N(c)^{-n}.$$

Let $x, y \in K$. Then

$$\begin{aligned} N_c(xy) &= \lim_{n\to\infty} N(xy c^{2n}) N(c)^{-2n} \\ &\leq \lim_{n\to\infty} N(xc^n) N(c)^{-n} N(yc^n) N(c)^{-n} = N_c(x) N_c(y). \end{aligned}$$

It is easy to see that

$$N_c(x+y) \leq N_c(x) + N_c(y)$$

and, if N is an ultranorm,
$$N_c(x+y) \leq \sup\{N_c(x), N_c(y)\}.$$

Consequently, as $N_c(1) = 1$, $\{x \in K : N_c(x) = 0\}$ is a proper ideal of K and hence is the zero ideal. Thus N_c is a[n] [ultra]norm. N_c is a spectral norm, for if $x \in K$ and $m \geq 1$,

$$N_c(x^m) = \lim_{n\to\infty} N(x^m c^{nm}) N(c)^{-nm} = \lim_{n\to\infty} N((xc^n)^m) N(c)^{-nm}$$
$$= \lim_{n\to\infty} N(xc^n)^m N(c)^{-nm} = N_c(x)^m.$$

Clearly (1) holds. To establish (2), let $x \in H$. Then for any $y \in K$,

$$N_c(xy) = \lim_{n\to\infty} N(xyc^n) N(c)^{-n}$$
$$= \lim_{n\to\infty} N(x) N(yc^n) N(c)^{-n} = N(x) N_c(y).$$

Choosing $y = 1$, we obtain $N_c(x) = N(x)$, so $N_c(xy) = N_c(x)N_c(y)$ for all $y \in K$. Thus x belongs to the core of N_c. (3) As N is a spectral norm, $N_c(c) = N(c)$. For each $y \in K$,

$$N_c(yc) = \lim_{n\to\infty} N(yc^{n+1}) N(c)^{-n}$$
$$= \lim_{n\to\infty} [N(yc^{n+1}) N(c)^{-n-1}] N(c) = N_c(y) N(c),$$

so as $N_c(c) = N(c)$, c belongs to the core of N_c. •

The following theorem, due to Aurora [1958], gives the fundamental relation between spectral norms and absolute values on a field:

14.10 Theorem. *Let N be a norm on a field K, and let H be its core. The following statements are equivalent:*

1° *N is a spectral [ultra]norm.*

2° *There is a family $(A_\lambda)_{\lambda \in L}$ of [nonarchimedean] absolute values on K such that*
$$N = \sup_{\lambda \in L} A_\lambda.$$

3° *There is a family $(A_c)_{c \in K^*}$ of [nonarchimedean] absolute values on K such that*
$$N = \sup_{c \in K^*} A_c.$$

and for each $c \in K^*$, $A_c(c) = N(c)$ and $A_c(x) = N(x)$ for all $x \in H$.

Proof. Clearly 3° implies 2°, and 2° implies 1°. Assume 1°. To prove 3°, it suffices to show that for each $c \in K^*$ there exists a [nonarchimedean] absolute value A_c such that $A_c \leq N$ and $A_c(x) = N(x)$ for all $x \in H \cup \{c\}$. Let $\mathcal{N}$ be the set of all spectral [ultra]norms P on K such that $P \leq N$. We order $\mathcal{N}$ by declaring $P \preceq Q$ if and only if $Q \leq P$, the core H_Q of Q contains the core H_P of P, and $Q(x) = P(x)$ for all $x \in H_P$. With the terminology of 14.9, let

$$\mathcal{N}_c = \{P \in \mathcal{N} : N_c \preceq P\}.$$

With its induced ordering, $\mathcal{N}_c$ is an inductive set. Indeed, if $\mathcal{C}$ is a chain in $\mathcal{N}_c$, $\mathcal{C}$ is totally ordered for $\leq$, and the infimum, P_0, of $\mathcal{C}$ for $\leq$ clearly satisfies (N 1)–(N 4) of Definition 1.2 [and (N 6) of Definition 6.5] and, for any $P \in \mathcal{C}$, $P_0(x) = P(x)$ for all $x \in H_P$; in particular, $P_0(c) = N_c(c) \neq 0$, so $\{x \in K : P_0(x) = 0\}$ is a proper ideal of K, thus the zero ideal, and therefore P_0 is a [nonarchimedean] norm. Consequently, $P_0 \in \mathcal{N}_c$ and P_0 is the supremum of $\mathcal{C}$ for $\preceq$. Therefore by Zorn's Lemma, $\mathcal{N}_c$ has a maximal member A_c. As $N_c \preceq A_c$ and as $N \preceq N_c$ by 14.9, $A_c \leq N_c \leq N$, and $A_c(x) = N_c(x) = N(x)$ for all $x \in H \cup \{c\}$. We have left to show that A_c is an absolute value, that is, that its core is K^*. Let $d \in K^*$. With the notation of 14.9, $(A_c)_d \succeq A_c \succeq N_c$, so by the maximality of A_c, $(A_c)_d = A_c$, and therefore d belongs to the core of A_c. •

14.11 Theorem. *If $\mathcal{T}$ is a locally bounded Hausdorff topology on a field K for which there is a nonzero topological nilpotent, then there is a proper absolute value on K whose topology is weaker than $\mathcal{T}$.*

Proof. By 14.4, $\mathcal{T}$ is the topology given by a norm N. By 14.8 there is a spectral norm N_s on K such that $N_s \leq N$, and by 14.10 there is an absolute value A on K such that $A \leq N_s$; consequently $A \leq N$, so the topology defined by A is weaker than $\mathcal{T}$. •

Finally, we obtain the following criterion for a Hausdorff ring topology on a field to be given by an absolute value:

14.12 Theorem. *A Hausdorff ring topology $\mathcal{T}$ on a field K is given by a proper absolute value if and only if $\mathcal{T}$ is locally retrobounded and there is a nonzero topological nilpotent for $\mathcal{T}$.*

Proof. The condition is sufficient by 13.7, 13.8, 13.3, and 14.11. It is necessary by 13.6. •

Exercises

14.1 A function N from a ring A to $\mathbb{R}_{\geq 0}$ is a *seminorm* if (N 1)–(N 4) of Definition 1.2 hold. (a) If N is a seminorm on A and if, for each $r \in \mathbb{R}_{>0}$,

$$V_r = \{x \in A : N(x) < r\},$$

then $(V_r)_{r>0}$ is a fundamental system of neighborhoods of zero for a locally bounded ring topology on A. (b) The topology defined by N is Hausdorff if and only if N is a norm.

14.2 Let N be a norm on a commutative ring A. (a) The function N_s, defined by

$$N_s(x) = \lim_{n \to \infty} N(x^n)^{1/n},$$

is a seminorm on A. (b) For any $x \in A$, $N_s(x) < 1$ if and only if x is a topological nilpotent.

14.3 Let A be a commutative Hausdorff ring with identity that contains an invertible topological nilpotent. The following statements are equivalent:

1° The set R of topological nilpotents is a bounded neighborhood of zero.

2° The topology of A is given by a spectral norm.

3° The topology of A is given by a spectral norm whose core contains an invertible topological nilpotent.

[Apply 14.4 where $V = R$ and 12.17.]

14.4 (Kowalsky [1953]) A field K is *rankfree* if for every nondiscrete locally retrobounded topology $\mathcal{T}$ on K there is a nonzero topological nilpotent for $\mathcal{T}$, and if each nonzero element of K is a topological nilpotent for at most finitely many locally retrobounded topologies. If K is rankfree and if $\mathcal{T}$ is a nondiscrete locally bounded ring topology on K, then $\mathcal{T}$ is the supremum of finitely many nondiscrete locally retrobounded topologies if and only if the set of elements that are topologically nilpotent for $\mathcal{T}$ is nonzero and bounded for $\mathcal{T}$. [Use 14.4, Exercise 14.3, and 14.10.]

14.5 (Arnautov [1965b]) Let A be a metrizable ring whose topology $\mathcal{T}$ is an ideal topology, and let $(V_n)_{n \geq 1}$ be a decreasing, fundamental system of ideal neighborhoods of zero. (a) For each $n \geq 1$, let U_n be the ideal generated by the union of all the sets $V_{i_1} V_{i_2} \ldots V_{i_r}$ such that $\sum_{k=1}^{r} i_k = n$. (a) For all $n \geq 1$, $V_n \subseteq U_n$ and $U_{n+1} \subseteq U_n$. (b) For all $n, m \geq 1$, $U_n U_m \subseteq U_{n+m}$, and U_{nm} is contained in the ideal generated by $V_m \cup V_1^n$. (c) $\mathcal{T}$ is defined by an ultranorm [norm] if and only if there is a neighborhood V of zero such that for every neighborhood W of zero, there exists $n \geq 1$ such that $V^n \subseteq W$. [Use (a) and 6.1.]

15 Finite-dimensional Vector Spaces

A *linear transformation* from an A-module E to an A-module F is a homomorphism from E to F, and a *linear operator* on E is simply an endomorphism of the A-module E.

An A-module E is the *direct sum* of submodules $(M_k)_{1\leq k\leq n}$ if the linear transformation s from $\prod_{k=1}^{n} M_k$ to E, defined by

$$(1) \qquad s(x_1,\ldots,x_n) = \sum_{k=1}^{n} x_k,$$

is an isomorphism. Clearly s is a linear transformation, so s is an isomorphism if and only if it is surjective and its kernel contains only $(0,\ldots,0)$. In this case, the *family of projections* associated to $(M_k)_{1\leq k\leq n}$ is the family $(p_k)_{1\leq k\leq n}$ of linear operators on E defined by

$$p_k = in_k \circ pr_k \circ s^{-1},$$

where pr_k is the canonical projection from $\prod_{i=1}^{n} M_i$ to M_k and in_k is the canonical injection from M_k to E; thus

$$p_k(x_1 + \cdots + x_n) = x_k$$

whenever $x_i \in M_i$ for all $i \in [1,n]$.

Similarly, a ring A is the *direct sum* of subrings $(B_k)_{1\leq k\leq n}$ if the function s from the ring $\prod_{k=1}^{n} B_k$ to A, defined by (1) is an isomorphism. Clearly A is the direct sum of subrings $(B_k)_{1\leq k\leq n}$ if and only if $B_1,\ldots,B_n$ are ideals of A such that $B_i B_j = \{0\}$ whenever $i \neq j$ and the A-module A is the direct sum of the submodules $(B_k)_{1\leq k\leq n}$.

These considerations may further be extended to any family of submodules of an A-module E. If $(M_\lambda)_{\lambda \in L}$ is a family of A-modules, we define $\bigoplus_{\lambda \in L} M_\lambda$, sometimes called *the outer direct sum* of $(M_\lambda)_{\lambda \in L}$, to be the submodule of $\prod_{\lambda \in L} M_\lambda$ consisting of all $(x_\lambda)_{\lambda \in L}$ such that $x_\lambda = 0$ for all but finitely many $\lambda \in L$. If each M_λ is a submodule of E, we define

$$\sum_{\lambda \in L} M_\lambda$$

to be the submodule of E generated by $\bigcup_{\lambda \in L} M_\lambda$ and say that E is the *direct sum* of $(M_\lambda)_{\lambda \in L}$ if the linear operator s from $\bigoplus_{\lambda \in L} M_\lambda$ to E, defined by

$$(2) \qquad s((x_\lambda)_{\lambda \in L}) = \sum_{\lambda \in L} x_\lambda,$$

is an isomorphism.

For example, $(b_\lambda)_{\lambda \in L}$ is a basis of a unitary A-module E if and only if E is the direct sum of $(Ab_\lambda)_{\lambda \in L}$ and for each $\lambda \in L$, $\{b_\lambda\}$ is linearly independent, that is, $\alpha.b_\lambda = 0$ only if $\alpha = 0$.

Similarly, if $(B_\lambda)_{\lambda \in L}$ is a family of rings, we define $\bigoplus_{\lambda \in L} B_\lambda$ to be the subring of $\prod_{\lambda \in L} B_\lambda$ consisting of all $(x_\lambda)_{\lambda \in L}$ such that $x_\lambda = 0$ for all but finitely many $\lambda \in L$. This notation is primarily useful when each B_λ is an ideal of a ring A_λ, in which case $\bigoplus_{\lambda \in L} B_\lambda$ is an ideal of $\prod_{\lambda \in L} A_\lambda$. If each B_λ is an ideal of a ring A, the ideal B generated by $\bigcup_{\lambda \in L} B_\lambda$ is the *direct sum* of $(B_\lambda)_{\lambda \in L}$ if the function s defined by (2) is an isomorphism from $\bigoplus_{\lambda \in L} B_\lambda$ to B.

15.1 Definition. *Let E be a topological module over a topological ring A. Then E is the **topological direct sum** of submodules $(M_k)_{1 \leq k \leq n}$ if the function s from $\prod_{k=1}^n M_k$ to E, defined by*

$$s(x_1, \ldots, x_n) = \sum_{i=1}^n x_i,$$

*is a topological isomorphism. Similarly, a topological ring A is the **topological direct sum** of subrings $(B_k)_{1 \leq k \leq n}$ if the function s from the ring $\prod_{k=1}^n B_k$ to A, defined by*

$$s(x_1, \ldots, x_n) = \sum_{k=1}^n x_k,$$

is a topological isomorphism.

15.2 Theorem. *Let E be a topological A-module [ring] that is the direct sum of submodules [subrings] $(M_k)_{1 \leq k \leq n}$. Then E is the topological direct sum of $(M_k)_{1 \leq k \leq n}$ if and only if each member of the associated family of projections is continuous.*

Proof. Let $(p_k)_{1 \leq k \leq n}$ be the associated family of projections, and let s be the isomorphism from $\prod_{k=1}^n M_k$ to E defined by

$$s(x_1, \ldots, x_n) = \sum_{k=1}^n x_k.$$

Since s is simply the restriction to $\prod_{k=1}^n M_k$ of addition on E^n, s is continuous. Thus s is a topological isomorphism if and only if s^{-1} is continuous. But

$$s^{-1}(x) = (p_1(x), \ldots, p_n(x))$$

for all $x \in E$, and hence s^{-1} is continuous if and only if each p_k is. •

If M is a submodule of a module E, a submodule N of E is a *supplement* (or, for emphasis, an *algebraic supplement*) of M if E is the direct sum of M and N.

15.3 Definition. *Let M be a submodule of a topological module E. A submodule N of E is a* **topological supplement** *of M if E is the topological direct sum of M and N.*

If an A-module E is the direct sum of submodules M and N and if p and q are the associated projections, p is called the *projection on M along N*, q the *projection on N along M*. Clearly $q = 1_E - p$ where 1_E is the identity map of E.

If M and N are supplementary submodules of E, the projection p on M along N is a linear operator on E satisfying $p \circ p = p$, and moreover, the range of p is M and its kernel is N. Conversely, if p is a linear operator on E such that $p \circ p = p$, then E is the direct sum of its range M and its kernel N, and p is the projection on M along N. Consequently, any linear operator on E that satisfies $p \circ p = p$ is called a *projection*.

15.4 Theorem. *Let M be a submodule of a topological A-module E. A supplement N of M in E is a topological supplement if and only if the projection on M along N is continuous, in which case the restriction ρ to N of the canonical epimorphism ϕ_M from E to E/M is a topological isomorphism. Moreover, M has a topological supplement if and only if there is a continuous projection on E whose range is M. If M has a topological supplement and if E is Hausdorff, then M is closed.*

Proof. Let E be the topological direct sum of M and N. If U is open in N, then as $s(M \times U) = M+U$, $M+U$ is open in E; thus as $\rho(U) = \phi_M(M+U)$, $\rho(U)$ is open in E/M. Clearly ρ is continuous and thus is a topological isomorphism. If E is Hausdorff, then M is closed as it is the kernel of the (continuous) projection on N along M. •

As we shall shortly see, if E is a Hausdorff finite-dimensional vector space over a complete straight division ring K, and if E is (algebraically) the direct sum of subspaces $M_1, \ldots M_n$, then E is the topological direct sum of $M_1, \ldots M_n$.

Henceforth, K is a division ring topologized by a Hausdorff ring topology.

A *linear form* on a K-vector space E is a linear transformation from E to the K-vector space K.

15.5 Theorem. *K is straight if and only if every linear form on a Hausdorff K-vector space whose kernel is closed is continuous.*

Proof. Necessity: Let u be a nonzero linear form on a Hausdorff K-vector space E whose kernel H is closed. Then there is an isomorphism v from the K-vector space E/H to K satisfying $u = v \circ \phi_H$ where ϕ_H is the canonical epimorphism from E to E/H. By (1) of 5.7, E/H is Hausdorff, and if $a = v^{-1}(1)$, $v^{-1}(\lambda) = \lambda.a$ for all $\lambda \in K$. As K is straight, v^{-1} is a homeomorphism, so v is continuous, and therefore u is also.

Sufficiency: Let a be a nonzero vector of a Hausdorff K-vector space E. If u_a, defined by $u_a(\lambda) = \lambda.a$, were not a homeomorphism from K to $K.a$, then u_a^{-1} would be a discontinuous linear form on $K.a$ with closed kernel (0), a contradiction. •

A *hyperplane* of a vector space E is a subspace H such that E/H is one-dimensional.

15.6 Corollary. *If K is straight and if E is Hausdorff K-vector space, every algebraic supplement D of a closed hyperplane H is a topological supplement.*

Proof. If $D = Ka$ and if p is the projection on D along H, then with the notation of 15.5 the linear form $u_a^{-1} \circ p$ is continuous by 15.5, so p is also continuous, and the assertion follows from 15.4. •

The *standard basis* of the K-vector space K^n is the basis $\{e_1, e_2, \ldots, e_n\}$ where for each $k \in [1, n]$, e_k is the n-tuple whose kth entry is 1 and whose remaining entries are 0.

15.7 Theorem. *Every linear form on the topological K-vector space K^n is continuous; hence every hyperplane of K^n is closed.*

Proof. Let u be a linear form on K^n, and let $u(e_k) = \alpha_k \in K$ for each $k \in [1, n]$. Then for any $(\lambda_1, \ldots, \lambda_n) \in K^n$,

$$u(\lambda_1, \ldots, \lambda_n) = u(\sum_{k=1}^n \lambda_k e_k) = \sum_{k=1}^n \lambda_k u(e_k) = \sum_{k=1}^n \alpha_k \lambda_k,$$

so u is continuous from K^n to K. •

Since every proper subspace of a vector space is an intersection of hyperplanes, the statements "Every hyperplane is closed" and "Every subspace is closed" about a topological vector space are equivalent.

15.8 Theorem. *The following statements are equivalent:*

$1°$ *K is straight.*

$2°$ *For each $n \geq 1$, every isomorphism from the K-vector space K^n to an n-dimensional Hausdorff K-vector space all of whose hyperplanes are closed is a topological isomorphism.*

3° For each $n \geq 1$, every n-dimensional Hausdorff K-vector space all of whose hyperplanes are closed is topologically isomorphic to K^n.

Proof. To prove 2° from 1°, we proceed by induction on n. Let S_n be the statement: Every isomorphism from the K-vector space K^n to an n-dimensional Hausdorff K-vector space all of whose hyperplanes are closed is a topological isomorphism. By the definition of straightness, S_1 holds. Let $m > 1$, assume that S_n holds whenever $n < m$, and let u be an isomorphism from K^m to an m-dimensional Hausdorff K-vector space E all of whose hyperplanes are closed. Let $a_k = u(e_k)$ for each $k \in [1, m]$, where $\{e_1, \ldots, e_m\}$ is the standard basis of K^m. Let F be the subspace generated by $a_1, \ldots a_{m-1}$. Then every hyperplane H of F is closed in F. Indeed, in the contrary case, the closure $\overline{H}$ of H in F would be F, so as $H + Ka_m$ is a hyperplane of E and hence is closed in E, $H + Ka_m$ would contain $\overline{H} + Ka_m = F + Ka_m = E$, a contradiction. Consequently by our inductive hypothesis, the linear transformation v from K^{m-1} to F, defined by

$$v(\lambda_1, \ldots, \lambda_{m-1}) = u(\lambda_1, \ldots, \lambda_{m-1}, 0) = \sum_{k=1}^{m-1} \lambda_k a_k,$$

is a topological isomorphism. As K is straight, therefore, the isomorphism

$$(\lambda_1, \ldots, \lambda_{m-1}, \lambda_m) \to (v(\lambda_1, \ldots, \lambda_{m-1}), \lambda_m a_m)$$

from K^m to $F \times Ka_m$ is a topological isomorphism. As F is a hyperplane of E, F is closed in E, so $(x, y) \to x + y$ is a topological isomorphism from $F \times Ka_m$ to E by 15.6. Thus as

$$u(\lambda_1, \ldots, \lambda_m) = v(\lambda_1, \ldots, \lambda_{m-1}) + \lambda_m a_m,$$

u is a topological isomorphism from K^m to E.

To show that 3° implies 1°, let a be a nonzero vector in a Hausdorff K-vector space. By 3° there is a topological K-isomorphism u from the topological K-vector space K to Ka. Clearly $u(\lambda) = \lambda b$, where $b = u(1)$. Let $\gamma \in K^*$ be such that $a = \gamma b$. Then $R_\gamma : \lambda \to \lambda \gamma$ is a homeomorphism from K to K, so $u \circ R_\gamma : \lambda \to \lambda a$ is a homeomorphism from K to Ka. •

15.9 Theorem. *The following statements are equivalent:*

1° K *is straight and complete.*

2° *For every $n \geq 1$, every isomorphism from the K-vector space K^n to an n-dimensional Hausdorff K-vector space is a topological isomorphism.*

3° *For every $n \geq 1$, every n-dimensional Hausdorff K-vector space is topologically isomorphic to K^n.*

Proof. To prove 2° from 1°, we proceed by induction on n. Let T_n be the statement: Every isomorphism from the K-vector space K^n to an n-dimensional Hausdorff K-vector space is a a topological isomorphism. By the definition of straightness, T_1 holds. Let $m > 1$, and assume that T_n holds for all $n < m$. To establish T_m, it suffices by 15.8 to show that if H is a hyperplane of a Hausdorff m-dimensional K-vector space E, then H is closed. But as H has dimension $m - 1$, H is topologically isomorphic to K^{m-1} by T_{m-1}, hence is complete, and thus is closed.

Finally, assume 3°. By 15.8, K is straight. Suppose that K were not complete. Let $a \in \widehat{K} \setminus K$, and let $E = K + Ka$, furnished with the topology inherited from $\widehat{K}$. Then E is a two-dimensional Hausdorff K-vector space, and K is a dense one-dimensional subspace of E. Consequently, E is not topologically isomorphic to K^2 by 15.7, a contradiction of our hypothesis. •

15.10 Corollary. *Let E be a finite-dimensional vector space over a complete straight division ring K. There is one and only one Hausdorff vector topology on E. For any basis $\{a_1, \ldots, a_n\}$ of E,*

$$u : (\lambda_1, \ldots, \lambda_n) \to \sum_{k=1}^{n} \lambda_k a_k$$

is a topological isomorphism from K^n to E, furnished with its unique Hausdorff vector topology.

15.11 Theorem. *Let K be a division ring furnished with a complete proper absolute value, and let E be a finite-dimensional K-vector space. The unique Hausdorff vector topology $\mathcal{T}$ on E is normable; indeed, if $\{b_1, \ldots b_n\}$ is a basis of E, then $\|..\|$, defined by*

$$\|\sum_{k=1}^{n} \lambda_k b_k\| = \sup_{1 \leq k \leq n} |\lambda_k|,$$

is a norm on E defining $\mathcal{T}$.

15.12 Theorem. *Let K be a complete straight division ring.*

(1) *Every Hausdorff finite-dimensional K-vector space is complete; hence every finite-dimensional subspace of a Hausdorff K-vector space is closed.*

(2) *Every linear transformation from a finite-dimensional Hausdorff K-vector space to a Hausdorff K-vector space is continuous.*

(3) *A Hausdorff finite-dimensional K-vector space that is the direct sum of a sequence of subspaces is the topological direct sum of those subspaces.*

(4) *Every linear transformation from a Hausdorff K-vector space to a finite-dimensional Hausdorff K-vector space whose kernel is closed is a topological homomorphism.*

(5) *If M is a closed subspace and N a finite-dimensional subspace of a Hausdorff K-vector space E, then $M + N$ is closed.*

Proof. (1) follows from 15.10 and 7.14, since K^n is complete for all $n \geq 1$ by (2) of 7.8.

To prove (2), let u be a linear transformation from a Hausdorff finite-dimensional K-vector space E to a Hausdorff K-vector space F. By (1), the kernel H of u is closed, so E/H is a Hausdorff finite-dimensional K-vector space. By 15.9, the isomorphism v from E/H to $u(E)$ satisfying $v \circ \phi_H = u$, where ϕ_H is the canonical epimorphism from E to E/H, is a topological isomorphism, so u is a topological homomorphism by the module analogue of 5.11. (3) follows from (2), and the proof of (4) is similar to that of (2).

(5) As M is closed, E/M is Hausdorff, so the finite-dimensional subspace $\phi_M(N)$ of E/M is closed by (1) (where ϕ_M is the canonical epimorphism from E to E/M), and therefore $\phi_M^{-1}(\phi_M(N))$ is closed. But $\phi_M^{-1}(\phi_M(N)) = M + N$. •

15.13 Definition. *Let A be a ring. A function u is A-multilinear if for some $n \geq 1$ the domain of u is the cartesian product of a sequence $(E_k)_{1 \leq k \leq n}$ of n A-modules, its codomain is an A-module F, and for each $k \in [1, n]$ and each sequence $c_1 \in E_1, \ldots, c_{k-1} \in E_{k-1}, c_{k+1} \in E_{k+1}, \ldots, c_n \in E_n$, the function $x \to u(c_1, \ldots, c_{k-1}, x, c_{k+1}, \ldots, c_n)$ is a linear transformation from E_k to F.*

For example, if E is an algebra over a commutative ring with identity A, multiplication is an A-multilinear transformation from $E \times E$ to E.

Theorem 15.14. *Let K be a complete straight division ring. Any K-multilinear transformation from the cartesian product of Hausdorff finite-dimensional K-vector spaces to a Hausdorff K-vector space is continuous.*

Proof. Let u be a multilinear transformation from the cartesian product E of Hausdorff finite-dimensional K-vector spaces $(E_k)_{1 \leq k \leq n}$ to a Hausdorff K-vector space F. For each $k \in [1, n]$, let $m(k)$ be the dimension of E_k, let $(e_{k,j})_{1 \leq j \leq m(k)}$ be a basis of E_k, and let pr_k be the canonical projection from E to E_k. For each $k \in [1, n]$ and each $i \in [1, m(k)]$, let $q_{k,i}$ be the linear form on E_k defined by

$$q_{k,i}(\sum_{j=1}^{m(k)} \lambda_j e_{k,j}) = \lambda_i.$$

By (2) of 15.12, each $q_{k,i}$ is continuous. Let $M = \prod_{k=1}^{n}[1, m(k)]$, and for each $i \in M$, let i_k be its kth component, so that $i = (i_1, \ldots, i_n)$. For each $i \in M$ and each $k \in [1, n]$, the function $q_{k,i_k} \circ pr_k$ is continuous from E to K, so the function P_i, defined by

$$P_i = \prod_{k=1}^{n}(q_{k,i_k} \circ pr_k)$$

is continuous from E to K. For each $i \in M$ let U_i be the function from K into F defined by

$$U_i(\lambda) = \lambda u(e_{i_1}, \ldots, e_{i_n}).$$

Then

$$u = \sum_{i \in M} U_i \circ P_i$$

and hence is continuous. •

15.15 Corollary. *If A is a finite-dimensional algebra over a complete straight field K, then the unique Hausdorff vector topology on A is an algebra topology, that is, multiplication is continuous from $A \times A$ to A.*

Exercises

Let E be a vector space over a field. We denote by E^* the vector space of all linear forms on E. A subspace E' of E^* is *total* if for each nonzero $x \in E$ there exists $u \in E^*$ such that $u(x) \neq 0$. If K is a Hausdorff topological field and if E' is a subspace of E, we denote by $\sigma_K(E, E')$ the weakest topology on E making each $u \in E'$ continuous, a vector topology by 2.17.

15.1 Let K be a Hausdorff field, E a K-vector space, E' a subspace of E^*. (a) $\sigma_K(E, E')$ is Hausdorff if and only if E' is a total subspace of E^*. (b) If the topology of K is given by a proper absolute value, then a linear form v on E is continuous for $\sigma_K(E, E')$ if and only if $v \in E'$. [Show that there exists a linearly independent sequence $(u_k)_{1 \leq k \leq n}$ in E' such that, if

$$H = \bigcap_{k=1}^{n} u_k^{-1}(0),$$

then $H \subseteq v^{-1}(0)$, and observe that each u_k induces a linear form $\bar{u}_k$ on E/H and that $(\bar{u}_k)_{1 \leq k \leq n}$ is a basis of $(E/H)^*$.]

15.2 (Warner [1956]) Let A be an algebra over a field K furnished with an absolute value, and let A' be a total subset of A^*. (a) If (TR 2) of Theorem

2.15 holds for $\sigma_K(A, A')$, then for every $v \in A'$, $v^{-1}(0)$ contains an ideal of finite codimension. [Let W be a neighborhood of zero such that

$$W \cup W^2 \cup W^3 \subseteq \{x \in A : |v(x)| \leq 1\},$$

let $(u_k)_{1 \leq k \leq n}$ be a sequence in A' such that

$$\{x \in A : |u_k(x)| \leq 1 \text{ for all } k \in [1, n]\} \subseteq W,$$

and let

$$J = \bigcap_{k=1}^{n} u_k^{-1}(0).$$

Show that the sets AJ, JA, and AJA are all contained in the kernel of v.]
(b) If K is complete and if the kernel of each $v \in A'$ contains an ideal of finite codimension, then (TR 2) holds for $\sigma_K(A, A')$. [Apply 15.13 to A/L, where L is a closed ideal of finite codimension contained in $v^{-1}(0)$.]

15.3 If L is a field that is an infinite-dimensional extension of a field K, furnished with a proper absolute value, then $\sigma_K(L, L^*)$ is a Hausdorff topology on the field L satisfying (TR 1) and (TR 2) of Definition 1.1 and (TR 5) of Theorem 2.15, but not (TR 4). [Use Exercise 15.2.]

15.4 Let L be a field furnished with a proper absolute value that is an infinite-dimensional extension of a subfield K, and let L' be the K-vector space of all continuous linear forms on the K-vector space L. The topology $\sigma_K(L, L')$ on L, regarded as a vector space over L, satisfies (TM 5) and (TM 6) of Theorem 2.16 but not (TM 4). [Modify the proof of Exercise 15.2(a).]

16 Topological Division Rings

Here we present some classical theorems concerning topological division rings. We begin with some theorems concerning locally compact division rings.

The hypothesis of 16.1 implies that multiplicative inversion is continuous by 11.11, but we do not need that fact in the proof.

16.1 Theorem. *If T is a locally compact ring topology on a division ring K, the set of all topological nilpotents is a neighborhood of zero.*

Proof. We may assume that T is not the discrete topology. Let U be a compact neighborhood of zero; as T is not discrete, there is a compact neighborhood V of zero such that $V \subset U$; let $W = \{x \in K : xU \subseteq V\}$. Then $WW \subseteq W$, for if $x, y \in W$, then $xyU \subseteq xV \subseteq xU \subseteq V$. Consequently, if $x \in W$, then by induction $x^n \in W$ for all $n \geq 1$. By 12.3, W is a neighborhood of zero; we shall show that each $a \in W$ is a topological nilpotent. As V is closed, so is W; as U contains a nonzero element d and as $Wd \subseteq V$, $W \subseteq Vd^{-1}$, a compact set; hence W is compact and contains a^n for all $n \geq 1$. Therefore to show that a is a topological nilpotent, it suffices to show that no nonzero element of K is an adherent point of the sequence $(a^n)_{n \geq 1}$.

Assume that b is a nonzero adherent point of $(a^n)_{n \geq 1}$. Then $b \notin bW$, since otherwise $1 \in W$ and hence $U \subseteq V$, a contradiction. As bW is compact, by (3) of 4.14 there is a neighborhood T of zero such that $b+T$ and $bW+T$ are disjoint, and by 12.3 there is a neighborhood S of zero such that $S \subseteq T$ and $SW \subseteq T$. As b is adherent to $(a^n)_{n \geq 1}$, there exist integers m and p such that $p > m$ and both a^m and a^p belong to $b+S$. But then

$$a^p = a^m a^{p-m} \in (b+S)W \subseteq bW + SW \subseteq bW + T,$$

a contradiction. Therefore a is a nonzero topological nilpotent. •

16.2 Theorem. *If E is a nonzero Hausdorff vector space over a nondiscrete topological division ring K that is straight and complete, then E is locally compact if and only if E is finite-dimensional and K is locally compact.*

Proof. Sufficiency: If E has dimension n, then E is topologically isomorphic to the K-vector space K^n by 15.9 and hence is locally compact.

Necessity: By hypothesis there is a nonzero vector $c \in E$. By (1) of 15.12, Kc is closed in E and hence is locally compact. As the K-vector space K is topologically isomorphic to Kc by hypothesis, K is locally compact and and E is not discrete. By 16.1 K has a nonzero topological nilpotent α. Let V be a compact neighborhood of zero in E. As αV is a neighborhood of zero, there exist $a_1, \ldots, a_n \in V$ such that

$$V \subseteq \bigcup_{k=1}^{n} (a_k + \alpha V).$$

Let M be the finite-dimensional subspace of E spanned by $a_1, \ldots, a_n$. Then M is closed in E by (1) of 15.12, so E/M is a Hausdorff K-vector space. Let $W = \phi_M(V)$, where ϕ_M is the canonical epimorphism from E to E/M. Then W is a compact neighborhood of zero in E/M, and $W \subseteq \alpha W$. By induction, $W \subseteq \alpha^n W$ for all $n \geq 1$, so

$$W \subseteq \bigcap_{n=1}^{\infty} \alpha^n W.$$

Let $w \in W$; then for each $n \geq 1$, $w = \alpha^n w_n$ for some $w_n \in W$. Hence as W is bounded by 12.3,
$$w = \lim_{n \to \infty} \alpha^n w_n = 0$$
by 12.12. Thus $W = \{0\}$, so $V \subseteq M$. For any $x \in E$, $\lim_{n \to \infty} \alpha^n x = 0$, so $\alpha^m x \in V$ for some $m \geq 1$, whence
$$x \in \alpha^{-m} V \subseteq \alpha^{-m} M = M.$$

Thus $E = M$. •

Once again, the hypothesis of the following theorem implies that multiplicative inversion is continuous by 11.11, but we shall obtain that conclusion by appealing to the much more elementary 11.12.

16.3 Theorem. *A locally compact ring topology T on a field K is defined by an absolute value.*

Proof. We may assume that T is not the discrete topology. By 16.1 there is a nonzero topological nilpotent $a \in K$. By 12.3, 14.4, and 7.7, T is defined by a complete norm, and hence inversion on K^* is continuous by 11.12. By 14.12 we need only show that T is locally retrobounded.

Let V be an open neighborhood of zero. Then $K \setminus V$ is closed, so $(K \setminus V)^{-1}$ is closed in K^*, and therefore $(K \setminus V)^{-1} \cup \{0\}$ is closed in K. By 12.3, therefore, as T is metrizable, we need only obtain a contradiction from the assumption that there is a sequence $(x_p)_{p \geq 1}$ in $K \setminus V$ such that $(x_p^{-1})_{p \geq 1}$ has no adherent point. In particular, zero is not an adherent point, so there is a compact neighborhood T of zero such that $x_p^{-1} \in K \setminus T$ for all $p \geq 1$.

For each $p \geq 1$, $\lim_{k \to \infty} a^k x_p^{-1} = 0$, so there is a smallest $n(p) \in \mathbb{N}$ such that $a^{n(p)} x_p^{-1} \in T$; moreover, $n(p) \geq 1$ since $x_p^{-1} \notin T$, so $a^{n(p)-1} x_p^{-1} \notin T$. If $n(p) \leq r$ for infinitely many $p \geq 1$, then for all such p,
$$x_p^{-1} \in \bigcup_{k=1}^{r} a^{-k} T,$$
a compact set, so $(x_p^{-1})_{p \geq 1}$ would have an adherent point, a contradiction. Thus $\lim_{p \to \infty} n(p) = +\infty$, so $\lim_{p \to \infty} a^{n(p)} = 0$. As $a^{n(p)} x_p^{-1} \in T$ for all $p \geq 1$, some subsequence of it converges; let
$$\lim_{k \to \infty} a^{n(p_k)} x_{p_k}^{-1} = b.$$

Then
$$\lim_{k \to \infty} a^{n(p_k)-1} x_{p_k}^{-1} = a^{-1} b,$$

so $a^{-1}b \neq 0$ as
$$a^{n(p_k)-1}x_{p_k}^{-1} \notin T$$
for all $k \geq 1$, and therefore $b \neq 0$. As inversion is continuous on K^*,
$$\lim_{k\to\infty} x_{p_k} a^{-n(p_k)} = b^{-1},$$
so
$$\lim_{k\to\infty} x_{p_k} = \lim_{k\to\infty} x_{p_k} a^{-n(p_k)} \lim_{k\to\infty} a^{n(p_k)} = b^{-1} \cdot 0 = 0,$$
a contradiction as $x_p \in K \setminus V$ for all $p \geq 1$. •

16.4 Theorem. (Frobenius) *If D is a division algebra over $\mathbb{R}$ every commutative division subalgebra of which has dimension at most 2, then D is isomorphic to $\mathbb{R}$, $\mathbb{C}$, or $\mathbb{H}$.*

Proof. We identify $\mathbb{R}$ with the division subalgebra $\mathbb{R}.1$ of D. For any commutative division subalgebra F properly containing $\mathbb{R}$, $\dim_\mathbb{R} F = 2$ by hypothesis, so F is $\mathbb{R}$-isomorphic to $\mathbb{C}$, and hence $F = \mathbb{R}(j)$ for some $j \in F$ satisfying $j^2 = -1$.

Case 1: The center Z of D properly contains $\mathbb{R}$. Then $Z = \mathbb{R}(i)$ for some $i \in Z$ satisfying $i^2 = -1$. If $x \in D \setminus Z$, $Z(x)$ is a commutative division subring properly containing Z, and hence $\dim_\mathbb{R} Z(x) > 2$, a contradiction. Therefore $D = Z = \mathbb{R}(i)$, so D is isomorphic to $\mathbb{C}$.

Case 2: $Z = \mathbb{R}$ and $D \neq Z$. Let $a \in D \setminus \mathbb{R}$. Then $\mathbb{R}(a)$ is a commutative division subalgebra properly containing $\mathbb{R}$, so $\mathbb{R}(a) = \mathbb{R}(i)$ for some $i \in \mathbb{R}(a)$ satisfying $i^2 = 1$. Let $D_+ = \{x \in D : ix = xi\}$, $D_- = \{x \in D : ix = -xi\}$. Then D_+ is a division subalgebra of D. As $D_+ D_- = D_-$, D_- is a D_+-vector space. Clearly $D_+ \cap D_- = \{0\}$; moreover, $D_+ + D_- = D$, for if $x \in D$,
$$x = \frac{1}{2}(x - ixi) + \frac{1}{2}(x + ixi) \in D_+ + D_-.$$

Now $D_+ = \mathbb{R}(i)$, for if $c \in D_+ \setminus \mathbb{R}(i)$, c would commute with each member of $\mathbb{R}(i)$, and hence $\mathbb{R}(i,c)$ would be a commutative division subalgebra whose dimension exceeds 2, a contradiction. Since $\mathbb{R}(i)$ is commutative, $\mathbb{R}(i) \neq D$, so there is a nonzero $b \in D_-$. As $b \notin D_+ = \mathbb{R}(i)$, $\mathbb{R}(b) \cap \mathbb{R}(i)$ is a proper division subalgebra of the 2-dimensional subalgebra $\mathbb{R}(i)$, so $\mathbb{R}(b) \cap D_+ = \mathbb{R}(b) \cap \mathbb{R}(i) = \mathbb{R}$. Consequently,
$$b^2 \in \mathbb{R}(b) \cap D_- D_- \subseteq \mathbb{R}(b) \cap D_+ = \mathbb{R}.$$

Moreover, $b^2 < 0$, for otherwise b^2 would have two square roots in $\mathbb{R}$ in addition to the square root b, so the field $\mathbb{R}(b)$ would contain three square roots of b^2, which is impossible. Consequently, $b^2 = -r$ for some $r > 0$; let $j = r^{-\frac{1}{2}}b \in D_-$; then $j^2 = -1$. If $x \in D_-$, then $xj^{-1} \in D_-D_- \subseteq D_+$, so $x = (xj^{-1})j \in D_+j$. Thus $\{j\}$ is a basis of the D_+-vector space D_-. Therefore as $D_+ = \mathbb{R}(i)$ and as $\{1, j\}$ is a basis of the D_+-vector space $D = D_+ + D_-$, $\{1, i, j, ij\}$ is a basis of the $\mathbb{R}$-vector space D. Let $k = ij$. It is easy to see that $k^2 = -1$, $jk = i$, $ki = j$, $ji = -k$, $kj = -i$, and $ik = -j$, so D is isomorphic to $\mathbb{H}$. •

16.5 Theorem. (Pontriagin [1931]) *If D is a division ring furnished with a connected locally compact ring topology, then D is topologically isomorphic to $\mathbb{R}$, $\mathbb{C}$, or $\mathbb{H}$.*

Proof. First, we observe that D cannot contain a closed subfield F whose topology is given by a proper nonarchimedean absolute value. For otherwise, as F is complete by 7.7, the left F-vector space D would be topologically isomorphic to F^n for some $n \geq 1$ by 16.2, 15.9, and 13.8 and hence would be totally disconnected, as F is.

As D is connected, it is not discrete, so by 16.1 D contains a nonzero topological nilpotent c. The set K of all elements of D commuting with c is easily seen to be a closed and hence locally compact division subring of D, and its center F is thus a closed and hence locally compact subfield. As $c \in F$, F is not discrete. By 16.3 the topology of F is defined by a proper absolute value A which, by the preceding, is archimedean. Hence F has characteristic zero by a remark following 1.12, so we may assume that F contains the rational field $\mathbb{Q}$. By 1.15 the topology induced on $\mathbb{Q}$ is that defined by the usual absolute value $|..|_\infty$, so the closure of $\mathbb{Q}$ in F is topologically isomorphic to $\mathbb{R}$. Consequently, we may regard D as a division algebra over $\mathbb{R}$ that has finite dimension by 16.2. By 16.4 there is an isomorphism from $\mathbb{R}$-division algebra D to either $\mathbb{R}$, $\mathbb{C}$, or $\mathbb{H}$, and by 15.9 that isomorphism is a topological isomorphism. •

We shall discuss totally disconnected locally compact division rings in §18.

To prove the Extension Theorem for absolute values, we need the following theorem concerning multilinear transformations on normed vector spaces.

16.6 Theorem. *Let $E_1, \ldots, E_n, F$ be vector spaces over a division ring K furnished with a proper absolute value $|..|$, let $N_1, \ldots, N_n, N$ be norms respectively on $E_1, \ldots, E_n, F$, and let u be a multilinear transformation from the cartesian product E of $(E_k)_{1 \leq k \leq n}$ to F. The following statements are equivalent:*

1° *u is continuous.*
2° *u is continuous at $(0, \ldots, 0)$.*

$3°$ There exists $c > 0$ such that
$$N(u(x_1,\ldots,x_n)) \le cN_1(x_1)\ldots N_n(x_n)$$
for all $(x_1,\ldots,x_n) \in E$.

Proof. Assume $2°$. Thus there exists $r > 0$ such that if $N_k(x_k) \le r$ for all $k \in [1,n]$, then $N(u(x_1,\ldots,x_n)) \le 1$. As $|..|$ is proper, there exists $\alpha \in K^*$ such that $|\alpha| < \inf\{1,r\}$; let $c = |\alpha|^{-2n}$. To establish $3°$, let $(x_1,\ldots,x_n) \in E$. If $x_i = 0$ for some $i \in [1,n]$, then $u(x_1,\ldots,x_n) = 0$, so we may assume that $x_i \ne 0$ for each $i \in [1,n]$. Let m_i be the integer such that
$$|\alpha|^{m_i+2} < N_i(x_i) \le |\alpha|^{m_i+1}.$$
Then
$$N_i(\alpha^{-m_i}x_i) = |\alpha|^{-m_i}N_i(x_i) \le |\alpha| \le r$$
for each $i \in [1,n]$, so
$$N(u(\alpha^{-m_1}x_1,\ldots,\alpha^{-m_n}x_n)) \le 1,$$
whence
$$N(u(x_1,\ldots,x_n)) \le |\alpha|^{m_1}\ldots|\alpha|^{m_n} < \prod_{i=1}^{n}(|\alpha|^{-2}N_i(x_i))$$
$$= cN_1(x_1)\ldots N_n(x_n).$$

Next, assume $3°$, let $(a_1,\ldots,a_n) \in E$, and let
$$M = \prod_{i=1}^{n}\sup\{1, N(a_k)\}.$$
Given $e > 0$, let
$$d = \inf\{1, e[(2^n-1)cM]^{-1}\}.$$
Let $(z_1,\ldots,z_n) \in E$ be such that $N_k(z_k) \le d$ for all $k \in [1,n]$. For each proper subset H of $[1,n]$, let $u_H = u(y_1,\ldots,y_n)$, where $y_i = a_i$ if $i \in H$, $y_i = z_i$ if $i \notin H$; then as u is multilinear,
$$u(a_1+z_1,\ldots,a_n+z_n) - u(a_1,\ldots,a_n) = \sum_{H \in \mathcal{P}} u_H,$$
where $\mathcal{P}$ is the set of all proper subsets of $[1,n]$. Given $H \in \mathcal{P}$, let $j \notin H$; then
$$N(u_H) \le cN_j(z_j)\prod_{i \ne j}\sup\{N(a_i), N(z_i)\} \le cdM.$$
Hence
$$N(u(a_1+z_1,\ldots,a_n+z_n) - u(a_1,\ldots,a_n)) \le \sum_{H \in \mathcal{P}} N(u_H) \le (2^n-1)cdM \le e.$$
Thus u is continuous at $(a_1,\ldots,a_n)$. ∎

16.7 Theorem. *Let A be a topological algebra over a field K, furnished with a proper absolute value. If N is a norm defining the topology of the underlying K-vector space, there is a K-algebra norm $\|..\|$ defining its topology. Furthermore, if A has an identity element e, there is a K-algebra norm $\|..\|_1$ defining the topology such that $\|e\|_1 = 1$.*

Proof. As multiplication is a K-bilinear, by 16.6 there exists $c > 0$ such that $N(xy) \leq cN(x)N(y)$ for all $x, y \in A$. Define $\|..\|$ by $\|x\| = cN(x)$. Clearly $\|..\|$ is a K-vector space norm defining the same topology as N, and

$$\|xy\| = cN(xy) \leq c^2 N(x)N(y) = \|x\|\|y\|.$$

Suppose, in addition, that A has an identity element e. For each $x \in A$, the K-linear mapping $L_x : t \to xt$ is continuous, and therefore by 16.6 there exists $c_x > 0$ such that for all $t \in A$, $\|xt\| \leq c_x \|t\|$. Thus $\{\|xt\|\|t\|^{-1} : t \in A^*\}$ is bounded, so $\|x\|_1$ is well defined by

$$\|x\|_1 = \sup_{t \neq 0} \frac{\|xt\|}{\|t\|}.$$

Clearly $\|..\|_1$ is a K-algebra norm satisfying $\|e\|_1 = 1$, for if $y \neq 0$,

$$\|xy\|_1 = \sup_{yt \neq 0} \frac{\|xyt\|}{\|t\|} \leq \sup_{yt \neq 0} \frac{\|xyt\|}{\|yt\|} \sup_{t \neq 0} \frac{\|yt\|}{\|t\|} \leq \|x\|_1 \|y\|_1.$$

Since

$$\|x\|\|e\|^{-1} \leq \|x\|_1 \leq \|x\|$$

for all $x \in A$, $\|..\|_1$ defines the same topology as $\|..\|$. •

16.8 Theorem. (Extension Theorem) *If A is a proper complete absolute value on a field K and if L is a finite-dimensional extension field of K, there is a unique absolute value A_L on L extending A. Moreover, for each $c \in L$,*

$$A_L(c) = A(\alpha_0)^{1/m}$$

where α_0 is the constant coefficient and m the degree of the minimal polynomial of c over K.

Proof. By 13.8 and 15.10 there is a unique K-vector topology $\mathcal{T}$ on L, and $\mathcal{T}$ is defined by a K-vector norm by 15.11. By 15.14 L is a topological K-algebra, so by 16.7 there is a K-algebra norm N defining $\mathcal{T}$ such that $N(1) = 1$ and hence

$$N(\lambda x) = A(\lambda)N(x) = A(\lambda)N(1)N(x) = N(\lambda 1)N(x) = N(\lambda)N(x)$$

for all $\lambda \in K$ and all $x \in L$. Thus K is contained in the core of N and hence in the core of the associated spectral norm N_s by 14.8. Choosing $x = 1$ in the above equalities yields $A(\lambda) = N(\lambda)$ and hence $N_s(\lambda) = A(\lambda)$ by 14.8. By 14.10 there is an absolute value A_L on L such that $A_L(\lambda) = N_s(\lambda) = A(\lambda)$ for all $\lambda \in K$. If B is an absolute value on L extending A, then both B and A_L define the unique K-vector topology on L, so there exists $r > 0$ such that $B^r = A_L$ by 1.10. Let $t \in K^*$ be such that $A(t) \neq 1$. Then

$$B(t) = A(t) = A_L(t) = B(t)^r,$$

so $r = 1$ and $B = A_L$.

Let $c \in L$, and let f be the minimal polynomial of c over K, and let

$$f = \prod_{k=1}^{m}(X - c_k)$$

in a splitting field Ω of f over L. The constant coefficient α_0 of f is $(-1)^m c_1 \ldots c_m$. For each $k \in [1, m]$ there is a K-automorphism σ_k of Ω such that $\sigma_k(c) = c_k$. As Ω is a finite-dimensional extension of K, there is a unique absolute value A_Ω on Ω extending A by what we have already proved, and the restriction of A_Ω to L is, of course, A_L. But for each $k \in [1, m]$, $A_\Omega \circ \sigma_k$ is clearly an absolute value on Ω extending A, so by the uniqueness of A_Ω, $A_\Omega \circ \sigma_k = A_\Omega$. Thus

$$A_\Omega(c_k) = A_\Omega(\sigma_k(c)) = A_\Omega(c)$$

for all $k \in [1, m]$. Consequently,

$$A(\alpha_0) = A((-1)^m c_1 \ldots c_m) = \prod_{k=1}^{m} A_\Omega(c_k) = A_\Omega(c)^m = A_L(c)^m,$$

so $A_L(c) = A(\alpha_0)^{1/m}$. ●

Let D be a finite-dimensional division algebra over a field K. For any $c \in D$, the norm $N_{D/K}(c)$ of c relative to K is the determinant of the linear operator $L_c : x \to cx$ on the K-vector space D. Let $X^m + \alpha_m X^{m-1} + \ldots + \alpha_1 X + \alpha_0$ be the minimal polynomial of c over K. Then $K(c)$ is an extension field of K, and $\{1, c, \ldots, c^{m-1}\}$ is a basis of the K-vector space $K(c)$. Let $\{e_1, \ldots, e_p\}$ be a basis of the (left) $K(c)$-vector space D. Then $\{e_1, ce_1, \ldots, c^{m-1}e_1, e_2, ce_2, \ldots, c^{m-1}e_2, \ldots, e_p, ce_p, \ldots, c^{m-1}e_p\}$

is a basis of the K-vector space D. Relative to this basis, the matrix of L_c is

$$\begin{bmatrix} A & 0 & \cdots & 0 \\ 0 & A & \cdots & 0 \\ \vdots & \vdots & \ddots & \vdots \\ 0 & 0 & \cdots & A \end{bmatrix}$$

where

$$A = \begin{bmatrix} 0 & 0 & 0 & \cdots & 0 & 0 & 0 & -\alpha_0 \\ 1 & 0 & 0 & \cdots & 0 & 0 & 0 & -\alpha_1 \\ 0 & 1 & 0 & \cdots & 0 & 0 & 0 & -\alpha_2 \\ \vdots & \vdots & \vdots & \ddots & \vdots & \vdots & \vdots & \vdots \\ 0 & 0 & 0 & \cdots & 1 & 0 & 0 & -\alpha_{m-3} \\ 0 & 0 & 0 & \cdots & 0 & 1 & 0 & -\alpha_{m-2} \\ 0 & 0 & 0 & \cdots & 0 & 0 & 1 & -\alpha_{m-1} \end{bmatrix}.$$

Developing $\det A$ by the minors of the first row, we obtain

$$\det A = (-1)^{1+m}(-\alpha_0) = (-1)^m \alpha_0.$$

Thus

$$N_{D/K}(c) = (\det A)^p = (-1)^{mp}\alpha_0^p = (-1)^n \alpha_0^{n/m}.$$

16.9 Theorem. *Let A be a proper complete [nonarchimedean] absolute value on a field K, and let D be an n-dimensional division algebra over K. There is a unique [nonarchimedean] absolute value A_D on D extending A. Moreover, for each $c \in D$,*

$$A_D(c) = A(N_{D/C}(c))^{1/n}.$$

Proof. For each $c \in D$, $K(c)$ is an extension field of K, so by 16.8 there is a unique absolute value $A_{K(c)}$ on $K(c)$ extending A; moreover, if α_0 is the constant coefficient and m the degree of the minimal polynomial of c over K,

$$A_{K(c)}(c)^n = A(\alpha_0)^{n/m} = A(\alpha_0^{n/m}) = A((-1)^n \alpha_0^{n/m}) = A(N_{D/K}(c)).$$

Thus the only possible absolute value on D extending A is the function A_D defined above, and the restriction of A_D to $K(c)$ is an absolute value on $K(c)$ for any $c \in D$.

To show that A_D is an absolute value, let $c, d \in D$. Since

$$N_{D/K}(cd) = \det L_{cd} = \det(L_c \circ L_d)$$
$$= \det(L_c)\det(L_d) = N_{D/K}(c)N_{D/K}(d),$$

$A_D(cd) = A_D(c)A_D(d)$. To show that $A_D(c+d) \leq A_D(c) + A_D(d)$, we may assume that $c \neq 0$. Then $1 + c^{-1}d \in K(c^{-1}d)$, so as the restriction of A_D to $K(c^{-1}d)$ is an absolute value,

$$A_D(1 + c^{-1}d) \leq A_D(1) + A_D(c^{-1}d) = 1 + A_D(c^{-1}d),$$

whence

$$A_D(c+d) = A_D(c)A_D(1+c^{-1}d) \leq A_D(c)[1 + A_D(c^{-1}d)] = A_D(c) + A_D(d).$$

By the remark after 1.12, A_D is nonarchimedean if and only if A is. •

16.10 Theorem. *If A is a proper absolute value on a field K and if L is a finite-dimensional extension field of K, there is an absolute value B on L extending A.*

Proof. Let Ω be the algebraic closure of $\widehat{K}$. There is a K-isomorphism σ from L to a subfield L' of Ω. As $\dim_K L' < +\infty$, there exist $x_1, \ldots, x_n \in L'$ such that $L' = K(x_1, \ldots, x_n)$. Then $x_1, \ldots, x_n$ are algebraic over $\widehat{K}$, so $\dim_{\widehat{K}} \widehat{K}(x_1, \ldots, x_n) < +\infty$. By 16.8 there is an absolute value B' on $\widehat{K}(x_1, \ldots, x_n)$ extending $\widehat{A}$, the unique absolute value on $\widehat{K}$ extending A and defining the topology of $\widehat{K}$. Then B, defined by $B(x) = B'(\sigma(x))$ for all $x \in L$, is an absolute value on L extending A. •

Lastly, we prove the impossibility of extending an archimedean absolute value or a norm relative to an archimedean absolute value on $\mathbb{R}$ to a field larger than $\mathbb{C}$:

16.11 Theorem. (Ostrowski [1915]) *If K is a field properly containing $\mathbb{C}$ and if $0 < r \leq 1$, there is no absolute value A on K extending $|\cdot|_\infty^r$.*

Proof. Assume that such an extension A exists. Let $a \in K \setminus \mathbb{C}$, and let

$$m = \inf_{\lambda \in \mathbb{C}} A(a - \lambda).$$

Since $\mathbb{C}$ is locally compact, by 7.7 and 7.5, $\mathbb{C}$ is a closed subfield of K, topologized by A. Therefore $m > 0$. Let $\lambda_n \in \mathbb{C}$ be such that

$$m \leq A(a - \lambda_n) \leq m + \frac{1}{n}$$

for each $n \geq 1$. Then

$$A(\lambda_n) \leq A(a - \lambda_n) + A(a) \leq m + 1 + A(a),$$

so
$$|\lambda_n| \leq [m+1+A(a)]^{1/r}.$$

Thus a subsequence of $(\lambda_n)_{n\geq 1}$ converges to some $\beta \in \mathbb{C}$, and $A(a-\beta) = m$. Let $b = a - \beta$. Then for all $\nu \in \mathbb{C}$,
$$A(b-\nu) \geq A(b)$$
since $A(b-\nu) = A(a-(\beta+\nu)) \geq m = A(b)$.

We shall show that if $c \in K^*$ satisfies $A(c-\nu) \geq A(c)$ for all $\nu \in \mathbb{C}$, then $A(c-\lambda) = A(c)$ for every $\lambda \in \mathbb{C}$ such that $A(\lambda) < A(c)$. Indeed, let ζ_n be a primitive nth root of unity in $\mathbb{C}$. By our assumption,
$$A(c - \zeta_n^k \lambda) \geq A(c)$$
for all $k \in [0, n-1]$, so
$$A(c^n - \lambda^n) = A(\prod_{k=0}^{n-1}(c - \zeta_n^k \lambda)) = \prod_{k=0}^{n-1} A(c - \zeta_n^k \lambda) \geq A(c-\lambda)A(c)^{n-1}.$$

Consequently,
$$A(c-\lambda)A(c)^{n-1} \leq A(c^n - \lambda^n) \leq A(c)^n + A(\lambda)^n,$$
so
$$A(c-\lambda)A(c)^{-1} \leq 1 + (A(\lambda)A(c)^{-1})^n.$$

Therefore
$$A(c-\lambda)A(c)^{-1} \leq \lim_{n\to\infty}[1 + (A(\lambda)A(c)^{-1})^n] = 1$$
as $A(\lambda)A(c)^{-1} < 1$. Hence $A(c-\lambda) \leq A(c)$, so by our assumption, $A(c-\lambda) = A(c)$. In addition, for any $\nu \in \mathbb{C}$, by our assumption
$$A((c-\lambda)-\nu) = A(c-(\lambda+\nu)) \geq A(c) = A(c-\lambda).$$

Let $\lambda \in \mathbb{C}^*$ be such that $|\lambda|_\infty^r = A(\lambda) < A(b)$. Applying the conclusion of the preceding paragraph successively to $b, b-\lambda, b-2\lambda, \ldots, b-(n-1)\lambda$, we conclude that
$$A(b) = A(b-\lambda) = A(b-2\lambda) = \ldots = A(b-n\lambda)$$
for all $n \geq 1$. Thus for each $n \geq 1$,
$$2A(b) = A(b-n\lambda) + A(b) \geq A(n\lambda) = |n\lambda|_\infty^r = n^r|\lambda|_\infty^r.$$

Consequently,
$$A(b) \geq \frac{1}{2}n^r|\lambda|_\infty^r$$
for all $n \geq 1$, an impossibility. ●

16.12 Theorem. *If D is a normed division algebra over $\mathbb{R}$, furnished with the absolute value $|..|_\infty^r$ where $0 < r \leq 1$, there is a topological isomorphism from D to one of the $\mathbb{R}$-algebras $\mathbb{R}$, $\mathbb{C}$, $\mathbb{H}$.*

Proof. By 16.7 there is a norm N on the algebra D that is equivalent to the given one and satisfies $N(1) = 1$. Consequently, for any $\lambda \in \mathbb{R}$,

$$N(\lambda.1) = |\lambda|_\infty^r N(1) = |\lambda|_\infty^r.$$

We identify $\mathbb{R}$ with $\mathbb{R}.1$. Thus N is a norm on the division ring D that extends $|..|_\infty^r$ and contains $\mathbb{R}$ in its core. To apply 16.4, let K be a commutative division subalgebra of D, N' the restriction of N to K. By 14.8 the corresponding spectral norm N'_s on K agrees with $|..|_\infty^r$ on $\mathbb{R}$ and contains $\mathbb{R}$ in its core. By 14.10 there is an absolute value A on K that agrees with N'_s and hence $|..|_\infty^r$ on $\mathbb{R}$. By 16.10 there is an absolute value A' extending A to $K(i)$, the field obtained by adjoining a root of $X^2 + 1$ to K. But $K(i) \supseteq \mathbb{R}(i) = \mathbb{C}$, so as $\mathbb{R}$ is complete for $|..|_\infty^r$, $A'(x) = |x|_\infty^r$ for all $x \in \mathbb{R}(i)$ by 16.8. Therefore $K(i) = \mathbb{C}$ by 16.11, so $\dim_\mathbb{R} K \leq 2$. By 16.4 there is an isomorphism σ from D to one of the $\mathbb{R}$-algebras $\mathbb{R}$, $\mathbb{C}$, $\mathbb{H}$. As these algebras are finite-dimensional, σ is a topological isomorphism by 15.10. •

16.13 Corollary. (Gel'fand-Mazur) *If D is a normed division algebra over $\mathbb{C}$, furnished with the absolute value $|..|_\infty^r$ where $0 < r \leq 1$, then D is one-dimensional.*

Proof. Restricting the scalar field to $\mathbb{R}$, we conclude from 16.12 that D is isomorphic to a subalgebra D' of $\mathbb{H}$ that contains $\mathbb{C}$ in its center. Consequently, $D' = \mathbb{C}$, that is, D is one-dimensional over $\mathbb{C}$. •

16.14 Theorem. (Ostrowski) *If A is an archimedean absolute value on a division ring [field] D, there exist $s \in (0,1]$ and an isomorphism σ from D to a division subring [subfield] of $\mathbb{H}$ [$\mathbb{C}$] such that*

$$A(x) = |\sigma(x)|_\infty^s$$

for all $x \in D$.

Proof. By a remark following 1.12, the characteristic of D is zero, so we may regard D as a $\mathbb{Q}$-algebra. By 1.15 there exists $s \in (0,1]$ such that $A(\lambda.1) = |\lambda|_\infty^s$ for all $\lambda \in \mathbb{Q}$. Consequently, A is a norm on the algebra D over $\mathbb{Q}$, furnished with the absolute value $|..|_\infty^s$. Therefore by 13.10 and 13.11, the unique absolute value $\widehat{A}$ on $\widehat{D}$ that extends A and defines the topology of $\widehat{D}$ is a norm on the algebra $\widehat{D}$ over $\widehat{Q} = \mathbb{R}$, furnished with $|..|_\infty^s$. Consequently by 16.12 there is an isomorphism σ from the $\mathbb{R}$-algebra $\widehat{D}$ to either $\mathbb{R}$, $\mathbb{C}$, or $\mathbb{H}$, and we may exclude $\mathbb{H}$, of course, if D and hence also $\widehat{D}$ are commutative. Now the functions $t \to \widehat{A}(\sigma^{-1}(t))$ and $|..|_\infty^s$ are absolute values on $\sigma(\widehat{D})$, and

$$|\lambda|_\infty^s = \widehat{A}(\lambda.1) = \widehat{A}(\sigma^{-1}(\lambda))$$

for all $\lambda \in \mathbb{R}$. Therefore by 1.10,

$$\widehat{A}(\sigma^{-1}(t)) = |t|_\infty^s$$

for all $t \in \sigma(\widehat{D})$, that is,

$$\widehat{A}(x) = |\sigma(x)|_\infty^s$$

for all $x \in \widehat{D}$. ∎

Exercises

16.1 Let A be a proper complete absolute value on a field K. If L is an algebraic extension of K, there is a unique [nonarchimedean] absolute value A_L on L that extends A. [Use 16.8.]

16.2 Derive the Fundamental Theorem of Algebra from 16.8 and 16.11.

16.3 (a) There are $2^{\text{card}(\mathbb{R})}$ subfields of $\mathbb{C}$ isomorphic to $\mathbb{R}$, but only one of them, $\mathbb{R}$, is closed. [Use the fact that there are $2^{\text{card}(\mathbb{R})}$ automorphisms of $\mathbb{C}$.] (b) If K is a locally compact proper subfield of $\mathbb{C}$, then $K = \mathbb{R}$.

16.4 (Baer and Hasse [1931]) Let $\mathbb{C}_\infty$ be the Riemann sphere, the Aleksandrov one-point compactification of $\mathbb{C}$. A theorem of Janiszewski [1915], generalizing the Jordan Curve Theorem, asserts that if F_1 and F_2 are closed connected subsets of $\mathbb{C}_\infty$, each containing more than one point, and if $F_1 \cap F_2$ is not connected, then $\mathbb{C}_\infty \setminus (F_1 \cup F_2)$ is not connected. Use this theorem to show that if K is a subfield of $\mathbb{C}$ that contains a closed connected subset of $\mathbb{C}$ containing more than one point, then $K = \mathbb{R}$ or $K = \mathbb{C}$. [Observe that if $F_1 \cup F_2 \subseteq K \cup \{\infty\}$ and if $a \in \mathbb{C} \setminus K$, then $a + (F_1 \cup F_2) \subseteq \mathbb{C}_\infty \setminus K$.]

16.5 Let A be an advertibly open topological algebra with identity over a complete straight field K. (a) Every maximal ideal of A is closed. (b) Every homomorphism from A to the K-algebra K is continuous. (c) If K is $\mathbb{C}$, furnished with $|..|_\infty^r$ for some $r \in (0, 1]$, then an element x of A is invertible if and only if $u(x) \neq 0$ for every nonzero homomorphism u from A to $\mathbb{C}$.

16.6 (Cantor [1883], Bendixson [1884]) Let X be a topological space. A subset A of X is *perfect* if A is closed and contains no isolated points. A point c is a *condensation point* of a subset A of X if every neighborhood of c contains uncountably many points of A. (a) The set of condensation points of a subset A of X is closed. (b) If X is a T_1-space and if every open subset of X is a Lindelöf space (that is, every open cover contains a countable subcover), then the set B of condensation points of a subset A of X is a perfect set, and $A \setminus B$ is countable. (c) If A is a nonempty perfect subset of a complete metric space X, then $\operatorname{card}(A) \geq \operatorname{card}(\mathbb{R})$. [Define recursively $A(a_0, \ldots, a_n)$ for all $n \in \mathbb{N}$, where each a_k is either 0 or 1, so that for each finite sequence $a_0, \ldots, a_m$ of 0's and 1's, $A(a_0, \ldots, a_m, 0)$ and $A(a_0, \ldots, a_m, 1)$ are disjoint infinite closed subsets of $A(a_0, \ldots, a_m)$ of diameter $\leq 1/(m+1)$; for each $a \in \{0, 1\}^{\mathbb{N}}$, let

$$x(a) \in \bigcap_{n=0}^{\infty} A(a(0), \ldots, a(n)).]$$

16.7 Let X be a connected complete separable metric space each point c of which has a fundamental system $\mathcal{V}_c$ of neighborhoods such that for all $V \in \mathcal{V}_c$, $V \setminus \{c\}$ is connected. (a) If G is a nonempty open nondense subset of X, the boundary of G contains a nonempty perfect subset. [To apply Exercise 16.6(b), observe that the boundary of G is a nonempty Baire space, and conclude that it is uncountable.] (b) (Livenson [1936]) If A is a dense subset of X that intersects nonvacuously each nonempty perfect subset of X, then A is connected.

16.8 (Dieudonné [1945]) Let $\mathfrak{c} = \operatorname{card}(\mathbb{R})$, and let γ be the smallest ordinal of cardinality $\mathfrak{c}$; thus $\operatorname{card}([0, \gamma)) = \mathfrak{c}$, and if $\beta < \gamma$, $\operatorname{card}([0, \beta)) < \mathfrak{c}$. (a) There is a bijection $\beta \to P_\beta$ from $[0, \gamma)$ to the set of all nonempty perfect subsets of $\mathbb{C}$ such that $0 \in P_0$. (b) There is an injection $\alpha \to K_\alpha$ from $[0, \gamma)$ to the set of all subfields of $\mathbb{C}$ such that $K_0 = \mathbb{Q}$, $K_\beta \subset K_\alpha$ whenever $\beta < \alpha$, and for all $\alpha \in (0, \gamma)$, if $K'_\alpha = \cup_{\beta < \alpha} K_\beta$, then $K_\alpha = K'_\alpha(u_\alpha)$ where u_α is transcendental over K'_α and $u_\alpha \in P_\alpha$. [Use Exercise 16.7] (c) Let $K = \cup_{\alpha < \gamma} K_\alpha$. Then K is a purely transcendental extension of $\mathbb{Q}$, and K is connected. [Use Exercise 16.7(b).] Also, K is locally connected. [Observe that if X is an open disk of center zero, every nonempty perfect subset of X contains a nonempty perfect subset of $\mathbb{C}$.] (d) There is a field K' containing K that is isomorphic to $\mathbb{R}$; with its induced topology, K' is connected and locally connected but not locally compact.

CHAPTER IV

REAL VALUATIONS

Some basic definitions and theorems concerning real valuations are given in §17. Discrete valuations, discussed in §18, are the principal real valuations we will encounter later. In §19 some subsequently needed theorems about extensions of real and discrete valuations are presented.

17 Real Valuations and Valuation Rings

We adjoin to $\mathbb{R}$ a new element, denoted by $+\infty$, and denote the set $\mathbb{R} \cup \{+\infty\}$ by $\mathbb{R}_\infty$. We extend addition on $\mathbb{R}$ to an associative, commutative composition on $\mathbb{R}_\infty$ by declaring, for all $\alpha \in \mathbb{R}$,

$$\alpha + (+\infty) = (+\infty) + \alpha = +\infty$$
$$(+\infty) + (+\infty) = +\infty.$$

We also define $\alpha \cdot (+\infty)$ and $(+\infty) \cdot \alpha$ to be $+\infty$ for all $\alpha \in \mathbb{R}_{>0}$. Finally, we extend the total ordering of $\mathbb{R}$ to one of $\mathbb{R}_\infty$ by declaring $\alpha \leq +\infty$ for all $\alpha \in \mathbb{R}_\infty$. Thus for all $\alpha, \beta, \gamma \in \mathbb{R}_\infty$, if $\alpha \leq \beta$, then $\alpha + \gamma \leq \beta + \gamma$.

17.1 Definition. *Let A be a ring with identity. A function v from A to $\mathbb{R}_\infty$ is a* **real valuation** *of A if for all $x, y \in A$,*

(V 1) $\qquad\qquad v(xy) = v(x) + v(y)$

(V 2) $\qquad\qquad v(x+y) \geq \inf\{v(x), v(y)\}$

(V 3) $\qquad\qquad v(1) = 0 \text{ and } v(0) = +\infty.$

Let v be a real valuation of A. If $z^n = 1$, then $0 = v(z^n) = n.v(z)$, so $v(z) = 0$. In particular, $v(-1) = 0$, so by (V 1), $v(-y) = v(y)$ for all $y \in A$. If $x \in A^\times$, then

$$0 = v(1) = v(xx^{-1}) = v(x) + v(x^{-1}),$$

so $v(x) \neq +\infty$, and $v(x^{-1}) = -v(x)$.

17.2 Theorem. *Let v be a real valuation of A, and let $x_1, \ldots, x_n \in A$. Then*

$$v(\sum_{i=1}^n x_i) \geq \inf_{1 \leq i \leq n} v(x_i),$$

and if there exists $r \in [1, n]$ such that $v(x_r) < v(x_i)$ for all $i \neq r$, then

$$v(\sum_{i=1}^n x_i) = v(x_r).$$

In particular, if $v(x) < v(y)$, then $v(x + y) = v(x)$.

Proof. The first assertion follows from (V 2) by induction. For the second, let $x = x_r$, $y = \sum_{i \neq r} x_i$. Then $v(y) > v(x)$, so $v(x + y) \geq v(x)$. If $v(x + y) > v(x)$, we would have

$$v(x) = v(((x+y)-y) \geq \inf\{v(x+y), v(-y)\} = \inf\{v(x+y), v(y)\} > v(x),$$

a contradiction. Hence $v(x + y) = v(x)$. •

17.3 Corollary. *If v is a real valuation of A and if $x_1, \ldots, x_n$ are elements of A^* such that $x_1 + \cdots + x_n = 0$, then there exist distinct integers r, s in $[1, n]$ such that*

$$v(x_r) = v(x_s) = \inf_{1 \leq i \leq n} v(x_i).$$

Henceforth, we shall consider only real valuations of division rings.

17.4 Theorem. *Let v be a real valuation of a division ring K, and let Γ be the commutator subgroup of K^* (the subgroup of the multiplicative group K^* generated by all elements of the form $xyx^{-1}y^{-1}$). Let*

$$A_v = \{x \in K : v(x) \geq 0\},$$
$$M_v = \{x \in K : v(x) > 0\}.$$

(1) $v(K^*)$ *is an additive subgroup of $\mathbb{R}$.*

(2) A_v *is a subring of K containing 1, M_v is a proper ideal of A_v containing every proper ideal of A_v, $A_v^\times = A_v \setminus M_v$, and A_v/M_v is a division ring.*

(3) $(K \setminus A_v)^{-1} = M_v$; *in particular, the smallest division subring containing A_v is K itself.*

(4) $\Gamma \subseteq A_v^\times$; $A_v t = t A_v$ *for every $t \in K$; hence every left or right ideal of A_v is an ideal of A_v.*

(5) For all $c, d \in K^*$, $v(c) \leq v(d)$ if and only if $d \in A_v c$.

Proof. By (V 1), the restriction of v to K^* is a homomorphism, so (1) holds. Clearly A_v is a subring of K containing 1, and M_v is a proper ideal of A_v. As $v(x) = 0$ if and only if $v(x^{-1}) = 0$, $A_v^\times = A_v \setminus M_v = \{x \in K : v(x) = 0\}$. In particular, every proper ideal of A_v is contained in M_v, and A_v/M_v is a division ring. Since $x \in K \setminus A_v$ if and only if $v(x) < 0$, or equivalently, $v(x^{-1}) > 0$, (3) holds.

If $x, y \in K^*$, then $xyx^{-1}y^{-1} \in A_v^\times$ since

$$v(xyx^{-1}y^{-1}) = v(x) + v(y) - v(x) - v(y) = 0.$$

Consequently, $\Gamma \subseteq A_v^\times$. Therefore $A_v t = t A_v$ for every $t \in K$, so every left or right ideal of A_v is an ideal.

(5) If $v(c) \leq v(d)$, then $dc^{-1} \in A_v$ since $v(dc^{-1}) = v(d) - v(c) \geq 0$, so $d = (dc^{-1})c \in A_v c$. Conversely, if $d = ac$ where $a \in A$, then $v(d) = v(a) + v(c) \geq v(c)$.

17.5 Definition. *Let v be a real valuation of a division ring K. The* **valuation ring** A_v *of v is the subring of K consisting of all $x \in K$ such that $v(x) \geq 0$, the* **valuation ideal** *of v is the ideal M_v of A consisting of all $x \in A$ such that $v(x) > 0$, and the* **value group** G_v *of v is the additive subgroup $v(K^*)$ of $\mathbb{R}$. The* **residue division ring** *(or* **residue field**, *if it is a field) of v is the quotient ring A_v/M_v.*

The only real valuations heretofore encountered (in §1) are the valuations of the quotient field of a principal ideal domain determined by primes of that domain. Let K be the quotient field of a principal ideal domain D, and let v_p be the valuation defined by a prime p of D. The valuation ring A of v_p is then the ring of all fractions a/s where $a \in D$, $s \in D^*$, and $p \nmid s$, and the valuation ideal M of v_p is pA. Furthermore, $M \cap D = pD$, for if $pa/s = b \in D$ where $p \nmid s$, then $pa = sb$, so $p \mid sb$, whence $p \mid b$ as $p \nmid s$, therefore $b = pt$ for some $t \in D$, and hence $a/s = t \in D$.

The discussion on page 8 of the absolute values on a field defined by a p-adic valuation is equally valid for arbitrary real valuations. If v is a real valuation of a division ring K and if $c > 1$, the defining properties of a real valuation imply that V_c, defined by

$$V_c(x) = c^{-v(x)}$$

for all $x \in K$ (where we adopt the convention $c^{-\infty} = 0$), is a nonarchimedean absolute value on K, called the *absolute value of v to base c*. An *absolute value of v* is simply an absolute value of v to base c for some $c > 1$. If $c > 1$ and $d > 1$, the absolute values of v to bases c and d are equivalent, since $V_d = V_c^s$ where $s = \log_c d$. Thus the absolute values of v all define the same topology, called the *topology defined by* v. The valuation ring A_v and maximal ideal M_v of v are then the closed unit ball and open unit ball of any absolute value V of v, that is,

$$A_v = \{x \in K : V(x) \leq 1\}$$
$$M_v = \{x \in K : V(x) < 1\}.$$

17.6 Definition. *The* **improper valuation** *of a division ring K is the real valuation v defined by $v(0) = +\infty$, $v(x) = 0$ for all $x \in K^*$. A real valuation of K is* **proper** *if it is not the improper valuation.*

17.7 Theorem. *The following statements about a real valuation v of a division ring K are equivalent:*

1° v is improper.
2° The valuation ring of v is K.
3° The value group of v is $\{0\}$.
4° The topology defined by v is the discrete topology.

The proof is easy.

17.8 Definition. *Real valuations v and w of a division ring K are* **equivalent** *if they define the same topology.*

17.9 Theorem. *Let v and w be proper real valuations of a division ring K with valuation rings A_v and A_w, valuation ideals M_v and M_w, and value groups G_v and G_w respectively. The following statements are equivalent:*

1° v and w are equivalent.
2° $A_v = A_w$.
3° $A_w \subseteq A_v$.
4° $M_v = M_w$.
5° $M_v \subseteq M_w$.
6° There exists $r \in \mathbb{R}_{>0}$ such that $w = rv$.
7° There is an increasing isomorphism ϕ from G_v to G_w such that $w(x) = (\phi \circ v)(x)$ for all $x \in K^*$.

Proof. By 1.10 applied to absolute values of v and w, 1°, 5°, and 6° are equivalent. Clearly 6° implies 4°, which implies 5°, and 6° implies 7°, which implies 2°, which implies 3°. We need only show, therefore, that 3° implies 5°. But if $A_w \subseteq A_v$, then $K \setminus A_v \subseteq K \setminus A_w$, so $(K \setminus A_v)^{-1} \subseteq (K \setminus A_w)^{-1}$, that is, $M_v \setminus \{0\} \subseteq M_w \setminus \{0\}$, and hence $M_v \subseteq M_w$. ∎

17.10 Theorem. *Let A be the valuation ring of a proper real valuation v of a division ring K. Every nonzero ideal of A is open and hence closed for the topology defined by v, and the nonzero principal ideals of A form a fundamental system of neighborhoods of zero. Furthermore, the restriction of v to K^* is continuous from K^* to the value group G of v, furnished with the discrete topology.*

Proof. A is a neighborhood of zero and hence is open by 4.9. If b is a nonzero element of K, then bA is open since $x \to bx$ is a homeomorphism from K to K. Consequently, every nonzero ideal is open by 4.9 and thus closed by 4.8. Let V be the absolute value defined by v to base $c > 1$, and let $r > 0$. Since v is proper, G is a nonzero subgroup of $\mathbb{R}$ under addition and hence is unbounded; therefore there exists $a \in K$ such that $c^{-v(a)} < \inf\{1, r\}$. Then $v(a) > 0$, so $a \in A$, and if $s = c^{-v(a)}$,

$$Aa = \{x \in K : v(x) \geq v(a)\} = \{x \in K : V(x) \leq s\}$$
$$\subseteq \{x \in K : V(x) < r\}.$$

Let $V_\alpha = \{x \in K : v(x) > \alpha\}$ and $W_\alpha = \{x \in K : v(x) \geq \alpha\}$ for each $\alpha \in G$. Both V_α and W_α are additive subgroups of K, so as $V_\alpha = \{x \in K : V(x) < c^{-\alpha}\}$, V_α is open and hence closed by 4.8, and W_α is open by 4.9. Consequently, as $v^{-1}(\alpha) = W_\alpha \setminus V_\alpha$, $v^{-1}(\alpha)$ is open. Thus v is continuous from K^* to the discrete group G. •

17.11 Definition. *A real valuation of a division ring is **complete** if the topology it defines is complete.*

17.12 Theorem. *Let v be a real valuation of a division ring K. The completion $\widehat{K}$ of K for the topology defined by v is a division ring, and there is a unique real valuation $\widehat{v}$ on $\widehat{K}$ that extends v and defines the topology of $\widehat{K}$. The value group of $\widehat{v}$ is the value group G of v, the valuation ring and ideal of $\widehat{v}$ are the closures in $\widehat{K}$ of the valuation ring A and valuation ideal M of v respectively, and consequently the residue division ring A/M of v is canonically isomorphic to the residue division ring $\widehat{A}/\widehat{M}$.*

Proof. Since $\widehat{K} = K$ if v is the improper valuation, we shall assume that v is proper. Let $c > 1$, and let V be the absolute value to base c defined by v. By 13.10 and 13.11, $\widehat{K}$ is a division ring whose topology is defined by a unique absolute value $\widehat{V}$ extending V. Let $\widehat{v} = -\log_c \widehat{V}$ (with the convention $-\log_c 0 = +\infty$). Clearly $\widehat{v}$ is a real valuation of $\widehat{K}$ that extends v, and $\widehat{V}$ is the absolute value of $\widehat{v}$ to base c. Thus the topology of $\widehat{K}$ defined by $\widehat{v}$ is its given topology. If w is a real valuation of $\widehat{K}$ that defines its topology and extends v, then w and $\widehat{v}$ are equivalent, so there exists

$r > 0$ such that $w = r\widehat{v}$. As v is proper, there exists $x \in K^*$ such that $v(x) \neq 0$, so
$$rv(x) = r\widehat{v}(x) = w(x) = v(x),$$
and therefore $r = 1$ and $w = \widehat{v}$.

As the valuation ring and ideal of $\widehat{v}$ are closed by 17.10, they contain $\widehat{A}$ and $\widehat{M}$ respectively. The statements that the value group of $\widehat{v}$ is G and that the valuation ring and ideal of $\widehat{v}$ are contained in $\widehat{A}$ and $\widehat{M}$ respectively all follow from the fact that for any $x \in \widehat{K}^*$ and any open neighborhood U of x, there exists $y \in U \cap K^*$ such that $\widehat{v}(x) = v(y)$. Indeed, by 17.10, $\widehat{v}^{-1}(\widehat{v}(x))$ is open in $\widehat{K}$, so there exists $y \in K \cap U \cap \widehat{v}^{-1}(\widehat{v}(x))$, and consequently $v(y) = \widehat{v}(y) = \widehat{v}(x)$. Finally, the function g from A/M to $\widehat{A}/\widehat{M}$ defined by $g(x + M) = x + \widehat{M}$ is an isomorphism by 5.14. •

17.13 Theorem. *If v is a proper complete real valuation of a field K with value group G and if D is an n-dimensional division algebra over K, there is a unique real valuation v_D of D extending v with value group $(1/n)G$.*

Proof. Let A be an absolute value of v. There is a unique nonarchimedean absolute value A_D on D extending A and consequently a valuation v_D of D extending v with value group $(1/n)G$ by 16.9. Arguing as in the proof of 17.12, we may conclude that v_D is the only valuation of D extending v. •

17.14 Theorem. *If A is the valuation ring of a proper real valuation v of a division ring K, then A is maximal in the set of all proper subrings of K, ordered by $\subseteq$.*

Proof. By 17.7, A is indeed a proper subring of K. Let B be a subring of K properly containing A. Then there exists $b \in B$ such that $v(b) < 0$. To show that $B = K$, let $x \in K \setminus A$. Then $v(x) < 0$, so by the archimedean property of $\mathbb{R}$, there exists $n \in \mathbb{N}$ such that $n.v(b) < v(x)$. Consequently $xb^{-n} \in A \subseteq B$ as $v(xb^{-n}) = v(x) - nv(b) \geq 0$, so $x = xb^{-n}b^n \in B$. •

17.15 Theorem. *If k is the residue division ring and G the value group of a real valuation v of a division ring K, then $\mathrm{card}(K) \leq \mathrm{card}(k^G)$.*

Proof. Let A be the valuation ring and M the maximal ideal of v. For each $\lambda \in G$, let $A_\lambda = \{x \in K : v(x) \geq \lambda\}$. Let $(c_\lambda)_{\lambda \in G}$ be a family of elements of K such that $v(c_\lambda) = \lambda$ for all $\lambda \in G$, and for each $\lambda \in G$ let B_λ be a subset of K such that B_λ contains precisely one member of each coset of A_λ in the additive group K (thus $K/A_\lambda = \{b + A_\lambda : b \in B_\lambda\}$). For each $x \in K$ and each $\lambda \in G$, let $b_{\lambda,x}$ be the unique member of B_λ such that $x + A_\lambda = b_{\lambda,x} + A_\lambda$. Then $v(x - b_{\lambda,x}) \geq \lambda$, so $c_\lambda^{-1}(x - b_{\lambda,x}) \in A$. For each

$x \in K$ and each $\lambda \in G$, let

$$\hat{x}(\lambda) = c_\lambda^{-1}(x - b_{\lambda,x}) + M,$$

an element of k. To show that $x \to \hat{x}$ is an injection from K to k^G, assume that x and y are distinct elements of K, and let $\delta = v(x-y)$. Then

$$b_{\delta,x} + A_\delta = x + A_\delta = y + A_\delta = b_{\delta,y} + A_\delta,$$

so $b_{\delta,x} = b_{\delta,y}$. Consequently

$$v(c_\delta^{-1}[x - b_{\delta,x}] - c_\delta^{-1}[y - b_{\delta,y}]) = v(c_\delta^{-1}(x-y)) = 0,$$

so $\hat{x}(\delta) \neq \hat{y}(\delta)$. •

Exercises

An *archimedean-ordered group* is a commutative group G furnished with a total ordering $\leq$ such that for all $x, y, z, \in G$, $x \leq y$ implies $x + z \leq y + z$, and for all $a, b \subset G$ such that $b > 0$ there exists $n \in \mathbb{N}$ such that $n.b \geq a$. In the proof of 17.14 we used the fact that the ordered group $\mathbb{R}$ is archimedean-ordered. A celebrated theorem of Baer [1928] states that if G is an archimedean-ordered group, then there is a strictly increasing monomorphism from G to the additive totally ordered group $\mathbb{R}$.

17.1 A subring A of a field K is a *valuation subring* of K if for all $x \in K^*$, either $x \in A$ or $x^{-1} \in A$. Let A be a valuation subring of a field K. (a) $1 \in A$, and if $M = A \setminus A^\times$, M is an ideal of A containing every proper ideal of A (and hence is called the *maximal ideal* of A). (b) The A-submodules of K are totally ordered by $\subseteq$. In particular, the ideals of A are totally ordered by $\subseteq$. (c) Let $G(A) = \{Ax : x \in K^*\}$. We define a composition on $G(A)$ by $Ax \cdot Ay = Axy$ and an ordering by declaring $Ax \preccurlyeq Ay$ if $Ax \supseteq Ay$. Under this multiplication, $G(A)$ is a commutative group, and for all $x, y, z \in K$, $Ax \preccurlyeq Ay$ implies $(Ax)(Az) \preccurlyeq (Ay)(Az)$. (d) Let v_A be the function from K^* to $G(A)$ defined by $v(x) = Ax$. For all $x, y \in K^*$, $v_A(xy) = v_A(x)v_A(y)$ and, if $x + y \neq 0$, $v_A(x+y) \geq \inf\{v_A(x), v_A(y)\}$.

17.2 Let A be a valuation subring of a field K that is maximal in the set of all proper subrings of K, ordered by $\subseteq$, and let M be the maximal ideal of A. (a) For each $b \in M$,

$$K = \bigcup_{n=0}^{\infty} Ab^{-n}.$$

(b) $G(A)$ (Exercise 17.1) is an archimedean-ordered group. (c) A is the valuation ring of a proper real valuation on K. [Use Exercise 17.1 and Baer's theorem, mentioned above.]

17.3 Prove directly that if G is a subgroup of the additive group $\mathbb{R}$ and if ϕ is an increasing isomorphism from G to a subgroup H of $\mathbb{R}$, then there exists $r \in \mathbb{R}_{>0}$ such that $\phi(x) = rx$ for all $x \in G$. [Consider separately the cases where $\mathbb{R}_{>0} \cap G$ does or does not have a smallest element. In the latter case, show that G is dense in $\mathbb{R}$ and that ϕ is continuous.]

18 Discrete Valuations

A nonzero subgroup G of $\mathbb{R}$ is clearly cyclic if and only if G is isomorphic to the additive group $\mathbb{Z}$. Furthermore, if G is a nonzero cyclic subgroup of $\mathbb{R}$, it has a a unique positive generator ζ, which is the smallest positive element of G.

18.1 Definition. *A* **discrete** *valuation of a division ring K is a proper real valuation of K whose value group is cyclic. If v is a discrete valuation of K and if ζ is the unique positive generator of its value group, any element u of K such that $v(u) = \zeta$ is called a* **uniformizer** *of v.*

If p is a prime of a principal ideal domain D, the p-adic valuation v_p of the quotient field K of D is an example of a discrete valuation, and p is a uniformizer of v_p.

18.2 Theorem. *Let u be a uniformizer of a discrete valuation v of a division ring K, and let A be the valuation ring of K. If M is a nonzero proper submodule of the A-module K, there is a unique $m \in \mathbb{Z}$ such that $M = Au^m$. Thus, every nonzero proper A-submodule of the A-module K is a member of the strictly decreasing sequence $(Au^n)_{n \in \mathbb{Z}}$ of A-submodules. In particular, if J is a nonzero ideal of A, there is a unique $m \in \mathbb{N}$ such that $J = Au^m$. Thus, every nonzero ideal of A is a member of the strictly decreasing sequence $(Au^n)_{n \in \mathbb{N}}$ of ideals.*

Proof. Let G be the value group of v, let M be a proper nonzero submodule of the A-module K, and let $H = v(M) \cap G$. If $\beta \in H$, then H contains every $\alpha \in G$ such that $\alpha \geq \beta$ by (5) of 17.4. If H were G, then, again by (5) of 17.4, M would be K, a contradiction. Consequently, H contains a smallest multiple $mv(u)$ of $v(u)$, and $H = \{nv(u) : n \geq m\}$. As $v(u^m) = mv(u)$, $M = Au^m$ by (5) of 17.4. •

18.3 Theorem. *If v is a proper real valuation of a division ring K, then v is a discrete valuation if and only if each ideal of its valuation ring A is a principal left ideal.*

Proof. The condition is necessary by 18.2. Sufficiency: Let M be the maximal ideal of v. By hypothesis, there exists $u \in A$ such that $M = Au$. Furthermore, $\bigcap_{n=1}^{\infty} Au^n$ is clearly an ideal of A, so by hypothesis there exists $t \in A$ such that
$$At = \bigcap_{n=1}^{\infty} Au^n.$$
Thus for each $n \in \mathbb{N}$ there exists $x_n \in A$ such that $t = x_n u^{n+1}$. Therefore for all $n \in \mathbb{N}$, $x_0 u = x_n u^{n+1}$, so $x_0 = x_n u^n \in Au^n$. Hence $x_0 \in At$, which is tA by (4) of 17.4, so $x_0 = ta$ for some $a \in A$. Thus $t = x_0 u = tau$. Consequently, if $t \neq 0$, $au = 1$ and hence u would be invertible in A, a contradiction. Thus $t = 0$ and
$$\bigcap_{n=1}^{\infty} Au^n = (0).$$
To show that u is a uniformizer of v, let $x \in A^*$. Then there is a largest $n \in \mathbb{N}$ such that $Ax \subseteq Au^n$ by what we have just proved. Consequently, $Axu^{-n} \subseteq A$, but if $Axu^{-n} \subseteq M = Au$, then $Ax \subseteq Au^{n+1}$, a contradiction; hence $Axu^{-n} = A$, so xu^{-n} is a unit of A. Therefore $v(xu^{-n}) = 0$, so $v(x) = nv(u)$. Finally, if $x \in K^* \setminus A$, then $x^{-1} \in A$, so $-v(x) = v(x^{-1}) = nv(u)$ for some $n \in \mathbb{N}$, whence $v(x) = -nv(u)$. •

18.4 Definition. *Let v be a proper real valuation of a division ring K, and let A and M be the valuation ring and ideal of v respectively. A subset S of A is a **representative set** for v (or for the residue division ring of v) if $0 \in S$ and the restriction to S of the canonical epimorphism from A to A/M is a bijection from S to A/M.*

18.5 Theorem. *Let v be a discrete valuation of a division ring K, let ζ be the the positive generator of its value group G, let A be the valuation ring of v, and let S be a representative set for v. For each $n \in \mathbb{Z}$, let $u_n \in K^*$ be such that $v(u_n) = n\zeta$.*

(1) For each $c \in K$ there is a unique family $(s_n)_{n \in \mathbb{Z}}$ of elements of S such that $s_n = 0$ for all but finitely many $n < 0$, $(s_n u_n)_{n \in \mathbb{Z}}$ is summable, and
$$c = \sum_{n \in \mathbb{Z}} s_n u_n;$$
moreover, if $v(c) = m\zeta$, then $s_n = 0$ for all $n < m$ and $s_m \neq 0$.

(2) If v is complete and if $(t_n)_{n \in \mathbb{Z}}$ is a family of elements of A such that $t_n = 0$ for all but finitely many $n < 0$, then $(t_n u_n)_{n \in \mathbb{Z}}$ is summable; if, moreover, $t_n = 0$ for all $n < m$ and $v(t_m) = 0$, then
$$v(\sum_{n \in \mathbb{Z}} t_n u_n) = m\zeta.$$

18 DISCRETE VALUATIONS

Proof. (1) Let $u = u_1$, a uniformizer of v. Then for each $n \in \mathbb{Z}$, $u_n u^{-n}$ is invertible in A, so $Au_n = Au^n$. We may assume that $c \neq 0$ and $v(c) = m\zeta$. A recursive argument establishes the existence of a sequence $(s_n)_{n \geq m}$ in S such that

$$c - \sum_{n=m}^{p} s_n u_n \in Au^{p+1}$$

for all $p \geq m$. Indeed, let $c = a_m u_m$ where $a_m \in A \setminus Au$; we define s_m to be the unique member of S such that $s_m - a_m \in Au$; then $s_m \neq 0$, and

$$c - s_m u_m = (a_m - s_m) u_m \in Auu_m = Au^{m+1}.$$

Similarly, if $s_m \ldots, s_p$ are defined so that

$$c - \sum_{n=m}^{p} s_n u_n = a_{p+1} u^{p+1}$$

where $a_{p+1} \in A$, we need only let s_{p+1} be the unique member of S such that $s_{p+1} - a_{p+1} \in Au$. Let $s_i = 0$ for all $i < m$. As $(Au^n)_{n \geq 1}$ is a fundamental system of neighborhoods of zero, $(s_n u_n)_{n \in \mathbb{Z}}$ is summable and

$$c = \sum_{n \in \mathbb{Z}} s_n u_n.$$

Uniqueness: Suppose that

$$\sum_{n \in \mathbb{Z}} s_n u_n = \sum_{n \in \mathbb{Z}} t_n u_n$$

where $s_n, t_n \in S$ for all $n \in \mathbb{Z}$ amd $s_n = t_n = 0$ for all but finitely many $n < 0$. If there were integers j such that $s_j \neq t_j$, there would be a smallest such integer m; but then $v(s_m - t_m) = 0$ as $s_m - t_m \in A \setminus Au$, so

$$v(\sum_{n \in \mathbb{Z}} (s_n - t_n) u_n) = m\zeta$$

by what we have just seen, a contradiction.

(2) Since $v(t_n u_n) \geq n\zeta$ for all $n \in \mathbb{Z}$, $(t_n u_n)_{n \in \mathbb{Z}}$ is summable by 10.5. If $v(t_m) = 0$ and $t_n = 0$ for all $n < m$, then

$$v(\sum_{n=m}^{p} t_n u_n) = m\zeta$$

for all $p \geq m$ by 17.2, and consequently

$$v(\sum_{n \in \mathbb{Z}} t_n u_n) = m\zeta$$

by 17.10. •

One choice of the sequence $(u_n)_{n \in \mathbb{Z}}$ is $(u^n)_{n \in \mathbb{Z}}$ where u is a uniformizer of v. Then for each $c \in A$ there is a unique sequence $(s_n)_{n \geq 0}$ of elements of S, called the *development* of c determined by S and u, such that

$$c = \sum_{n=0}^{\infty} s_n u^n.$$

18.6 Theorem. *Let v be a real valuation of a division ring K, let F be a subfield of its center such that K is n-dimensional over F, and let w be the restriction of v to F.*

(1) *v is improper if and only if w is improper.*
(2) *v is discrete if and only if w is discrete.*

Proof. Let $c \in K^*$. Then c is algebraic over F, so there is a nonconstant polynomial f over F of degree $\leq n$ such that $f(c) = 0$. Consequently by 17.3 there exist $a, b \in F^*$ and distinct $r, s \in [0, n]$ such that $v(ac^r) = v(bc^s)$, whence $w(a) + rv(c) = w(b) + sv(c)$.

Assume, first, that w is improper. Then $w(a) = w(b) = 0$, so $rv(c) = sv(c)$ and hence $v(c) = 0$. Thus v is improper.

Assume, next, that w is discrete, and let G be its (cyclic) value group. Then

$$v(c) = \frac{w(b) - w(a)}{r - s} \in \frac{1}{n!}G.$$

Thus the value group of v is a subgroup of the cyclic group $(1/n!)G$ and hence is cyclic. •

If the topology defined by a proper real valuation of a division ring is locally compact, that valuation is necessarily discrete:

18.7 Theorem. *Let v be a proper real valuation of a division ring K. The topology defined by v is locally compact if and only if the following conditions hold:*

1° *v is complete.*
2° *v is discrete.*
3° *The residue division ring of v is finite.*

These conditions hold if and only if the valuation ring A of v is compact.

Proof. Let G be the value group, A the valuation ring, and M the valuation ideal of v. Let $c \geq 1$, let V be the absolute value to base c defined by v, and for each $\alpha \in G$ let

$$W_\alpha = \{x \in K : v(x) \geq \alpha\} = \{x \in K : V(x) \leq c^{-v(\alpha)}\}$$

for each $\alpha \in G$. Necessity: 1° holds by 7.7. By hypothesis, W_β is compact for some $\beta \in G$. Then as $A = b^{-1}W_\beta$ where b is any element of K such that $v(b) = \beta$, A is also compact. Consequently, as M is open, A/M is a compact discrete space and hence is finite. Thus 3° holds.

To prove 2°, it suffices by 18.3 it suffices to show that if J is a proper nonzero ideal of A, then J is a principal left ideal. Let b be a nonzero element of J. We may assume that $J \setminus Ab \neq \emptyset$. As J and Ab are both open and hence closed, $J \setminus Ab$ is a closed subset of compact A and hence is compact. Therefore by 17.10, $v(J \setminus Ab)$ is a nonempty compact subset of the discrete space G, so $v(J \setminus Ab)$ is finite and hence has a smallest element γ. Let $c \in J \setminus Ab$ be such that $v(c) = \gamma$. If $v(c) \geq v(b)$, then $c \in Ab$ by (5) of 17.4, a contradiction. Thus $v(b) \geq v(c)$, so $b \in Ac$, and hence $Ab \subseteq Ac$. If $x \in J \setminus Ab$, then $v(x) \geq v(c)$, so $x \in Ac$ by (5) of 17.4. Thus $J = Ac$.

Sufficiency: Let u be a uniformizer of v. For each $n \in \mathbb{N}$, Au^n is open, hence closed, and therefore complete by 1°. Consequently by 18.2 and (2) of 8.5, A is topologically isomorphic to $\varprojlim_{n \geq 1}(A/Au^n)$. By 5.24, therefore, it suffices to show that each A/Au^n is finite. By 3°, A/Au is finite as $Au = M$. Assume A/Au^m is finite. Now $x \to xu^m + Au^{m+1}$ is an epimorphism from the additive group A to the additive group Au^m/Au^{m+1} whose kernel is Au, so Au^m/Au^{m+1} is isomorphic to A/Au and hence is also finite. Therefore as $(A/Au^{m+1})/(Au^m/Au^{m+1})$ is isomorphic to A/Au^m, A/Au^{m+1} is also finite. Thus by induction, A/Au^n is finite for all $n \geq 1$. •

18.8 Theorem. *Let p be a prime in a principal ideal domain D, and let A_p be valuation ring of the valuation v_p of the quotient field K of D defined by p. Then D is dense in A_p, pD is dense in the maximal ideal pA_p of v_p, and consequently the restriction to D of the canonical epimorphism from A_p to the residue field A_p/pA_p is an epimorphism with kernel pD.*

Proof. As $x \to px$ is a homeomorphism from K to K, the second assertion follows from the first, and the third follows from the first by 5.14 and the remark following 17.5. By 18.2, it suffices to show that for any $n \in \mathbb{N}$ and any $a, b \in D$ such that $b \neq 0$ and p does not divide b, there exists $s \in D$ such that

$$s - \frac{a}{b} \in p^n A_p.$$

As b and p^n are relatively prime, $Db + Dp^n = D$, so there exist $s, t \in D$ such that $sb + tp^n = a$, whence

$$s - \frac{a}{b} = \frac{-tp^n}{b} \in p^n A_p. \bullet$$

18.9 Definition. *Let p be a prime integer. The completion of $\mathbb{Q}$ for the p-adic valuation v_p is called the **p-adic number field** and is denoted by $\mathbb{Q}_p$. The valuation $\hat{v}_p$ of $\mathbb{Q}_p$ is called the **p-adic valuation** of $\mathbb{Q}_p$; its valuation ring is called the **ring of p-adic integers** and is denoted by $\mathbb{Z}_p$. The absolute value $|..|_p$ of $\hat{v}_p$ to base p is called the **p-adic absolute value** on $\mathbb{Q}_p$.*

18.10 Theorem. *Let p be a prime integer.*

(1) $\mathbb{Q}_p$ is a locally compact field.

(2) The compact valuation ring $\mathbb{Z}_p$ of v_p is the closure of $\mathbb{Z}$ in $\mathbb{Q}_p$.

(3) The only nonzero proper closed additive subgroups of $\mathbb{Q}_p$ are the compact groups $p^n \mathbb{Z}_p$, where $n \in \mathbb{Z}$, and each $p^n \mathbb{Z}_p$ is the closure in $\mathbb{Q}$ of $p^n \mathbb{Z}$.

(4) The nonzero ideals of $\mathbb{Z}$ are the ideals $p^n \mathbb{Z}_p$ where $n \in \mathbb{N}$; in particular, $p\mathbb{Z}_p$ is the maximal ideal of $\mathbb{Z}_p$.

(5) The restriction to $\mathbb{Z}$ of the canonical epimorphism from $\mathbb{Z}_p$ to $\mathbb{Z}_p/p\mathbb{Z}_p$ is an epimorphism with kernel $p\mathbb{Z}$, so the residue field of $\hat{v}_p$ is the finite field of p elements, and $\{0, 1, \ldots p-1\}$ is a representative set for $\hat{v}_p$.

Proof. By 18.8, $\mathbb{Z}$ is dense in the valuation ring A_p of v_p, $p\mathbb{Z}$ is dense in its maximal ideal pA_p, and the restriction to $\mathbb{Z}$ of the canonical epimorphism from A_p to the residue field A_p/pA_p is an epimorphism with kernel $p\mathbb{Z}$. Consequently (5) holds by 5.14. The value group of $\hat{v}_p$ is $\mathbb{Z}$ by 17.12. Therefore (1) and (2) follow from 18.7. Since p is a uniformizer of $\hat{v}_p$, (3) and (4) follow from 18.2. •

18.11 Theorem. *If K is a division ring of characteristic zero furnished with a nondiscrete Hausdorff ring topology, then K is locally compact and totally disconnected if and only if K is a finite-dimensional extension of $\mathbb{Q}_p$ for some prime p, in which case its topology is given by a unique valuation extending the p-adic valuation of $\mathbb{Q}$.*

Proof. The condition is sufficient by 13.8, 18.10, and 16.2. Necessity: By 4.21, there is a compact open subring A of K, and A contains a nonzero element a as K is not discrete. The sequence $(2^k a)_{k \geq 1}$ lies in A and therefore has an adherent point b. Then ba^{-1} is an adherent point of $(2^k)_{k \geq 1}$. If the topology induced on $\mathbb{Q}$ were discrete, then $\mathbb{Q}$ would be closed by 4.13, so no sequence of distinct rationals would have an adherent point in K.

Consequently, the topology induced on $\mathbb{Q}$ is not discrete. The center F of K is closed by 4.4 and hence is a locally compact field containing $\mathbb{Q}$. Consequently, as F is not discrete, its topology is given by a proper absolute value V by 16.3. Moreover, K is a finite-dimensional over F by 16.2, and hence by 15.10 F cannot be connected, as otherwise K would be. If V were archimedean, then as F is complete, F would be isomorphic to either $\mathbb{R}$ or $\mathbb{C}$ by 16.14 and hence would be connected. Thus V is nonarchimedean. Therefore by 1.14 the topology induced on $\mathbb{Q}$ is given by the p-adic valuation for some prime p. The closure of $\mathbb{Q}$ in K is therefore $\mathbb{Q}_p$, and K is finite-dimensional over $\mathbb{Q}_p$ by 16.2. The final assertion follows from 17.12 and 17.13. •

Let K be a commutative ring with identity. The *ring [K-algebra] of formal power series over* K is the ring [K-algebra] $S(K, \mathbb{Z})$ of all sequences $(a_n)_{n \in \mathbb{Z}}$ of elements of K such that $a_n = 0$ for all but finitely many $n < 0$, where addition is defined componentwise, multiplication by

$$(a_n)_{n \in \mathbb{Z}} (b_n)_{n \in \mathbb{Z}} = \left(\sum_{i+j=n} a_i b_j \right)_{n \in \mathbb{Z}},$$

(a definition which makes sense as for each $n \in \mathbb{Z}$ there are only finitely many couples (i, j) such that $i + j = n$ and $a_i b_j \neq 0$), and scalar multiplication by

$$c(a_n)_{n \in \mathbb{Z}} = (ca_n)_{n \in \mathbb{Z}}.$$

It is easy to verify that $S(K, \mathbb{Z})$ is, indeed a commutative ring [K-algebra]. Let $\delta_{i,j} = 0$ if $i \neq j$ and $\delta_{i,j} = 1$ if $i = j$. The identity element of $S(K, \mathbb{Z})$ is then $(\delta_{0,n})_{n \in \mathbb{Z}}$. It is easy to see that if K is an integral domain, $S(K, \mathbb{Z})$ is also. Furthermore, if K is a field, $S(K, \mathbb{Z})$ is a field, for an inductive argument establishes that if $(a_n)_{n \in \mathbb{Z}} \in S(K, \mathbb{Z})^*$, there exists $(b_n)_{n \in \mathbb{Z}} \in S(K, \mathbb{Z})$ such that

$$\sum_{i+j=n} a_i b_j = \delta_{0,n}$$

for all $n \in \mathbb{Z}$. We denote by X the sequence $(\delta_{1,n})_{n \in \mathbb{Z}}$. An inductive argument then establishes that $X^m = (\delta_{m,n})_{n \in \mathbb{Z}}$ for all $m \in \mathbb{Z}$.

We define the *order* of each nonzero $(a_n)_{n \in \mathbb{Z}} \in S(K, Z)$ to be the smallest of the integers m such that $a_m \neq 0$, and we denote it by $\mathrm{ord}((a_n)_{n \in \mathbb{Z}}$. We also define the *order* of the zero sequence to be $+\infty$. It is easy to see that for any $f, g \in S(K, \mathbb{Z})$,

$$\mathrm{ord}(f + g) \geq \inf\{\mathrm{ord}(f), \mathrm{ord}(g)\},$$
$$\mathrm{ord}(fg) \geq \mathrm{ord}(f) + \mathrm{ord}(g).$$

Consequently, if $c > 1$ and
$$\|f\| = c^{-\mathrm{ord}(f)}$$
for all $f \in S(K, \mathbb{Z})$, $\|..\|$ is a norm on $S(K, \mathbb{Z})$ for which $(V_m)_{m \geq 0}$ is a fundamental system of neighborhoods of zero, where for each $m \in \mathbb{Z}$,
$$V_m = \{f \in S(K, \mathbb{Z}) : \mathrm{ord}(f) \geq m\}.$$
We furnish $S(K, \mathbb{Z})$ with the topology defined by this norm, called the *order topology*. It is easy to see that for any $(a_n)_{n \in \mathbb{Z}} \in S(K, \mathbb{Z})$, $(a_n X^n)_{n \in \mathbb{Z}}$ is summable, and
$$(a_n)_{n \in \mathbb{Z}} = \sum_{n \in \mathbb{Z}} a_n X^n.$$
Consequently, we commonly denote $S(K, \mathbb{Z})$ by $K((X))$ and the subring V_0 by $K[[X]]$.

If K is an integral domain, then ord is a real valuation of $K((X))$. If K is a field, the valuation ring of ord is $K[[X]]$ ($= V_0$), and the valuation ideal of ord is (X) ($= V_1$).

18.12 Theorem. *If K is a commutative ring with identity, $K((X))$ is complete for the order topology.*

Proof. Let $(f_m)_{m \geq 0}$ be a Cauchy sequence in $K((X))$, and for each $m \geq 0$, let
$$f_m = \sum_{n \in \mathbb{Z}} a_{m,n} X^n.$$
For each $k \in \mathbb{Z}$, the additive homomorphism pr_k, defined by
$$pr_k(\sum_{n \in \mathbb{Z}} c_n X^n) = c_k,$$
is continuous from $K((X))$ to the discrete group $\mathbb{Z}$, since $pr_k(V_{k+1}) = \{0\}$. Consequently, $(pr_k(f_m))_{m \geq 0}$ is a Cauchy sequence in $\mathbb{Z}$ by 7.12 and 7.11. Thus there exists $m_k \in \mathbb{Z}$ and $b_k \in K$ such that $a_{m,k} = b_k$ for all $m \geq m_k$. Suppose that $b_k \neq 0$ for infinitely many $k < 0$. Then there would exist a strictly decreasing sequence $(k_r)_{r \geq 1}$ of integers such that for all $r \geq 1$,
$$k_{r+1} < \inf\{\mathrm{ord}(f_{m_{k_r}}), k_r\}$$
and $b_{k_{r+1}} \neq 0$, whence
$$\mathrm{ord}(f_{m_{k_{r+1}}}) \leq k_{r+1} < \mathrm{ord}(f_{m_{k_r}})$$
and therefore
$$\mathrm{ord}(f_{m_{k_{r+1}}} - f_{m_{k_r}}) = \mathrm{ord}(f_{m_{k_{r+1}}}) \leq k_{r+1},$$
a contradiction of our hypothesis that $(f_m)_{m \geq 0}$ is a Cauchy sequence. Therefore $(b_k)_{k \in \mathbb{Z}}$ belongs to $K((X))$, and it is easy to see that $(f_m)_{m \geq 0}$ converges to $\sum_{k \in \mathbb{Z}} b_k X^k$. •

18.13 Corollary. *If K is a field, then $K((X))$ is locally compact if and only if K is finite.*

Proof. With the terminology of the proof of 18.12, the restriction of pr_0 to the valuation ring $K[[X]]$ of ord is an epimorphism from the ring $K[[X]]$ to K whose kernel is the maximal ideal (X) of $K[[X]]$, so the residue field $K[[X]]/(X)$ of ord is isomorphic to K. The assertion therefore follows from 18.12 and 18.7. •

18.14 Theorem. *If K is a nondiscrete locally compact field of prime characteristic p, then there is a finite field k such that K is topologically isomorphic to $k((X))$.*

Proof. By 16.3, the topology of K is given by a proper absolute value, which is necessary nonarchimedean by the remark following 1.12. Consequently by 18.7, the topology of K is defined by a proper valuation v whose value group is $\mathbb{Z}$, whose valuation ring A is compact, and whose residue field k_v is finite. By the theory of finite fields, the order q of k_v is a power of p, and the multiplicative group k_v^* is cyclic. Let α be a generator of k_v^*, and let $a \in A$ be such that $\phi(a) = \alpha$, where ϕ is the canonical epimorphism from A to k_v. Then $\phi(a^q - a) = \alpha^q - \alpha = 0$, so $v(a^q - a) \geq 1$, and therefore

$$\lim_{n \to \infty} (a^{q^{n+1}} - a^{q^n}) = \lim_{n \to \infty} (a^q - a)^{q^n} = 0.$$

As A is compact, some subsequence $(a^{q^{n_k}})$ of $(a^{q^n})_{n \geq 1}$ converges to a point b of A. Then

$$b^q - b = \lim_{k \to \infty} (a^{q^{n_k+1}} - a^{q^{n_k}}) = \lim_{k \to \infty} (a^q - a)^{q^{n_k}} = 0,$$

so b is algebraic over the prime field P of K, and hence the smallest subfield $P(b)$ of K containing b is contained in A. Moreover,

$$\phi(b) = \lim_{k \to \infty} \phi(a^{q^{n_k}}) = \lim_{k \to \infty} \alpha^{q^{n_k}} = \alpha.$$

Let $k = P(b)$. The restriction to k of ϕ is therefore an isomorphism from k to k_v. In particular, k is a representative set for v. Let u be a uniformizer for v. By 18.7 and the remark following, for each element x of K there is a unique sequence $(c_n)_{n \in \mathbb{Z}}$ in k such that $c_n \neq 0$ for only finitely many $n < 0$ and

$$x = \sum_{n \in \mathbb{Z}} c_n u^n,$$

and furthermore, if $x \neq 0$, $v(x)$ is the smallest of the integers m such that $c_m \neq 0$; furthermore, for any sequence $(c_n)_{n \in \mathbb{Z}}$ of elements of k such that $c_n = 0$ for all but finitely many $n < 0$, $(c_n u^n)_{n \in \mathbb{Z}}$ is summable and $\sum_{n \in \mathbb{Z}} c_n u^n \in K$. Therefore K is topologically isomorphic to $k((X))$. •

To describe nondiscrete locally compact division rings of prime characteristic, we shall use the following theorem, due to Artin and Whaples [1942]:

18.15 Theorem. *Let D be a division ring, C its center. If F is a division subring of D such that D, regarded as a left vector space over F, has finite dimension, and if F' is the division ring consisting of all elements of D commuting with each element of F, then $\dim_C F' \leq \dim_F D$.*

18.16 Theorem. *If K is a division ring of prime characteristic furnished with a nondiscrete locally compact ring topology, then its center C is a nondiscrete locally compact field and hence is topologically isomorphic to $k((X))$ for some finite field k, and K is a finite-dimensional division algebra over C.*

Proof. By 16.1, D has a nonzero topological nilpotent c. As in the proof of 16.5, the set K of all elements of D commuting with c is a closed and hence locally compact division subring of D, and its center F is thus a closed and hence locally compact subfield. As $c \in F$, F is not discrete. The topology of F is thus given by a discrete valuation v by 16.3, the remark following 1.12, and 18.7, so D, regarded as a topological left vector space over F, is finite-dimensional by 16.2, 7.7, and 13.8. By 18.15, the division subring F' of all elements of D commuting with each element of F is finite-dimensional over C. But as F is commutative, $F' \supseteq F$, and therefore F, which clearly contains C, is finite-dimensional over C. Consequently by 18.6, the topology of C is given by a discrete valuation and hence is nondiscrete, and

$$\dim_C K = (\dim_F K)(\dim_C F) < +\infty.$$

The theorem now follows from 18.14. •

18.17 Theorem. *A nondiscrete locally compact division ring is finite-dimensional over its center, a nondiscrete locally compact field, and its topology is given by a proper, complete absolute value.*

Proof. The assertion follows from 16.5, 18.11, 18.16, 16.9, and 7.7. •

Exercises

If p is a prime of $\mathbb{Z}$, the p-adic development of $c \in \mathbb{Z}_p$ is the sequence $(s_n)_{n \geq 0}$ in $\{0, 1, \ldots, p-1\}$ such that

$$c = \sum_{n=0}^{\infty} s_n p^n.$$

18.1 Let p be a prime of $\mathbb{Z}$. Use geometric series to establish the following equalities in $\mathbb{Q}_p$:

$$(a)\ -1 = \sum_{n=0}^{\infty}(p-1)p^n. \qquad (b)\ -\frac{k}{p-1} = \sum_{n=0}^{\infty}kp^n$$

for each $k \in [1, p-1]$.

18.2 Find the 5-adic development of $2/3$.

18.3 Let p be a prime of $\mathbb{Z}$, and let $(s_n)_{n \geq 0}$ be the p-adic development of $c \in \mathbb{Z}_p$. If there exist $k \geq 1$ and $m \geq 0$ such that $s_{n+k} = s_n$ for all $n \geq m$, then $c \in \mathbb{Q}$.

18.4 Let p be a prime of $\mathbb{Z}$, and let $(s_n)_{n \geq 0}$ be the p-adic development of the rational a/b, where $a, b \in \mathbb{Z}$, $b > 0$, and $p \nmid b$. (a) For each $k \geq 1$, let

$$q_k = \sum_{n=0}^{k-1} s_n p^n.$$

Show that $0 \leq q_k \leq p^k - 1$ and that $a - bq_k = a_k p^k$ where $a_k \in \mathbb{Z}$. (b) There exist $m \geq 1$ and $k \geq 1$ such that $a_{m+k} = a_m$. [Show first that $ap^{-k} - b \leq a_k < ap^{-k}$.] (c) For all $n \geq m$, $a_{n+1+k} = a_{n+1}$ and $s_{n+k} = s_n$. [Show that $a_{n+1}p = a_n - bs_n$ for all $n \geq 1$, and use induction.]

18.5 Let G be a nonzero subgroup of $\mathbb{R}$. If A and B are well ordered subsets of G, then $A + B$ is well ordered, and for each $\gamma \in A + B$, there are only finitely many $(\alpha, \beta) \in A \times B$ such that $\gamma = \alpha + \beta$. [If a nonempty subset M of $B + C$ had no smallest element, show that there would exist a strictly decreasing sequence $(\beta_n + \gamma_n)_{n \geq 1}$ of elements of M such that for each $n \geq 1$, $\beta_n \in B$, $\gamma_n \in C$, and γ_n is the smallest of the elements $\gamma \in C$ such that $b_n + \gamma \in M$; extract a strictly increasing subsequence $(\beta_{n_k})_{k \geq 1}$ of $(\beta_n)_{n \geq 1}$, and consider $\{\gamma_{n_k} : k \geq 1\}$.]

18.6 Let G be a nonzero subgroup of $\mathbb{R}$, and let K be a commutative ring with identity. The *support* of $f \in K^G$, denoted by $\operatorname{supp}(f)$, is $\{\alpha \in G : f(\alpha) \neq 0\}$. (a) The subset $S(K, G)$, consisting of all functions f from G to K such that $\operatorname{supp}(f)$ is a well-ordered subset of G, is a K-submodule of the K-module K^G. (b) Under multiplication defined by

$$(fg)(\alpha) = \sum_{\beta + \gamma = \alpha} f(\beta)g(\gamma),$$

for all $\alpha \in G$ (cf. Exercise 18.5), $S(K, G)$ is a K-algebra. (c) If K is an integral domain, so is $S(K, G)$. (d) If K is a field, $S(K, G)$ is a K-division algebra. [Suppose that 0 is the smallest element in $\operatorname{supp}(f)$; to show that

f is invertible, consider the set of all $\alpha \geq 0$ for which there exists a unique $g_\alpha \in S(K,G)$ such that $\mathrm{supp}(g_\alpha) \subseteq [0,\alpha]$, $(fg_\alpha)(0) = 1$, and $(fg_\alpha)(\beta) = 0$ for all $\beta \in (0, \alpha]$.]

18.7 Let G be a nonzero subgroup of $\mathbb{R}$, and let K be a commutative ring with identity. For each nonzero $f \in S(K,G)$, the *order of f*, denoted by $\mathrm{ord}(f)$, is the smallest element in $\mathrm{supp}(f)$, and that of the zero function is defined to be $+\infty$. (a) Show that for all $f, g \in S(K,G)$,

$$\mathrm{ord}(f+g) \geq \inf\{\mathrm{ord}(f), \mathrm{ord}(g)\},$$
$$\mathrm{ord}(fg) \geq \mathrm{ord}(f) + \mathrm{ord}(g).$$

(b) If $c > 1$ and $\|..\|$ is defined by

$$\|f\| = c^{-\mathrm{ord}(f)}$$

for all $f \in S(K,G)$, $\|..\|$ is a norm on the ring $S(K,G)$ for which $(V_\alpha)_{\alpha \in G}$ is a fundamental system of neighborhoods of zero, where for each $\alpha \in G$, $V_\alpha = \{f \in S(K,G) : \mathrm{ord}(f) \geq \alpha\}$. The topology defined by this norm is called the *order topology* of $S(K,G)$. (c) If K is an integral domain, ord is a real valuation of $S(K,G)$. If K is a field, then the valuation ring of ord is V_0, its maximal ideal is $\{f \in S(K,G) : \mathrm{ord}(f) > 0\}$, and its residue field is isomorphic to K. (d) For each $\alpha \in G$, let X^α be the function from G to K taking α into 1 and β into 0 for all $\beta \neq \alpha$. Then $X^{\alpha+\beta} = X^\alpha X^\beta$ for all $\alpha, \beta \in G$. (e) For each $f \in S(K,G)$, $(f(\alpha)X^\alpha)_{\alpha \in G}$ is summable, and

$$f = \sum_{\alpha \in G} f(\alpha) X^\alpha.$$

(f) If $(f_n)_{n \geq 1}$ is a Cauchy sequence in $S(K,G)$, then for each $\alpha \in G$ there exists $n_\alpha \geq 1$ such that $f_m(\beta) = f_p(\beta)$ for all $m, p \geq n_\alpha$ and all $\beta \leq \alpha$. (g) $S(K,G)$ is complete for the order topology.

19 Extensions of Real Valuations

We begin by describing all ring topologies on a simple algebraic extension L of a field K that induce on K the topology given by a proper absolute value. For this, we need a preliminary theorem:

19.1 Theorem. *Let E be a finite-dimensional Hausdorff vector space over a straight division ring K. If B is a finite set of generators of the K-vector space E, B also generates the $\widehat{K}$-vector space $\widehat{E}$. In particular,*

$$\dim_{\widehat{K}} \widehat{E} \leq \dim_K E.$$

Proof. By 13.4, $\widehat{K}$ is straight, so by (1) of 15.12, the subspace $\sum_{b \in B} \widehat{K}b$ is closed in $\widehat{E}$. As $\sum_{b \in B} \widehat{K}b$ contains the dense subspace E,

$$\widehat{E} = \sum_{b \in B} \widehat{K}b. \bullet$$

19.2 Theorem. *Let K be a field, $\mathcal{T}_V$ the topology on K given by a proper absolute value V of K. Let L be a simple algebraic extension of K, c an element of L such that $L = K(c)$, and f the minimal polynomial of c over K. There is a bijection $g \to \mathcal{T}_g$ from the set $D(f)$ of all monic divisors of f in $\widehat{K}[X]$ to the set of all ring topologies on L inducing $\mathcal{T}_V$ on K such that for all g, $h \in D(f)$, $g \mid h$ if and only if $\mathcal{T}_g \subseteq \mathcal{T}_h$. For each $g \in D(f)$, the completion $\widehat{L}_g$ of L for $\mathcal{T}_g$ is a $\widehat{K}$-algebra generated by 1 and c, and g is the minimal polynomial of c in $\widehat{L}_g$. In particular,*

$$\dim_{\widehat{K}} \widehat{L}_g = \deg g.$$

Each ring topology on L inducing $\mathcal{T}_V$ on K is normable and hence is a field topology. The topologies on L defined by proper absolute values extending V are precisely the topologies $\mathcal{T}_p$ where p is a prime polynomial in $\widehat{K}[X]$ belonging to $D(f)$.

Proof. For each $g \in D(f)$, let A_g be the $\widehat{K}$-algebra $\widehat{K}[X]/(g)$, and let $c_g = X + (g) \in A_g$. Clearly $A_g = \widehat{K}[c_g]$, and the minimal polynomial over $\widehat{K}$ of c_g is g. Since $g \mid f$ in $\widehat{K}[X]$, $f(c_g) = 0$; but as f is a prime polynomial over K, f is the minimal polynomial of c_g over K. Thus there is a unique K-isomorphism u_g from L to $K[c_g]$ satisfying $u_g(c) = c_g$. By 15.15 there is a unique Hausdorff topology on A_g making it a $\widehat{K}$-topological algebra; that topology is complete by (1) of 15.12, defined by a $\widehat{K}$-algebra norm by 16.7, and hence has continuous inversion by 11.12. We define $\mathcal{T}_g$ to be the topology on L making u_g a homeomorphism from L to $K[c_g]$, furnished with the topology it inherits from A_g, and we shall denote by L_g the field L furnished with topology $\mathcal{T}_g$. Clearly $\mathcal{T}_g$ is a field topology that induces $\mathcal{T}_V$ on K. By 19.1, $K[c_g]$ is dense in A_g, so there is by 8.7 and 8.4 there is a unique topological $\widehat{K}$-isomorphism $\widehat{u}_g$ from $\widehat{L}_g$ to A_g extending u_g. Since $\widehat{u}_g(c) = c_g$, the minimal polynomial over $\widehat{K}$ of c is g. Consequently,

$$\deg_{\widehat{K}} \widehat{L}_g = \deg g.$$

Assume that $\mathcal{T}_g \subseteq \mathcal{T}_h$. The identity mapping from L_h to L_g is then continuous and hence is the restriction of a continuous $\widehat{K}$-homomorphism w from $\widehat{L}_h$ to $\widehat{L}_g$. Thus k, defined by

$$k = \widehat{u}_g \circ w \circ \widehat{u}_h^{-1},$$

is a continuous $\widehat{K}$-homomorphism from A_h to A_g taking c_h into c_g. Consequently, as $h(c_h) = 0$,

$$0 = k(h(c_h)) = h(k(c_h)) = h(c_g),$$

so the minimal polynomial g of c_g divides h. In particular, if $\mathcal{T}_g = \mathcal{T}_h$, then $g = h$.

Conversely, assume that $g \mid h$. The canonical epimorphism from $A_h = \widehat{K}[X]/(h)$ to $A_g = \widehat{K}[X]/(g)$ is $\widehat{K}$-linear and hence continuous by (2) of 15.12, and takes c_h into c_g. Its restriction q to the subfield $K[c_h]$ of A_h is therefore a continuous isomorphism from $K[c_h]$ to $K[c_g]$ satisfying $q(c_h) = c_g$. Hence $u_g^{-1} \circ q \circ u_h$ is the identity map of L and is continuous from L_h to L_g. Thus $\mathcal{T}_g \subseteq \mathcal{T}_h$.

Let $\mathcal{T}$ be a ring topology on L inducing $\mathcal{T}_V$ on K, and let $\widehat{L}$ be the completion of L for $\mathcal{T}$. Then $\widehat{L}$ is a topological $\widehat{K}$-algebra, and by 19.1, $\widehat{L} = \widehat{K}[c]$. The minimal polynomial g of c over $\widehat{L}$ divides f in $\widehat{K}[X]$ and hence belongs to $D(f)$. Thus there is a unique $\widehat{K}$-isomorphism from $\widehat{L}$ into A_g taking c to c_g, and that isomorphism is a topological isomorphism by (2) of 15.12; its restriction to L is clearly u_g, so $\mathcal{T} = \mathcal{T}_g$.

Since $\widehat{L}_g$ is $\widehat{K}$-isomorphic to $A_g = \widehat{K}[X]/(g)$, $\widehat{L}_g$ is a field if and only if g is a prime factor of f in $\widehat{K}[X]$. Thus a topology on L defined by an absolute value extending V is necessarily one of the topologies $\mathcal{T}_p$ where p is a prime factor of f in $\widehat{K}[X]$, by 13.10. Conversely, if p is a prime factor of f in $\widehat{K}[X]$, the unique Hausdorff topology on A_p making it a $\widehat{K}$-topological algebra is given by an absolute value extending $\widehat{V}$ by 16.8. •

If L is an extension field of a field K, we shall frequently denote $\dim_K L$ by $[L : K]$.

19.3 Theorem. *Let L be a simple algebraic extension of a field K, and let V be a proper absolute value on K [a proper real valuation of K]. There are only finitely many absolute values [real valuations] $V_1, \ldots, V_m$ on L extending V, any two of them are inequivalent, and*

$$(1) \qquad \sum_{k=1}^{m} [\widehat{L}_k : \widehat{K}] \leq [L : K]$$

where $\widehat{L}_k$ is the completion of L for V_k. If L is a separable extension of K, and if $\mathcal{T}_i$ is the topological defined by V_i for each $i \in [1, m]$, then equality holds in (1), and for each ring topology $\mathcal{T}$ on L that induces the topology $\mathcal{T}_V$ defined by V on K, there is a unique nonempty subset J of $[1, m]$ such that

$$\mathcal{T} = \sup_{k \in J} \mathcal{T}_k.$$

Proof. The assertion for a real valuation V follows directly from that for an absolute value by replacing V with an absolute value of V. Consequently, we shall consider only the absolute value case. Let $c \in L$ be such that $L = K(c)$, and let f be the minimal polynomial of c over K. Let $(p_k)_{1 \le k \le m}$ be the distinct prime polynomials in $\widehat{K}[X]$ that divide f in $\widehat{K}[X]$. Then their product divides f in $\widehat{K}[X]$, so by 19.2

$$(1) \qquad \sum_{k=1}^{m} [\widehat{L}_k : \widehat{K}] = \sum_{k=1}^{m} \deg p_k \le \deg f = [L : K].$$

Equivalent absolute values A and B on a field L that induce the same proper absolute value V on a subfield K are identical, however, for there exists $t \in K$ such that $V(t) > 1$, and by 1.10 there exists $r > 0$ such that $B = A^r$, so

$$A(t)^r = B(t) = V(t) = A(t),$$

whence $r = 1$. Thus there are exactly m absolute values that extend V.

Assume henceforth that L is a separable extension of K. Then f is the product of $(p_k)_{1 \le k \le m}$ of distinct prime polynomials in $\widehat{K}[X]$, so equality holds in (1).

If H and J are distinct nonempty subsets of $[1, m]$, then

$$\sup_{k \in H} \mathcal{T}_{p_k} \ne \sup_{k \in J} \mathcal{T}_{p_k}.$$

Indeed, if they were identical and if, for example, $r \in H \setminus J$, there would exist for each $k \in J$ an open neighborhood U_k of zero for $\mathcal{T}_{p_k}$ such that the unit ball B_r of L for $\mathcal{T}_{p_r}$ would contain $\cap_{k \in J} U_k$, and consequently any nonempty subset U_r of L open for $\mathcal{T}_{p_r}$ and disjoint from B_r would be disjoint from $\cap_{k \in J} U_k$, in contradiction to (1) of 13.12.

Consequently, the topologies $\sup_{k \in J} \mathcal{T}_{p_k}$ where J is a nonempty subset of $[1, m]$ are $2^m - 1$ in number and thus, by 19.2, are all the ring topologies on L extending $\mathcal{T}_V$. •

Since an inseparable finite-dimensional extension L of a field K of prime characteristic is a purely inseparable extension of the separable closure (or largest separable extension) of K in L, a discussion of the extensions to L of a real valuation of K is reduced to the separable case by virtue of the following theorem:

19.4 Theorem. *If L is a finite-dimensional purely inseparable extension of a field K of prime characteristic p and if v is a real valuation of K, there is exactly one real valuation w of L extending v, defined by*

$$w(x) = p^{-n} v(x^{p^n}),$$

where $p^n = [L:K]$, for all $x \in L$.

Proof. The function w is well defined since $x^{p^n} \in K$ for all $x \in L$, and clearly w extends v. Moreover, for all $x, y \in L$,

$$w(x+y) = p^{-n}v((x+y)^{p^n}) = p^{-n}v(x^{p^n} + y^{p^n})$$
$$\geq p^{-n}\inf\{v(x^{p^n})v(y^{p^n})\} = \inf\{w(x), w(y)\}$$
$$w(xy) = p^{-n}v((xy)^{p^n}) = p^{-n}v(x^{p^n}y^{p^n})$$
$$= p^{-n}(v(x^{p^n}) + v(y^{p^n})) = w(x) + w(y). \bullet$$

To illustrate 19.2 and 19.3 let $K = \mathbb{Q}$ furnished with the restriction to $\mathbb{Q}$ of $|..|_\infty$, and let $L = \mathbb{Q}(\sqrt{2})$. Then $f = X^2 - 2$, and $f = pq$ in $\mathbb{R}[X] = \widehat{\mathbb{Q}}[X]$, where $p = X - \sqrt{2}$ and $q = X + \sqrt{2}$. Thus $u_p^{-1}(c_p) = \sqrt{2}$ and $u_q^{-1}(c_q) = -\sqrt{2}$, so the topologies T_p and T_q are defined respectively by absolute values $|..|_p$ and $|..|_q$, where $|a + b\sqrt{2}|_p = |a + b\sqrt{2}|_\infty$ and $|a + b\sqrt{2}|_q = |a - b\sqrt{2}|_\infty$ for all $a, b \in \mathbb{Q}$. Consequently, T_f is defined by the norm $\sup\{|..|_p, |..|_q\}$.

Let v be a real valuation of a field K with valuation ring A_v, maximal ideal M_v, and residue field k_v ($= A_v/M_v$). Let v' an extension of v to a larger field L, with $A_{v'}$, $M_{v'}$, and $k_{v'}$ similarly defined. Then $\phi_{v',v} : k_v \to k_{v'}$, defined by

$$\phi_{v',v}(x + M_v) = x + M_{v'}$$

for all $x \in A_v$, is a monomorphism from k_v to $k_{v'}$, called the *canonical embedding* of k_v in $k_{v'}$. We also regard $k_{v'}$ as a vector space over k_v under the scalar multiplication

$$(x + M_v).(y' + M_{v'}) = xy' + M_{v'}$$

for all $x \in A_v$ and all $y' \in A_{v'}$. It is easy to verify that the scalar multiplication is, indeed, well defined and converts $k_{v'}$ into a k_v-vector space.

19.5 Definition. *Let v be a real valuation of a field K, v' a real valuation of a larger field K' that extends v. Let k and k' be respectively the residue fields of v and v', and let G and G' be respectively the value groups of v and v'. The index $(G':G)$ of G in G' is called the* **ramification index** *of v' over v and is denoted by $e(v'/v)$. The dimension of the k-vector space k' is called the* **residue class degree** *of v' over v and is denoted by $f(v'/v)$.*

19.6 Theorem. *Let K, K', and K'' be fields such that $K \subseteq K' \subseteq K''$, and let v'' be a real valuation of K'', v' and v its restrictions to K' and K respectively. Then*

$$e(v''/v')e(v'/v) = e(v''/v),$$
$$f(v''/v')f(v'/v) = f(v''/v).$$

The equalities are apparent.

19.7 Theorem. *Let v be a real valuation of a field K, $\hat{v}$ its continuous extension to $\hat{K}$. Then*
$$e(\hat{v}/v) = 1 = f(\hat{v}/v).$$
If v' is a real valuation extending v to a larger field L and $\hat{v}'$ its continuous extension to $\hat{L}$, then
$$e(\hat{v}'/\hat{v}) = e(v'/v)$$
$$f(\hat{v}'/\hat{v}) = f(v'/v).$$

Proof. The first equality is a consequence of 17.12. The other two follow from the first and 19.6, for
$$e(v'/v) = e(\hat{v}'/v')e(v'/v) = e(\hat{v}'/v)$$
$$= e(\hat{v}'/\hat{v})e(\hat{v}/v) = e(\hat{v}'/\hat{v}),$$
and similarly $f(v'/v) = f(\hat{v}'/\hat{v})$. •

Let v be a real valuation of a field K with valuation ring A and residue field k. For any $x \in A$ we shall frequently denote by $\bar{x}$ its image under the canonical epimorphism from A to k, and if $f = \sum_{k=0}^{n} a_k X^k \in A[X]$, we shall similarly denote by $\bar{f}$ the polynomial $\sum_{k=0}^{n} \bar{a}_k X^k \in k[X]$.

19.8 Theorem. *Let K be a field, L a finite-dimensional extension of K, v a real valuation of K, v' a real valuation of L extending K. Then $e(v'/v)$ and $f(v'/v)$ are finite, and*
$$e(v'/v)f(v'/v) \leq [L:K].$$

Proof. Let $n = [L:K]$. Let A_v and $A_{v'}$ be the valuation rings of v and v' respectively, k_v and $k_{v'}$ their residue fields, G and G' their value groups. Let r and s be any positive integers not exceeding $e(v'/v)$ and $f(v'/v)$ respectively. It suffices to show that $rs \leq n$.

There exist $x_1, \ldots, x_r \in L^*$ such that $v(x_i) - v(x_j) \notin G$ whenever $i \neq j$, and there exist $y_1, \ldots, y_s \in A_{v'}$ such that their images in $k_{v'}$ form a linearly independent set over k_v (in particular, $v'(y_k) = 0$ for all $k \in [1, s]$). To show that $rs \leq n$, we need only show that $\{x_i y_j : i \in [1, r], j \in [1, s]\}$ is linearly independent over K. Assume that

(2) $$\sum_{i=1}^{r}\sum_{j=1}^{s} a_{ij} x_i y_j = 0$$

where $a_{ij} \in K$ but not all $a_{ij} = 0$. Let $p \in [1, r]$ and $q \in [1, s]$ be such that

$$v'(a_{pq}x_p) = \inf\{v'(a_{ij}x_i) : i \in [1,r], j \in [1,s]\}.$$

By our assumption, $a_{pq} \neq 0$. If $i \neq p$, then

$$v'(a_{pq}x_p) < v'(a_{ij}x_i)$$

for each $j \in [1, s]$, for otherwise

$$v'(x_i) - v'(x_p) = v(a_{pq}) - v(a_{ij}) \in G,$$

a contradiction. Let

$$b_{ij} = (a_{ij}x_i)(a_{pq}x_p)^{-1}.$$

From (2) we obtain

$$\sum_{j=1}^{s} b_{pj}y_j + \sum_{i \neq p}\sum_{j=1}^{s} b_{ij}y_j = 0,$$

and

$$v'(b_{ij}y_j) = v'(b_{ij}) > 0$$

if $i \neq p$,

$$v'(b_{pj}y_j) = v'(b_{pj}) \geq 0$$

for all $j \in [1, s]$. Therefore in $k_{v'}$

$$\sum_{j=1}^{s} \overline{b}_{pj}\overline{y}_j = \overline{0},$$

whereas $\overline{b}_{pq} = \overline{1}$, in contradiction to the linear independence of $\overline{y}_1, \ldots, \overline{y}_s$. ●

In an important case, the inequality of 19.8 is an equality:

19.9 Theorem. *Let v be a complete discrete valuation of a field K, let L be an extension field of K, and let v' be an extension of v to L such that $e(v'/v) < +\infty$ and $f(v'/v) < +\infty$. Then L is a finite-dimensional extension of K, and*

$$e(v'/v)f(v'/v) = [L : K].$$

Proof. We abbreviate $e(v'/v)$ and $f(v'/v)$ to e and f respectively. We may assume that the value group of v is $\mathbb{Z}$; let G' be the value group of v'. As $(G' : \mathbb{Z}) = e$, $eG' \subseteq \mathbb{Z}$, and hence $G' \subseteq (1/e)\mathbb{Z}$. Thus $G' = (1/e)\mathbb{Z}$ as no other group H between $\mathbb{Z}$ and $(1/e)\mathbb{Z}$ satisfies $(H:\mathbb{Z})=e$. In particular, v' is discrete.

Let u and t be uniformizers of v' and v respectively, and for each $n \in \mathbb{Z}$, let $z_n = t^q u^r$ where $n = qe + r$ and $0 \le r < e$. Then $v'(z_n) = n/e$.

Let $A_{v'}$ and $M_{v'}$ be the valuation ring and ideal respectively of v'. By hypothesis there exist $b_1, \ldots, b_f \in A_{v'} \setminus M_{v'}$ whose images in the residue field $k_{v'}$ of v' form a basis of $k_{v'}$ over the residue field k_v of v. Let S be a representative set for v. Then S', defined by

$$S' = \{\sum_{i=1}^{f} s_i b_i : s_i \in S \text{ for all } i \in [1, f]\},$$

is a representative set for v'. By 19.8 it suffices to show that $\{b_i u^r : 1 \le i \le f, 0 \le r < e\}$ generates the K-vector space L.

Let $x \in L$. By (1) of 18.5 there exists a family $(c_n)_{n \in \mathbb{Z}}$ of elements of S' such that $c_n = 0$ for all but finitely many $n < 0$ and

$$x = \sum_{n \in \mathbb{Z}} c_n z_n.$$

By the definition of S', for each $q \in \mathbb{Z}$ and each $r \in [0, e-1]$ there exists a sequence $(s_{q,r,i})_{1 \le i \le f}$ in S such that

$$c_{qe+r} = \sum_{i=1}^{f} s_{q,r,i} b_i,$$

and moreover, for all but finitely many $q < 0$, $s_{q,r,i} = 0$ for all $r \in [0, e-1]$ and all $i \in [1, f]$. Thus by (2) of 18.5, for each such r and i, $(s_{q,r,i} t^q)_{q \in \mathbb{Z}}$ is summable in K; let

$$a_{r,i} = \sum_{q \in \mathbb{Z}} s_{q,r,i} t^q \in K.$$

By 10.12, 10.8, and 10.2,

$$\sum_{r=0}^{e-1} \sum_{i=1}^{f} a_{r,i} b_i u^r = \sum_{r=0}^{e-1} \sum_{i=1}^{f} (\sum_{q \in \mathbb{Z}} s_{q,r,i} t^q) b_i u^r = \sum_{q \in \mathbb{Z}} \sum_{r=0}^{e-1} \sum_{i=1}^{f} (\sum s_{q,r,i} b_i) t^q u^r$$

$$= \sum_{q \in \mathbb{Z}} \sum_{r=0}^{e-1} c_{qe+r} z_{qe+r} = \sum_{n \in \mathbb{Z}} c_n z_n = x. \bullet$$

19.10 Theorem. *Let v be a discrete valuation of a field K, and let L be a simple algebraic extension of K. There are only finitely many valuations $v'_1, \ldots, v'_m$ of L extending v, and*

$$(3) \qquad \sum_{k=1}^{m} e(v'_k/v) f(v'_k/v) \leq [L:K]$$

Further, if L is a separable extension of K, equality holds in (3).

Proof. By 19.3 there are only finitely many extensions. Let $\widehat{L}_k$ be the completion of L for v'_k. Then

$$\sum_{k=1}^{m} e(v'_k/v) f(v'_k/v) = \sum_{k=1}^{m} e(\hat{v}'_k/\hat{v}) f(\hat{v}'_k/\hat{v}) = \sum_{k=1}^{m} [\widehat{L}_k : \widehat{K}]$$

by Theorems 19.7 and 19.9, so the assertions hold by 19.3. •

A real valuation v of a field K induces in a natural way a real valuation $\bar{v}$ of the field $K(X)$ of fractions over K:

19.11 Theorem. *If v is a real valuation of a field K, there is a valuation $\bar{v}$ of $K(X)$ satisfying*

$$\bar{v}(\sum_{k=0}^{n} a_k X^k) = \inf\{v(a_k) : k \in [0, n]\}$$

for every polynomial $\sum_{k=0}^{n} a_k X^k$ over K.

Proof. The function w with domain $K[X]$ defined by

$$w(\sum_{k=0}^{n} a_k X^k) = \inf\{v(a_k) : k \in [0, n]\}$$

is easily seen to be a real valuation of $K[X]$. We extend it to a function $\bar{v}$ on $K(X)$ by

$$\bar{v}(f/g) = w(f) - w(g)$$

for all polynomials f, g over K such that $g \neq 0$, and it is easy to see that $\bar{v}$ is a real valuation of $K(X)$. •

19.12 Theorem. *Let v be a real valuation of a field K, and let A be its valuation ring. If f is a monic polynomial over A and if $f = g_1 \ldots g_n$ where $g, \ldots, g_n$ are polynomials over K, then there exist $a_1, \ldots, a_n \in K^*$ such that if $g'_i = a_i g_i$ for each $i \in [1,n]$, then each g'_i is a polynomial over A, and $f = g'_1 \ldots g'_n$.*

Proof. For each $i \in [1, n-1]$, let $b_i \in K^*$ be such that $v(b_i) = \overline{v}(g_i)$. We need only let $a_i = b_i^{-1}$ for all $i \in [1, n-1]$ and $a_n = b_1 b_2 \ldots b_{n-1}$. Then

$$v(a_n) = \sum_{i=0}^{n-1} \overline{v}(g_i) = \overline{v}(f) - \overline{v}(g_n) = -\overline{v}(g_n),$$

so $\overline{v}(g_i) = 0$ and hence $g_i \in A[X]$ for all $i \in [1,n]$. •

19.13 Corollary. *Let v be a real valuation of a field K, and let A be its valuation ring. If f is a monic polynomial over A that is an irreducible element of $A[X]$, then f is a prime polynomial over K.*

Proof. If f were not a prime polynomial over K, then there would exist nonconstant polynomials g and h over K such that $f = gh$, and hence by 19.12 there would exist nonconstant polynomials $g', h' \in A[X]$ such that $f = g'h'$, a contradiction of our hypothesis that f is irreducible in $A[X]$.•

19.14 Theorem. *Let v be a valuation of a field K. If k'_1 is a finite-dimensional [separable] extension of the residue field k of v, there exist a finite-dimensional [separable] extension K' of K and a valuation v' of K' extending v such that the residue field k' of v' is k-isomorphic to k'_1, $f(v'/v) = [K' : K]$, and $e(v'/v) = 1$.*

Proof. By induction and the theorem of the primitive element, we may assume that $k'_1 = k[\alpha]$ where α is [separable] algebraic over k. Let f be a monic polynomial over the valuation ring A of v such that $\overline{f}$ is the minimal polynomial of α. In particular, $\overline{f}$ is a prime polynomial over k. If $f = gh$ where g and h are nonunits of $A[X]$, then their leading coefficients are invertible elements of A, so neither is a constant polynomial; hence $\overline{f} = \overline{g}\overline{h}$ where neither $\overline{g}$ nor $\overline{h}$ is a constant polynomial, a contradiction. Thus f is an irreducible element of $A[X]$ and hence, by 19.13, a prime polynomial over K. Moreover, if $\overline{f}$ is separable, then its derivative $D\overline{f} \neq 0$, so $\overline{Df} = D\overline{f} \neq 0$, hence $Df \neq 0$, and therefore f is separable.

Let $K' = K[a]$ where a is a root of f. It follows at once from 16.10 (or from 19.2) that there is a real valuation v' of K' extending v; let A' be its valuation ring, k' its residue field. If $v'(a) < 0$, then $v'(g(a)) = nv'(a) < 0$ for any monic polynomial g of degree n over A by 17.2; consequently as

$f(a) = 0$, $v'(a) \geq 0$ and thus $a \in A'$. As $\bar{f}$ is prime, $\bar{f}$ is the minimal polynomial of $\bar{a} \in k'$ since $\bar{f}(\bar{a}) = \overline{f(a)} = \bar{0}$. Therefore there is a k-isomorphism σ from $k[\bar{a}]$ to $k[\alpha]$, and

$$[k[\bar{a}] : k] = \deg \bar{f} = \deg f = [K' : K].$$

Consequently by 19.8, $k[\bar{a}] = k'$, and $e(v'/v) = 1$. •

Our final result extends 19.14:

19.15 Theorem. *Let v be a proper real valuation of a field K with residue field k_v, and let ψ be a monomorphism from k_v to a field Ω. There exist an extension field H of K, a real valuation w of H extending v, and an isomorphism Ψ from the residue field k_w of w to Ω such that $e(w/v) = 1$ and $\Psi \circ \phi_{w,v} = \psi$.*

Proof. Let G be the value group of v, and let E be a set such that $\text{card}(E) > \text{card}(\Omega^G)$. Let $\mathcal{L}$ be the set of all (F, u, σ) such that F is an extension field of K, the set F is a subset of E, u is a real valuation of F extending v whose value group is G, and σ is an monomorphism from the residue field k_u of u to Ω such that $\sigma \circ \phi_{u,v} = \psi$. Clearly $\mathcal{L}$ is a set, and $(K, v, \psi) \in \mathcal{L}$. The relation $\preccurlyeq$ on $\mathcal{L}$ satisfying

$$(F_1, u_1, \sigma_1) \preccurlyeq (F_2, u_2, \sigma_2)$$

if and only if F_2 is an extension field of F_1, u_2 is an extension of u_1, and $\sigma_2 \circ \phi_{u_2,u_1} = \sigma_1$, is easily seen to be an ordering. To show that $(\mathcal{L}, \preccurlyeq)$ is inductive, let $\{(F_\lambda, u_\lambda, \sigma_\lambda) : \lambda \in L\}$ be a totally ordered subset of $\mathcal{L}$, indexed by a set L, let A_λ and M_λ be the valuation ring and maximal ideal of each u_λ, and let $\leq$ be the total ordering on L satisfying $\alpha \leq \beta$ if and only if $(F_\alpha, u_\alpha, \sigma_\alpha) \preccurlyeq (F_\beta, u_\beta, \sigma_\beta)$. Let

$$F = \bigcup_{\lambda \in L} F_\lambda,$$

a subset of E. There are unique compositions, addition and multiplication, on F that are extensions of addition and multiplication on each F_λ, and with them F is a field. Since G is the value group of each u_λ, there is a unique real valuation u of F that extends each u_λ, and its value group is G. The valuation ring A and maximal ideal M of u are then given by

$$A = \bigcup_{\lambda \in L} A_\lambda \qquad M = \bigcup_{\lambda \in L} M_\lambda,$$

and moreover, $M \cap A_\lambda = M_\lambda$ for all $\lambda \in L$. We define $\sigma : A/M \to \Omega$ as follows: For each $x \in A$, let $\lambda \in L$ be such that $x \in A_\lambda$, and define $\sigma(x+M)$ to be $\sigma_\lambda(x+M_\lambda)$. This is well defined, for if also $x \in A_\mu$ where $\lambda < \mu$, then

$$\sigma_\lambda(x+M_\lambda) = (\sigma_\mu \circ \phi_{v_\mu, v_\lambda})(x+M_\lambda) = \sigma_\mu(x+M_\mu).$$

Clearly σ is a monomorphism from A/M to Ω, and $\sigma \circ \phi_{u,u_\lambda} = \sigma_\lambda$ for all $\lambda \in L$. In particular, let $\lambda \in L$; then

$$\sigma \circ \phi_{u,v} = \sigma \circ \phi_{u,u_\lambda} \circ \phi_{u_\lambda,v} = \sigma_\lambda \circ \phi_{u_\lambda,v} = \psi.$$

Thus $(F, u, \sigma) \in \mathcal{L}$, and clearly

$$(F, u, \sigma) = \sup_{\lambda \in L}(F_\lambda, u_\lambda, \sigma_\lambda).$$

By Zorn's Lemma, therefore, $\mathcal{L}$ contains a maximal member (H, w, Ψ). Let A_w and k_w be respectively the valuation ring and residue field of w. We need only show that $\Psi(k_w) = \Omega$. By 17.15,

$$\text{card}(H) \leq \text{card}(k_w^G) \leq \text{card}(\Omega^G) < \text{card}(E)$$

and hence $\text{card}(E \setminus H) = \text{card}(E)$. Suppose there exists $t \in \Omega \setminus \Psi(k_w)$.

Case 1: t is transcendental over $\Psi(k_w)$. Now

$$\text{card}(H(X) \setminus H) = \aleph_0 \text{card}(H) = \sup\{\aleph_0, \text{card}(H)\} < \text{card}(E \setminus H).$$

Consequently, by a set-theoretic "push-out," there is a field extension $H(x)$ of H such that the set $H(x)$ is contained in E and x is transcendental over H. By 19.11 there is a valuation $\overline{w}$ of $H(x)$ satisfying

$$\overline{w}(\sum_{j=0}^{n} a_j x^j) = \inf\{w(a_j) : j \in [0, n]\}$$

for all $a_0, a_1, \ldots, a_n \in H$. The value group of $\overline{w}$ is thus G. By definition, $\overline{w}(x) = 0$, so $\overline{x} \neq \overline{0}$ in the residue field $k_{\overline{w}}$ of $\overline{w}$. Moreover, $\overline{x}$ is transcendental over the subfield $\phi_{\overline{w},w}(k_w)$ of $k_{\overline{w}}$. Indeed, let $a_0, a_1, \ldots, a_n \in A_w$. If

$$\sum_{j=0}^{n} \phi_{\overline{w},w}(\overline{a}_j)\overline{x}^j = \overline{0},$$

then

$$\sum_{j=0}^{n} a_j x^j = \overline{0},$$

so
$$\overline{w}(\sum_{j=0}^{n} a_j x^j) > 0,$$

thus $w(a_j) > 0$ for all $j \in [0,n]$, and hence $\overline{a}_j = \overline{0}$ for all $j \in [0,n]$. It is easy to see that $k_{\overline{w}}$ is the field $[\phi_{\overline{w},w}(k_w)](\overline{x})$ of rational functions in the transcendental element $\overline{x}$ over $\phi_{\overline{w},w}(k_w)$. As $\Psi \circ \phi_{\overline{w},w}^{-1}$ is an isomorphism from $\phi_{\overline{w},w}(k_w)$ to $\Psi(k_w)$, there is a unique isomorphism $\overline{\Psi}$ from $k_{\overline{w}}$ to $[\Psi(k_w)](t)$ satisfying $\overline{\Psi} \circ \phi_{\overline{w},w} = \Psi$ and $\overline{\Psi}(\overline{x}) = t$. Thus $(H(x), \overline{w}, \overline{\Psi}) \in \mathcal{L}$ and $(H(x), \overline{w}, \overline{\Psi}) \succ (H, w, \Psi)$, a contradiction.

Case 2: t is algebraic over $\Psi(k_w)$. If H_0 is an n-dimensional field extension of H, then as in Case 1 $\mathrm{card}(H_0 \setminus H) < \mathrm{card}(E \setminus H)$, so there is a bijection τ from H_0 to a subset H_1 of E containing H such that $\tau(x) = x$ for all $x \in H$, and consequently H_1 may be made into a field such that τ is an H-isomorphism from H_0 to H_1. Therefore by 19.4 there exist a field extension H_1 of H such that $H_1 \subset E$, a real valuation w_1 of H_1 extending w whose value group is G, and an isomorphism Ψ_1 from the residue field k_{w_1} of w_1 to $\Psi(k_w)(t)$ such that $\Psi_1 \circ \phi_{w_1,w} = \Psi$. Consequently, $(H_1, w_1, \Psi_1) \in \mathcal{L}$ and $(H_1, w_1, \Psi_1) \succ (H, w, \Psi)$, a contradiction. •

Exercises

In Exercises 19.1–19.3 A is a proper absolute value of a field K, $\mathcal{T}$ is the topology defined by A, L is the field extension $K[c]$ where c is algebraic over K, $n = [L:K]$, and f is the minimal polynomial of c over K.

19.1 The following statements are equivalent:

1° There are n absolute values on L extending A.
2° There are $2^n - 1$ ring topologies on L inducing $\mathcal{T}$ on K.
3° L is a separable extension of K, and $\widehat{K}$ contains a splitting field of f.

19.2 The following statements are equivalent:

1° There are n ring topologies on L inducing $\mathcal{T}$ on K, but there is only one absolute value on L extending A.
2° There are n ring topologies on L inducing $\mathcal{T}$ on K, and they are totally ordered by inclusion.
3° c is purely inseparable over $\widehat{K}$.

19.3 The completion of L for some ring topology on L inducing $\mathcal{T}$ on K has a nonzero nilpotent if and only if there is a prime polynomial p over $\widehat{K}$ such that $p^2 \mid f$ in $\widehat{K}[X]$.

19.4 If v is a complete proper real valuation of a field K and if L is an algebraic extension of K, there is a unique real valuation of L extending v. [Use Exercise 16.1.]

19.5 If v is a real valuation of an algebraically closed field K, then the residue field of v is algebraically closed and the value group of v is the additive group of a $\mathbb{Q}$-vector space.

19.6 (Ostrowski [1932]) Let Ω be an algebraic closure of the 2-adic field $\mathbb{Q}_2$, and let $(c_n)_{n\geq 0}$ be a sequence of elements of Ω satisfying $c_0 = 2$, $c_n^2 = c_{n-1}$ for all $n \geq 1$. Let $K_n = \mathbb{Q}_2(c_n)$ for each $n \in \mathbb{N}$, and let

$$K = \bigcup_{n=0}^{\infty} K_n.$$

(a) For all $n \in \mathbb{N}$, $[K : \mathbb{Q}_2] = 2^n$. (b) There is a unique real valuation v of K extending the 2-adic valuation of $\mathbb{Q}_2 = K_0$. [Use Exercise 19.4.] The value group of the restriction v_n of v to K_n is $2^n.\mathbb{Z}$, and the value group G of v satisfies $2.G = G$. (c) $[K(\sqrt{3}) : K] = 2$. [Use (b) in showing $[K_n(\sqrt{3}) : K_n] = 2$ for all $n \in \mathbb{N}$.] (d) There is a unique valuation w of $K(\sqrt{3})$ extending v_0, and for all $a, b \in K$, $w(a+b\sqrt{3}) = w(a-b\sqrt{3})$. (e) $e(w/v) = 1$. (f) $w(s_n - \sqrt{3}) = 1 - 2^{-(n+1)}$ where $s_n = 1 + 2(c_1^{-1} + c_2^{-1} + \ldots + c_n^{-1})$. [Calculate $v(s_n^2 - 3)$ by expanding s_n^2, and use (d).] (g) $f(w, v) = 1$. [If $w(a + b\sqrt{3}) \geq 0$ where $a, b \in K_n$, use (d) to show that $w(a) \geq -1$ and $w(b) \geq -1$. If the restriction w_n of w to $K_n(\sqrt{3})$ satisfies $f(w_n/v_n) = 2$, use (b), (d), and (f) to show that $w((a+b\sqrt{3})-(a-b\sqrt{3})) > 0$, and consider $a + bs_{n+1}$.]

CHAPTER V

COMPLETE LOCAL RINGS

Here we investigate an area of commutative algebra in which the topology determined by an ideal of a commutative ring, for which the powers of the ideal form a fundamental system of neighborhoods of zero, plays a crucial role. After an introductory discussion of local and noetherian rings in §20, we establish in §21 I. S. Cohen's fundamental theorem that a local noetherian ring complete for the topology determined by its maximal ideal contains a special type of subring, called here a Cohen subring. This theorem is applied in §22 to describe complete discrete valuation rings: such a ring is either the valuation ring of a formal power series field or is finitely generated as a module over a Cohen subring, which, in turn, is completely determined by its characteristic and residue field. In §23 we give characterizations of complete local noetherian rings, and after a general discussion of the topologies determined by an ideal, we show in §24 that a complete semilocal noetherian ring is the topological direct sum of complete local noetherian rings.

20 Noetherian Modules and Rings

Here we shall give some basic properties of noetherian rings and modules.

20.1 Definition. *Let A be a ring. An A-module E is* **noetherian** *if every submodule M of E is finitely generated, that is, if there exist $x_1, \ldots, x_n \in M$ such that $M = \mathbb{Z}.x_1 + Ax_1 + \ldots + \mathbb{Z}.x_n + Ax_n$. A ring is* **noetherian** *if it is noetherian as a left module over itself, that is, if every left ideal is finitely generated.*

If A is a ring with identity and E a unitary A-module, then E is noetherian if and only if for each submodule M of E there exist $x_1, \ldots, x_n \in M$ such that $M = Ax_1 + \cdots + Ax_n$.

20.2 Theorem. *If E is an A module, the following statements are equivalent:*

1° E is noetherian.

2° If $(M_n)_{n \geq 1}$ is any increasing sequence of submodules of E, there exists $q \geq 1$ such that $M_n = M_q$ for all $n \geq q$.

3° *Every nonempty set of submodules of E, ordered by inclusion, contains a maximal element.*

Proof. Assume 1°, let $(M_n)_{n \geq 1}$ be an increasing sequence of submodules of E, and let
$$M = \bigcup_{n=1}^{\infty} M_n.$$
Then M is a submodule, so by 1°, M is generated by a finite subset $\{x_1, \ldots, x_m\}$. Consequently, there exists $q \geq 1$ such that $\{x_1, \ldots, x_m\} \subseteq M_q$, so $M = M_q$, and hence $M_n = M_q$ for all $n \geq q$.

To show that 2° implies 3°, assume that $\mathcal{M}$ is a nonempty family of submodules of E that, ordered by inclusion, contains no maximal element. Then for each $M \in \mathcal{M}$, the set $\mathcal{S}_M$ of all submodules in $\mathcal{M}$ strictly containing M is nonempty. Consequently, by the Axiom of Choice, there is a function c from $\mathcal{M}$ into itself such that for each $M \in \mathcal{M}$, $c(M) \in \mathcal{S}_M$, whence $c(M) \supset M$. Let $M_1 \in \mathcal{M}$. The sequence $(c^n(M_1))_{n \geq 0}$ is thus a strictly increasing sequence of submodules, so 2° does not hold.

Finally, 3° implies 1°, for we need only apply 3° to the set of all finitely generated submodules of a given submodule of E. •

Thus, a ring is noetherian if every increasing sequence of left ideals is eventually stationary, or equivalently, if every nonempty set of left ideals, ordered by inclusion, contains a maximal member.

The statement of 2° is frequently called the *Ascending Chain Condition*.

20.3 Theorem. *If E is an A-module and F a submodule of E, then E is noetherian if and only if both F and E/F are noetherian.*

Proof. Necessity: Clearly every submodule of F is finitely generated. If Q is a submodule of E/F and if ϕ_F is the canonical epimorphism from E to E/F, then $\phi^{-1}(Q)$ is generated by finitely many elements $x_1, \ldots, x_q$, and hence Q, which is $\phi_F(\phi_F^{-1}(Q))$, is generated by $x_1 + F, \ldots, x_q + F$.

Sufficiency: Let M be a submodule of E. By hypothesis, $(M + F)/F$ and $M \cap F$ are finitely generated, so there exist $x_1, \ldots, x_n \in M$ and $x_{n+1}, \ldots, x_p \in M \cap F$ such that

$$\sum_{i=1}^{n} Ax_i + \mathbb{Z}.x_i + F = M + F, \qquad \sum_{i=n+1}^{p} Ax_i + \mathbb{Z}.x_i = M \cap F.$$

Then
$$\sum_{i=1}^{p} (Ax_i + \mathbb{Z}.x_i) = M,$$

for if $z \in M$, there exist $a_i, \ldots, a_n \in A$ and $q_1, \ldots, q_n \in \mathbb{Z}$ such that $z - \sum_{i=1}^{n}(a_i x_i + q_i.x_i)$ belongs to F and hence to $M \cap F$, and therefore there exist $a_{n+1}, \ldots, a_p \in A$ and $q_{n+1}, \ldots, q_p \in \mathbb{Z}$ such that

$$z - \sum_{i=1}^{n}(a_i x_i + q_i.x_i) = \sum_{i=n+1}^{p}(a_i x_i + q_i.x_i),$$

and consequently z belongs to the submodule generated by $x_1, \ldots, x_p$. •

20.4 Corollary. *If A is a noetherian ring and if J is an ideal of A, then A/J is a noetherian ring.*

Proof. The assertion follows from 20.3, since the left ideals of A/J are precisely the submodules of the A-module A/J. •

20.5 Corollary. *The sum of finitely many noetherian submodules of an A-module E is noetherian.*

Proof. By induction, it suffices to prove that the sum of two noetherian submodules is noetherian. If F_1 and F_2 are noetherian submodules, then $F_1/(F_1 \cap F_2)$ is noetherian by 20.3, so as $(F_1 + F_2)/F_2$ and $F_1/(F_1 \cap F_2)$ are isomorphic A-modules, $(F_1 + F_2)/F_2$ also is noetherian; therefore by 20.3, $F_1 + F_2$ is noetherian. •

20.6 Corollary. *The cartesian product of finitely many noetherian A-modules is noetherian.*

Proof. By induction, it suffices to show that if E and F are noetherian A-modules, so is $E \times F$. As the submodules $E \times \{0\}$ and $\{0\} \times F$ of $E \times F$ are respectively isomorphic to E and F, they are noetherian, and hence their sum $E \times F$ is noetherian. •

20.7 Theorem. *The cartesian product of finitely many noetherian rings is noetherian.*

Proof. By induction it suffices to show that if A and B are noetherian rings, so is $A \times B$. We make A and B into $(A \times B)$-modules by defining $(a,b)x = ax$ and $(a,b)y = by$ for all $a, x \in A$ and all $b, y \in B$. The left ideals of A and B respectively are then precisely the submodules of the $(A \times B)$-modules A and B. Consequently by 20.6 the $(A \times B)$-module $A \times B$ is noetherian, that is, the ring $A \times B$ is noetherian •

20.8 Theorem. *If A is a noetherian ring with identity, a unitary A-module E is noetherian if and only if E is finitely generated.*

Proof. Sufficiency: By 20.5 it suffices to show that if $x \in E$, then Ax is a noetherian A-module. But if $N = \{a \in A : ax = 0\}$, then Ax is isomorphic to the A-module A/N, which is noetherian by 20.3. •

We recall that an ideal P of a commutative ring with identity A is *prime* if P is a proper ideal and for all $a, b \in A$, if $ab \in P$, then either $a \in P$ or $b \in P$; equivalently, an ideal P is prime if and only if A/P is an integral domain.

20.9 Theorem. *If every prime ideal of a commutative ring with identity A is finitely generated, then A is noetherian.*

Proof. Suppose that the set $\mathcal{J}$ of ideals of A that are not finitely generated is nonempty. Ordered by inclusion, $\mathcal{J}$ is inductive, for if the union S of a totally ordered subset S of T were generated by a finite set F, some member of S would contain F, hence be identical with S and thus be finitely generated, a contradiction. By Zorn's Lemma, there is an ideal M that is maximal in $\mathcal{J}$ for the ordering defined by inclusion. To obtain a contradiction, we need only show that M is a prime ideal. Suppose that $ab \in M$ but that $a \notin M$ and $b \notin M$. By the maximality of M, $M + Ab$ is finitely generated, say, by $m_1 + x_1 b, \ldots, m_n + x_n b$ where $m_1, \ldots, m_n \in M$ and $x_1, \ldots, x_n \in A$. Let $J = \{x \in A : xb \in M\}$. Then J contains both M and a and hence is finitely generated, so $Am_1 + \cdots + Am_n + Jb$ is also. To show that M is contained in the latter ideal, let $z \in M \subset M + Ab$. Then there exist $y_1, \ldots, y_n \in A$ such that

$$z = \sum_{i=1}^{n} y_i(m_i + x_i b).$$

As

$$\sum_{i=1}^{n} y_i x_i b = z - \sum_{i=1}^{n} y_i m_i \in M,$$

$y_1 x_1 + \cdots + y_n x_n \in J$. Thus $z \in Am_1 + \cdots + Am_n + Jb$. But by definition of J, $Am_1 + \cdots + Am_n + Jb \subseteq M$. Therefore $M = Am_1 + \cdots + Am_n + Jb$, a finitely generated ideal. •

Henceforth we shall use the following notational convention: Let A be a ring, E an A-module. For any additive subgroup J of A and any additive subgroup F of E, we shall denote by JF the additive subgroup of E generated by all cx, where $c \in J$ and $x \in F$. Thus

$$JF = \{\sum_{i=1}^{n} c_i x_i : n \geq 1, \text{ and for all } i \in [1, n], c_i \in J \text{ and } x \in F\}.$$

This convention is applicable, in particular, to additive subgroups of A, regarded as a left module over itself. Thus, if I and J are additive subgroups of A, IJ is the additive subgroup of A generated by the elements ab, where $a \in I$ and $b \in J$. This notational convention for products of additive subgroups of a ring is, of course, in conflict with that introduced on page 13 where the composition $*$ is the multiplicative composition of a ring, but in any context it will be clear which convention is meant.

Let I, J, and K be additive subgroups of A, F and G additive subgroups of E. Clearly

$$(IJ)F = I(JF)$$
$$(I+J)F = IF + JF$$
$$I(F+G) = IF + IG.$$

If I is a left ideal, IF is a submodule. If $(I_\lambda)_{\lambda \in L}$ is a family of additive subgroups of A and F an additive subgroup of E, then

$$\left(\bigcap_{\lambda \in L} I_\lambda\right) F \subseteq \bigcap_{\lambda \in L} I_\lambda F$$

and if I is an additive subgroup of A and $(F_\lambda)_{\lambda \in L}$ a family of additive subgroups of E, then

$$I\left(\bigcap_{\lambda \in L} F_\lambda\right) \subseteq \bigcap_{\lambda \in L} IF_\lambda.$$

In particular, $(IJ)K = I(JK)$, IJ is a left ideal if I is, IJ is a right ideal if J is, and thus IJ is an ideal if I is a left ideal, J a right ideal.

20.10 Theorem. *Let A be a commutative ring with identity. If F is a finitely generated unitary A-module and if J is an ideal of A such that $JF = F$, then J contains an element a such that $(1-a)F = \{0\}$.*

Proof. Let $F = Ax_1 + \ldots + Ax_n$, let $F_k = Ax_k + \ldots Ax_n$ for each $k \in [1, n]$, and let $F_{n+1} = \{0\}$. We shall show inductively that for each $k \in [q, n+1]$ there exists $a_k \in J$ such that $(1-a_k)F \subseteq F_k$; then a_{n+1} is the desired a. Let $a_1 = 0$, and assume that there exists $a_k \in J$ such that $(1-a_k)F \subseteq F_k$ where $k \in [1, n]$. By hypothesis,

$$(1-a_k)F = (1-a_k)JF = J(1-a_k)F \subseteq JF_k,$$

so there exist $a_{kk}, \ldots a_{kn} \in J$ such that

$$(1-a_k)x_k = \sum_{j=k}^{n} a_{kj} x_j.$$

Thus $(1 - a_k - a_{kk})x_k \in F_{k+1}$, so we need only let

$$a_{k+1} = a_k + (a_k + a_{kk}) - (a_k + a_{kk})a_k \in J,$$

for then

$$(1-a_{k+1})F = (1 - a_k - a_{kk})(1-a_k)F \subseteq (1-a_k-a_{kk})F_k \subseteq F_{k+1}. \bullet$$

20.11 Theorem. *Let A be a commutative noetherian ring with identity, and let E be a finitely generated unitary A-module. If F is a submodule of E and J an ideal of A, there exists $n \geq 1$ such that $J^n E \cap F \subseteq JF$.*

Proof. The set $\mathcal{A}$ of submodules G of E such that $G \cap F = JF$ is nonempty since $JF \in \mathcal{A}$. Consequently, as E is noetherian by 20.8, $\mathcal{A}$ contains a maximal member H. We need only show that $J^n E \subseteq H$ for some $n \geq 1$. If $J = Aa_1 + \cdots + Aa_r$, and if $a_j^m E \subseteq H$ for each $j \in [1,r]$, then clearly $J^{rm} E \subseteq H$. Thus it suffices to show that if $a \in J$, then $a^m E \subseteq H$ for some $m \geq 1$.

Let $D_r = \{x \in E : a^r x \in H\}$ for each $r \geq 1$. Then $(D_r)_{r \geq 1}$ is an increasing sequence of submodules, so there exists $m \geq 1$ such that $D_{m+1} = D_m$. Clearly $(a^m E + H) \cap F \supseteq H \cap F = JF$. Conversely, let $y = a^m x + h \in F$ where $x \in E$, $h \in H$. Then

$$ay = a^{m+1} x + ah \in aF \subseteq JF \subseteq H,$$

so $a^{m+1} x \in H$. Thus $x \in D_{m+1} = D_m$, so $a^m x \in H$ and hence $y \in H$. Therefore $y \in H \cap F = JF$. Consequently, $(a^m E + H) \cap F = JF$, so by the maximality of H, $a^m E \subseteq H$. •

20.12 Corollary. *Let A be a commutative noetherian ring with identity, E a finitely generated unitary A-module. If J is an ideal of A and if $F = \bigcap_{n=1}^{\infty} J^n E$, then $JF = F$.*

Proof. By 20.11 there exists $n \geq 1$ such that

$$F = J^n E \cap F \subseteq JF \subseteq F,$$

so $F = JF$. •

20.13 Corollary. *Let A be a commutative noetherian ring with identity, and let E be a finitely generated unitary A-module. If J is an ideal of A, there exists $a \in J$ such that*

$$(1-a)\left(\bigcap_{n=1}^{\infty} J^n E\right) = \{0\}.$$

Conversely, if $x \in E$ satisfies $(1-a)x = 0$ for some $a \in J$, then

$$x \in \bigcap_{n=1}^{\infty} J^n E.$$

Proof. The first assertion follows from 20.12 and 20.10. If $(1-a)x = 0$ where $a \in J$, then $x = ax$, so by induction $x = a^n x \in J^n E$ for all $n \geq 1$. •

20.14 Definition. A **local ring** is a commutative ring with identity having only one maximal ideal, and a **local domain** is an integral domain that is a local ring. If A is a local ring with maximal ideal M, the **residue field** of A is the field A/M, and the **natural topology** of A is the ring topology for which the ideals $(M^n)_{n \geq 1}$ form a fundamental system of neighborhoods of zero.

If M is a proper ideal of a commutative ring with identity A, clearly A is a local ring with maximal ideal M if and only $A^\times = A \setminus M$.

A field is a local ring. More generally, the valuation ring of a real valuation of a field is a local ring by (2) of 17.4.

20.15 Theorem. *If J is a proper ideal of a local ring A, then A/J is a local ring whose natural topology is the quotient topology induced by the natural topology of A.*

Proof. Let M be the maximal ideal of A. Clearly M/J is the unique maximal ideal of A/J. Since $(M/J)^n = (M^n + J)/J$ for all $n \geq 1$, the final assertion follows. •

20.16 Theorem. *The natural topology of a local noetherian ring A is Hausdorff, and each of its ideals is closed for that topology.*

Proof. Let M be the maximal ideal of A, and let $x \in \cap_{n=1}^\infty M^n$. By 20.13 applied to $E = A$, there exists $a \in M$ such that $(1-a)x = 0$, whence $x = 0$ as $1 - a$ is invertible in A. Thus A is Hausdorff by 3.4. If J is a proper ideal of A, then A/J is Hausdorff for its natural topology by 20.4 and what we have just proved, so J is closed by 20.15 and 5.7. •

We conclude by characterizing the local domains that are discrete valuation rings:

20.17 Theorem. *Let A be a local domain distinct from its quotient field K, and let M be its maximal ideal. The following statements are equivalent:*
 1° *A is the valuation ring of a discrete valuation.*
 2° *A is noetherian and the valuation ring of a real valuation.*
 3° *A is a principal ideal domain.*
 4° *A is noetherian, and M is a principal ideal.*
 5° *M is a principal ideal, and $\cap_{n=1}^\infty M^n = (0)$.*

Proof. By 18.2, 1° implies 2° and 3°. Assume 2°. By (5) of 17.4, the principal ideals of A are totally ordered by inclusion. Consequently, if $M = Ax_1 + \cdots + Ax_n$, then there exists $k \in [1, n]$ such that $Ax_k \supseteq Ax_j$ for all $j \in [1, n]$, whence $M = Ax_k$. Thus 4° holds. Clearly 3° implies 4°, and 4° implies 5° by 20.16.

Assume, finally, 5°, and let $M = Ap$. As $D \neq K$ and hence $p \neq 0$, $(Ap^n)_{n \in \mathbb{Z}}$ is a strictly decreasing sequence of A-submodules of K whose intersection is (0) by hypothesis and whose union in K; indeed, for each $a \in A^*$ there is a unique $n \in \mathbb{N}$ such that $a \in Ap^n \setminus Ap^{n+1}$, so $a = up^n$ where $u \in A \setminus Ap$ and hence is a unit of A, and therefore $a^{-1} = u^{-1}p^{-n} \in Ap^{-n}$. For each $x \in K^*$, let $v(x)$ be the unique $n \in \mathbb{Z}$ such that $x \in Ap^n \setminus Ap^{n+1}$, and let $v(0) = +\infty$. Clearly $v(a + b) \geq \inf\{v(a), v(b)\}$ for all $a, b \in K$. If $v(a) = n$ and $v(b) = m$, then $a = tp^n$ and $b = up^m$ where $t, u \in A \setminus Ap$ and hence are units of A; thus $ab = tup^{n+m}$ where tu is a unit of A, so $ab \in Ap^{n+m} \setminus Ap^{n+m+1}$, and hence $v(ab) = n + m = v(a) + v(b)$. Thus v is a discrete valuation of K, and clearly A is its valuation ring. •

Exercises

20.1 (a) If A is a finite-dimensional algebra over a field, the ring A is noetherian. (b) Let K be an infinite field, and let A be the K-algebra $K \times K$ where addition is defined componentwise and multiplication by

$$(s, x)(t, y) = (st, sy + tx)$$

for all $s, t, x, y \in K$. Let $J = \{0\} \times K$. The ring A is noetherian and J is an ideal of A, but J is not a noetherian ring.

20.2 If E is a vector space over a division ring K, the ring A of all linear operators on E is noetherian if and only if E is finite-dimensional.

20.3 Let D be a noetherian domain, and let a be a noninvertible element of D^*. (a) There is an irreducible element p of D such that $p \mid a$. [Consider the family of proper principal ideals containing (a).] (b) a is the product of irreducible elements. [Consider the family of principal ideals (a/d) where d is a product of irreducible elements and $d \mid a$.]

20.4 Let A be a commutative ring with identity. (a) A proper ideal P of A is a prime ideal if and only if for all ideals I, J of A, if $IJ \subseteq P$, then either $I \subseteq P$ or $J \subseteq P$. (b) If L is a proper ideal of A that is not prime, then there exist ideals I and J properly containing L such that $IJ \subseteq L$.

20.5 If A is a commutative noetherian ring with identity, every ideal of A contains a product of prime ideals. [Use Exercise 20.4(b) in considering the set of all proper ideals of A that do not contain a product of prime ideals.]

21 Cohen Subrings of Complete Local Rings

The completeness of the natural topology of a local noetherian ring has important consequences, which we shall investigate in the remainder of this chapter and in Chapter 9.

21.1 Definition. *A local ring is* **equicharacteristic** *if it has the same characteristic as its residue field.*

We shall extend to local rings the notational convention introduced on page 158: Let M is the maximal ideal of a local ring A, ϕ_M the canonical epimorphism from A to A/M. For any x in A we shall denote $\phi_M(x)$ by $\bar{x}$, and for any $f \in A[X]$ we shall denote by $\bar{f}$ its image under the epimorphism from $A[X]$ to $(A/M)[X]$ induced by ϕ_M.

21.2 Theorem. *Let p be the characteristic of the residue field of a local ring A. If $p = 0$, then A has characteristic zero and, in particular, is equicharacteristic. If p is a prime, then the characteristic of A is either zero or p^r for some $r \geq 1$.*

Proof. The first assertion is obvious. Assume that the characteristic q of A is not zero, let M be the maximal ideal of A, and let $q = p^r s$ where $p \nmid s$. Then $s.\bar{1} \neq \bar{0}$, so $s.1 \notin M$ and hence $s.1$ is invertible. Therefore as $0 = q.1 = (p^r.1)(s.1)$, $p^r.1 = 0$ and $s = 1$. •

21.3 Definition. *A* **complete local ring** *is a local ring that is Hausdorff and complete for its natural topology. A ring C is a* **Cohen ring** *if C is a complete local ring whose maximal ideal is the principal ideal generated by $p.1$, where p is the characteristic of its residue field. If A is a local ring, a subring C of A is a* **Cohen subring** *(or a* **Cohen subfield** *if it is a field) if C is a Cohen ring and the restriction to C of the canonical epimorphism from A to its residue field k is an epimorphism from C to k.*

For example, for each prime p the ring $\mathbb{Z}_p$ of p-adic integers is a Cohen ring.

Let C be a local ring whose maximal ideal is the principal ideal generated by $p.1$, where p is the characteristic of its residue field. If $p = 0$, or if the characteristic of C is p, then C is a field and its natural topology is the discrete topology and, in particular, is complete. By virtue of the following theorem, if the characteristic of C is a prime power p^r, then the natural topology of C is again discrete. Consequently, the requirement that a Cohen ring be complete and Hausdorff for its natural topology is substantive only if it has characteristic zero and its residue field has prime characteristic.

A ring (or ideal) A is *nilpotent* if for some $n \geq 1$, $A^n = \{0\}$, or equivalently, if zero is the only product of n terms of A. A is *nilpotent of index n* if n is the smallest natural number such that $A^n = \{0\}$.

21.4 Theorem. *Let A be a local ring whose maximal ideal is a principal ideal Ab satisfying $\cap_{n=1}^{\infty} Ab^n = (0)$. If b is nilpotent of index n, then Ab is nilpotent of index n, hence the natural topology of A is discrete, for every nonzero $x \in A$ there is a unique unit u and a unique $r \in [0, n-1]$ such* that $x = ub^r$, and $(Ab^k)_{0 \le k \le n}$ is a strictly decreasing sequence consisting of all ideals of A. If b is not nilpotent, then A is an integral domain and, furthermore, is the valuation ring of a discrete valuation of its quotient field.

Proof. For each nonzero x, there is a largest integer r such that $x \in Ab^r$, so $x = ub^r$ where $u \in A \setminus Ab$ and hence is a unit, as $x \notin Ab^{r+1}$; clearly $r < n$ if $b^n = 0$. If J is a nonzero ideal, clearly $J = Ab^k$ where k is the smallest of the integers r such that for some unit u, $ub^r \in J$. Assume, finally, that b is not nilpotent. If $x = ub^r$ and $y = vb^s$ where u and v are units, then $xy = uvb^{r+s} \ne 0$. Hence A is an integral domain, so the final assertion follows from 20.17. •

21.5 Lemma. *Let A be a local ring, M its maximal ideal. Let K be a subfield of A, let $g \in K[X]$, and let $a \in A$ be such that $\bar{a}$ is a simple root of $\bar{g}$ in A/M. If Q and N are proper ideals of A such that $g(a) \in Q$ and $g(a)^{n+1} \in N$ where $n \ge 1$, there exists $a_1 \in A$ such that $a - a_1 \in Q$ and $g(a_1)^n \in N$; moreover $\bar{a}_1$ is a simple root of $\bar{g}$.*

Proof. As $\bar{a}$ is a simple root of $\bar{g}$, $D\bar{g}(\bar{a}) \ne \bar{0}$, where $D\bar{g}$ is the derivative of $\bar{g}$. Consequently, as $D\bar{g}(\bar{a}) = \overline{Dg(a)}$, $Dg(a) \notin M$ and therefore is invertible. Let
$$b = Dg(a)^{-1}(g(a) - 1), \qquad h = bg(a), \qquad a_1 = a + h.$$
Then $a - a_1 \in Q$. The set of all $f \in K[X]$ such that $f(a+h) - f(a) - hDf(a) \in Ah^2$ is clearly a subspace of $K[X]$ containing all monomials and hence is $K[X]$; consequently, there exists $c \in A$ such that
$$\begin{aligned} g(a_1) &= g(a) + Dg(a)bg(a) + cb^2 g(a)^2 \\ &= g(a)[1 + Dg(a)b] + cb^2 g(a)^2 \\ &= g(a)^2 (1 + cb^2). \end{aligned}$$
Hence $g(a_1)^n$ is a multiple of $g(a)^{2n}$ and thus belongs to N. The final assertion follows since $\bar{a}_1 = \bar{a}$ as $a - a_1 \in Q \subseteq M$. •

21.6 Lemma. *Let A be a local ring furnished with a complete metrizable ideal topology for which every element of the maximal ideal M of A is a topological nilpotent. Let K be a subfield of A, let $g \in K[X]$, and let $a \in A$. If $\bar{a}$ is a simple root of $\bar{g}$ in A/M, then there is a root b of g in A such that $\bar{b} = \bar{a}$.*

Proof. As the topology is a Hausdorff ideal topoloy, M is open and thus closed. Let $(N_k)_{k \ge 0}$ be a decreasing fundamental system of ideal neighborhoods of zero such that $N_0 = M$. We shall show that there is a sequence $(a_k)_{k \ge 0}$ of elements of A such that $a_0 = a$, $g(a_k) \in N_k$ for all $k \ge 0$, and

$a_k - a_{k-1} \in N_{k-1}$ for all $k \geq 1$. Indeed, as $\overline{g(a)} = \bar{g}(\bar{a}) = \bar{0}$ by hypothesis, $g(a_0) \in M = N_0$. Assume that $k \geq 0$, that $g(a_k) \in N_k \subseteq M$, and, if $k \geq 1$, that $a_k - a_{k-1} \in N_{k-1}$. Then $\lim_{n\to\infty} g(a_k)^n = 0$ by hypothesis, so there exists $m \geq 1$ such that $g(a_k)^m \in N_{k+1}$; by 21.5 applied $m-1$ times, there exists $a_{k+1} \in A$ such that $g(a_{k+1}) \in N_{k+1}$ and $a_{k+1} - a_k \in N_k$. Consequently, $\lim_{k\to\infty}(a_{k+1} - a_k) = 0$, so as the topology is a complete ideal topology, $(a_k)_{k\geq 0}$ converges to an element $b \in A$. Also $\lim_{k\to\infty} g(a_k) = 0$, so as g is a polynomial and therefore continuous,

$$0 = \lim_{k\to\infty} g(a_k) = g(b).$$

As M is closed and as

$$a - a_k = \sum_{j=0}^{k-1}(a_j - a_{j+1}) \in \sum_{j=0}^{k-1} N_j = M$$

for all $k \geq 1$, we conclude that $a - b \in M$ and hence that $\bar{b} = \bar{a}$. •

21.7 Theorem. *Let A be an equicharacteristic local ring furnished with a complete metrizable ideal topology for which every element of the maximal ideal M of A is a topological nilpotent. Then A contains a subfield K such that the residue field k of A is a purely inseparable algebraic extension of the image $\overline{K}$ of K under the canonical epimorphism from A to k; indeed, if K is any maximal subfield of A, k is a purely inseparable algebraic extension of $\overline{K}$.*

Proof. If A has prime characteristic, clearly A contains a prime subfield. If A has characteristic zero and if $n \in \mathbb{Z}^*$, then $n.\bar{1} \neq \bar{0}$ since by hypothesis k has characteristic zero, so $n.1 \notin M$ and thus is invertible in A. Therefore if A has characteristic zero, A contains a subfield isomorphic to $\mathbb{Q}$. Consequently by Zorn's Lemma, A contains maximal subfields, so it remains to show that if K is a maximal subfield of A, k is purely inseparable over $\overline{K}$.

If $\bar{a} \in k$ were transcendental over $\overline{K}$, then for every nonzero $g \in K[X]$, we would have $\bar{g}(\bar{a}) \neq \bar{0}$, whence $g(a) \notin M$, and thus $g(a)$ would be invertible in A; the set of all $f(a)/g(a)$, where $f, g \in K[X]$ and $g \neq 0$, would then be a subfield of A properly containing K, a contradiction. Thus k is an algebraic extension of $\overline{K}$. If $\bar{a}$ is separable over $\overline{K}$ and if $g \in K[X]$ is such that $\bar{g}$ is the minimal polynomial of a over $\overline{K}$, then by 21.6 there is a root $b \in A$ of g such that $\bar{b} = \bar{a}$; as $\bar{g}$ and hence also g are irreducible, $K[b]$ is a subfield of A containing K, so by maximality $K[b] = K$ and thus $\bar{a} = \bar{b} \in \overline{K}$. Therefore k is a purely inseparable extension of $\overline{K}$. •

21.8 Corollary. *If A is an equicharacteristic local ring of characteristic zero furnished with a complete metrizable ideal topology for which every element of the maximal ideal of A is a topological nilpotent, then A contains a Cohen subfield.*

21.9 Definition. *An ideal J of a ring A is a **nil** ideal if every element of J is nilpotent; if $r \geq 1$, J is a **nil ideal of index r** if $x^r = 0$ for all $x \in J$ but $x^{r-1} \neq 0$ for some $x \in J$; J is a **nil ideal of bounded index** if J is a nil ideal of index r for some $r \geq 1$.*

Applying 21.8 to the discrete topology, we obtain:

21.10 Theorem. *If A is an equicharacteristic local ring of characteristic zero whose maximal ideal is a nil ideal, then A contains a Cohen subfield, and moreover, every maximal subfield of A is a Cohen subfield.*

21.11 Theorem. *If A is a local ring whose characteristic is a prime p and whose maximal ideal M is a nil ideal of index $\leq p$, then A has a Cohen subfield, and moreover, the Cohen subfields of A are precisely the maximal subfields of A containing the image $\sigma(A)$ of A under the endomorphism $\sigma : x \to x^p$ of A.*

Proof. To show that $\sigma(A)$ is a field, let $x \in \sigma(A)$, $x \neq 0$. Then there exists $y \in A$ such that $y^p = x$. If $y \in M$, then $0 = y^p = x$, a contradiction; hence y is invertible in A, and $y^{-p} \in \sigma(A)$ is clearly the inverse of x.

The set $\mathcal{F}$ of all subfields of A containing $\sigma(A)$, ordered by inclusion, is clearly inductive and hence contains maximal members; but a maximal member of $\mathcal{F}$ is maximal in the set of all subfields of A. Therefore, to complete the proof, we need only show that if C is a subfield of A, then C is a Cohen subfield if and only if C is a maximal subfield and $C \supseteq \sigma(A)$. Necessity: A Cohen subfield C is clearly maximal. To show that $C \supseteq \sigma(A)$, let $x \in \sigma(A)$, and let $y \in A$ be such that $y^p = x$. There exists $z \in C$ such that $z - y \in M$, whence

$$z^p - y^p = (z-y)^p = 0$$

by hypothesis. Thus $x = y^p = z^p \in C$.

Sufficiency: Let C be a maximal subfield of A that contains $\sigma(A)$, and let $a \in A$. By 21.7 applied to the discrete topology of A, the residue field k of A is a purely inseparable algebraic extension of the image $\overline{C}$ of C under the canonical epimorphism from A to k. Assume that $\bar{a} \notin \overline{C}$. Then since $\bar{a}^p \in \overline{C}$, the minimal polynomial of $\bar{a}$ over $\overline{C}$ is $X^p - \bar{a}^p$. As $X^p - \bar{a}^p$ is thus irreducible over $\overline{C}$, $X^p - a^p$ is irreducible over C. Therefore $C(a)$ is a subfield of A properly containing C, as $\bar{a} \notin \overline{C}$, a contradiction. Hence $\overline{C} = k$, so C is a Cohen subfield of A. ●

21.12 Lemma. *Let A and A' be local rings whose characteristics are powers of a prime p, let f be an epimorphism from A to A', and let C' be a Cohen subring of A'. Then $f^{-1}(C')$ is a local ring whose maximal ideal is contained in the maximal ideal M of A. If C is a Cohen subring of $f^{-1}(C')$, then C is a Cohen subring of A, and $f(C) = C'$.*

Proof. Let M' be the maximal ideal of A'. As the residue field of A' has characteristic p, $p.A' \subseteq M'$. As C' is a local ring whose maximal ideal is $p.C'$, clearly $f^{-1}(C')$ is a local ring whose maximal ideal is $f^{-1}(p.C')$, and

$$f^{-1}(p.C') \subseteq f^{-1}(p.A') \subseteq f^{-1}(M') = M.$$

Let C be a Cohen subring of $f^{-1}(C')$. To show that C is a Cohen subring of A, let $a \in A$. As f is an epimorphism and C' a Cohen subring, there exists $b \in f^{-1}(C')$ such that $f(a) - f(b) \in M'$, whence $a - b \in f^{-1}(M') = M$; also there exists $c \in C$ such that $b - c \in f^{-1}(p.C') \subseteq M$; thus $a - c = (a - b) + (b - c) \in M$.

To show that $f(C) \supseteq C'$, let p^r be the characteristic of A', and let $y \in C'$. We shall show by induction that there is a sequence $x_1, \ldots, x_r$ of elements of C such that for each $n \in [1, r]$,

(1) $$y - f(x_1 + p.x_2 + \ldots + p^{n-1}.x_n) \in p^n C'.$$

Indeed, let $x \in f^{-1}(C')$ be such that $f(x) = y$; there exists $x_1 \in C$ such that $x - x_1 \in f^{-1}(p.C')$, so $y - f(x_1) \in p.C'$. If $x_1, \ldots, x_n$ are elements of C satisfying (1) where $n < r$, there exists $z \in C'$ such that

$$y - f(x_1 + p.x_2 + \cdots + p^{n-1}.x_n) = p^n.z.$$

Let $u \in f^{-1}(C')$ be such that $f(u) = z$. There exists $x_{n+1} \in C$ such that $u - x_{n+1} \in f^{-1}(p.C')$. Then

$$\begin{aligned} y - f(x_1 + p.x_2 + \ldots + p^n.x_{n+1}) &= p^n.z - p^n.f(x_{n+1}) \\ &= p^n.f(u - x_{n+1}) \in p^{n+1}.C'. \end{aligned}$$

Thus such a sequence exists. By (1) where $n = r$,

$$y = f(x_1 + p.x_2 + \ldots + p^{r-1}.x_r) \in f(C). \bullet$$

21 COHEN SUBRINGS OF COMPLETE LOCAL RINGS

21.13 Lemma. *Let A be a local ring furnished with a complete Hausdorff ring topology, and let $(J_n)_{n\geq 1}$ be a decreasing sequence of proper closed ideals of A that converges to zero. For all $m, n \in \mathbb{N}$ such that $m \geq n \geq 1$, let $f_{n,m}$ be the canonical epimorphism from A/J_m to A/J_n, let $A_0 = \varprojlim_{n\geq 1}(A/J_n)$, and let p be the characteristic of the residue field of A. If for all $n \geq 1$ the local ring A/J_n contains a Cohen subring C_n, if $f_{n,n+1}(C_{n+1}) = C_n$ for all $n \geq 1$, and if*

$$C_0 = (\prod_{n=1}^\infty C_n) \cap A_0, \qquad M_0 = (\prod_{n=1}^\infty p.C_n) \cap A_0,$$

then C_0 is a local ring whose maximal ideal is M_0, for all $m \geq 1$ the restriction to C_0 of the canonical projection pr_m from $\prod_{n=1}^\infty C_n$ to C_m is an epimorphism, and the restriction to C_0 of the canonical epimorphism from A_0 to its residue field k_0 is an epimorphism.

Proof. By (2) of 8.5, the canonical homomorphism g from A to A_0 is a topological isomorphism, so in particular, A_0 is a local ring. If $m \geq n \geq 1$, then since

$$f_{n,m} = f_{n,n+1} \circ f_{n+1,n+2} \circ \cdots \circ f_{m-1,m},$$

clearly $f_{n,m}(C_m) = C_n$. To show that the restriction to C_0 of pr_m is surjective, let $y_m \in C_m$. For each $j \in [1, m-1]$ let $y_j = f_{j,m}(y_m) \in C_j$. By hypothesis there is a sequence $(y_k)_{k\geq m+1}$ such that for all $k \geq m+1$, $y_k \in C_k$ and $f_{k-1,k}(y_k) = y_{k-1}$. Let $y = (y_k)_{k\geq 1}$; clearly $y \in C_0$ and $pr_m(y) = y_m$.

Since $p.C_n$ is the maximal ideal of C_n for all $n \geq 1$, M_0 is a proper ideal of C_0. Let $(c_n)_{n\geq 1} \in C_0 \setminus M_0$. Then for some $m \geq 1$, $c_m \notin p.C_m$, so c_m has an inverse $c_m^{-1} \in C_m$. Consequently, if $j \in [1, m-1]$, then c_j, which is $f_{j,m}(c_m)$, is invertible in C_j. If $k \geq m+1$, then $f_{m,k}(c_k) = c_m \notin p.C_m$, so as $f_{m,k}(p.C_k) = p.C_m$, $c_k \notin p.C_k$ and therefore c_k is invertible in C_k. Clearly $(c_n^{-1})_{n\geq 1} \in A_0$, so $(c_n^{-1})_{n\geq 1}$ is the inverse of $(c_n)_{n\geq 1}$ in C_0. Therefore C_0 is a local ring whose maximal ideal is M_0.

Let M be the maximal ideal of A. By hypothesis, the restriction q_1 to C_1 of the canonical epimorphism from A/J_1 to its residue field $(A/J_1)/(M/J_1)$ is surjective. Let r be the canonical isomorphism from $(A/J_1)/(M/J_1)$ to A/M, $\bar{g}$ the isomorphism from A/M to k_0 induced by the isomorphism g from A to A_0. Let

$$h = \bar{g} \circ r \circ q_1 \circ \pi_1$$

where π_1 is the restriction to C_0 of pr_1. Since π_1 is an epimorphism from C_0 to C_1, h is an epimorphism from C_0 to k_0. To see that h is the restriction to C_0 of the canonical epimorphism from A_0 to k_0, let $c \in C_0$. Since g is an

isomorphism, there exists $b \in A$ such that $g(b) = c$. Then $c = (b + J_n)_{n \geq 1}$. Consequently,
$$h(c) = (\bar{g} \circ r \circ q_1)(b + J_1) = \bar{g}(b + M) = g(b) + M = c + M. \bullet$$

21.14 Theorem. *Let A be a local ring of prime characteristic p, furnished with a complete Hausdorff ideal topology, let σ be the endomorphism of A defined by $\sigma(x) = x^p$ for all $x \in A$, and let M be the maximal ideal of A. If the filter base $(\sigma^n(M))_{n \geq 0}$ converges to zero, then A contains a Cohen subfield.*

Proof. For each $n \in \mathbb{N}$ let J_n be the closure of the ideal Q_n generated by $\sigma^n(M)$. Since the given topology is an ideal topology and is, by 3.4, regular, the filter base $(J_n)_{n \geq 0}$ also converges to zero. Furthermore, as the topology is an ideal topology, the maximal ideal M is necessarily open and hence closed, so $J_0 = M$. Let $A_0 = \varprojlim_{n \geq 0}(A/J_n)$. By (2) of 8.5 A is topologically isomorphic to A_0, so we need only show that A_0 has a Cohen subfield.

For each $n \in \mathbb{N}$ let f_n be the canonical epimorphism from A/J_{n+1} to A/J_n. We shall first show that if C_n is a Cohen subfield of A/J_n, then there is a Cohen subfield C_{n+1} of A/J_{n+1} such that $f_n(C_{n+1}) = C_n$. By 21.12 it suffices to show that the local ring $f_n^{-1}(C_n)$ contains a Cohen subfield, and for this, it suffices by 21.11 to show that the maximal ideal $f_n^{-1}(0) = J_n/J_{n+1}$ of $f_n^{-1}(C_n)$ is a nil ideal of index $\leq p$. If $a \in Q_n$, then
$$a = \sum_{i=1}^{s} a_i m_i^{p^n}$$
where $a_i \in A$ and $m_i \in M$ for all $i \in [1, s]$, so
$$a^p = \sum_{i=1}^{s} a_i^p m_i^{p^{n+1}} \in Q_{n+1}.$$
Thus $\sigma(Q_n) \subseteq Q_{n+1}$. Since σ is continuous, $\sigma(J_n) \subseteq J_{n+1}$, so J_n/J_{n+1} is a nil ideal of A/J_{n+1} of index $\leq p$.

In particular, since A/J_0 is a field and hence is a Cohen subfield of itself, by the preceding and induction there exists $(C_n)_{n \geq 1}$ such that for all $n \geq 1$, C_n is a Cohen subfield of A/J_n and $f_n(C_{n+1}) = C_n$. Let
$$C_0 = (\prod_{n=1}^{\infty} C_n) \cap A_0.$$
By 21.13, the maximal ideal M_0 of C_0 is the zero ideal, since each C_n has characteristic p. Thus C_0 is a subfield of A_0, and by 21.13 it is a Cohen subfield. $\bullet$

21.15 Theorem. *Let A be a local ring of prime characteristic p, furnished with a complete Hausdorff ideal topology, let σ be the endomorphism of A defined by $\sigma(x) = x^p$ for all $x \in A$, let M be the maximal ideal of A, and let k_0 be a perfect subfield of the residue field k of A. If $(\sigma^n(M))_{n \geq 0}$ converges to zero, then A contains a unique subfield K_0 such that the restriction to K_0 of the canonical epimorphism from A to k is an isomorphism from K_0 to k_0.*

Proof. Since A has a Cohen subfield C by 21.14, $\{x \in C : \bar{x} \in k_0\}$ is such a field K_0. To establish its uniqueness, it suffices to show that for each $\alpha \in k_0$, the element a of C such that $\bar{a} = \alpha$ is the only element x of A such that $\bar{x} = \alpha$ and $x^{p^{-n}} \in A$ for all $n \in \mathbb{N}$. Indeed, for each $n \in \mathbb{N}$, since

$$\bar{x}^{p^{-n}} = \alpha^{p^{-n}} = \bar{a}^{p^{-n}},$$

$x^{p^{-n}} - a^{p^{-n}} \in M$, so

$$x - a = (x^{p^{-n}} - a^{p^{-n}})^{p^n} \in \sigma^n(M).$$

Consequently,

$$x - a \in \bigcap_{n=0}^{\infty} \sigma^n(M) = (0)$$

by hypothesis. •

21.16 Corollary. *If A is a local ring of prime characteristic p whose maximal ideal M is a nil ideal of bounded index, then A contains a Cohen subfield; if, moreover, k_0 is a perfect subfield of the residue field k of A, then there is a unique subfield K_0 of A such that the restriction to K_0 of the canonical epimorphism from A to k is an isomorphism from K_0 to k_0.*

The assertion follows from 21.14 and 21.15 for the case where the topology is the discrete topology.

21.17 Theorem. *A local ring whose maximal ideal is nilpotent contains a Cohen subring.*

Proof. The assertion follows from 21.10 if the characteristic of the residue field is zero. Consequently, we need only prove by induction on n the assertion that a local ring whose residue field has prime characteristic p and whose maximal ideal is nilpotent of index $\leq n$ contains a Cohen subring.

First, we consider the case $n = 2$: Let A be a local ring whose residue field has characteristic p and whose maximal ideal M satisfies $M^2 = (0)$. Then $p.1 \in M$, so the characteristic of A is p or p^2. Moreover, $A/p.A$ is a local ring whose characteristic is p and whose maximal ideal $M/p.A$ is nilpotent of index not exceeding 2. As $2 \leq p$, $A/p.A$ has a Cohen subfield F by 21.11. Let $C = \psi^{-1}(F)$, where ψ is the canonical epimorphism from A to $A/p.A$. Then C is a local ring whose maximal ideal, $\psi^{-1}(0)$, is $p.A$. To show that C is a Cohen subring of A, it therefore suffices to show that if $x \in A$, there exists $y \in C$ such that $x - y \in M$ and $p.x = p.y$. As the maximal ideal of $A/p.A$ is $M/p.A$, there exists $y \in C$ such that

$$\psi(x) - \psi(y) \in M/p.A,$$

whence $x - y \in M$; moreover,

$$p.x - p.y \in (p.1)M \subseteq M^2 = (0),$$

so $p.x = p.y$.

Finally, assume the truth of the assertion if $n < m$, where $m \geq 3$, and let A be a local ring whose residue field has prime characteristic p and whose maximal ideal M satisfies $M^m = (0)$. As $p.A \subseteq M$, $p^{m-1}.A \subseteq M^{m-1}$; let f be the canonical epimorphism from $A/(p^{m-1}.A)$ to A/M^{m-1}. The index of nilpotency of the maximal ideal M/M^{m-1} of A/M^{m-1} clearly does not exceed $m-1$, so by our inductive hypothesis, A/M^{m-1} has a Cohen subring C'. Then $f^{-1}(C')$ is a local ring whose maximal ideal is $f^{-1}(p.C'')$. It is easy to verify that

$$f^{-1}(p.C') \subseteq (p.A + M^{m-1})/(p^{m-1}.A).$$

Now

$$(p.A + M^{m-1})^{m-1} \subseteq p^{m-1}.A + \sum_{k=1}^{m-1} p^{m-1-k}.M^{(m-1)k}$$

$$\subseteq p^{m-1}.A + \sum_{k=1}^{m-1} M^{m-1-k+(m-1)k} = p^{m-1}.A$$

since, as $m \geq 3$, $m-1-k+(m-1)k \geq m$ for all $k \in [1, m-1]$. Therefore the index of nilpotency of $f^{-1}(p.C)$ does not exceed $m-1$, so by our inductive hypothesis, $f^{-1}(C')$ contains a Cohen subring C. The characteristic of $A/(p^{m-1}.A)$ is clearly p^r for some $r \in [1, m-1]$, so by 21.12, C is a Cohen subring of $A/(p^{m-1}.A)$.

Let ϕ be the canonical epimorphism from A to $A/(p^{m-1}.A)$. If $x \in A$, there exists $y \in \phi^{-1}(C)$ such that

$$\phi(x) - \phi(y) \in M/(p^{m-1}.A),$$

whence $x - y \in M$. Therefore, since $\phi^{-1}(C)$ is a local ring whose maximal ideal is $\phi^{-1}(p.C)$, to show that $\phi^{-1}(C)$ is a Cohen subring of A it suffices to show that $p.\phi^{-1}(C) = \phi^{-1}(p.C)$. Clearly $p.\phi^{-1}(C) \subseteq \phi^{-1}(p.C)$. Conversely, let $x \in \phi^{-1}(p.C)$; then there exist $y \in \phi^{-1}(C)$ and $z \in A$ such that

$$x = p.y + p^{m-1}.z.$$

As C is a Cohen subring of $A/(p^{m-1}.A)$, there exists $t \in \phi^{-1}(C)$ such that

$$\phi(t) - \phi(z) \in M/(p^{m-1}.A),$$

whence $t - z \in M$, and therefore

$$p^{m-1}.t - p^{m-1}.z = p^{m-1}.(t-z) \in M^m = (0).$$

Thus

$$x = p.y + p^{m-1}.t = p.(y + p^{m-2}.t) \in p.\phi^{-1}(C). \bullet$$

21.18 Theorem. *Let A be a local ring of characteristic zero with maximal ideal M whose residue field has prime characteristic p, and let C be a subring of A that is the valuation ring of a real valuation v of some field. If $\mathcal{T}$ is a Hausdorff ideal topology on A for which $(M^r)_{r \geq 1}$ converges to zero, then the topology on C induced by $\mathcal{T}$ is the topology defined by v, for which all nonzero ideals of C form a fundamental system of neighborhoods of zero.*

Proof. Since the ideals of C are totally ordered and since $\mathcal{T}$ induces on C a Hausdorff ideal topology, every nonzero ideal of C is open for the topology induced by $\mathcal{T}$. Consequently, we need only show that the zero ideal is not open for that topology. In the contrary case, $J \cap C = (0)$ for some ideal J of A that is open for $\mathcal{T}$. By hypothesis, $M^n \subseteq J$ for some $n \geq 1$, and $p.1 \in M$. Hence

$$p^n.1 \in M^n \cap C \subseteq J \cap C = (0),$$

in contradiction to our hypothesis that A has characteristic zero. $\bullet$

21.19 Theorem. *If A is a local ring furnished with a complete Hausdorff ring topology that is weaker than its natural topology, then A contains a Cohen subring.*

Proof. Let M be the maximal ideal of A. For each $n \geq 1$, the closure Q_n of M^n is an ideal, and $(Q_n)_{n \geq 1}$ converges to zero since the topology is regular by 3.4. The ring topology $\mathcal{T}$ on A for which $(Q_n)_{n \geq 1}$ is a fundamental system of neighborhoods of zero is then stronger than the given complete topology and hence is also complete by 7.21. Replacing the given topology by $\mathcal{T}$, therefore, we may assume that the topology of A is a complete metrizable ideal topology for which $(M^n)_{n\geq 1}$ converges to zero.

If the residue field of A has characteristic zero, then A contains a Cohen subfield by 21.8. Therefore we may assume that the residue field has prime characteristic p.

Since M is necessarily open, there is a fundamental decreasing sequence $(J_n)_{n\geq 1}$ of neighborhoods of zero such that each J_n is an open ideal and $J_1 = M$. The maximal ideal M/J_n of each A/J_n is then nilpotent and the characteristic of A/J_n is a power of p; indeed, J_n contains M^r for some $r \geq 1$ by hypothesis, so $(M/J_n)^r = (0)$, and as $p.1 \in M$, $p^r.1 \in M^r \subseteq J_n$, so A/J_n has characteristic p^s for some $s \in [1, r]$. Let $A_0 = \varprojlim_{n\geq 1}(A/J_n)$. By (2) of 8.5 we need only show that A_0 contains a Cohen subring. For each $n \geq 1$, let pr_n be the canonical projection from A_0 onto A/J_n, and let $f_{m,n}$ be the canonical epimorphism from A/J_n to A/J_m whenever $n \geq m \geq 1$. By the definition of projective limit, $f_{m,n} \circ pr_n = pr_m$.

We shall first show that if C_n is a Cohen subring of A/J_n, then there is a Cohen subring C_{n+1} of A/J_{n+1} such that $f_{n,n+1}(C_{n+1}) = C_n$. By 21.12 the maximal ideal of $f_{n,n+1}^{-1}(C_n)$ is contained in that of A/J_{n+1} and hence is nilpotent, so by 21.17, $f_{n,n+1}^{-1}(C_n)$ contains a Cohen subring C_{n+1}, and $f_{n,n+1}(C_{n+1}) = C_n$ by 21.12.

Let C_1 be the field A/J_1. By the preceding and induction, there exists $(C_n)_{n\geq 1}$ such that for each $n \geq 1$, C_n is a Cohen subring of A/J_n and $f_{n,n+1}(C_{n+1}) = C_n$. Let

$$C_0 = (\prod_{n=1}^{\infty} C_n) \cap A_0, \qquad M_0 = (\prod_{n=1}^{\infty} p.C_n) \cap A_0.$$

By 21.13, we need only show that $M_0 = p.C_0$ and, if A has characteristic zero, the discrete valuation ring C_0 is complete. For each $n \geq 1$, let p^{r_n} be the characteristic of A/J_n and hence of C_n. Since C_n is an epimorphic image of C_{n+1}, $r_n \leq r_{n+1}$.

Case 1: The characteristic of A is p^r for some $r \geq 1$. Then there exists s such that $p^{r-1}.1 \notin J_s$, so $r_s = r$ and hence $r_n = r$ for all $n \geq s$. By 21.13, $pr_s(C_0) = C_s$. To show that the restriction π_s of pr_s to C_0 is an isomorphism from C_0 to C_s, let $(x_n)_{n\geq 1} \in C_0$ be such that $x_s = 0$. If $j \in [1, s-1]$, $x_j = f_{j,s}(x_s) = 0$. Suppose that $x_k \neq 0$ for some $k > s$; by 21.4, $x_k = p^t.u_k$ where u_k is invertible in C_k and $t \in [0, r-1]$. Since

$$0 = x_s = f_{s,k}(x_k) = p^t.f_{s,k}(u_k)$$

and since $f_{s,k}(u_k)$ is invertible in C_s, $t \geq r_s = r$, a contradiction. Thus $x_n = 0$ for all $n \geq 1$. Therefore π_s is an isomorphism from C_0 to C_s, so the maximal ideal M_0 of C_0 is $\pi_s^{-1}(p.C_s) = p.C_0$.

Case 2: The characteristic of A is zero. Then $(r_n)_{n \geq 1}$ is an increasing sequence diverging to $+\infty$, since for any $m \geq 1$ there exists $k \geq 1$ such that $p^m.1 \notin J_k$, whence $r_k \geq m$. Let $y \in M_0$. Then $y = (p.y_n)_{n \geq 1}$ where $y_n \in C_n$ for all $n \geq 1$ and $f_{m,n}(p.y_n) = p.y_m$ whenever $n \geq m \geq 1$. By 21.13, for each $n \geq 1$ there exists $z_n \in C_0$ such that $pr_n(z_n) = y_n$. We shall show that $(z_n)_{n \geq 1}$ is a Cauchy sequence in C_0, or equivalently by 7.8. that $(pr_m(z_n))_{n \geq 1}$ is an eventually stationary sequence in C_m for each $m \geq 1$, since A/J_m is discrete. Let $q > m$ be such that $r_q > r_m$, and let $s > q$. Then

$$p.pr_q(z_q) = p.y_q = f_{q,s}(p.y_s) = f_{q,s}(p.pr_s(z_s))$$
$$= p.f_{q,s}(pr_s(z_s)) = p.pr_q(z_s).$$

Therefore $p.(pr_q(z_q - z_s)) = 0$, so $pr_q(z_q - z_s) \in p^{r_q - 1}.C_q$ by 21.4, and consequently

$$pr_m(z_q) - pr_m(z_s) = pr_m(z_q - z_s) = f_{m,q}(pr_q(z_q - z_s))$$
$$\in p^{r_q - 1}.C_m \subseteq p^{r_m}.C_m = (0).$$

Since each A/J_n is discrete, C_0 is closed in A_0 and thus complete for the topology it inherits from A_0. Therefore $(z_n)_{n \geq 1}$ has a limit $z \in C_0$, and consequently $\lim_{n \to \infty} p.z_n = p.z$. If $n > m$,

$$pr_m(z_n) = f_{m,n}(pr_n(z_n)) = f_{m,n}(y_n).$$

Hence

$$pr_m(p.z) = \lim_{n \to \infty} pr_m(p.z_n) = \lim_{n \to \infty} p.pr_m(z_n) = \lim_{n \to \infty} p.f_{m,n}(y_n)$$
$$= \lim_{n \to \infty} f_{m,n}(p.y_n) = p.y_m.$$

Therefore $p.z = (p.y_m)_{m \geq 1} = y$. Consequently C_0 is a discrete valuation ring that is complete for the topology it inherits from A_0, which by 21.4 and 21.18 is its valuation topology. •

21.20 Corollary. *A complete local ring contains a Cohen subring.*

Exercises

21.1 Let A be the valuation ring, M the valuation ideal of a proper real valuation v of a field K. (a) If v is not discrete, then $M^2 = M$, and hence the natural topology of A is not Hausdorff. (b) If Q is an ideal of A properly contained in M, then $\cap_{n=1}^{\infty} Q^n = (0)$.

In the following exercises, F is a field, f is a prime polynmomial in $F[X]$, v_f is the valuation of $F(X)$ determined by f, and A_f is the valuation ring of $\hat{v}_f$. We recall that a *stem field* of f is a field $F(c)$ generated by F and a root c of f.

21.2 (a) The residue field of $\hat{v}_f$ is isomorphic to a stem field of f over F. (b) If f is separable, A_f contains a stem field of f over F. [Use 21.7.]

21.3 (I. S. Cohen [1945]) Let F be an imperfect field of prime characteristic p, and let f be the prime polynomial $X^p - a$ where $a \in F$ has no pth root in F. (a) A_f contains no root of f. [If c were a root, calculate $\hat{v}_f(X - c)$, and use 17.12.] (b) a does not belong to any Cohen subfield of A_f.

21.4 For every prime polynomial $f \in F[X]$, there is a Cohen subfield of A_f containing F if and only if F is perfect.

22 Complete Discretely Valued Fields

From the results of §21 we may derive a classical description of all complete discretely valued fields and their valuation rings. The equicharacteristic case is immediate:

22.1 Theorem. *Let v be a complete discrete valuation of a field K whose value group is $\mathbb{Z}$ and whose residue field k has the same characteristic as K, and let u be a uniformizer of v. The valuation ring A of v contains a Cohen subfield. If C is a Cohen subfield of A and ϕ the restriction to C of the canonical epimorphism from A to k, then the function Φ from $k((X))$ to K defined by*

$$\Phi(\sum_{n \in \mathbb{Z}} c_n X^n) = \sum_{n \in \mathbb{Z}} \phi^{-1}(c_n) u^n$$

is an isomorphism satisfying $v \circ \Phi = \mathrm{ord}$, the canonical valuation of $k((X))$; in particular, $\Phi(k[[X]]) = A$.

Proof. The first assertion is a consequence of 21.8 and 21.14. Let C be a Cohen subfield of A. By 18.5, for each family $(c_n)_{n \in \mathbb{Z}}$ of elements of k such that $c_n = 0$ for all but finitely many $n < 0$, the family $(\phi^{-1}(c_n) u^n)_{n \in \mathbb{Z}}$ of elements of K is summable, and Φ is a bijection from $k((X))$ to K satisfying $v \circ \Phi = \mathrm{ord}$. In particular, $\Phi(k[[X]]) = A$. By 10.12 and 10.16, Φ is a homomorphism. •

Thus a complete discrete valuation whose valuation ring is equicharacteristic is completely determined by its residue field.

The nonequicharacteristic case requires a further concept:

22.2 Definition. *Let A be a local domain. An **Eisenstein polynomial** over A is a monic polynomial whose nonleading coefficients all belong to the* maximal ideal M of A and whose constant coefficient does not belong to M^2. An **Eisenstein polynomial relative to** v, where v is a real valuation of a field, is an Eisenstein polynomial over the valuation ring of v.

22.3 Theorem. *Let v be a discrete valuation of a field K. Every Eisenstein polynomial relative to v is a prime polynomial over K. If u is a root in an extension field of K of an Eisenstein polynomial relative to v of degree m, then v has a unique extension v' to $K(u)$, $e(v'/v) = m$, $f(v'/v) = 1$, and u is a uniformizer of v'.*

Proof. We may assume that $\mathbb{Z}$ is the value group of v. Let u be a root of an Eisenstein polynomial g relative to v, and let

$$g(x) = X^m + a_{m-1}X^{m-1} + \ldots + a_1 X + a_0.$$

Let v' be an extension of v to $K(u)$. By hypothesis, $v(a_j) \geq 1$ for all $j \in [0, m-1]$, and $v'(u^m) = v'(\sum_{j=0}^{m-1} a_j u^j)$. If $v'(u) \leq 0$, then for each $j \in [0, m-1]$,

$$v'(a_j u^j) \geq 1 + jv'(u) > mv'(u) = v'(u^m),$$

so

$$v'(\sum_{j=0}^{m-1} a_j u^j) \geq \inf\{v'(a_j u^j) : j \in [0, m-1]\} > v'(u^m),$$

a contradiction. Hence $v'(u) > 0$. Thus if $j \in [0, m-1]$,

$$v'(a_j u^j) \geq 1 + jv'(u) > 1,$$

but $v'(a_0) = 1$ by hypothesis, so

$$mv'(u) = v'(u^m) = v'(\sum_{j=0}^{m-1} a_j u^j) = \inf\{v'(a_j u^j) : j \in [0, m-1]\} = 1$$

by 17.2, and consequently $v'(u) = 1/m$.

The value group of v' therefore contains $(1/m)\mathbb{Z}$, so $e(v'/v) \geq m$. Since g is a multiple of the minimal polynomial of u over K, $m \geq [K(u):K]$. By 19.8,

$$[K(u):K] \geq e(v'/v)f(v'/v) \geq e(v'/v).$$

Therefore $m = e(v'/v)$, so the value group of v' is $(1/m)\mathbb{Z}$, and hence u is a uniformizer of v'; deg $g = m = [K(u):K]$, so g is a prime polynomial; v' is the only extension of v to $K(u)$ by 19.10; and $f(v'/v) = 1$. •

The preceding theorem admits a converse:

22.4 Theorem. *Let v be a discrete valuation of a field K, let v' be a valuation of a finite-dimensional extension field K' of K that extends v, and let $m = [K' : K]$. If $e(v'/v) = m$, then v' is the only extension of v to K', and for each uniformizer u of v', $K' = K(u)$, the minimal polynomial of u over K is an Eisenstein polynomial relative to v, and $A' = A + Au + \ldots + Au^{m-1}$, where A and A' are respectively the valuation rings of v and v'.*

Proof. We may assume that the value group of v is $\mathbb{Z}$, so by hypothesis the value group of v' is $(1/m)\mathbb{Z}$. Let u be a uniformizer of v'; then $v'(u) = 1/m$. If $a, b \in K^*$ and if $0 \le j < k \le m-1$, then $v(a) \in \mathbb{Z}$, $v(b) \in \mathbb{Z}$, so

$$v'(au^j) = v(a) + \frac{j}{m} \ne v(b) + \frac{k}{m} = v'(bu^k).$$

Consequently, by 17.2, if $a_0, a_1, \ldots, a_{m-1} \in K$, then

$$v'\Big(\sum_{j=0}^{m-1} a_j u^j\Big) = \inf\{v(a_j) + \frac{j}{m} : j \in [0, m-1]\}.$$

In particular, $\{1, u, u^2, \ldots, u^{m-1}\}$ is linearly independent. Consequently, the minimal polynomial g of u cannot have degree $< m$, so $\deg g = m$ and hence $K' = K[u]$. Therefore by 19.10, v' is the only valuation of K' that extends v. Let

$$g = X^m + b_{m-1}X^{m-1} + \ldots b_1 X + b_0.$$

Then

$$1 = v'(u^m) = v'\Big(\sum_{j=0}^{m-1} b_j u^j\Big) = \inf\{v(b_j) + \frac{j}{m} : j \in [0, m-1]\}.$$

If $j \in [1, m-1]$, then $v(b_j) + (j/m) \notin \mathbb{Z}$, so $1 = v(b_0) < v(b_j) + (j/m)$ and consequently $1 = v(b_0) \le v(b_j)$. Therefore g is an Eisenstein polynomial relative to v. Finally, let $c \in A'$. As $K' = K(u)$, $c = \sum_{j=0}^{m-1} a_j u^j$ where $a_0, a_1, \ldots, a_{m-1} \in K$. Then

$$0 \le v'(c) = \inf\{v(a_j) + \frac{j}{m} : j \in [0, m-1]\}.$$

Thus for each $j \in [0, m-1]$, $v(a_j) + (j/m) \ge 0$, so $v(a_j) \ge 0$ as $v(a_j) \in \mathbb{Z}$; consequently, $c \in A + Au + \ldots + Au^{m-1}$. ∎

22.5 Definition. *Let v be a real valuation of a field K. A finite-dimensional extension field K' of K is an **Eisenstein extension relative to** v if there exists $u \in K'$ such that $K' = K(u)$ and the minimal polynomial of u over K is an Eisenstein polynomial relative to v.*

From 22.3 and 22.4, we conclude:

22.6 Theorem. *Let v be a discrete valuation of a field K. A finite-dimensional extension field K' of K is an Eisenstein extension relative to v if and only if there is a valuation v' of K' extending v such that $e(v'/v) = [K' : K]$; in this case, v' is the only valuation of K' extending v, the uniformizers of v' are precisely the elements u of K' such that $K' = K(u)$, the minimal polynomial of u over K is an Eisenstein polynomial relative to v, and the valuation ring A' of v' is a finitely generated module over the valuation ring A of v.*

The *codimension* of a subfield K of a field K' is $[K' : K]$. The following theorem reduces the problem of describing complete discrete valuations in the nonequicharacteristic case to that of characterizing those whose valuation rings are Cohen rings:

22.7 Theorem. *If v is a complete discrete valuation of a field K of characteristic zero whose residue field has prime characteristic p, there is a finite-codimensional subfield K_0 of K such that the restriction v_0 of v to K_0 is a complete discrete valuation whose valuation ring is a Cohen ring, and K is an Eisenstein extension of K_0 relative to v_0.*

Proof. By 20.17 and 21.20 the valuation ring A of v contains a Cohen subring A_0; let K_0 be the quotient field in K of A_0. The restriction v_0 of v to K_0 is then a complete discrete valuation by 21.18 and 7.6. We may assume that the value group of v is $\mathbb{Z}$. The value group of v_0 is then a nonzero subgroup of $\mathbb{Z}$ and hence is $m.\mathbb{Z}$ for some $m \geq 1$. Consequently, $e(v/v_0) = (\mathbb{Z} : m.\mathbb{Z}) = m < +\infty$. By the definition of a Cohen subring, $f(v/v_0) = 1$. Therefore by 19.9 $[K : K_0] = e(v/v_0) < +\infty$, so by 22.6, K is an Eisenstein extension of K_0 relative to v_0. •

An investigation of complete discretely valued fields therefore devolves upon nonequicharacteristic Cohen rings of characteristic zero. The main result is that for any field F of prime characteristic there is one and, to within isomorphism, only one Cohen ring of characteristic zero whose residue field is isomorphic to F.

22.8 Theorem. *Let F be a field of prime characteristic p. If q is either zero or a power of p, there is a Cohen ring of characteristic q whose residue field is isomorphic to F.*

Proof. Assume first that $q = 0$. The ring $\mathbb{Z}_p$ of p-adic integers is a Cohen ring of characteristic zero whose residue field is the prime field of p elements. By 19.15, there is a real valuation w of a field K extending the p-adic valuation $\hat{v}_p$ of $\mathbb{Q}_p$ such that the value group of w is $\mathbb{Z}$ and the residue field of w is isomorphic to F. The valuation ring $A_{\hat{w}}$ of $\hat{w}$ is then the desired Cohen ring of characteristic p whose residue field is isomorphic to F by 19.7. If $q = p^n$, then $A_{\hat{w}}/p^n A_{\hat{w}}$ is a Cohen ring of characteristic q whose residue field is isomorphic to F. •

Our first step in establishing the uniqueness of Cohen rings is to show that a Cohen ring is the only Cohen subring of itself, an assertion that is clearly true if the Cohen ring is a field.

22.9 Theorem. *If B is a Cohen subring of a Cohen ring C whose residue field has prime characteristic p, then $B = C$.*

Proof. Since pC is the maximal ideal of C, for each $x \in C$ there exists $y \in B$ such that $y \equiv x \pmod{pC}$; consequently, $C = B + pC$. If $C = B + p^m C$, then $pC = pB + p^{m+1}C$, so $C = B + pC = B + p^{m+1}C$. An inductive argument thus establishes that $C = B + p^n C$ for all $n \geq 1$. If C has characteristic p^r, then, in particular, $C = B + p^r C = B$. If C has characteristic zero, then

$$C = \bigcap_{n=1}^{\infty}(B + p^n C) = \overline{B}$$

by (3) of 3.3, but as B is a complete subring of C, $\overline{B} = B$. •

22.10 Lemma. *Let p be a prime, let C_1 and C_2 be Cohen rings of characteristic p^m, let K_1 and K_2 be Cohen rings of characteristic p^n where $1 \leq n < m$, let f_1 and f_2 be epimorphisms respectively from C_1 to K_1 and from C_2 to K_2, and let g be an isomorphism from K_1 to K_2. Then there is an isomorphism h from C_1 to C_2 such that $f_2 \circ h = g \circ f_1$.*

Proof. By 21.4, the only ideals of C_1 are the ideals $p^k C_1$ where $k \in [0, m]$, and the characteristic of $C_1/p^k C_1$ is p^k. Since the characteristic of K_1 is p^n, therefore, the kernel of f_1 is $p^n C_1$. Similarly, the kernel of f_2 is $p^n C_2$. Let

$$A = \{(x_1, x_2) \in C_1 \times C_2 : g(f_1(x_1)) = f_2(x_2)\}.$$

Clearly A is a subring of $C_1 \times C_2$ containing $p^n C_1 \times p^n C_2$. Let q_1 and q_2 be the restrictions to A of the canonical projections from $C_1 \times C_2$ to C_1 and C_2 respectively. Each is surjective since both $g \circ f_1$ and f_2 are surjective. Let

$$M = (f_2 \circ q_2)^{-1}(pK_2).$$

As q_2 and f_2 are surjective, and as pK_2 is the maximal ideal of K_2, M is a maximal ideal of A. Also, M is nilpotent, for by the definition of A,

$$(g \circ f_1 \circ q_1)(M^n) \subseteq (f_2 \circ q_2)(M^n) = (pK_2)^n = (0)$$

and hence $(f_1 \circ q_1)(M^n) = (0)$, so $M^n \subseteq q_1(M^n) \times q_2(M^n) \subseteq p^n C_1 \times p^n C_2$, and therefore

$$M^{nm} \subseteq p^{nm}C_1 \times p^{nm}C_2 = (0).$$

Since a prime ideal of a commutative ring with identity necessarily contains every nilpotent and since a maximal ideal is prime, M is the only maximal ideal of A.

Consequently by 21.17, A contains a Cohen subring C. Again by 21.4, the kernel of q_1 is $p^k C$ for some $k \in [0, m]$. As the characteristic of $C/p^k C$ is p^k and as the characteristic of the subring $q_1(C)$ of C_1 is p^m, therefore, $k = m$, so $p^k C = (0)$ and q_1 is injective. To show that $q_1(C)$ is a Cohen subring of C_1, let $x_1 \in C_1$. As q_1 is surjective, there exists $x_2 \in C_2$ such that $(x_1, x_2) \in A$. Consequently, there exists $(y_1, y_2) \in C$ such that $(y_1, y_2) - (x_1, x_2) \in M$, so

$$(g \circ f_1)(y_1 - x_1) = f_2(y_2 - x_2) = (f_2 \circ q_2)((y_1, y_2) - (x_1, x_2)) \in pK_2,$$

the maximal ideal of K_2, and thus $y_1 - x_1$ belongs to the maximal ideal pC_1 of C_1. Therefore $q_1(C)$ is a Cohen subring of C_1, so $q_1(C) = C_1$ by 22.9, and thus q_1 is an isomorphism from C to C_1. Similarly, q_2 is an isomorphism from C to C_2. Let $h = q_2 \circ q_1^{-1}$, an isomorphism from C_1 to C_2. As $f_2 \circ q_2 = g \circ f_1 \circ q_1$, $f \circ h = f_2 \circ q_2 \circ q_1^{-1} = g \circ f_1$. •

22.11 Theorem. *Let C_1 and C_2 be Cohen rings of the same characteristic, and let f_1 and f_2 be the canonical epimorphisms from C_1 and C_2 to their residue fields k_1 and k_2 respectively. If g_1 is an isomorphism from k_1 to k_2, there is an isomorphism h from C_1 to C_2 such that $f_2 \circ h = g_1 \circ f_1$.*

Proof. If the common characteristic is p^m, we need only apply 22.10 where $K_1 = k_1$, $K_2 = k_2$. Therefore we may assume that the characteristic of C_1 and C_2 is zero and the characteristic of k_1 and k_2 is a prime p. For each $n \geq 1$, let $f_{n,1}$ and $f_{n,2}$ be the canonical epimorphisms from $C_1/p^{n+1}C_1$ to $C_1/p^n C_1$ and from $C_2/p^{n+1}C_2$ to $C_2/p^n C_2$ respectively. By 22.10 there is an isomorphism g_2 from $C_1/p^2 C_1$ to $C_2/p^2 C_2$ such that $f_{1,2} \circ g_2 = g_1 \circ f_{1,1}$, and in general, by an inductive argument, there is a sequence $(g_n)_{n \geq 1}$ such that each g_n is an isomorphism from $C_1/p^n C_1$ to $C_2/p^n C_2$ and $f_{n,2} \circ g_{n+1} = g_n \circ f_{n,1}$ for all $n \geq 1$. For each $(x_n)_{n \geq 1} \in \varprojlim_{n \geq 1}(C_1/p^n C_1)$, $(g_n(x_n))_{n \geq 1} \in \varprojlim_{n \geq 1}(C_2/p^n C_2)$, for if $m \geq 1$, $f_{m,1}(x_{m+1}) = x_m$, so

$$f_{n,2}(g_{m+1}(x_{m+1})) = g_m(f_{m,1}(x_{m+1})) = g_m(x_m).$$

The function g from $\varprojlim_{n\geq 1}(C_1/p^n C_1)$ to $\varprojlim_{n\geq 1}(C_2/p^n C_2)$, defined by

$$g((x_n)_{n\geq 1}) = (g_n(x_n))_{n\geq 1},$$

is easily seen to be an isomorphism. By (2) of 8.5, the canonical homomorphisms $h_1 : x \to (x + p^n C_1)_{n\geq 1}$ from C_1 to $\varprojlim_{n\geq 1}(C_1/p^n C_1)$ and $h_2 : x \to (x + p^n C_2)_{n\geq 1}$ from C_2 to $\varprojlim_{n\geq 1}(C_2/p^n C_2)$ are isomorphisms. Thus h, defined to be $h_2^{-1} \circ g \circ h_1$, is an isomorphism from C_1 to C_2. Let π_1 be the restriction to $\varprojlim_{n\geq 1}(C_2/p^n C_2)$ of the canonical projection pr_1 from $\prod_{n=1}^{\infty}(C_2/p^n C_2)$ to C_2/pC_2. Then $g_1 \circ f_1 = \pi_1 \circ g \circ h_1$ and $f_2 = \pi_1 \circ h_2$, so

$$f_2 \circ h = f_2 \circ h_2^{-1} \circ g \circ h_1 = \pi_1 \circ g \circ h_1 = g_1 \circ f_1. \bullet$$

Thus a Cohen ring is completely determined by its residue field and its characteristic.

23 Complete Local Noetherian Rings

Let A be a commutative ring with identity. By the definition on page 148, $A[[X]]$ is the ring or A-algebra of all sequences $(a_n)_{n\in\mathbb{Z}}$ of elements of A such that $a_n = 0$ for all $n < 0$. For notational convenience, we shall henceforth omit reference to the terms of negative index, and denote any such $(a_n)_{n\in\mathbb{Z}}$ by $(a_n)_{n\in\mathbb{N}}$ or $(a_n)_{n\geq 0}$.

For every $(a_n)_{n\in\mathbb{N}} \in A[[X]]$, $(a_n X^n)_{n\in\mathbb{N}}$ is summable and its sum

$$\sum_{n\in\mathbb{N}} a_n X^n = (a_n)_{n\in\mathbb{N}}$$

for the order topology. Here we shall be concerned with Hausdorff ring topologies strictly weaker than the order topology, but any family of elements summable in $A[[X]]$ for the order topology is a *fortiori* summable (with the same sum) for any such weaker topology. In any case, even if we have no specific topology in mind, we shall usually denote the sequence $(a_n)_{n\in\mathbb{N}}$ by $\sum_{n\in\mathbb{N}} a_n X^n$. Consequently, the elements of $A[[X]]$ are frequently called *formal power series*. If $a \in A$, the *constant formal power series* determined by a is $\sum_{n\in\mathbb{N}} a_n X^n$ where $a_0 = a$ and $a_n = 0$ for all $n \geq 1$; we shall frequently identify a with the constant formal power series it defines. The *constant term* of a formal power series $\sum_{n\in\mathbb{N}} a_n X^n$ is the element a_0 of A.

The principal ideal of $A[[X]]$ generated by X consists of all formal power series $\sum_{n\in\mathbb{N}} a_n X^n$ such that $a_0 = 0$. It is easy to see, by an inductive argument, that $\sum_{n\in\mathbb{N}} a_n X^n$ is invertible in $A[[X]]$ if and only if a_0 is invertible in A. Indeed, if $\sum_{n\in\mathbb{N}} a_n X^n$ is a formal power series such that a_0 is invertible, then $\sum_{n\in\mathbb{N}} b_n X^n$ is the inverse of $\sum_{n\in\mathbb{N}} a_n X^n$ where $(b_n)_{n\geq 0}$ is defined recursively by $b_0 = a_0^{-1}$,

$$b_{n+1} = -a_0^{-1}(a_1 b_n + \ldots a_{n+1} b_0).$$

Consequently, we have:

23.1 Theorem. *If A is a local ring with maximal ideal M, then $A[[X]]$ is a local ring with maximal ideal $M + (X)$.*

23.2 Theorem. *If A is a commutative noetherian ring with identity, so is $A[[X]]$. If A is an integral domain, so is $A[[X]]$.*

Proof. Let $B = A[[X]]$. The second statement is easy to verify. For the first, it suffices by 20.9 to show that each prime ideal P of B is finitely generated. The constant terms of formal power series belonging to P form an ideal Q of A, so there exist $a_1, \ldots, a_m \in Q$ such that $Q = Aa_1 + \ldots + Aa_m$. If $X \in P$, the constant power series determined by each $q \in Q$ clearly belongs to P, so $P = Ba_1 + \ldots + Ba_m + BX$. We need only show, therefore, that if $X \notin P$ and if, for each $j \in [1,m]$, f_j is a member of P whose constant term is a_j, then $P = Bf_1 + \ldots + Bf_m$.

Let $g \in P$. We define recursively sequences $(b_{1,n})_{n\geq 0}, \ldots, (b_{m,n})_{n\geq 0}$ such that for each $n \in \mathbb{N}$,

$$(1) \qquad g - \sum_{j=1}^{m} \left(\sum_{k=0}^{n-1} b_{j,k} X^k \right) f_j \in BX^n.$$

Indeed, if (1) holds, then

$$g - \sum_{j=1}^{m} \left(\sum_{k=0}^{n-1} b_{j,k} X^k \right) f_j = hX^n$$

for some $h \in B$. Since P is a prime ideal and $X \notin P$, $h \in P$, so its constant term c belongs to Q. Thus there exist $b_{1,n}, \ldots, b_{m,n} \in Q$ such that $c = \sum_{j=1}^{m} b_{j,n} a_j$. The constant term of $h - \sum_{j=1}^{m} b_{j,n} f_j$ is then zero, so

$$g - \sum_{j=1}^{m} \left(\sum_{k=0}^{n} b_{j,k} X^k \right) f_j = hX^n - \sum_{j=1}^{m} b_{j,n} X^n f_j \in X^{n+1} B.$$

Let

$$h_j = \sum_{k\in\mathbb{N}} b_{k,j} X^k$$

for each $j \in [1,m]$. Then by (1), $g - \sum_{j=1}^{m} h_j f_j \in BX^n$ for all $n \in \mathbb{N}$, so $g = \sum_{j=1}^{m} h_j f_j$. ∎

23.3 Theorem. *Let A be a local ring.* (a) *If the natural topology of A is Hausdorff, the natural topology of $A[[X]]$ is Hausdorff.* (b) *If A is a complete local ring, so is $A[[X]]$.*

Proof. Let $B = A[[X]]$, and let M be the maximal ideal of A. By 23.1, the maximal ideal N of B is $M + BX$. Consequently

$$N^k = M^k + M^{k-1}X + \ldots + MX^{k-1} + BX^k$$

for all $k \geq 1$. It readily follows that if $\cap_{k=0}^\infty M^k = \{0\}$, then $\cap_{k=0}^\infty N^k = \{0\}$ also. Thus (a) holds.

(b) Let $(f_n)_{n \geq 1}$ be a Cauchy sequence in B. To show that $(f_n)_{n \geq 1}$ converges to an element of B, it suffices by 7.2 to show that some $f \in B$ is adherent to $(f_n)_{n \geq 1}$, and thus we may assume, by extracting a subsequence if necessary, that for each $k \geq 1$, $f_n - f_k \in N^k$ for all $n \geq k$. Let

$$f_n = \sum_{j \in \mathbb{N}} a_{n,j} X^j.$$

If $n \geq k \geq j$, then $a_{n,j} - a_{k,j} \in M^{k-j}$, so $(a_{n,j})_{n \geq 1}$ is a Cauchy sequence in A. Let $a_j = \lim_{n \to \infty} a_{n,j}$ for each $j \in \mathbb{N}$, and let

$$f = \sum_{j \in \mathbb{N}} a_j X^j.$$

Since $a_{n,j} - a_{k,j} \in M^{k-j}$ whenever $n \geq k \geq j$ and since M^{k-j} is open and thus closed for the natural topology of A, $a_j - a_{k,j} \in M^{j-k}$ for all $k \geq j$, and hence

$$f - f_k = \sum_{j \in \mathbb{N}} (a_j - a_{k,j}) X^j \in M^k + M^{k-1}X + \ldots + MX^{k-1} + BX^k = N^k$$

for all $k \geq 1$. Thus $f = \lim_{k \to \infty} f_k$. •

There is a natural extension of the definition of the power series ring (in one variable) over a commutative ring A with identity to one of the power series ring in several variables over A: Let $m \geq 1$, and let $\mathbb{N}^m$ be the cartesian product of m copies of the additive semigroup $\mathbb{N}$. The *ring of formal power series in m variables over A* is the set $S[A, \mathbb{N}^m]$ of all families of elements of A indexed by $\mathbb{N}^m$ (or equivalently, the set of all functions from $\mathbb{N}^m$ into A), where addition is defined componentwise and multiplication by

$$(a_n)_{n \in \mathbb{N}^m} (b_n)_{n \in \mathbb{N}^m} = \Big(\sum_{i+j=n} a_i b_j \Big)_{n \in \mathbb{N}^m}.$$

For each $n \in \mathbb{N}^m$, let n_i be the ith component of n for all $i \in [1, m]$, so that $n = (n_1, \ldots, n_m)$. It is customary to choose some capital letter, say X, and denote by X_i the element $(a_n)_{n \in \mathbb{N}^m}$ where $a_n = 0$ unless $n_i = 1$ and $n_j = 0$ for all other $j \in [1, m]$, in which case $a_n = 1$. From the definition of multiplication, it then follows that for any $r \in \mathbb{N}^m$, the element $(a_n)_{n \in \mathbb{N}^m}$, where $a_r = 1$ and $a_n = 0$ for all $n \neq r$, is simply $X^{r_1} X^{r_2} \ldots X^{r_m}$. Consequently, if $(a_n)_{n \in \mathbb{N}^m}$ is any element all but finitely many of whose terms are zero,

$$(a_n)_{n \in \mathbb{N}^m} = \sum_{(n_1, n_2, \ldots, n_m) \in \mathbb{N}^m} a_{(n_1, n_2, \ldots, n_m)} X_1^{n_1} X_2^{n_2} \ldots X_n^{n_m}.$$

For this reason, we denote any element $(a_n)_{n \in \mathbb{N}^m}$ by

$$\sum_{(n_1, n_2, \ldots, n_m) \in \mathbb{N}^m} a_{(n_1, n_2, \ldots, n_m)} X_1^{n_1} X_2^{n_2} \ldots X_m^{n_m},$$

or, for short,

$$\sum_{n \in \mathbb{N}^m} a_n X^n$$

where X^n stands for $X_1^{n_1} X_2^{n_2} \ldots X_m^{n_m}$. Consequently, we customarily denote the ring $S[A, \mathbb{N}^m]$ by $A[[X_1, X_2, \ldots, X_m]]$.

The ring $(A[[X_1, X_2, \ldots, X_{m-1}]])[[X_m]]$ is naturally isomorphic to the formal power series ring $A[[X_1, X_2, \ldots, X_m]]$ under the isomorphism Φ defined by

$$\Phi(\sum_{k \in \mathbb{N}} (\sum_{(n_1, n_2, \ldots, n_{m-1}) \in \mathbb{N}^{m-1}} a_{(n_1, n_2, \ldots, n_{m-1}, k)} X_1^{n_1} \ldots X_{m-1}^{n_{m-1}}) X_m^k) =$$

$$\sum_{(n_1, n_2, \ldots, n_{m-1}, k) \in \mathbb{N}^m} a_{(n_1, n_2, \ldots, n_{m-1}, k)} X_1^{n_1} X_2^{n_2} \ldots X_{m-1}^{n_{m-1}} X_m^k.$$

Consequently we obtain by induction from Theorems 23.1-23.3:

23.4 Theorem. *Let A be a commutative ring with identity, $m \geq 1$. (a) If A is a local ring with maximal ideal M, $A[[X_1, \ldots, X_m]]$ is a local ring with ideal $M + (X_1) + \ldots + (X_m)$. (b) If A is noetherian [an integral domain], so is $A[[X_1, \ldots, X_m]]$. (c) If A is a local ring that is Hausdorff for its natural topology, then so is $A[[X_1, \ldots, X_m]]$. (d) If A is a complete local ring, so is $A[[X_1, \ldots, X_m]]$.*

23.5 Theorem. *Let A be a commutative ring with identity that is Hausdorff and complete for the ring topology for which $(J^n)_{n\geq 1}$ is a fundamental system of neighborhoods of zero, where J is an ideal of A. Let C be a subring of A such that the restriction to C of the canonical epimorphism from A to A/J is a surjection, and let $x_1, \ldots, x_m \in J$. For any family $(c_n)_{n\in\mathbb{N}^m}$ of elements of C indexed by $\mathbb{N}^m$, the family $(c_n x_1^{n_1} \ldots x_m^{n_m})_{(n_1,\ldots,n_m)\in\mathbb{N}^m}$ is summable in A, and the function S from $C[[X_1, \ldots, X_m]]$ to A defined by*

$$S(\sum_{n\in\mathbb{N}^m} c_n X_1^{n_1} \ldots X_m^{n_m}) = \sum_{n\in\mathbb{N}^m} c_n x_1^{n_1} \ldots x_m^{n_m}$$

is a homomorphism. If $\{x_1, \ldots, x_m\}$ generates J, then S is an epimorphism.

Proof. For each $n \in \mathbb{N}^m$ we shall denote $n_1 + \ldots + n_m$ by $|n|$, the monomial $X_1^{n_1} \ldots X_m^{n_m}$ by X^n, and the element $x_1^{n_1} \ldots x_m^{n_m}$ of A by x^n. For any family $(c_n)_{n\in\mathbb{N}^m}$ of elements of C, $(c_n x^n)_{n\in\mathbb{N}^m}$ is summable by 10.5, since $c_n x^n \in J^k$ whenever $|n| \geq k$. By 10.12 and 10.16, S is a homomorphism.

Assume further that $J = Ax_1 + \cdots + Ax_m$. To show that S is surjective, let $y \in A$. We shall define c_n inductively for all $n \in \mathbb{N}^m$ so that

$$y - \sum_{|n|<k} c_n x^n \in J^k.$$

Indeed if that statement holds, then

$$y - \sum_{|n|<k} c_n x^n = \sum_{|p|=k} a_p x^p$$

where $a_p \in A$ for all $p \in \mathbb{N}^m$ such that $|p| = k$. By hypothesis, for each such $p \in \mathbb{N}^m$ there exists $c_p \in C$ such that $c_p - a_p \in J$, so

$$y - \sum_{|n|<k+1} c_n x^n = \sum_{|p|=k} (a_p - c_p)x^p \in JJ^k = J^{k+1}.$$

Clearly
$$y = \sum_{n\in\mathbb{N}_m} c_n x^n. \ \bullet$$

We may now characterize complete local noetherian rings:

23.6 Theorem. *Let A be a commutative ring with identity. The following statements are equivalent:*

$1°$ *A is a complete local ring whose maximal ideal is finitely generated.*

2° A is a complete local noetherian ring.

3° There exist a Cohen ring C and $m \geq 0$ such that A is an epimorphic image of $C[[X_1, \ldots, X_m]]$.

Proof. Clearly 2° implies 1°. If 1° holds, then A contains a Cohen subring by 21.20, so 3° holds by 23.5. Assume 3°. Then A is isomorphic to $C[[X_1, \ldots, X_m]]/J$ for some proper ideal J of $C[[X,, \ldots, X_m]]$. Consequently, 1° follows by 23.4, 20.16, 20.15, 6.4, and 6.12. •

Exercise

23.1 (Nagata [1949]) Let K be a field, and let $A = \{f/g \in K(X)$: the constant coefficient of g is not $0\}$. (a) A is a subring of $K[[X]]$ and hence of $K[[X, Y]]$. (b) Let $B = A + K[[X, Y]]Y$, $N = AX + K[[X, Y]]Y$. Show that B is a local ring whose maximal ideal is N, and that $N = BX + BY$. (c) Let M be the maximal ideal of $K[[X, Y]]$. Show that $N = M \cap B$ and, more generally, that $N^n = M^n \cap B$ for all $n \geq 1$. (d) B, furnished with its natural topology, is a proper dense subring of $K[[X, Y]]$, furnished with its natural topology. (e) There exists $z \in K[[X]] \setminus B$, and $zY \in B$. (f) The ideal BY of B is not closed in B. [Use (d) and (e).] (g) B is a local nonnoetherian ring whose maximal ideal is finitely generated and whose natural topology is Hausdorff. [Use 20.16.]

24 Complete Semilocal Noetherian Rings

Theorems relating properties of a commutative noetherian ring to those of its completion for the topology for which the powers of an ideal form a fundamental system of neighborhoods of zero have important applications in commutative algebra. Here we will present a few basic examples of such theorems.

24.1 Definition. *Let J be an ideal of a ring A. The J-topology on A is the ring topology for which $(J^n)_{n \geq 1}$ is a fundamental system of neighborhoods of zero. If E is an A-module, the J-topology on E is the additive group topology for which $(J^n E)_{n \geq 1}$ is a fundamental system of neighborhoods of zero.*

24.2 Theorem. *let E be an A-module, J an ideal of A. If A is furnished with its J-topology, then E, furnished with its J-topology, is a topological module over A. If M is a submodule of E, the quotient topology induced on E/M by the J-topology of E is the J-topology of E/M.*

Proof. Clearly $\{J^n E : n \geq 1\}$ satisfies (TMN 1)-(TMN 3) of 3.6 if A is furnished with the J-topology. The second assertion follows from the identity $J^n(E/M) = (J^n E + M)/M$. •

24.3 Theorem. *If E is a finitely generated unitary module over a commutative noetherian ring A with identity and if J is an ideal of A, then for each submodule F of E, the topology induced on F by the J-topology of E is the J-topology of F.*

The assertion is a consequence of 20.11.

Let E be an A-module, J an ideal of A, and assume that the J-topologies of A and E are Hausdorff. By 24.2 and 8.6 we may regard the completion $\widehat{E}$ of E for the J-topology as a module over the completion $\widehat{A}$ of A for its J-topology. If F is a submodule of E, we shall denote its closure in $\widehat{E}$ by $\widehat{F}$, since that is the completion of F for the topology induced on F by the J-topology of E. If the hypotheses of 24.3 hold, $\widehat{F}$ is also the completion of F for its J-topology.

24.4 Theorem. *Let E be a unitary A-module, and let $x_1, \ldots, x_n \in E$ be such that $E = Ax_1 + \ldots Ax_n$. If J is an ideal of A and if the J topologies of both A and E are Hausdorff, then*

$$\widehat{E} = \widehat{A}x_1 + \ldots + \widehat{A}x_n = \widehat{A}E.$$

Proof. Let $y \in \widehat{E}$. There is a Cauchy sequence $(y_k)_{k \geq 1}$ in E such that $\lim_{k \to \infty} y_k = y$. By extracting a subsequence if necessary, we may assume that $y_{k+1} - y_k \in J^k E$ for all $k \geq 1$. Thus for each $k \geq 1$, there exist $c_{k,1}, \ldots, c_{k,s(k)} \in J^k$ and $z_{k,1}, \ldots, z_{k,s(k)} \in E$ such that

$$y_{k+1} - y_k = \sum_{i=1}^{s(k)} c_{k,i} z_{k,i}.$$

For each $i \in [1, s(k)]$ let

$$z_{k,i} = \sum_{j=1}^{n} d_{k,i,j} x_j$$

where $d_{k,i,j} \in A$ for all $j \in [1, n]$. Then

$$y_{k+1} - y_k = \sum_{j=1}^{n} a_{k,j} x_j$$

where

$$a_{k,j} = \sum_{i=1}^{s(k)} c_{k,i} d_{k,i,j} \in J^k.$$

Let
$$y_1 = \sum_{j=1}^{n} b_{1,j} x_j$$

where $b_{1,1}, \ldots b_{1,n} \in A$. For each $j \in [1,n]$ we define $(b_{k,j})_{k \geq 1}$ recursively by $b_{k+1,j} = b_{k,j} + a_{k,j}$. An inductive argument establishes that

$$y_k = \sum_{j=1}^{n} b_{k,j} x_j$$

for all $k \geq 1$. Since $b_{k+1,j} - b_{k,j} = a_{k,j} \in J^k$, $(b_{k,j})_{k \geq 1}$ is a Cauchy sequence in A and hence has a limit b_j in $\widehat{A}$. Clearly

$$y = \lim_{k \to \infty} y_k = \lim_{k \to \infty} \sum_{j=1}^{n} b_{k,j} x_j = \sum_{j=1}^{n} b_j x_j. \bullet$$

24.5 Corollary. *Let E be a finitely generated unitary A-module, J a finitely generated ideal of A. If the J-topologies of A and E are Hausdorff, then*

$$\widehat{J^n E} = \widehat{J^n} E = \widehat{A} J^n E$$

for all $n \geq 1$, and in particular, $\widehat{J^n} = \widehat{A} J^n$. If, moreover, A is commutative, then

$$(\widehat{J})^n = \widehat{J^n},$$

and the topologies of $\widehat{E}$ and $\widehat{A}$ are their $\widehat{J}$-topologies.

Proof. The topology induced on $J^n E$ by the J topology of E is the J-topology of $J^n E$, since $J^m(J^n E) = J^{m+n} E \cap J^n E$ for all $m \geq 1$. Therefore $\widehat{J^n E}$ is also the completion, for the J-topology, of $J^n E$, which is finitely generated as both E and J are. Consequently by 24.4, $\widehat{J^n E} = \widehat{A} J^n E$. In particular, $\widehat{J^n} = \widehat{A} J^n$.

Assume further that A is commutative. Then $\widehat{A}$ is commutative by 8.3, so

$$(\widehat{J})^n = (\widehat{A} J)^n = \widehat{A} J^n = \widehat{J^n}.$$

Consequently,

$$\widehat{J^n E} = \widehat{A} J^n E = (\widehat{J})^n E = (\widehat{J})^n \widehat{A} E = (\widehat{J})^n \widehat{E}.$$

As $(\widehat{J^n E})_{n \geq 1}$ is a fundamental system of neighborhoods of zero for the topology of $\widehat{E}$ by 4.22, therefore, the topology of $\widehat{E}$ is its $\widehat{J}$-topology. In particular, the topology of $\widehat{A}$ is its $\widehat{J}$-topology. $\bullet$

24.6 Corollary. *Let A be a commutative noetherian ring with identity, and let E be a finitely generated unitary A-module. If J is an ideal of A such that the J-topologies of both A and E are Hausdorff, then for any $x_1, \ldots, x_n \in E$, if $F = Ax_1 + \ldots + Ax_n$, the closure $\widehat{F}$ of F in $\widehat{E}$ is $\widehat{A}F = \widehat{A}x_1 + \ldots \widehat{A}x_n$.*

Proof. By 24.3, $\widehat{F}$ is also the completion of F for its J-topology, so by 28.4, $\widehat{F} = \widehat{A}F = \widehat{A}x_1 + \ldots + \widehat{A}x_n$. •

24.7 Theorem. *Let A be a commutative ring with identity furnished with the M-topology, where M is an ideal of A and the M-topology of A is Hausdorff. (1) If A is noetherian, then $\widehat{A}$ is noetherian, and the topology of $\widehat{A}$ is its $\widehat{M}$-topology. (2) If M is a finitely generated maximal ideal, then $\widehat{A}$ is a local noetherian ring whose maximal ideal is $\widehat{M}$, and the topology of $\widehat{A}$ is its natural topology.*

Proof. In both cases, the topology of $\widehat{A}$ is its $\widehat{M}$-topology by 24.5. (1) By 5.14 applied to the dense subring A of $\widehat{A}$ and its open ideal M and by 23.5, $\widehat{A}$ is an epimorphic image of $A[[X_1, \ldots, X_m]]$ for some $m \geq 0$, so $\widehat{A}$ is noetherian by 23.4 and 20.4. (2) If $x \in \widehat{M}$, then x is a topological nilpotent, so $1 - x$ is invertible by 11.16. Thus each element of $1 + \widehat{M}$ is invertible in $\widehat{A}$. By 5.14, A/M and $\widehat{A}/\widehat{M}$ are isomorphic, so M is a maximal ideal of $\widehat{A}$. Therefore if $y \in \widehat{A} \setminus \widehat{M}$, there exists $z \in \widehat{A} \setminus \widehat{M}$ such that $1 - yz \in \widehat{M}$; by the preceding, yz is invertible in $\widehat{A}$, so y is also. Thus $\widehat{A}$ is a local ring whose maximal ideal is $\widehat{M}$. By 24.5, $\widehat{M} = \widehat{A}M$ and hence $\widehat{M}$ is a finitely generated ideal of $\widehat{A}$. Therefore by 23.6, $\widehat{A}$ is noetherian. •

24.8 Corollary. *The completion of a local noetherian ring, furnished with its natural topology, is a complete local noetherian ring, furnished with its natural toology.*

For our subsequent discussion, we shall need the following definitions and three theorems of algebra:

24.9 Definition. *Let A be a ring. Ideals I and J are **relatively prime** if $I + J = A$. A family $(I_\lambda)_{\lambda \in L}$ of ideals is **pairwise relatively prime** if I_λ and I_μ are relatively prime whenever $\lambda, \mu \in L$ and $\lambda \neq \mu$.*

The hypothesis $(A/I)^2 = A/I$ of the following theorem, which is equivalent to the statement $A^2 + I = A$, is always satisfied if A has an identity element.

24.10 Theorem. *Let $I, J_1, \ldots, J_n$ be ideals of a ring A such that $(A/I)^2 = A/I$. If I and J_k are relatively prime for each $k \in [1, n]$, then*

I and $J_1 J_2 \ldots J_n$ are relatively prime, and a fortiori I and $J_1 \cap J_2 \cap \cdots \cap J_n$ are relatively prime. In particular, if I and J are relative prime, then for any $n \geq 1$, I and J^n are relatively prime.

Proof. By induction, $(A/I)^m = A/I$ for all $m \geq 1$, that is, $A^m + I = A$. In particular

$$A = A^n + I = (I + J_1)(I + J_2) \ldots (I + J_n) + I = I + J_1 J_2 \ldots J_n.$$

Since $J_1 J_2 \ldots J_n \subseteq J_1 \cap J_2 \cap \cdots \cap J_n$, the second assertion also holds. ●

24.11 Theorem. *Let $J_1, J_2, \ldots, J_n$ be pairwise relatively prime ideals of a ring A such that $(A/J_k)^2 = A/J_k$ for each $k \in [1, n-1]$. Then $\Phi : x \to (x+J_1, x+J_2, \ldots, x+J_n)$ is an epimorphism from A to $\prod_{k=1}^n (A/J_k)$, that is, for any sequence $a_1, a_2, \ldots, a_n \in A$ there exists $c \in A$ such that $c \equiv a_k \pmod{J_k}$ for each $k \in [1, n]$. If, moreover, A is a commutative ring with identity, then $J_1 \cap J_2 \cap \cdots \cap J_n = J_1 J_2 \ldots J_n$.*

Proof. We shall show by induction that if there exists $b \in A$ such that $b \equiv a_k \pmod{J_k}$ for all $k \in [m, n]$, where $1 < m \leq n$, then there exists $c \in A$ such that $c \equiv a_k \pmod{J_k}$ for all $k \in [m-1, n]$. By 24.10, J_{m-1} and $\cap_{k=m}^n J_k$ are relatively prime, so there exist $x \in J_{m-1}$ and $y \in \cap_{k=m}^n J_k$ such that $x + y = b - a_{m-1}$. Let $c = x + a_{m-1} = b - y$. Then $c \equiv a_{m-1} \pmod{J_{m-1}}$, and for each $k \in [m, n]$, $c \equiv b \equiv a_k \pmod{J_k}$. Thus by induction, the first assertion holds.

For the second assertion, assume that $J_m \cap \cdots \cap J_n = J_m \ldots J_n$ where $1 < m \leq n$. We shall show that $J_{m-1} \cap J_m \cap \cdots \cap J_n = J_{m-1} J_m \ldots J_n$. Let $J = J_m \cap \cdots \cap J_n = J_m \ldots J_n$. By 24.10, J_{m-1} and J are relatively prime, so there exist $e \in J_{m-1}$ and $f \in J$ such that $e + f = 1$. If $x \in J_{m-1} \cap J$, then $x = xe + xf \in JJ_{m-1} + J_{m-1}J$. Thus $J_{m-1} \cap J \subseteq J_{m-1}J \subseteq J_{m-1} \cap J$, so

$$J_{m-1} \cap J_m \cap \cdots \cap J_n = J_{m-1} J_m \ldots J_n.$$

An inductive argument thus establishes the assertion. ●

24.12 Theorem. *Let $(A_\lambda)_{\lambda \in L}$ be a family of rings with identity, let A be a subring of $\prod_{\lambda \in L} A_\lambda$ containing $\bigoplus_{\lambda \in L} A_\lambda$, and for each $\mu \in L$ let pr_μ be the canonical projection from $\prod_{\lambda \in L} A_\lambda$ to A_μ, defined by $pr_\mu((x_\lambda)_{\lambda \in L}) = x_\mu$. If J is a left or right ideal of A, then $\bigoplus_{\lambda \in L} pr_\lambda(J) \subseteq J$; in particular, if L is finite, $J = \prod_{\lambda \in L} pr_\lambda(J)$.*

Proof. For each $\mu \in L$ let in_μ be the canonical injection from A_μ to $\prod_{\lambda \in L} A_\lambda$ (so that $pr_\mu \circ in_\mu$ is the identity mapping of A_μ), and let $e_\mu = in_\mu(1_\mu)$, where 1_μ is the identity element of A_μ. Then $e_\mu \in \bigoplus_{\lambda \in L} A_\lambda \subseteq A$, and for any $z \in A$,

$$in_\mu(pr_\mu(z)) = ze_\mu = e_\mu z.$$

Let $(x_\lambda)_{\lambda \in L} \in \bigoplus_{\lambda \in L} pr_\lambda(J)$; then for each $\lambda \in L$ there exists $y_\lambda \in J$ such that $x_\lambda = pr_\lambda(y_\lambda)$, and there is a finite subset M of L such that $x_\lambda = 0$ for all $\lambda \in L \setminus M$. Therefore

$$(x_\lambda)_{\lambda \in L} = \sum_{\mu \in M} in_\mu(x_\mu) = \sum_{\mu \in M} in_\mu(pr_\mu(y_\mu)) = \sum_{\mu \in M} e_\mu y_\mu \in J$$

if J is a left ideal, and similarly $(x_\lambda)_{\lambda \in L} = \sum_{\mu \in M} y_\mu e_\mu \in J$ if J is a right ideal. •

24.13 Definition. *Let A be a commutative ring with identity. The* **radical** *of A is the intersection of the maximal ideals of A.*

24.14 Theorem. *Let A be a commutative noetherian ring with identity, and let J be an ideal of A. The following statements are equivalent:*

$1°$ *Every finitely generated unitary A-module is Hausdorff for the J-topology.*
$2°$ *Every submodule of every finitely generated unitary A-module is closed for the J-topology.*
$3°$ *Every ideal of A is closed for the J-topology.*
$4°$ *J is contained in the radical of A.*

In particular, if R is the radical of A, the R-topology of A or of any finitely generated unitary A-module is Hausdorff.

Proof. To show that $1°$ implies $2°$, let F be a submodule of a finitely generated unitary A-module E. By $1°$, E/F is Hausdorff for its J-topology, which by 24.2 is quotient topology induced on E/F by the J-topology of E. Therefore by 5.7, F is closed in E.

To show that $3°$ implies $4°$, assume that J is not contained in the radical of A. Then there is a maximal ideal M of A such that $J \not\subseteq M$, so $M + J = A$. By 24.10, $M + J^n = A$ for all $n \geq 1$, so the closure of M for the J-topology is A by (3) of 3.3.

Finally, to show that $4°$ implies $1°$, let E be a finitely generated unitary A-module, and let $x \in \cap_{n=1}^{\infty} J^n E$. By 20.13 there exists $a \in J$ such that $(1-a)x = 0$. Since a is contained in the radical of A, $1-a$ belongs to no maximal ideal of A and hence is invertible, so $x = 0$. •

24.15 Definition. *A* **semilocal** *ring is a commutative ring with identity that has only finitely many maximal ideals. The* **natural topology** *of a semilocal ring A or of a unitary A-module E is its R-topology, where R is the radical of A. A* **complete semilocal ring** *is a semilocal ring that is Hausdorff and complete for its natural topology.*

24.16 Theorem. *Let A be a semilocal ring with radical R.*
(1) A/R has only finitely many ideals.
(2) The natural topology of A is the supremum of its M-topologies, where M is a maximal ideal of A.
(3) R is a nilpotent ideal if and only if (0) is a product of maximal ideals.
(4) If A is noetherian, then A is Hausdorff and each of its ideals is closed for the natural topology; more generally, if E is a finitely generated unitary A-module, then E is Hausdorff and each of its submodules is closed for the natural topology.

Proof. Let $M_1, \ldots, M_r$ be the maximal ideals of A. By 24.11, A/R is isomorphic to $\prod_{k=1}^r (A/M_k)$, the cartesian product of r fields, so A/R has 2^r ideals by 24.12. By 24.10 and 24.11,

$$R^n = (M_1 \cap \cdots \cap M_r)^n = (M_1 \ldots M_r)^n = M_1^n \ldots M_r^n = M_1^n \cap \cdots \cap M_r^n.$$

Thus (2) and (3) hold, and (4) follows from 24.14. ∎

24.17 Theorem. *Let A be a semilocal ring, R its radical. (1) If R is finitely generated, so is each maximal ideal of A. (2) If the natural topology of A is Hausdorff and if each maximal ideal of A is finitely generated, then the completion $\widehat{A}$ of A for the natural topology is a semilocal noetherian ring whose maximal ideals are the closures in $\widehat{A}$ of the maximal ideals of A, $\widehat{A}$ is the topological direct sum of finitely many complete local noetherian rings, and the topology of $\widehat{A}$ is its natural topology.*

Proof. By (1) of 24.16 and 20.2, A/R is noetherian, so if M is a maximal ideal of A, there exist $x_1, \ldots, x_s \in M$ such that

$$M/R = (A/R)(x_1 + R) + \cdots + (A/R)(x_s + R),$$

whence $M = Ax_1 + \ldots, +Ax_s + R$. Thus if R is finitely generated, so is M.

Let $M_1, \ldots, M_r$ be the maximal ideals of A. Then for any $n \geq 1$, $M_1^n, \ldots, M_r^n$ are pairwise relatively prime by 24.10, so

$$R^n = (\bigcap_{k=1}^r M_k)^n = (M_1 \ldots M_r)^n = M_1^n \ldots M_r^n = \bigcap_{k=1}^r M_k^n$$

by 24.11. To prove (2), for each $k \in [1,r]$ let $N_k = \bigcap_{n=1}^\infty M_k^n$, and let $A_k = A/N_k$, furnished with its (M_k/N_k)-topology. By (2) of 24.7, $\widehat{A}_k$ is a local noetherian ring whose topology is its natural topology. By hypothesis,

$$\bigcap_{k=1}^r N_k = \bigcap_{k=1}^r (\bigcap_{n=1}^\infty M_k^n) = \bigcap_{n=1}^\infty (\bigcap_{k=1}^r M_k^n) = \bigcap_{n=1}^\infty R^n = (0).$$

Therefore the function Δ from A to $\prod_{k=1}^{n} A_k$, defined by

$$\Delta(x) = (x + N_1, \ldots, x + N_r),$$

is a monomorphism. Moreover, for any $n \geq 1$,

$$\Delta(R^n) = \prod_{k=1}^{r}(M_k^n/N_k) \cap \Delta(A),$$

for $x \in R^n$ if and only if for all $i \in [1, r]$, $x \in M_i^n$, or equivalently, $x + N_i \in M_i^n/N_i = (M_i/N_i)^n$. Therefore Δ is a topological isomorphism from A to its range A'. Furthermore, A' is dense in $\prod_{k=1}^{r} A_k$, for if $n \geq 1$ and if $a_1, \ldots, a_r \in A$, there exists $x \in A$ such that $x \equiv a_k \pmod{M_k^n}$ for all $k \in [1, r]$ by 24.11. By 8.4 there is a topological isomorphism $\widehat{\Delta}$ from $\widehat{A}$ to $\prod_{k=1}^{r} \widehat{A}_k$ that extends Δ. Thus $\widehat{A}$ is the topological direct sum of finitely many complete local noetherian rings. In particular, $\widehat{A}$ is noetherian by 20.7, and its topology is its natural topology. •

24.18 Theorem. *If A is a semilocal ring, then A is noetherian if and only if each maximal ideal of A is finitely generated and each ideal of A is closed for the natural topology.*

Proof. The condition is necessary by 24.14. Sufficiency: Since the zero ideal is closed, the natural topology of A is Hausdorff. By 24.17, $\widehat{A}$ is noetherian. If $(J_n)_{n \geq 1}$ is an increasing sequence of ideals in A, their closures $(\widehat{J}_n)_{n \geq 1}$ in $\widehat{A}$ form an increasing sequence of ideals, so there exists $q \geq 1$ such that $\widehat{J}_n = \widehat{J}_q$ for all $n \geq q$, whence

$$J_n = \widehat{J}_n \cap A = \widehat{J}_q \cap A = J_q$$

for all $n \geq q$, as each ideal of A is closed. •

24.19 Theorem. *If A is a commutative ring with identity, then A is a complete semilocal noetherian ring if and only if A is the topological direct sum of finitely many complete local noetherian rings.*

Proof. The condition is necessary by (2) of 24.17. Sufficiency: Let $A = \prod_{i=1}^{n} A_i$ where each A_i is a complete local noetherian ring with maximal ideal M_i. The maximal ideals of A are clearly the ideals $pr_i^{-1}(M_i)$ where $i \in [1, n]$ and pr_i is the canonical projection of A on A_i. Thus A is semilocal and its radical R satisfies

$$R = \bigcap_{i=1}^{n} pr_i^{-1}(M_i) = \prod_{i=1}^{n} M_i.$$

By 20.7 A is noetherian, and clearly $R^k = \prod_{i=1}^n M_i^k$ for all $k \geq 1$, so the cartesian product topology on A determined by the natural topologies of $A_1, \ldots, A_n$ is the natural topology of A. By 7.8, A is complete for that topology. •

Exercises

In these exercises, all rings are commutative rings with identity.

24.1 A ring topology on a ring A is a *Zariski topology* is it is an ideal topology for which every ideal is closed. For example (Zariski [1945]), if A is a noetherian ring and if J is an ideal of A, the J-topology is a Zariski topology if and only if J is contained in the radical of A (Theorem 24.14). (a) A Zariski topology on a ring is Hausdorff. (b) If J is an ideal of ring A, the topology induced on A/J by a Zariski topology on A is a Zariski topology. (c) If A is complete for a Zariski topology, it is complete for any stronger ideal topology. [Use 7.21.] (d) (Chevalley [1943]) If $\hat{A}$ is the completion of A for a Zariski topology and if c is a cancellable element of A, then c is a cancellable element of $\hat{A}$. [Use 7.20.]

24.2 (Zariski [1945]) If a noetherian ring A is Hausdorff and complete for the J-topology, where J is an ideal of A, then the J-topology is a Zariski topology, and consequently A is also complete for any stronger ideal topology. [Use 24.6.]

24.3 (Lafon [1955]) If $J_1, \ldots, J_r$ are ideals of a noetherian ring A and if E is a finitely generated A-module, then the supremum of the J_k-topologies, $k \in [1, r]$, is the $(J_1 \ldots J_n)$-topology on E. [Use 20.11.]

24.4 Let $M_1, \ldots, M_n$ be distinct maximal ideals of a commutative ring A with identity. If J is an ideal of A contained in $\bigcup_{i=1}^n M_i$, then for some $k \in [1, n]$, $J \subseteq M_k$. [Use 24.11 and 24.12.]

CHAPTER VI

PRIMITIVE AND SEMISIMPLE RINGS

This chapter is mostly devoted to fundamental concepts occurring in the theory of noncommutative rings. In §25 we discuss primitive rings, in §26 the radical of an arbitrary ring, and in §27 artinian rings and modules, where we conclude with the celebrated Artin-Wedderburn theorem.

25 Primitive Rings

If E is a commutative group, $\text{End}(E)$ is the ring of all endomorphisms of E, and if E is a K-module, $\text{End}_K(E)$ is the ring of all linear operators on E.

25.1 Definition. *If E is a commutative group, a subring A of $\text{End}(E)$ is a* **primitive ring of endomorphisms** *of E if for all x, $y \in E$ such that $x \neq 0$ there exists $a \in A$ such that $a(x) = y$.*

If E is a vector space, then A is a primitive ring of endomorphisms of (the additive group) E if and only if A is 1-fold transitive in the following sense:

25.2 Definition. *Let E be a vector space over a division ring K. A subring A of $\text{End}(E)$ is* **n-fold transitive** *if for every linearly independent sequence $x_1, \ldots, x_n$ of n vectors of E and every sequence $y_1, \ldots, y_n$ of vectors of E there exists $a \in A$ such that $a(x_i) = y_i$ for all $i \in [1, n]$. The ring A is a* **dense ring of linear operators** *on E if A is n-fold transitive for all $n \geq 1$.*

For example, any subring of $\text{End}_K(E)$ that contains all linear operators whose range is finite-dimensional is a dense ring of linear operators.

We shall need an extension of Definition 25.2 to one for left ideals of a ring of linear operators. If L is a subset of $\text{End}_K(E)$, we shall call the subspace of all $x \in E$ such that $v(x) = 0$ for all $v \in L$ the annihilator of L and denote it by $\text{Ann}_E(L)$.

25.3 Definition. *Let A be a ring of linear operators on a vector space E over a division ring K such that $\text{Ann}_E(A) = \{0\}$. If L is a left ideal of A and if $M = \text{Ann}_E(L)$, then L is* **n-fold transitive** *if for every sequence*

$x_1, \ldots, x_n$ of vectors of E such that $x_1 + M, \ldots, x_n + M$ is a linearly independent sequence of vectors in E/M and for every sequence $y_1, \ldots, y_n$ of vectors of E there exists $a \in L$ such that $a(x_i) = y_i$ for all $i \in [1, n]$.

25.4 Theorem. *Let E be a vector space over a division ring K, let A be a 1-fold transitive ring of linear operators on E, let L be a left ideal of A, and let $M = \mathrm{Ann}_E(L)$. For any $n \geq 1$, L is n-fold transitive if and only if for every sequence $x_1, \ldots, x_n$ of n vectors of E such that $x_1 + M, \ldots, x_n + M$ is a linearly independent sequence of n vectors of E/M there exists $a \in L$ such that $a(x_n) \neq 0$ and $a(x_i) = 0$ for all $i < n$.*

Proof. Sufficiency: Let $y_1, \ldots, y_n \in E$. For each $j \in [1, n]$, the sequence $x_{j+1} + M, \ldots, x_n + M, x_1 + M, \ldots, x_j + M$ is a linearly independent sequence, so there exists $a_j \in L$ such that $a_j(x_i) = 0$ if $i \neq j$ and $a_j(x_j) \neq 0$. As A is 1-fold transitive, for each $i \in [1, n]$ there exists $b_i \in A$ such that $b_i(a_i(x_i)) = y_i$. Let

$$a = \sum_{j=1}^{n} b_j a_j \in L.$$

Clearly $a(x_i) = y_i$ for all $i \in [1, n]$. •

25.5 Theorem. *Let E be a vector space over a division ring K. If a subring A of $\mathrm{End}(E)$ is 1-fold transitive and if a left ideal L of A is 2-fold transitive, then L is n-fold transitive for all $n \geq 1$.*

Proof. Let $M = \mathrm{Ann}_E(L)$. First, L is 1-fold transitive, for if $x \in E \setminus M$ and if $y \in E$, there exists $a \in L$ such that $a(x) \neq 0$, and hence there exists $b \in A$ such that $b(a(x)) = y$; then $ba \in L$ and $(ba)(x) = y$.

Assume that L is n-fold transitive for all $n < m$, where $m \geq 3$. To show that L is m-fold transitive, it suffices by 25.4 to show that if $x_1, \ldots, x_m$ are vectors of E such that $x_1 + M, \ldots, x_m + M$ is a linearly independent sequence of vectors of E/M, there exists $c \in L$ such that $c(x_i) = 0$ for all $i < m$ and $c(x_m) \neq 0$. By assumption, for each $i \in [1, m-1]$ there exists $a_i \in L$ such that $a_i(x_i) = x_i$ and $a_i(x_j) = 0$ for all $j \in [1, m-1]$ such that $j \neq i$. Let

$$a = \sum_{i=1}^{m-1} a_i \in L.$$

Case 1: $a(x_m) - x_m \notin M$. By hypothesis there exists $b \in L$ such that $b(a(x_m) - x_m) \neq 0$. Thus if $c = ba - b \in L$, then c has the desired properties.

Case 2: $a(x_m) - x_m \in M$. Suppose that for all $i < m$, $x_i + M$ and $a_i(x_m) + M$ were linearly dependent vectors in E/M. Then for each $i < m$

there would exist $\lambda_i \in K$ such that $\lambda_i x_i - a_i(x_m) \in M$. Consequently,

$$\sum_{i=1}^{m-1} \lambda_i x_i - x_m = \sum_{i=1}^{m-1} (\lambda_i x_i - a_i(x_m)) + (a(x_m) - x_m) \in M,$$

a contradiction of the linear independence of $x_1 + M, \ldots, x_m + M$. Thus there exists $j < m$ such that $x_j + M$ and $a_j(x_m) + M$ are linearly independent. As L is 2-fold transitive, there exists $b \in L$ such that $b(x_j) = 0$ and $b(a_j(x_m)) \neq 0$. If $c = ba_j \in L$, then c has the desired properties. •

25.6 Theorem. (Density Theorem) *Let A be a primitive ring of endomorphisms of a nonzero commutative group E. The set D of all endomorphisms of E that commute with each member of A is a division subring of $\mathrm{End}(E)$. Under scalar multiplication defined by $\lambda.x = \lambda(x)$ for all $\lambda \in D$ and all $x \in E$, E is a vector space over D, A is a dense ring of linear operators on the D-vector space E, and more generally, every left ideal L of A is n-fold transitive for all $n \geq 1$.*

Proof. Clearly D is a ring with identity. Assume that $\lambda \in D^*$. Then there exists $x \in E$ such that $\lambda(x) \neq 0$. For any nonzero $y \in E$ there exist $a, b \in A$ such that $a(\lambda(x)) = y$ and $b(y) = x$, whence

$$\lambda(a(x)) = a(\lambda(x)) = y,$$

so λ is surjective, and

$$b(\lambda(y)) = \lambda(b(y)) = \lambda(x) \neq 0,$$

so λ is injective. Therefore λ is an automorphism of E. As λ commutes with each member of A, so does λ^{-1}. Thus D is a division ring.

Clearly E is a D-vector space under the indicated scalar multiplication, and each $a \in A$ is a linear operator on the D-vector space E, since for all $\lambda \in D$ and all $x \in E$,

$$a(\lambda.x) = a(\lambda(x)) = \lambda(a(x)) = \lambda.a(x).$$

Let L be a left ideal of A, and let $M = \mathrm{Ann}_E(L)$. As A is primitive, A is 1-fold transitive. To show that L is n-fold transitive for all $n \geq 1$, it suffices by 25.5 and 25.4 to show that if $x_1 + M$ and $x_2 + M$ are linearly independent vectors of E/M, there exists $a \in L$ such that $a(x_1) = 0$ and $a(x_2) \neq 0$. Suppose, on the contrary, that for all $a \in L$, $a(x_1) = 0$ implies $a(x_2) = 0$. Then $\mu : a(x_1) \to a(x_2)$ for all $a \in L$ is a well-defined function from E into E. To establish this, we first note that for any $x \in E$ there exists $a \in L$ such that $a(x_1) = x$. Indeed, as $x_1 \notin M$, there exists $c \in L$ such that $c(x_1) \neq 0$, so as A is primitive, there exists $b \in A$ such that $b(c(x_1)) = x$; thus if $a = bc \in L$, then $a(x_1) = x$. Moreover, if $b(x_1) = c(x_1)$ where $b, c \in L$, then $(b-c)(x_1) = 0$, so by assumption $(b-c)(x_2) = 0$, that is, $b(x_2) = c(x_2)$. Thus μ is well defined, and μ is clearly an endomorphism of the additive group E. If $a \in L$ and $b \in A$, then $ba \in L$, so

$$(b\mu)(a(x_1)) = b(\mu(a(x_1))) = b(a(x_2)) = (ba)(x_2)$$
$$= \mu((ba)(x_1)) = \mu(b(a(x_1))) = (\mu b)(a(x_1)).$$

Hence $\mu \in D$. For any $a \in L$,

$$a(\mu.x_1) = a(\mu(x_1)) = \mu(a(x_1)) = a(x_2),$$

so $a(\mu.x_1 - x_2) = 0$; hence $\mu.x_1 - x_2 \in M$, a contradiction of the linear independence of $x_1 + M$ and $x_2 + M$. Therefore there exists $a \in L$ such that $a(x_1) = 0$ and $a(x_2) \neq 0$.

Applying this result to the case $L = A$, we conclude that A is a dense ring of linear operators on E. •

Conversely, if A is a dense ring of linear operators on a nonzero K-vector space E, then K is in a natural way isomorphic to the ring of all endomorphisms of the commutative group E that commute with each member of A:

25.7 Theorem. *If A is a dense ring of linear operators on a nonzero K-vector space E, and if for each $\lambda \in K$, $\hat{\lambda}$ is the endomorphism of the commutative group E defined by $\hat{\lambda}(x) = \lambda.x$ for all $x \in E$, then $\lambda \to \hat{\lambda}$ is an isomorphism from K to the division ring D of all endomorphisms of the commutative group E that commute with each member of A.*

Proof. The only nontrivial verification is to show that if $v \in D$, then $v = \hat{\lambda}$ for some $\lambda \in K$. For any nonzero $x \in E$, if x and $v(x)$ were linearly independent, then there would exist $u \in A$ such that $u(x) = x$ and $u(v(x)) = x$, whence $(uv)(x) = x \neq v(x) = (vu)(x)$, a contradiction. Thus for each nonzero $x \in E$ there exists $\lambda_x \in K$ such that $v(x) = \lambda_x x$. We need only show, therefore, that if x and y are nonzero vectors of E, then $\lambda_x = \lambda_y$. There exists $u \in A$ such that $u(x) = y$, so

$$\lambda_y y = v(y) = (vu)(x) = (uv)(x) = u(\lambda_x x) = \lambda_x u(x) = \lambda_x y,$$

whence $\lambda_y = \lambda_x$. •

25.8 Corollary. *Let A be a dense ring of linear operators on a K-vector space E, and let L be a left ideal of A. If $F = \mathrm{Ann}_E(L) + N$ where N is a finite-dimensional subspace, then for any $u \in A$ satisfying $u(y) = 0$ for all $y \in \mathrm{Ann}_E(L)$ there exists $v \in L$ such that $v(x) = u(x)$ for all $x \in F$.*

Proof. Let $M = \mathrm{Ann}_E(L)$, and let $\{x_1, \ldots, x_n\}$ be a basis of a supplement of M in F. Then $\{x_1 + M, \ldots, x_n + M\}$ is a basis of F/M, so by 25.6 and 25.7 there exists $v \in L$ such that $v(x_i) = u(x_i)$ for all $i \in [1,n]$. Consequently, $v(x) = u(x)$ for all $x \in F$. •

If A is a ring of linear operators on a vector space E and if F is a subspace of E, we shall call the *annihilator* of F in A the left ideal of all $u \in A$ such that $u(F) = (0)$ and denote it by $\mathrm{Ann}_A(F)$.

25.9 Corollary. *Let A be the ring of all linear operators on a nonzero finite-dimensional K-vector space E. Then $L \to \mathrm{Ann}_E(L)$ is an order-inverting bijection from the set of all left ideals of A to the set of all subspaces of E, and its inverse is the bijection $F \to \mathrm{Ann}_A F$. Thus the zero ideal is the only proper ideal of A.*

Proof. By 25.8, if L is a left ideal of A, $\mathrm{Ann}_A(\mathrm{Ann}_E(L)) = L$, and clearly for any subspace F of E, $\mathrm{Ann}_E(\mathrm{Ann}_A(F)) = F$. •

25.10 Theorem. *Let A be a dense ring of linear operators on a K-vector space E. (1) A nonzero ideal J of A is also a dense ring of linear operators on E. (2) If e is a nonzero idempotent of A with range M and if for each $v \in eAe$, v_M is the function obtained by restricting the domain and codomain of v to M, then $v \to v_M$ is an isomorphism from eAe to a dense ring of linear operators on M.*

Proof. (1) By hypothesis there exist $u \in J$ and nonzero vectors a and b such that $u(a) = b$. Let $x_1, \ldots, x_n$ be a linearly independent sequence of vectors, and let $y_1, \ldots, y_n \in E$. For each $i \in [1,n]$ there exist $v_i, w_i \in A$ such that $v_i(x_i) = a$, $v_i(x_j) = 0$ if $j \neq i$, and $w_i(b) = y_i$. Let

$$t = \sum_{j=1}^m w_j u v_j \in J.$$

Clearly $t(x_i) = y_i$ for all $i \in [1,n]$.

(2) Let $x_1, \ldots, x_n$ be a linearly independent sequence of vectors in M, and let $y_1, \ldots, y_n \in M$. There exists $u \in A$ such that $u(x_i) = y_i$ for all $i \in [1,n]$. But then, as $e(x) = x$ for all $x \in M$, $eue_M(x_i) = eue(x_i) = y_i$ for all $i \in [1,n]$. If $v \in eAe$ is such that $v(M) = \{0\}$ and if N is the kernel of e, then $v(N) = \{0\}$ as $v = eve$, so $v = 0$ as E is the direct sum of M and N. Thus $v \to v_M$ is an isomorphism from eAe to a dense ring of linear operators on M. •

25.11 Corollary. *Let A be a dense ring of linear operators on a K-vector space E. If I and J are ideals such that $IJ = (0)$, then either $I = (0)$ or $J = (0)$. In particular, A is not the direct sum of two proper ideals.*

Proof. Suppose $I \neq (0)$ and $J \neq (0)$, and let a be a nonzero vector in E. By (1) of 25.10 there exist $u \in I$ and $v \in J$ such that $u(a) = a = v(a)$, whence $uv \neq 0$. •

25.12 Definition. *A ring A is **left primitive**, or simply **primitive**, if it is isomorphic to a primitive ring of endomorphisms of a nonzero commutative group; A is **right primitive** if A is anti-isomorphic to a left primitive ring.*

Primitivity may be expressed in terms of modules: An A-module E is *simple* if E contains no proper nonzero submodules, and E is *faithful* if $\text{Ann}_A(E) = (0)$. Let $\phi : a \to \hat{a}$ be an isomorphism from a ring A to a ring $\hat{A}$ of endomorphisms of a nonzero commutative group E. We may make E into an A-module by defining $a.x = \hat{a}(x)$ for all $a \in A$, $x \in E$. If $\hat{A}$ is a primitive ring of endomorphisms of E, then $A.x = E$ for all nonzero $x \in E$, and hence E is a simple nontrivial A-module; as ϕ is injective, E is faithful. Thus a primitive ring admits a faithful simple nontrivial module.

Conversely, assume that E is a faithful simple nontrivial A-module, and for each $a \in A$, let $\hat{a}$ be the endomorphism of the commutative group E defined by $\hat{a}(x) = a.x$. Since E is faithful, $\phi : a \to \hat{a}$ is an isomorphism from A to a subring $\hat{A}$ of $\text{End}(E)$. For each $x \in E$, Ax is either E or (0) as E is simple, but as E is nontrivial, $\{x \in E : Ax = (0)\}$ is a proper submodule of E and hence is the zero submodule; thus $Ax = E$ for all nonzero $x \in E$, so $\hat{A}$ is a primitive ring of endomorphisms of E.

In sum, a ring is primitive if and only if it admits a faithful simple nontrivial module. Similarly, a ring is right primitive if and only if it admits a faithful simple nontrivial right module.

25.13 Theorem. *A ring A is primitive if and only if A is isomorphic to a dense ring of linear operators on a nonzero vector space.*

The assertion follows from 25.6.

25.14 Corollary. *A commutative primitive ring is a field.*

25.15 Corollary. *Let A be a primitive ring. Every nonzero ideal of A is a primitive ring, and for every nonzero idempotent e of A, eAe is a primitive ring. Moreover, A is not the direct sum of two proper ideals.*

The assertions follow from 25.13, 25.10, and 25.11.

A left [right] ideal I of a ring A is a *minimal left [right] ideal* if I is minimal in the set of all nonzero left [right] ideals of A, ordered by inclusion.

25.16 Theorem. *If I is a minimal left ideal of a ring A, then either $I^2 = (0)$ or there is an idempotent e such that $I = Ae$.*

Proof. Assume that $I^2 \neq (0)$. Then there exists $b \in I$ such that $Ib \neq (0)$. Since $\{x \in I : xb = 0\}$ is therefore a left ideal properly contained in I, it is the zero ideal, so $xb \neq 0$ for all nonzero $x \in K$. As Ib is a nonzero left ideal contained in I, $Ib = I$, and hence there exists $e \in I$ such that $eb = b$. Then $(e^2 - e)b = 0$, so $e^2 - e = 0$. In particular, $e \in Ae$, so Ae is a nonzero left ideal contained in I, whence $Ae = I$. •

25.17 Theorem. *Let A be a ring having no nonzero nilpotent ideals. Then A has no nonzero left or right nilpotent ideals, and consequently if I is a minimal left ideal of A, there is an idempotent e such that $I = Ae$.*

Proof. Let I be a left ideal, and let $J = I + IA$, the ideal of A generated by I. If $J^k = I^k + I^k A$, then

$$J^{k+1} = (I + IA)(I^k + I^k A) = I^{k+1} + I(AI^k) + II^k A + I(AI^k)A$$
$$= I^{k+1} + I^{k+1} A.$$

Hence if $I^n = (0)$, then $J^n = (0)$, so $J = (0)$ and thus $I = (0)$. The final assertion therefore follows from 25.16. •

25.18 Theorem. *Let e be an idempotent of a ring A having no nonzero nilpotent ideals, and for each $a \in A$, let a_L be the endomorphism of the commutative group Ae defined by $a_L(x) = ax$ for all $x \in Ae$. (1) Ae is a minimal left ideal of A if and only if eAe is a division ring. (2) If Ae is a minimal left ideal, then Ae is a right vector space over eAe under the scalar multiplication defined by $a.c = ac$ for all $a \in Ae$, $c \in eAe$, and $\lambda : a \to a_L$ is an epimorphism from A to a dense ring A_L of linear operators on the right eAe-vector space Ae; furthermore, if A is primitive, λ is an isomorphism.*

Proof. If $J = \{x \in A : Ax = (0)\}$, then J is an ideal satisfying $J^2 = (0)$, so $J = (0)$. Thus for every nonzero $x \in A$, $Ax \neq (0)$. Assume first that Ae is a minimal left ideal. By the remark just made, $Ax = Ae$ for every nonzero $x \in Ae$. Therefore A_L is a primitive ring of endomorphisms of the commutative group Ae. Let K be the ring of all endomorphisms of Ae that commute with each member of A_L. By 25.6, K is a division ring. For each $c \in eAe$, let c_R be the endomorphism of Ae defined by $c_R(x) = xc$ for all $x \in Ae$. Clearly $c_R \in K$. If $c_R = 0$, then $Aec = (0)$, so $c = ec = e^2 c = 0$. Let $\beta \in K$, and let $c = \beta(e) \in Ae$. Then

$$ec = e_L(c) = e_L(\beta(e)) = \beta(e_L(e)) = \beta(e^2) = \beta(e) = c,$$

so $c = ec \in eAe$. For any $a \in Ae$, $ae = a$, so

$$\beta(a) = \beta(ae) = \beta(a_L(e)) = a_L(\beta(e)) = a_L(c) = ac = c_R(a),$$

and therefore $\beta = c_R$. If $c, d \in eAe$, clearly $(cd)_R = d_R c_R$. Therefore $\beta : c \to c_R$ is an anti-isomorphism from eAe to K, so eAe is also a division ring. Clearly λ is an epimorphism from A to A_L, and by 25.6, A_L is a dense ring of linear operators on the right eAe-vector space Ae. The kernel L of λ satisfies $LAe = (0)$ and hence $L(Ae + AeA) = (0)$, so if A is primitive, $L = (0)$ by 25.11.

Finally, assume that eAe is a division ring, and let I be a nonzero left ideal contained in Ae. Then $Ie = I$ as e is a right identity of Ae. Hence if $eI = (0)$, then $I^2 = IeI = (0)$, a contradiction of our hypothesis by 25.17. Therefore there exists $u \in I$ such that $eu \neq 0$. As $u \in Ae$, $u = ue$, so $eue \neq 0$. Therefore there exists $x \in A$ such that $(exe)(eue) = e$, so $e = (exe)u \in I$, whence $Ae \subseteq I$. Thus Ae is a minimal left ideal. •

If A is a ring, the ring *opposite* A, or the *opposite ring* of A, is the ring obtained from A by replacing the multiplicative composition of A with the composition $*$, defined by $x * y = yx$ for all $x, y \in A$. The identity mapping is thus an anti-isomorphism from a ring to its opposite. If $\mathcal{T}$ is a ring topology on A, $\mathcal{T}$ is also a ring topology on its opposite by (2) of 2.11, and if A has an identity element, A and its opposite have the same multiplicative inversion, which therefore is continuous on A for $\mathcal{T}$ if and only if it is on the ring opposite A. Clearly A and its opposite have the same ideals, the same nilpotent ideals and the same idempotents. Moreover, if e is an idempotent, $e * A * e = eAe$, and eAe is a division subring of A if and only if its opposite is a division subring of the opposite ring of A. Therefore by applying 25.14 to the opposite ring of A, we conclude:

25.19 Corollary. *Let e be an idempotent of a ring A having no nonzero nilpotent ideals, and for each $a \in A$, let a_R be the endomorphism of the commutative group eA defined by $a_R(x) = xa$ for all $x \in eA$. (1) eA is a minimal right ideal of A if and only if eAe is a division ring. (2) If eA is a minimal right ideal, then eA is a left vector space over eAe under the scalar multiplication defined by $c.a = ca$ for all $c \in eAe$, $a \in eA$, and $\rho : a \to a_R$ is an anti-isomorphism from A to a dense ring A_R of linear operators on the left eAe-vector space eA; furthermore, if A is primitive, ρ is an anti-isomorphism.*

25.20 Theorem. *If A is a dense ring of linear operators on a K-vector space E, then I is a minimal left ideal of A if and only if there is an idempotent $e \in A$ such that $I = Ae$ and e is a projection on a one-dimensional subspace of E.*

Proof. Necessity: By 25.11 and 25.17, $I = Ae$ where e is an idempotent of A. Thus e is a projection on a subspace M of E. Suppose that M contained linearly independent vectors a and b. Then there would exist $u \in A$ such that $u(a) = a$ and $u(b) = 0$. Thus $eue \neq 0$, so $Aue \neq (0)$ and hence $Aue = Ae$. Consequently, there would exist $v \in A$ such that $vue = e$, whence $b = e(b) = vue(b) = vu(b) = 0$, a contradiction. Therefore M is one-dimensional.

Sufficiency: Let $e \in A$ be a projection on a one-dimensional subspace $K.a$ of E, and let $u \in Ae$, $u \neq 0$. As $u = ue$, the kernel of u contains that of e, a subspace supplementary to $K.a$, so $u(a) \neq 0$. Therefore there exists $v \in A$ such that $v(u(a)) = a$. Since the kernel of vu contains that of e, therefore, $vu = e$, and hence $e \in Au$, so $Ae \subseteq Au$. Thus Ae is a minimal left ideal. •

25.21 Theorem. *Let A be a dense ring of linear operators on a K-vector space E that has a minimal left ideal. (1) Every nonzero left ideal of A contains a minimal left ideal. (2) The ideal of all linear operators in A of finite-dimensional range contains a projection on each finite-dimensional subspace of E, is the smallest nonzero ideal of A, and is the sum of all the minimal left ideals of A.*

Proof. By 25.20, A contains a projection e on a one-dimensional subspace $K.c$. (1) Let L be a nonzero left ideal of A. There exist nonzero $a, b \in E$ and $t \in L$ such that $t(a) = b$. There exist $r, s \in A$ such that $r(c) = a$ and $s(b) = c$. If $f = rest \in L$, then f is a projection on $K.a$, so Af is a minimal left ideal contained in L by 25.20. (2) Let $\{a_1, \ldots, a_n\}$ be a basis of a subspace M. For each $i \in [1, n]$ there exist $u_i, v_i \in A$ such that $u_i(a_i) = c$, $u_i(a_k) = 0$ for all $k \neq i$, and $v_i(c) = a_i$. Then $\sum_{i=1}^{n} v_i e u_i$ is a projection on M. Finally, let J be a nonzero ideal of A, let w be a linear operator in A with n-dimensional range, and let $a_1, \ldots, a_n$ be such that $\{w(a_1), \ldots, w(a_n)\}$ is a basis of $w(E)$. By 25.10, for each $i \in [1, n]$ there exists $u_i \in J$ such that $u_i(w(a_i)) = a_i$ and $u_i(w(a_j)) = 0$ for all $j \neq i$. Let $e_i = u_i w \in J$. Then e_i is a projection on $K.a_i$, so Ae_i is a minimal left ideal by 25.16. Moreover,

$$w = \sum_{i=1}^{n} we_i,$$

so w belongs to $Ae_1 + \cdots + Ae_n$ and to J. •

We conclude with an application to topological rings:

25.22 Theorem. *If A is a Hausdorff primitive topological ring with a minimal left ideal Ae, where e is an idempotent, then A is topologically [anti-]isomorphic to a topological dense ring A_L [A_R] of continuous linear* operators containing nonzero linear operators of finite rank on the straight Hausdorff right [left] vector space Ae [eA] over the division ring eAe, where Ae [eA] and eAe are topologized as subsets of A, $(u, x) \to u(x)$ is continuous from $A_L \times Ae$ to Ae [$A_R \times eA$ to eA], and the additive groups Ae, [eA], and eAe are topological epimorphic images of the additive group A.

Proof. With the notation of (2) of 25.18, let A_L be furnished with the topology making $\lambda : a \to a_L$ a topological isomorphism. Since scalar multiplication of the right eAe-vector space Ae is simply the restriction to $Ae \times eAe$ of multiplication on A, Ae is a topological eAe-vector space. Also $(u, x) \to u(x)$ from $A_L \times Ae$ to Ae is simply the mapping $(u, x) \to (\lambda^{-1}(u), x)$ from $A_L \times Ae$ to $A \times Ae$ followed by the restriction of multiplication to $A \times Ae$ and hence is continuous. In particular, each $u \in A_L$ is continuous.

To show that Ae is a straight vector space over eAe, we first observe that $\lambda \to e.\lambda$ from eAe to $e.eAe$ is simply the identity mapping of eAe and hence is a homeomorphism. For any nonzero $c \in Ae$ there exist $v, w \in A$ such that $v_L(e) = c$ and $w_L(c) = e$, so the restriction of v_L to $e.eAe$ and that of w_L to $c.eAe$ are continuous maps that are inverses of each other; hence $e.\lambda \to c.\lambda$ is a homeomorphism. Consequently, $\lambda \to c.\lambda$ is the composite of two homeomorphisms and hence is a homeomorphism.

Finally, for any subset U of A, $U \cap Ae \subseteq Ue$ and $U \cap eAe \subseteq eUe$. Thus $x \to xe$ and $x \to exe$ are respectively continuous open epimorphisms from the additive group A to the additive groups Ae and eAe. The ring A_R and the anti-isomorphism from A to A_R are defined in 25.19, and the analogous statements concerning them are similarly proved. •

25.23 Theorem. *A topological ring A is a locally compact, connected, primitive ring with a minimal left ideal if and only if it is topologically isomorphic to the ring of all linear operators on a nonzero finite-dimensional right vector space E over $\mathbb{R}$, $\mathbb{C}$, or $\mathbb{H}$, furnished with the unique Hausdorff topology making it a topological algebra over $\mathbb{R}$.*

Proof. Sufficiency: By 25.13 and 25.20, A is a primitive ring with a minimal left ideal, and A is locally compact and connected since, as a topological $\mathbb{R}$-vector space, it is by 15.10 topologically isomorphic to $\mathbb{R}^{n^2}$ where $n = \dim_{\mathbb{R}} E$.

Necessity: By 25.13 and 25.11, A has no nonzero nilpotent ideals, so by 25.16 A has an idempotent e such that Ae is a minimal left ideal. Therefore by 25.22 we shall regard A as a locally compact dense ring of continuous linear operators on a right Hausdorff vector space E over a division ring K furnished with a Hausdorff ring topology such that the additive groups E and K are topological epimorphic images of A. Consequently, both E and

K are connected and locally compact, so K is topologically isomorphic to $\mathbb{R}$, $\mathbb{C}$, or $\mathbb{H}$ by 16.5 and E is finite-dimensional by 16.2. •

Exercises

25.1 If A is a dense ring of linear operators on a vector space E and if A contains a minimal left ideal, then for any finite-dimensional subspace F of E, A contains a projection on F.

25.2 A primitive ring that contains a nonzero finite left or right ideal is finite.

25.3 Let E be a Hausdorff vector space over a complete, straight division ring K. If every finite-dimensional subspace of E has a topological supplement, then the ring A of all continuous linear operators on E is a dense ring of linear operators containing a minimal left ideal. [Use 15.2.] (Corollaries of the Hahn-Banach theorem imply that the hypothesis holds if K is $\mathbb{R}$ or $\mathbb{C}$ and E is a normed (or, more generally, a Hausdorff locally convex) space.)

25.4 Let E be the topological vector space $\mathbb{Q} + \mathbb{Q}\sqrt[3]{2}$ over $\mathbb{Q}$, where both E and $\mathbb{Q}$ are given the topology induced from $\mathbb{R}$. The ring of all continuous linear operators on E is not a primitive ring of endomorphisms of the additive group E. [Use 8.7.]

26 The Radical of a Ring

In §24 we defined the radical of a commutative ring with identity to be the intersection of its maximal ideals. Here we extend that definition to one for arbitrary rings.

26.1 Definition. *An ideal J of a ring A is* **left primitive**, *or simply* **primitive**, *if A/J is a primitive ring, and J is* **right primitive** *if A/J is a right primitive ring.*

26.2 Definition. *A left ideal J of a ring A is* **regular** *if there exists $e \in A$ such that $x - xe \in J$ for all $x \in A$, and a right ideal J of A is* **regular** *if there exists $e \in A$ such that $x - ex \in A$ for all $x \in A$. A* **maximal regular left [right] ideal** *of A is a left [right] ideal that is maximal in the set of all proper regular left [right] ideals of A, ordered by inclusion.*

Clearly any left [right] ideal containing a regular left [right] ideal is again regular, and if A has an identity element, every left or right ideal is regular. Consequently, a maximal regular left [right] ideal is actually maximal in the set of all proper left [right] ideals of A, and thus is a *regular maximal left [right] ideal*, and conversely, a regular maximal left [right] ideal is clearly a maximal regular left [right] ideal.

26.3 Theorem. *A proper regular left [right] ideal of a ring A is contained in a maximal regular left [right] ideal.*

Proof. Let J be a proper regular left ideal. Then there exists $e \in A$ such that $x - xe \in J$ for all $x \in A$. As J is proper, $e \notin J$. The set of all ideals of A containing J but not e, ordered by inclusion, is clearly inductive, and so by Zorn's Lemma contains a maximal member M. Any left ideal of A properly containing M would therefore contain e and hence would be A; thus M is a maximal regular left ideal. •

If J is a regular left ideal of a ring A, we define $P(J)$ and $D(J)$ by

$$P(J) = \{a \in A : aA \subseteq J\} \qquad D(J) = \{d \in A : Jd \subseteq J\}.$$

26.4 Theorem. *If J is a regular left ideal of a ring A, $P(J)$ is the largest ideal of A contained in J, and $D(J)$ is the largest subring of A in which J is an ideal.*

Proof. There exists $e \in A$ such that $x - xe \in J$ for all $x \in A$. If $x \in P(J)$, then $x = (x - xe) + xe \in J + J = J$. Thus $P(J) \subseteq J$, and hence $P(J)$ is the largest ideal of A contained in J. •

26.5 Theorem. *Let A be a ring. (1) If M is a regular maximal left ideal of A and if, for each $a \in A$ and each $d \in D(M)$, $\hat{a}$ and $\tilde{d}$ are the endomorphisms of the commutative group A/M well defined by*

$$\hat{a}(x + M) = ax + M \qquad \tilde{d}(x + M) = xd + M$$

for all $x \in A$, then $\phi : a \to \hat{a}$ is an epimorphism from A to a primitive ring $\hat{A}$ of endomorphisms of A/M whose kernel, $P(M)$, is thus a primitive ideal, $\psi : d \to \tilde{d}$ is an anti-epimorphism with kernel M from $D(M)$ to the division ring D of all endomorphisms of A/M that commute with each member of $\hat{A}$, A/M is a right vector space over $D(M)/M$ under the well defined scalar multiplication

$$(x + M).(d + M) = xd + M$$

for all $x \in A$, $d \in D(M)$, and $\hat{A}$ is a dense ring of linear operators on the $D(M)/M$-vector space A/M. (2) If P is the kernel of an epimorphism ψ from A to a primitive ring $\hat{A}$ of endomorphisms of a nonzero commutative group E, then for each nonzero $x \in E$, the left ideal M_x, defined by

$$M_x = \{a \in A : \hat{a}(x) = 0\},$$

is a regular maximal left ideal of A, and $P = P(M_x)$. (3) If P is a primitive ideal of A, then P is the intersection of all the regular maximal left ideals M such that $P = P(M)$.

Proof. Let $e \in A$ be such that $z - ze \in M$ for all $z \in A$. (1) Clearly $\hat{a}$ and $\tilde{d}$ are well defined. To show that $\hat{A}$ is a primitive ring of endomorphisms of A/M, let $x \in A \setminus M$ and $y \in A$. Now $Ae \not\subseteq M$, since otherwise $z = (z - ze) + ze \in M$ for all $z \in A$, a contradiction. Thus $\{d \in A : Ad \subseteq M\}$ is a proper left ideal of A containing M and hence is M. Consequently, $Ax \not\subseteq M$, so $Ax + M = A$. In particular, there exists $a \in A$ such that $ax - y \in M$, so $\hat{a}(x + M) = y + M$. Thus $\hat{A}$ is a primitive ring of endomorphisms of A/M, so the kernel of ϕ, which is clearly $P(M)$, is a primitive ideal.

If $d \in D(M)$, then for all $a, x \in A$,
$$(\hat{a} \circ \tilde{d})(x + M) = \hat{a}(xd + M) = axd + M = \tilde{d}(ax + M) = (\tilde{d} \circ \hat{a})(x + M),$$
so $\tilde{d} \in D$. Conversely, let $\delta \in D$, and let $d \in A$ be such that $d + M = \delta(e + M)$. Then for any $x \in A$, $x + M = xe + M$ since $x - xe \in M$, so
$$\delta(x + M) = \delta(xe + M) = \delta(\hat{x}(e + M)) = \hat{x}(\delta(e + M)) = \hat{x}(d + M) = xd + M.$$
In particular, if $x \in M$, then $M = \delta(0 + M) = \delta(x + M) = xd + M$, so $xd \in M$. Thus $d \in D(M)$, and for each $x \in A$, $\delta(x + M) = xd + M = \tilde{d}(x + M)$, so $\delta = \tilde{d}$. Clearly, if $c, d \in B(M)$, $\widetilde{cd} = \tilde{d} \circ \tilde{c}$. Moreover,
$$\psi^{-1}(0) = \{d \in A : Ad \subseteq M\} \cap D(M) = M \cap D(M) = M$$
as we saw above. Therefore $D(M)/M$ is isomorphic to the division ring opposite D, and thus $\hat{A}$ is a dense ring of linear operators on the right $D(M)/M$-vector space A/M by 25.6.

(2) Clearly each M_x is a proper left ideal containing P, and
$$P = \bigcap \{M_x : x \in E, x \neq 0\}.$$
It therefore suffices to show that each M_x is a regular maximal left ideal and that $P = P(M_x)$. Since $\hat{A}$ is primitive, there exists $e \in A$ such that $\hat{e}(x) = x$, so $a - ae \in M_x$ for all $a \in A$, and thus M_x is a regular left ideal. To show that for any $c \in A \setminus M_x$, $A = Ac + M_x$, let $d \in A$. As $\hat{c}(x) \neq 0$, there exists $a \in A$ such that
$$\hat{a}(\hat{c}(x)) = \hat{d}(x),$$
so $ac - d \in M_x$, and thus $d \in Ac + M_x$. Hence M_x is a regular maximal left ideal of A. Since $P(M_x)$ is the largest ideal of A contained in M_x by 26.4, $P \subseteq P(M_x)$. Conversely, let $a \in P(M_x)$, and let $y \in E$. Then there exists $b \in A$ such that $\hat{b}(x) = y$, so as $ab \in M_x$,
$$\hat{a}(y) = \hat{a}(\hat{b}(x)) = (\widehat{ab})(x) = 0.$$
Thus $\hat{a} = 0$, so $a \in P$. Clearly (3) follows from (2). •

26.6 Theorem. *If M is a maximal ideal of a ring A that is also a regular left ideal, then M is primitive.*

Proof. By 26.3, M is contained in a maximal regular left ideal N. By 26.4, $M = P(N)$ and hence by (1) of 26.5, M is primitive. •

26.7 Definition. Let A be a ring. The **radical** of A is A if A has no proper regular left ideals, the intersection of all regular maximal left ideals of A if A has a proper regular left ideal. The ring A is **semisimple** if its radical is $\{0\}$, a **radical ring** if its radical is A.

Thus, if A is a commutative ring with identity, the radical of A is the intersection of all the maximal ideals of A, in accordance with Definition 24.13.

26.8 Theorem. *The radical of a ring is the intersection of all its primitive ideals and thus is an ideal.*

The assertion follows from 26.5 and 26.4.

26.9 Theorem. *Let R be the radical of a ring A. (1) Each element of R is advertible. (2) For any $c \in A$, the following statements are equivalent:*

$1°$ ac is left advertible for all $a \in A$.
$2°$ ca is left advertible for all $a \in A$.
$3°$ acb is left advertible for all $a, b \in A$.
$4°$ $c \in R$.

Proof. (1) Let $c \in A$ have no left adverse, and let $J = \{x - xc : x \in A\}$. Clearly J is a regular left ideal, and J is proper since $-c \notin J$. By 26.3, J is contained in a maximal regular left ideal M, and $c \notin M$, since otherwise, for any $x \in A$, $x = (x - xc) + xc \in J + M = M$, a contradiction. Therefore $c \notin R$.

Let $b \in R$. We have just seen that b has a left adverse a, and $a = ab - b \in R$ by 26.8, and hence a also has a left adverse c. Therefore as $\circ$ is associative, $b = c$, and a is the adverse of b.

(2) By (1) and 26.8, $4°$ implies each of $1° - 3°$. Suppose that $c \notin R$. By 26.8, there is a primitive ideal P of A such that $c \notin P$. By definition, P is the kernel of an epimorphism $a \to \hat{a}$ from A to a primitive ring of endomorphisms of a nonzero commutative group E. As $\hat{c} \neq 0$, there exists $x \in E$ such that $\hat{c}(x) \neq 0$. Therefore there exist $a, b \in A$ such that $\hat{a}(\hat{c}(x)) = x$ and $\hat{b}(x) = x$. For any $u \in A$, $[\widehat{u \circ ac}](x) = x \neq 0$, so ac is not left advertible, $[\widehat{u \circ ca}](\hat{c}(x)) = \hat{c}(x) \neq 0$, so ca is not left advertible, and $[\widehat{u \circ acb}](x) = x \neq 0$, so acb is not left advertible. •

26.10 Corollary. *If R is the radical of a ring A, then the ring opposite R is the radical of the ring opposite A.*

Proof. By (1) and 3° of (2) of 26.9, $c \in R$ if and only if acb is advertible for all $a, b \in A$. Since a ring and its opposite clearly have the same advertible elements, the assertion follows.

26.11 Theorem. *Let R be the radical of a ring A. (1) R is the intersection of the right primitive ideals of A. (2) For any $c \in A$, the following statements are equivalent:*

1° ac is right advertible for all $a \in A$.
2° ca is right advertible for all $a \in A$.
3° acb is right advertible for all $a, b \in A$.
4° $c \in R$.

Proof. The statements follow from 26.8, 26.9, and 26.10. For example, 1° of 26.9, applied to the ring opposite A, becomes 2° of this theorem. •

26.12 Theorem. *If A is a primitive or right primitive ring, then A is semisimple. In particular, a dense ring of linear operators on a vector space is semisimple.*

The assertion follows from 26.8 and 26.10.

26.13 Theorem. *Let J be a left or right ideal of a ring A. If every element of J is left [right] advertible, then J is contained in the radical of A.*

The assertion follows from 26.9 and 26.11.

Extending Definition 21.9, we shall say that a left or right ideal of a ring is *nil* if each of its elements is nilpotent.

26.14 Corollary. *A nil left or right ideal of a ring is contained in its radical.*

Proof. If $x^n = 0$, then the adverse of x is $-\sum_{k=1}^{n-1} x^n$. •

26.15 Theorem. *Let A and A' be rings with radicals R and R' respectively, and let h be an epimorphism from A to A'. (1) $h(R) \subseteq R'$. (2) If $h^{-1}(0) \subseteq R$, then $h(R) = R'$.*

Proof. (1) By (1) of 26.9, $h(R)$ is an ideal of A' all of whose elements are advertible, so $h(R) \subseteq R'$ by 26.13. (2) If Q is a primitive ideal of A', then $h^{-1}(Q)$ is a primitive ideal of A since h induces an isomorphism from $A/h^{-1}(Q)$ to A'/Q. If P is a primitive ideal of A, then $h^{-1}(0) \subseteq R \subseteq P$, so $h^{-1}(h(P)) = P$, and consequently $h(P)$ is a primitive ideal of A' since $A'/h(P)$ is isomorphic to $A/h^{-1}(h(P)) = A/P$. Thus $P \to h(P)$ is a bijection from the set $\mathcal{P}$ of primitive ideals of A to the set $\mathcal{Q}$ of primitive ideals of A', and its inverse is $Q \to h^{-1}(Q)$. Hence by 26.8,

$$h^{-1}(R') = h^{-1}(\bigcap_{Q \in \mathcal{Q}} Q) = \bigcap_{Q \in \mathcal{Q}} h^{-1}(Q) = \bigcap_{P \in \mathcal{P}} P = R,$$

so $R' = h(h^{-1}(R')) = h(R)$. •

26.16 Corollary. *If J is an ideal of a ring A contained in its radical R, then R/J is the radical of A/J. In particular, A/R is a semisimple ring.*

26.17 Theorem. *If e is an idempotent in a ring A with radical R, then eRe is the radical of eAe.*

Proof. Let S be the radical of eAe. As eRe is an ideal of eAe, to show that $eRe \subseteq S$, is suffices by 26.13 to show that each $c \in eRe$ has an adverse in eAe. As $eRe \subseteq R$ and as $c = ece$, c has an adverse c^a in A by 26.9 and 26.11. Thus

$$c^a ece - ece = c^a = ecec^a - ece,$$

so $c^a \in Ae \cap eA = eAe$.

Conversely, let $c \in S$. To show that $c \in R$ (and hence in eRe), it suffices by 26.11 to show that for any $a \in A$, ca has a right adverse in A. Now $ceae$ has an adverse b in eAe by 26.9 and 26.11. As $c = ce$ and $b = eb$,

$$ca \circ b = cea + b - ceab = cea + b - ceaeb$$
$$= [ceae + (cea - ceae)] + b - ceaeb$$
$$= (ceae \circ b) + (cea - ceae) = cea - ceae.$$

As $c = ec$, $(cea - ceae)^2 = 0$, and hence if $d = -(cea - ceae)$, d is the adverse of $cea - ceae$. Thus $ca \circ b \circ d = (cea - ceae) \circ d = 0$, so ca is right advertible in A. •

26.18 Theorem. *If R is the radical of a ring A and if J is an ideal of A, then the radical $R(J)$ of the ring J is $R \cap J$.*

Proof. Let $\mathcal{P}$ be the set of primitive ideals of A, and let $\mathcal{Q}$ be the subset of those primitive ideals P not containing J. If $P \in \mathcal{Q}$, then $(J + P)/P$ is a nonzero ideal of the primitive ring A/P and hence is a primitive ring by 25.15. As $J/(P \cap J)$ is isomorphic to $(J+P)/P$, therefore,

$$R \cap J = (\bigcap_{P \in \mathcal{P}} P) \cap J = \bigcap_{P \in \mathcal{P}} (P \cap J) = \bigcap_{P \in \mathcal{Q}} (P \cap J) \supseteq R(J).$$

But also, $R \cap J$ is an ideal of J each of whose elements is an advertible element of A and hence also an advertible element of J, so $R \cap J \subseteq R(J)$ by 26.13. •

26.19 Corollary. *If R is the radical of a ring A, then R is a radical ring and is the radical of the ring A_1 obtained by adjoining an identity element to A.*

Proof. A is an ideal of A_1, and A_1/A is isomorphic to the semisimple ring $\mathbb{Z}$, so by (1) of 26.15, applied to the canonical epimorphism from A_1 to A_1/A, the radical R_1 of A_1 is contained in A. Thus by 26.18, $R_1 = R_1 \cap A = R$. •

26.20 Theorem. *If A is a semisimple ring and if J is a left ideal of A, the radical $R(J)$ of the ring J is $\{x \in J : Jx = (0)\}$.*

Proof. Let $I = \{x \in J : Jx = (0)\}$. $JR(J)$ is a left ideal of A contained in $R(J)$, so each of its elements is advertible by (1) of 26.9. Consequently, $JR(J)$ is contained in the radical of A by 26.13, so $JR(J) = (0)$ by hypothesis, and thus $R(J) \subseteq I$. But also I is an ideal of J satisfying $I^2 = (0)$, so $I \subseteq R(J)$ by 26.14. Thus $R(J) = I$. •

We shall call a left or right ideal of a ring *advertible* if each of its elements is advertible. By 26.9 and 26.13, the radical of a ring is an advertible ideal and is the largest advertible left or right ideal of the ring.

26.21 Theorem. *If $(A_\lambda)_{\lambda \in L}$ is a family of rings and if R_λ is the radical of A_λ for each $\lambda \in L$, then $\prod_{\lambda \in L} R_\lambda$ is the radical of $\prod_{\lambda \in L} A_\lambda$; in particular, if each A_λ is semisimple, so is $\prod_{\lambda \in L} A_\lambda$.*

Proof. Let $A = \prod_{\lambda \in L} A_\lambda$, and let R be the radical of A. For each $\mu \in L$, let pr_μ be the canonical projection from A to A_μ. By 26.15, $pr_\lambda(R) \subseteq R_\lambda$ for each $\lambda \in L$, so $R \subseteq \prod_{\lambda \in L} R_\lambda$. But $\prod_{\lambda \in L} R_\lambda$ is clearly an advertible ideal of A since each R_λ is an advertible ideal of A_λ. Therefore $\prod_{\lambda \in L} R_\lambda \subseteq R$ by 26.13. •

From 26.21 and 26.18 we obtain:

26.22 Theorem. *If $(A_\lambda)_{\lambda \in L}$ is a family of rings and if R_λ is the radical of A_λ for each $\lambda \in L$, then $\bigoplus_{\lambda \in L} R_\lambda$ is the radical of $\bigoplus_{\lambda \in L} A_\lambda$.*

If A is an algebra over a commutative ring with identity K, an ideal of the underlying ring A need not be an ideal of the algebra A.

26.23 Theorem. *Let A be an algebra over a commutative ring K with identity. (1) Every regular maximal left ideal of the ring A is a left ideal of the algebra A. (2) Every primitive ideal of the ring A is an ideal of the algebra A. (3) The radical of the ring A is an ideal of the algebra A.*

Proof. By (3) of 26.5 and 26.8, we need only prove (1). Let M be a regular maximal left ideal of A, and let $e \in A$ be such that $x - xe \in M$ for all $x \in A$. Suppose that M were not a left ideal of the algebra A. Then the algebra left ideal $K.M$ generated by M would strictly contain M and hence be A, so there would exist $m_1, \ldots, m_n \in M$ and scalars $\lambda_1, \ldots, \lambda_n$ such that $e = \sum_{j=1}^n \lambda_j m_j$. But then, for any $x \in A$, $xe = \sum_{j=1}^n (\lambda_j x) m_j \in M$, so $x = (x - xe) + xe \in M$, and hence $M = A$, a contradiction. •

26.24 Theorem. *Let A be a primitive ring that is an algebra over a field F. There exist a regular maximal left ideal M of A, an isomorphism $\phi: a \to \hat{a}$ from A to a dense ring $\hat{A}$ of linear operators on a K-vector space A/M, where K is the division ring of all endomorphisms of A/M commuting with each $\hat{a} \in \hat{A}$, and an isomorphism $\psi: r \to \hat{r}$ from F to a subfield $\hat{F}$ of the center of K such that $\hat{A}$ is a subalgebra of the $\hat{F}$-algebra of all linear operators on A/M, and $\widehat{r.a} = \hat{r}.\hat{a}$ for all $r \in F$ and all $a \in A$.*

Proof. As the zero ideal of A is a primitive ideal, by (3) and (1) of 26.5 there is a regular maximal ideal M of A such that $\phi: a \to \hat{a}$ is an isomorphism from A onto a primitive ring $\hat{A}$ of endomorphisms of A/M, where $\hat{a}(x + M) = ax + M$ for all $a, x \in A$. By 26.23, M is an algebra left ideal, so for each $r \in F$, $\hat{r}: x + M \to r.x + M$ is a well-defined endomorphism of the additive group A/M. For each $r \in F$, $\hat{r} \in K$ since for all $a, x \in A$,

$$(\hat{r} \circ \hat{a})(x + M) = r.(ax) + M = a(r.x) + M = (\hat{a} \circ \hat{r})(x + M).$$

Clearly $\psi: r \to \hat{r}$ is an epimorphism from F to its range $\hat{F}$ in K; since $\hat{1}$ is the identity linear operator on E, ψ is an isomorphism from F to $\hat{F}$. Also, for all $r \in F$, $a \in A$, $\widehat{r.a} = \hat{r} \circ \hat{a}$, since for any $x \in A$,

$$\widehat{r.a}(x + M) = (r.a)x + M = r.(ax) + M = (\hat{r} \circ \hat{a})(x + M).$$

To show that $\hat{F}$ is a subfield of the center of K, let $r \in F$, $\lambda \in K$, and let $x \in A \setminus M$. As $\hat{A}$ is a primitive ring of endomorphisms, there exists $a \in A$ such that $\hat{a}(x + M) = x + M$. Then by the preceding,

$$\lambda \circ \hat{r} \circ \hat{a} = \lambda \circ \widehat{r.a} = \widehat{r.a} \circ \lambda = \hat{r} \circ \hat{a} \circ \lambda = \hat{r} \circ \lambda \circ \hat{a},$$

and consequently $(\lambda \circ \hat{r})(x + M) = (\lambda \circ \hat{r} \circ \hat{a})(x + M) = (\hat{r} \circ \lambda \circ \hat{a})(x + M) = (\hat{r} \circ \lambda)(x + M)$. By the definition of a scalar multiple of a linear operator, where the scalar belongs to the center of the coefficient division ring,

$$(\hat{r}.\hat{a})(x + M) = \hat{r}.(ax + M) = r.(ax) + M = (r.a)x + M = \widehat{r.a}(x + M)$$

for all $r \in F$, $a, x \in A$. Thus $\hat{r}.\hat{a} = \widehat{r.a} \in \hat{A}$, so $\hat{A}$ is a subalgebra of the $\hat{F}$-algebra of all linear operators on the K-vector space A/M. •

By set-theoretic considerations, we may construct a field K_1 containing F as a subfield, an isomorphism τ from K_1 to K such that $\tau(r) = \hat{r}$ for all $r \in F$ and a scalar multiplication from $K_1 \times (A/M)$ to A/M such that $t.(x + M) = \tau(t)(x + M)$ for all $t \in K_1$ and all $x \in A$. Consequently, we obtain:

26.25 Corollary. *A primitive ring A that is an algebra over a field F is isomorphic to a dense F-algebra of linear operators on a vector space E over a division ring K containing F in its center.*

We shall conclude by applying these concepts to obtain some information about advertibly open rings.

26.26 Theorem. (1) *If A is an advertibly open ring, then every quotient ring of A and every left or right ideal of A is advertibly open.* (2) *If J is an ideal of a topological ring A, then A is advertibly open if and only if J and A/J are advertibly open.*

Proof. (1) Clearly every epimorphic image of an advertible element is advertible. Consequently, if J is an ideal of A and ϕ_J the canonical epimorphism from A to A/J, then $\phi_J(A^a)$ is an open subset of $(A/J)^a$, so A/J is advertibly open by 11.8. Also, if x is advertible, then as $x^a = xx^a - x = x^a x - x$, x^a belongs to every left or right ideal I that x does. Consequently, $A^a \cap I \subseteq I^a$, so I is advertibly open by 11.8. (2) Assume that J and A/J are advertibly open, where J is an ideal of A. Then $\phi_J^{-1}((A/J)^a)$ is open, and each of its elements is advertible: Indeed, if $b+J$ is the adverse of $a+J$ in A/J, then $a \circ b$ and $b \circ a$ belong to J and hence have adverses c and d respectively. Thus $a \circ (b \circ c) = (a \circ b) \circ c = 0 = d \circ (b \circ a) = (d \circ b) \circ a$, so a is left and right advertible and hence advertible as $\circ$ is associative. •

26.27 Theorem. *If A is an advertibly open ring, then every regular maximal left ideal of A, every primitive ideal of A, and the radical of A are closed.*

Proof. Let M be a regular maximal left ideal, and let $e \in A$ be such that $x - xe \in M$ for all $x \in A$. Suppose that M is not closed. Then as $\overline{M}$ is a left ideal properly containing M, $\overline{M} = A$. Consequently, as A^a is an open neighborhood of zero, there exists $m \in M$ such that $e - m \in A^a$. Let q be the adverse of $e - m$. Then $e = m - (q - qe) - qm \in M$, and hence $M = A$, a contradiction. Therefore M is closed. Similarly, every regular maximal right ideal is closed. By (3) of 26.5, every primitive ideal is therefore closed, and also the radical of A is closed by 26.7. •

26.28 Theorem. *If e is an idempotent of a Hausdorff ring A, then Ae is a closed left ideal, eA is a closed right ideal, and eAe is a closed subring.*

Proof. The function f from A to A, defined by $f(x) = x - xe$, is continuous, and clearly $f^{-1}(0) = Ae$. Similarly, eA is closed. Therefore as $eAe = eA \cap Ae$, eAe is closed. •

26.29 Theorem. *If A is a Hausdorff, advertibly open, semisimple ring, a left [right] ideal I is a minimal closed left [right] ideal of A (that is, I is minimal in the set of all nonzero closed left [right] ideals of A, ordered by inclusion) if and only if I is a minimal left [right] ideal of A, in which case there is an idempotent e such that $I = Ae$ $[I = eA]$.*

Proof. The condition is sufficient by 26.14, 25.17, and 26.28. Necessity: Let I be a minimal closed left ideal of A. First, we shall show that if J is a nonzero proper closed left ideal of the ring I, then $IJ = \{0\}$. Indeed, suppose that $IJ \neq \{0\}$. Then there exists $c \in J$ such that $Ic \neq \{0\}$. As Ic is a left ideal of A, therefore, $\overline{Ic} = I$. Thus as J is closed in A, $I = \overline{Ic} \subseteq J \subset I$, a contradiction. Thus $IJ = \{0\}$, and, in particular, $J^2 = \{0\}$.

By 26.26 I is advertibly open, so by 26.27 its radical $R(I)$ is closed. Now I is not a nilpotent left ideal by 26.14, so by 26.20 $R(I)$ is a proper ideal of I; consequently by the preceding, $IR(I) = \{0\}$, and $I/R(I)$ is a nonzero Hausdorff ring that is semisimple and advertibly open by 26.16 and 26.26. By the preceding and 26.14, $R(I)$ contains every proper closed left ideal of the ring I. Consequently, $I/R(I)$ contains no proper nonzero closed left ideals, and hence by 26.27 the zero ideal of $I/R(I)$ is the only regular maximal left ideal of $I/R(I)$. Let $D = I/R(I)$. Then D has a right identity ϵ and no proper nonzero left ideals. Consequently, the right annihilator of D in D is (0), the left annihilator of each nonzero τ in D is (0), $\alpha - \epsilon\alpha = 0$ for each $\alpha \in D$, and finally, D is a division ring.

Let J be a nonzero left ideal of A contained in I. If $J \subseteq R(I)$, then $\overline{J} \subseteq R(I) \subset I$, as $R(I)$ is closed, contradicting the minimality of I. Thus $J \not\subseteq R(I)$ and hence, as D is a division ring, J is mapped surjectively to D by the canonical epimorphism from I to D. In particular, J contains an element f mapped onto the identity of D, so $f^2 - f \in R(I)$. Consequently

$$f^3 - f^2 = f(f^2 - f) \in JR(I) \subseteq IR(I) = \{0\}.$$

Therefore if $e = f^2$, e is an idempotent of J mapped onto the identity of D and hence is nonzero. By 26.28, Ae is closed, so $I = Ae \subseteq J \subseteq I$. Therefore I is a minimal left ideal and $I = Ae$. •

Exercises

26.1 Let E be a K-vector space having a countably infinite basis, let $A = \text{End}_K(E)$, and let P be the ideal of all $u \in A$ such that $\dim_K u(E) < +\infty$. (a) P is a maximal ideal of A. (b) P is a primitive ideal of A. (c) A/P is a ring with identity that has no proper nonzero ideals and no minimal left ideals. [If $u \in A \setminus P$, construct $v \in A$ such that $u \notin Avu + P$.]

26.2 Let A be a topological ring, M a regular maximal left ideal of A. With the notation of the proof of (1) of 26.5, let g be the isomorphism from $A/P(M)$ to $\hat{A}$ defined by $g(a + P(M)) = \hat{a}$ for all $a \in A$. If $\hat{A}$ is equipped with the topology making g a topological isomorphism, then $\hat{A}$ is a topological ring, and $(u, x) \to u(x)$ is continuous from $\hat{A} \times (A/M)$ to A/M.

26.3 Let A be a Hausdorff ring such that adversion is uniformly continuous on the radical R of A (a condition holding, for example, if A is bounded (Exercise 12.8)). (a) $\hat{R}$ is a radical ring. (b) $\hat{R}$ is contained in the radical of $\hat{A}$. (c) If R is open in A, then $\hat{R}$ is the radical of $\hat{A}$. [Use 5.14.] (d) (Kurke [1967]) If A is complete, R is closed.

26.4 (Yood [1962]) Let A be a dense subring of an advertibly open ring B. The following statements are equivalent:

1° A is advertibly open.

2° Every regular maximal left ideal of A is closed.

3° Every regular maximal right ideal of A is closed.

[If $a \in A$ has an adverse in B and if $J_a = \{x - xa : x \in A\}$, show that J_a is dense in B.]

26.5 A Hausdorff ring is a radical ring if and only if it is advertibly open and has no proper closed regular left ideals.

26.6 A *Gel'fand* ring is an advertibly open, Hausdorff ring with continuous adversion. (Kaplansky [1947c]) If J is a closed ideal of a Hausdorff ring A, then A is a Gel'fand ring if and only if J and A/J are Gel'fand rings.

26.7 (Correl [1958]) Let A be a Gel'fand ring whose completion $\hat{A}$ is locally compact. (a) $\hat{A}$ is a Gel'fand ring. [If U is a symmetric neighborhood of zero in the topological group A^a and if the closure $\hat{U}$ of U in $\hat{A}$ is compact, show that $\hat{U} \subseteq (\hat{A})^a$. (b) If A is a field, there is a Hausdorff field topology $\mathcal{S}$ weaker than the given topology of A such that the completion A_1 of A for $\mathcal{S}$ is a locally compact topological field, and A_1 is a continuous epimorphic image of $\hat{A}$. [Use (a), 26.27, and 11.11 in considering $\hat{A}/M$, where M is a maximal ideal of $\hat{A}$.]

27 Artinian Modules and Rings

Here we shall give some basic properties of artinian rings, algebras, and modules.

27.1 Definition. *Let A be a ring. An A-module E is **artinian** if every nonempty set of submodules of E, ordered by inclusion, contains a minimal element. A ring is **artinian** if it is artinian as a left module over itself, that is, if every nonempty set of left ideals, ordered by inclusion, contains a minimal element. If A is an algebra over a commutative ring with identity K, A is an **artinian K-algebra** if every nonempty set of left (K-algebra) ideals, ordered by inclusion, contains a minimal element.*

If A is a K-algebra such that $A^2 = A$ (in particular, if A has an identity element), then A is an artinian ring if and only if it is an artinian K-algebra, simply because every left ideal of the ring A is also a left ideal of the K-algebra A. If A is a finite-dimensional algebra over a field K, then A is an artinian K-algebra, since if I and J are left ideals such that $I \subset J$, then $\dim_K I < \dim_K J$. The ring of all linear operators on a finite-dimensional vector space over a division ring is also artinian by 25.9.

An argument similar to that in the proof of 20.2 establishes the following equivalent formulation:

27.2 Theorem. *An A-module E is artinian if and only if for every decreasing sequence $(M_n)_{n \geq 1}$ of submodules of E, there exists $q \geq 1$ such that $M_n = M_q$ for all $n \geq q$. A K-algebra A is artinian if and only if for every decreasing sequence $(M_n)_{n \geq 1}$ of (K-algebra) left ideals, there exists $q \geq 1$ such that $M_n = M_q$ for all $n \geq q$.*

The condition of 27.2 is frequently called the *Descending Chain Condition*.

27.3 Theorem. *If E is an A-module and F a submodules of E, then E is artinian if and only if both F and E/F are artinian.*

Proof. Necessity: Clearly F is artinian. If $(M_n)_{n \geq 1}$ were a strictly decreasing sequence of submodules of E/F, then $(\phi_F^{-1}(M_n))_{n \geq 1}$ would be a strictly decreasing sequence of submodules of E, where ϕ_F is the canonical epimorphism from E to E/F. Sufficiency: Let $(M_n)_{n \geq 1}$ be a decreasing sequence of submodules of E. By hypothesis, there exists $p \geq 1$ such that $M_n \cap F = M_p \cap F$ for all $n \geq p$, and there exists $q \geq p$ such that $(M_n + F)/F = (M_q + F)/F$ for all $n \geq q$. Then $M_n = M_q$ for all $n \geq q$. Indeed, let $n \geq q$ and let $x \in M_q$. Then $x + F \in (M_n + F)/F$, so there exists $y \in M_n$ such that $x - y \in F$. Then $x - y \in F \cap M_q = F \cap M_p = F \cap M_n$. Consequently, $x = (x - y) + y \in M_n$. •

The proofs of the following five statements are similar to the proofs of 20.4-20.8.

27.4 Corollary. *If A is an artinian ring [K-algebra] and if J is an ideal of A, then A/J is an artinian ring [K-algebra].*

27.5 Corollary. *The sum of finitely many artinian submodules of an A-module E is artinian.*

27.6 Corollary. *The cartesian product of finitely many artinian A-modules is artinian.*

27.7 Theorem. *The cartesian product of finitely many artinian rings [K-algebras] is artinian.*

27.8 Theorem. *If A is an artinian ring with identity and if E is a finitely generated unitary A-module, then E is artinian.*

27.9 Theorem. *If $(M_n)_{0 \leq n \leq k}$ is a decreasing sequence of submodules of an A-module E such that $M_0 = E$ and $M_k = \{0\}$, then E is noetherian [artinian] if and only if M_{n-1}/M_n is noetherian [artinian] for each $n \in [1, k]$.*

Proof. An inductive argument based on 20.3 [27.3] establishes that E/M_n is noetherian [artinian] for each $n \in [1, k]$. •

27.10 Theorem. *If J is a proper ideal of a ring A with identity such that A/J is noetherian [artinian] ring, and if $(M_n)_{n \geq 0}$ is a decreasing sequence of finitely generated submodules of a unitary A-module E such that $M_0 = E$ and $JM_{n-1} \subseteq M_n$ for all $n \geq 1$, then E/M_k is a noetherian [artinian] A-module for all $k \geq 1$.*

Proof. For each $n \geq 1$ we may regard M_{n-1}/M_n as an A/J-module having the same submodules as the A-module M_{n-1}/M_n. By hypothesis, M_{n-1}/M_n is finitely generated unitary A-module, hence a finitely generated unitary A/J-module, therefore a noetherian [artinian] A/J-module by 20.8 [27.8], and hence also a noetherian [artinian] A-module. Therefore by 27.9, the A-module E/M_k is noetherian [artinian]. •

27.11 Definition. *A ring [K-algebra] A is **simple** if A has no proper nonzero [K-algebra] ideals and A is not a radical ring.*

27.12 Theorem. *The following statements about a ring A [F-algebra A, where F is a field] are equivalent:*

1° *A is primitive and is an artinian ring [F-algebra].*

2° *A is isomorphic to the ring [F-algebra] of all linear operators on a nonzero finite-dimensional vector space over a division ring K [that contains F in its center].*

3° *A is a simple artinian ring [F-algebra].*

Proof. Assume 1°. By 25.13 [26.25] we may suppose that A is a dense ring [F-algebra] of linear operators on a K-vector space E, where K is a division ring [containing F in its center]. Suppose that $(x_n)_{n \geq 1}$ were a linearly independent sequence in E. For each $n \geq 1$, let

$$J_n = \{u \in A : u(x_i) = 0 \text{ for all } i \in [1, n]\}.$$

Clearly $(J_n)_{n \geq 1}$ is a strictly decreasing sequence of left [F-algebra] ideals, a contradiction. Thus E is finite-dimensional, and hence A is the ring [F-algebra] of all linear operators on E.

If 2° holds, then A is artinian by 25.9, not a radical ring as it has an identity element, and hence a simple ring by (2) of 25.21.

If 3° holds, then A contains a primitive [F-algebra] ideal by 27.11 [and 26.23], so the zero ideal is primitive as A has no other proper [F-algebra] ideals, and consequently A is primitive. •

27.13 Corollary. *An ideal P of an artinian ring or F-algebra A, where F is a field, is primitive if and only if it is a maximal ideal and A/P has an identity element.*

Proof. We need only apply 27.12 to A/P. •

27.14 Theorem. (Artin-Wedderburn) *The following statements about a nonzero ring A [F-algebra A, where F is a field] are equivalent:*

1° *A is a semisimple artinian ring [F-algebra].*
2° *A is the direct sum of finitely many rings [F-algebras], each isomorphic to the ring [F-algebra] of all linear operators on a nonzero finite-dimensional vector space over a division ring [that contains F in its center].*
3° *A is the direct sum of finitely many simple artinian rings [F-algebras].*

Proof. The following argument for a ring A is equally valid if A is an F-algebra, since every primitive ideal of the ring A is also an algebra ideal by 26.23. Assume 1°. As A is not a radical ring, it contains primitive ideals. As A is artinian, the set $\mathcal{Q}$ of all finite intersections of primitive ideals, ordered by inclusion, has a minimal element; let $P_1, \ldots, P_n$ be primitive ideals such that $P_1 \cap \cdots \cap P_n$ is minimal in $\mathcal{Q}$. If $P_1 \cap \cdots \cap P_n$ contained a nonzero element a, there would be a primitive ideal Q such that $a \notin Q$ as A is semisimple, and hence $P_1 \cap \cdots \cap P_n \cap Q \subset P_1 \cap \cdots \cap P_n$, a contradiction of the minimality of $P_1 \cap \cdots \cap P_n$. Thus $P_1 \cap \cdots \cap P_n = (0)$. Moreover, any two primitive ideals are relatively prime as they are maximal ideals by 27.13. Consequently by 24.11, as $P_1 \cap \cdots \cap P_n = (0)$, A is isomorphic to $\prod_{i=1}^{n} (A/P_i)$. Therefore 2° holds by 27.12, and by that same theorem, 2° and 3° are equivalent. If 3° holds, then A is semisimple by 26.21 as the radical of a simple ring is the zero ideal, and A is artinian by 27.7. •

27.15 Theorem. *The radical R of an artinian ring [K-algebra] A is a nilpotent ideal.*

Proof. If A is a K-algebra, by 26.23 its radical is an algebra ideal. Since $(R^k)_{k \geq 1}$ is a decreasing sequence of ideals, there exists $q \geq 1$ such that $R^{q+k} = R^q$ for all $k \geq 0$. Let $N = R^q$, and suppose that $N \neq (0)$. The set $\mathcal{L}$ of nonzero left [algebra] ideals L such that $NL \neq (0)$ is nonempty, since $N^2 = R^{2q} = R^q \neq (0)$ and thus $N \in \mathcal{L}$. Therefore $\mathcal{L}$ has a minimal element M. Since $NM \neq (0)$, there exists $b \in M$ such that $Nb \neq (0)$. As

$$N(Nb) = N^2 b = Nb \neq (0)$$

and $Nb \subseteq M$, $Nb = M$ by the minimality of M. Thus there exists $n \in N$ such that $nb = b$. As $n \in N \subseteq R$, n is advertible, so

$$b = b - (n^a \circ n)b = b - n^a b - nb + n^a nb = (b - nb) - n^a(b - nb) = 0,$$

a contradiction. Thus $N = (0)$. •

We next apply these results to the commutative case:

27.16 Theorem. *If A is a commutative ring with identity and if the zero ideal of A is a product of finitely many maximal ideals, then A is artinian if and only if A is noetherian.*

Proof. Let $(0) = M_1 \ldots M_s$ where $(M_k)_{1 \leq k \leq s}$ is a sequence of (not necessarily distinct) maximal ideals. Let $A_0 = A$ and $A_k = M_1 \ldots M_k$ for each $k \in [1, s]$. Then for each such k, A_{k-1}/A_k is a noetherian A-module if and only if it is an artinian A-module. Indeed, as $A_k = A_{k-1} M_k$, A_{k-1}/A_k is a vector space over the field A/M_k under the well-defined scalar multiplication $(a + M_k).(x + A_k) = ax + A_k$ for all $a \in A$ and all $x \in A_{k-1}$, and the submodules of the A-module A_{k-1}/A_k are identical with the subspaces of the (A/M_k)-vector space A_{k-1}/A_k. Consequently, if A is artinian [noetherian], then the A-module A_{k-1}/A_k is artinian [noetherian] by 27.8 [20.3], so the (A/M_k)-vector space A_{k-1}/A_k is artinian [noetherian], hence finite-dimensional [by 20.2], therefore noetherian [artinian], and so the A-module A_{k-1}/A_k is noetherian [artinian]. Consequently, by 27.9, A is an artinian ring if and only if it is a noetherian ring. •

27.17 Theorem. *If A is a commutative ring with identity with radical R, then A is an artinian ring if and only if A is a semilocal noetherian ring and R is nilpotent, or equivalently, if and only if A is the direct sum of finitely many local noetherian rings whose maximal ideals are nilpotent.*

Proof. Necessity: By 26.16 and 27.14, A/R is the direct sum of finitely many fields, so A/R is a semilocal ring, and consequently A is also. By 27.15 and (3) of 24.16, (0) is the product of finitely many maximal ideals, so by 27.16, A is a noetherian ring and its natural topology is discrete and thus complete, again by 27.15. Consequently, A has the desired descriptions by 24.19. The condition is sufficient by 27.16. •

We conclude with an application to topological rings.

By the *minimum condition* on a class $\mathcal{Q}$ of subrings of a ring or submodules of a module, we mean the statement that every nonempty subset of $\mathcal{Q}$, ordered by inclusion, contains a minimal element. This statement implies and, by the Axiom of Choice, is implied by the *descending chain condition* on $\mathcal{Q}$: There is no strictly decreasing sequence $(Q_n)_{n \geq 0}$ of members of $\mathcal{Q}$.

27.18 Theorem. *If A is a Hausdorff, advertibly open, primitive ring satisfying the minimum condition on closed left ideals, then A is a simple artinian ring.*

Proof. By hypothesis, the set of nonzero closed left ideals of A contains a minimal member I. By 26.29, I is a minimal left ideal of A and there is an idempotent e in A such that $I = Ae$. Consequently by 25.22 we may regard A as a topological dense ring of continuous linear operators containing nonzero linear operators of finite rank on a Hausdorff right vector space E over a division ring K furnished with a ring topology such that $(u, x) \to u(x)$ is continuous from $A \times E$ to E. In particular, for each $x \in E$, $u \to u(x)$ is continuous from A to E. Suppose that E had an infinite sequence $(x_n)_{n \geq 1}$ of linearly idependent vectors. If $J_n = \{u \in A : u(x_i) = 0 \text{ for all } i \in [1, n]\}$, then $(J_n)_{n \geq 1}$ would be a strictly decreasing sequence of closed left ideals, a contradiction. Thus E is finite-dimensional, so A is isomorphic to the ring of all linear operators on a finite-dimensional vector space and hence is a simple artinian ring by 27.12. •

The discrete case of Theorem 27.18 is Theorem 27.12.

Exercises

27.1 Let A be a ring. (a) A is simple if and only if A is primitive and A has no proper nonzero ideals. (b) A is a simple ring with a minimal left ideal if and only if A is isomorphic to a dense ring of linear operators of finite rank on a vector space over a division ring.

27.2 Let E be a K-vector space having a countably infinite basis, let A be the ring of all linear operators on E, and let P be the ideal of all linear operators on E of finite rank. (a) P is a maximal ideal of A. (b) P is a primitive ideal of A [Use 26.6.] (c) A/P is a simple ring that has no minimal left ideals. [If $u \in A \setminus P$, construct $v \in A$ such that $u \notin Avu + P$.]

27.3 If J is a finitely generated ideal of a commutative ring with identity A that is contained in its radical and if A/J is artinian, then A is semilocal and the J-topology is the natural topology of A [Use 27.17 and 27.15.]

27.4 (Kaplansky [1947c]) A Hausdorff, semisimple, advertibly open ring satisfying the minimum condition on closed left ideals is an artinian ring. [Use 27.18 in arguing as in the proof of 27.14.]

27.5 If A is a trivial ring that is a nonzero, finite-dimensional vector space over a field F of characteristic zero, then the F-algebra A is artinian, but A is not an artinian ring.

CHAPTER VII

LINEAR COMPACTNESS AND SEMISIMPLICITY

Linearly compact rings include compact, totally disconnected rings and discrete artinian rings, and hence offer a natural domain for generalizations of theorems concerning those two subjects. In this chapter, we shall primarily be concerned with semisimple rings. We conclude with a discussion of connected locally compact rings and present some informaton about semisimple locally compact rings.

28 Linearly Compact Rings and Modules

28.1 Definition. *A topological A-module E is* **linearly topologized**, *and its topology is a* **linear topology**, *if the open submodules of E form a fundamental system of neighborhoods of zero. A* [closed] **linear filter base** *on E is a filter base consisting of cosets of* [closed] *submodules of E, and a* **linear filter** *is a filter having a linear filter base.*

If M is a submodule of a linearly topologized A-module E, the topology of E clearly induces linear topologies on M and E/M. The cartesian product of a family of linearly topologized A-modules is also a linearly topologized A-module.

28.2 Definition. *A topological A-module E is* **linearly compact**, *and its topology is a* **linearly compact topology**, *if E is Hausdorff and linearly topologized, and if every linear filter on E has an adherent point.*

Thus a Hausdorff linear topology on E is linearly compact if and only if every closed linear filter base has a nonempty intersection. If $\mathcal{T}$ is a linear [linearly compact] topology on an A-module E, and if the topology of A is replaced by a stronger ring topology (for example, the discrete topology), then $\mathcal{T}$ is still a linear [linearly compact] topology on E.

28.3 Theorem. *If u is a continuous homomorphism from a linearly compact A-module E to a Hausdorff linearly topologized A-module F, then $u(E)$ is linearly compact.*

Proof. If $\mathcal{F}$ is a closed linear filter base on $u(E)$, then $u^{-1}(\mathcal{F})$ is a closed linear filter base on E, so there exists $b \in E$ such that $b \in u^{-1}(F)$ for all $F \in \mathcal{F}$, whence $u(b) \in u(u^{-1}(F)) = F$ for all $F \in \mathcal{F}$. •

28.4 Corollary. *If $\mathcal{T}$ is a linearly compact topology on a module, so is every weaker Hausdorff linear topology.*

28.5 Theorem. *A linearly compact module is complete.*

Proof. Let $\mathcal{F}$ be a Cauchy filter on a linearly compact A-module E, and let $\mathcal{V}$ be the set of open submodules of E. For each $V \in \mathcal{V}$, let $F_V \in \mathcal{F}$ be V-small, and let $a_V \in F_V$; then $F_V \subseteq a_V + V$, an open and hence closed set. As $\mathcal{F}$ is a filter, $\{a_V + V : V \in \mathcal{V}\}$ is a closed linear filter base, so by hypothesis there exists

$$a \in \bigcap_{V \in \mathcal{V}} (a_V + V).$$

Consequently, for each $V \in \mathcal{V}$, $a + V = a_V + V$, so $F_V \subseteq a + V$. Thus $\mathcal{F}$ converges to a. •

28.6 Theorem. *Let E be a Hausdorff linearly topologized A-module. (1) If a submodule M of E is linearly compact for its induced topology, then M is closed. (2) If E is linearly compact, then a submodule of E is linearly compact if and only if it is closed. (3) If E is linearly compact and if M and N are closed submodules of E, then $M + N$ is closed.*

Proof. (1) follows from 28.5. (2) If M is a closed submodule of a linearly compact module E, an adherent point of a filter base of subsets of M must belong to M, so M is also linearly compact. (3) E/M is Hausdorff and hence linearly compact by 28.3. Let ϕ_M be the canonical epimorphism from E to E/M. By (2), N is linearly compact, so $\phi_M(N)$ is linearly compact and hence closed by 28.3 and (1). Therefore as $M + N = \phi_M^{-1}(\phi_M(N))$, $M + N$ is closed. •

The use of ultrafilters in proving Tikhonov's theorem in topology has a counterpart in proving that the cartesian product of linearly compact modules is linearly compact. A *maximal linear filter* on an A-module E is a linear filter on E maximal for the ordering $\subseteq$ on the set of all linear filters on E. The set of all linear filters on E containing a given linear filter $\mathcal{F}$ is clearly inductive, since the supremum of a totally ordered set Γ of such filters is its union $\bigcup_{\mathcal{C} \in \Gamma} \mathcal{C}$. Consequently by Zorn's Lemma: *A linear filter is contained in a maximal linear filter*, for a linear filter maximal in the set of all linear filters containing a given linear filter $\mathcal{F}$ is clearly maximal in the set of all linear filters. Furthermore: *If c is adherent to a maximal linear filter $\mathcal{U}$ for a linear topology on E, then $\mathcal{U}$ converges to*

c. Indeed, let V be an open coset containing c. Then $V \cap F \neq \emptyset$ for all $F \in \mathcal{U}$, so $\{V \cap F : F$ is a coset in $\mathcal{U}\}$ is a linear filter base for a linear filter containing and hence identical with $\mathcal{U}$; therefore $V \in \mathcal{U}$. Thus $\mathcal{U}$ converges to c. Consequently: A Hausdorff linear topology on E is linearly compact if and only if every maximal linear filter on E converges to a point of E. Necessity: A maximal linear filter has, by hypothesis, an adherent point and hence converges to it. Sufficiency: If $\mathcal{F}$ is a linear filter, $\mathcal{F}$ is contained in a maximal linear filter $\mathcal{U}$, which converges to some $c \in E$ by hypothesis, and hence c is adherent to $\mathcal{U}$ and a fortiori to $\mathcal{F}$. Finally: If f is an epimorphism from an A-module E to an A-module F and if $\mathcal{U}$ is a maximal linear filter on E, then $f(\mathcal{U})$ is a maximal linear filter on F. If not, there would exist a coset $V \notin f(\mathcal{U})$ intersecting nonvacuously $f(U)$ for every $U \in \mathcal{U}$, so as $\emptyset \neq f^{-1}(V \cap f(U)) = f^{-1}(V) \cap U$, $f^{-1}(V)$ would be a coset in E not belonging to $\mathcal{U}$ (as otherwise $V = f(f^{-1}(V)) \in f(\mathcal{U})$) that intersects nonvacuously each member of $\mathcal{U}$, a contradiction of the maximality of $\mathcal{U}$.

28.7 Theorem. *The cartesian product E of a family $(E_\lambda)_{\lambda \in L}$ of linearly compact A-modules is linearly compact.*

Proof. For each $\lambda \in L$, let pr_λ be the projection from E to E_λ. Let $\mathcal{U}$ be a maximal linear filter on E. Then for each $\lambda \in L$, $pr_\lambda(\mathcal{U})$ is a maximal linear filter on E_λ, so $pr_\lambda(\mathcal{U})$ converges to some $c_\lambda \in E_\lambda$, and therefore $\mathcal{U}$ converges to $(c_\lambda)_{\lambda \in L}$. ∎

28.8 Definition. *A topological ring A is **linearly compact** if A, regarded as a left module over itself, is a linearly compact A-module.*

For example, a totally disconnected compact ring is linearly topologized and hence linearly compact by 4.20. Linear compactness distinguishes an important class of real valuations, as is shown by the following theorem, which we shall not use and hence will omit the proof:

28.9 Theorem. *Let K be a field topologized by a real valuation v, and let A be the valuation ring of v. The following statements are equivalent:*

1° *A is a linearly compact ring.*

2° *K is a linearly compact A-module.*

3° *If w is a real valuation of an extension field L of K extending v such that $e(w/v) < +\infty$ and $f(w/v) < +\infty$, then $[L : K] = e(w/v)f(w/v)$, and w is the only real valuation of L extending v.*

4° *If w is a real valuation of an extension field L of K extending v such that $e(w/v) = 1 = f(w/v)$, then $L = K$.*

A proof may be found, for example, in *Topological Fields*, 31.12 - 31.21.

For example, the valuation ring of a complete discrete valuation is linearly compact by 19.9, a fact also established by the equivalence of 8° and 5° of 36.33.

28.10 Definition. *If E is a topological A-module, E is **strictly linearly compact** and its topology is a **strictly linearly compact topology** if E is linearly compact and every continuous epimorphism from E to a Hausdorff linearly topologized A-module is a topological epimorphism. A topological ring A is a **strictly linearly compact ring** if the (left) A-module A is strictly linearly compact.*

For example, a totally disconnected compact ring A is a strictly linearly compact ring and is, moreover, a strictly linearly compact module over any of its subrings. Indeed, the open ideals of A form a fundamental system of neighborhoods of zero by 4.20, and a theorem of topology establishes that if f is a continuous epimorphism from the additive group A to a Hausdorff topological group B with kernel K, then the induced isomorphism $\bar{f}$ from A/K to B is a homeomorphism, so f is a topological epimorphism.

28.11 Theorem. *If u is a continuous homomorphism from a strictly linearly compact A-module E to a Hausdorff linearly topologized A-module F, then $u(E)$ is strictly linearly compact.*

Proof. By 28.3, $u(E)$ is linearly compact. If v is a continuous epimorphism from $u(E)$ to a Hausdorff linearly topologized A-module F, then $v \circ u$ is a continuous epimorphism from E to F, so $v \circ u$ is open by hypothesis. Consequently, if O is open in $u(E)$, then as $v(O) = (v \circ u)(u^{-1}(O))$, $v(O)$ is open in F. •

28.12 Definition. *A Hausdorff linear topology T on an A-module E is **minimal** if there are no Hausdorff linear topologies on E strictly weaker than T, that is, if T is minimal in the set of all Hausdorff linear topologies on E, ordered by inclusion.*

28.13 Theorem. *A strictly linearly compact topology on an A-module E is minimal.*

The assertion is evident.

28.14 Theorem. *Let E be an A-module. (1) If E is artinian, the discrete topology is the only Hausdorff linear topology on E. (2) E is strictly linearly compact for the discrete topology if and only if E is artinian.*

Proof. (1) By hypothesis, the filter base of open submodules for a linear topology on E contains a minimal member, which is actually the smallest member.

(2) Necessity: Let $(M_n)_{n\geq 1}$ be a decreasing sequence of submodules of E, and let

$$M = \bigcap_{n=1}^{\infty} M_n.$$

Then $(M_n/M)_{n\geq 1}$ is a fundamental system of neighborhoods of zero for a Hausdorff linear topology on E/M. By hypothesis and 28.11, E/M is strictly linearly compact for the discrete topology, so in particular, the discrete topology is minimal by 28.13. Consequently, there exists $q \geq 1$ such that for all $n \geq q$, $M_q/M_n = (0)$, or equivalently, $M_q = M_n$.

Sufficiency: Let $\mathcal{F}$ be a linear filter base on an artinian A-module E. Then $\mathcal{F}$ contains a minimal and hence a smallest member M. Let $a \in E$ be such that $a + M \in \mathcal{F}$. Then each member of $\mathcal{F}$ contains $a + M$, so a is adherent to $\mathcal{F}$. Thus E is linearly compact. Let u be an epimorphism from E, furnished with the discrete topology, to a Hausdorff linearly topologized A-module F. Then F is isomorphic to a quotient module of E and hence is artinian by 27.3. Consequently by (1), the topology of F is the discrete topology, so u is a topological epimorphism. •

28.15 Theorem. *Let E be a Hausdorff linearly topologized A-module, and let $(U_\lambda)_{\lambda \in L}$ be a fundamental system of neighborhoods of zero consisting of open submodules. (1) E is linearly compact if and only if E is complete and for each $\lambda \in L$, E/U_λ is linearly compact for the discrete topology. (2) E is strictly linearly compact if and only if E is complete and for each $\lambda \in L$, E/U_λ is artinian.*

Proof. (1) The condition is necessary by 28.5 and 28.3. Sufficiency: By the module analogue of 5.22, E is topologically isomorphic to $\varprojlim_{\lambda \in L}(E/U_\lambda)$, where each E/U_λ has the discrete topology. By 5.20, $\varprojlim_{\lambda \in L}(E/U_\lambda)$ is a closed submodule of $\prod_{\lambda \in L}(E/U_\lambda)$, which is linearly compact by 28.7. Thus by 28.6, E is linearly compact.

(2) The condition is necessary by 28.5, 28.11, and 28.14. Sufficiency: By (1) and 28.14, E is linearly compact. To establish that E is strictly linearly compact, let u be a continuous epimorphism from E to a Hausdorff linearly topologized A-module F, and let O be an open submodule of E. Then O is closed, so by 28.3 and 28.6, $u(O)$ is closed in F, and thus the induced topology of $F/u(O)$ is Hausdorff. The kernel H of the epimorphism $x \to u(x) + u(O)$ from E to $F/u(O)$ contains O, hence is open, and therefore contains U_λ for some $\lambda \in L$. Consequently, E/H is isomorphic to a quotient module of the artinian module E/U_λ and hence is artinian by 27.3. Therefore as E/H is isomorphic to $F/u(O)$, the discrete topology is the only Hausdorff linear topology on $F/u(O)$ by 28.14. Consequently, the quotient topology of $F/u(O)$ is the discrete topology, so $u(O)$ is open. •

28.16 Theorem. *Let E be a Hausdorff linearly topologized A-module, and let F be a closed submodule of E. (1) E is linearly compact if and only if F and E/F are linearly compact. (2) E is strictly linearly compact if and only if F and E/F are strictly linearly compact.*

Proof. Let ϕ be the canonical epimorphism from E to E/F. (1) The condition is necessary by 28.3 and 28.6. Sufficiency: Let $(F_\lambda)_{\lambda \in L}$ be a linear filter base on E, and for each $\lambda \in L$ let $F_\lambda = z_\lambda + M_\lambda$, where M_λ is a submodule of E. Since E/F is linearly compact, there exists $z \in E$ such that for each $\lambda \in L$, $\phi(z) \in \overline{\phi(z_\lambda + M_\lambda)}$, whence as ϕ is a topological epimorphism,

$$z \in \phi^{-1}(\overline{\phi(z_\lambda + M_\lambda)}) = \overline{\phi^{-1}(\phi(z_\lambda + M_\lambda))} = \overline{z_\lambda + M_\lambda + F}.$$

Consequently by (3) of 3.3, for each open submodule V of E and for each $\lambda \in L$, $z \in z_\lambda + M_\lambda + F + V$, so

$$((z - z_\lambda) + M_\lambda + V) \cap F \neq \emptyset.$$

Therefore the set of all $((z - z_\lambda) + M_\lambda + V) \cap F$ such that $\lambda \in L$ and V is an open submodule of E is a filter base of cosets of open and hence closed submodules of F. Consequently, as F is linearly compact, there exists $a \in F$ such that for all $\lambda \in L$ and all open submodules V of E, $a \in (z-z_\lambda)+M_\lambda+V$. Thus for each $\lambda \in L$, $z - a \in z_\lambda + M_\lambda + V$ for all open submodules V of E, so $z - a \in \overline{z_\lambda + M_\lambda}$ by (3) of 3.3.

(2) Necessity: By 28.11, E/F is strictly linearly compact, and by (1), F is linearly compact and hence complete. Consequently, by (2) of 28.15, to show that F is strictly linearly compact, it suffices to show that if U is an open submodule of E, then $F/(F \cap U)$ is artinian. But as $F/(F \cap U)$ is isomorphic to $(F + U)/U$, a submodule of E/U, which is artinian by (2) of 28.15, $F/(F \cap U)$ is artinian by 27.3.

Sufficiency: By (1), E is linearly compact and hence complete, so by (2) of 28.15, it suffices to show that if U is an open submodule of E, then E/U is artinian. Since $(U+F)/F$ is an open submodule of E/F, $(E/F)/((U+F)/F$ is artinian by (2) of 28.15, so its isomorphic copy $E/(U+F)$ is also artinian. Now $(U+F)/U$ is isomorphic to $F/(U \cap F)$, which is artinian by (2) of 28.15 since F is strictly linearly compact. Therefore as $(E/U)/((U + F)/U)$ is isomorphic to $E/(U + F)$, E/U is artinian by 27.3. •

28.17 Theorem. *If E is the cartesian product of a family $(E_\lambda)_{\lambda \in L}$ of strictly linearly compact A-modules, E is strictly linearly compact.*

Proof. By 28.7, E is linearly compact and hence complete. Consequently, it suffices by (2) of 28.15 to show that if U_λ is an open submodule of E_λ for each $\lambda \in L$ and if $U_\lambda = E_\lambda$ for all $\lambda \in L \setminus Q$, where Q is a finite subset of L, then $E/(\prod_{\lambda \in L} U_\lambda)$ is an artinian module. But $E/(\prod_{\lambda \in L} U_\lambda)$ is isomorphic to $\prod_{\lambda \in Q}(E_\lambda/U_\lambda)$, which is artinian by (2) of 28.15 and 27.6. •

28.18 Theorem. *If E is a Hausdorff linearly topologized module over a [strictly] linearly compact ring A and if $x_1, \ldots, x_n \in E$, then $Ax_1 + \cdots + Ax_n$ is a [strictly] linearly compact, hence complete and thus closed submodule of E.*

Proof. $Ax_1 + \cdots + Ax_n$ is the image of the A-module A^n under the homomorphism $u : (a_1, \ldots, a_n) \to a_1 x_1 + \cdots + a_n x_n$. By 28.7 [28.17], A^n is a [strictly] linearly compact A-module, so as u is continuous, the assertion follows from 28.3 [28.11] and 28.5. •

28.19 Theorem. *If $\mathcal{T}$ is a linearly compact topology on an A-module E for which every submodule is closed, then E is linearly compact for the discrete topology.*

Proof. The adherence for $\mathcal{T}$ of any linear filter base on E is simply its intersection. Hence E is linearly compact for the discrete topology. •

28.20 Theorem. *Let $\mathcal{N}$ be a linear filter base on a linearly compact A-module E. (1) If u is a continuous homomorphism from A to a Hausdorff linearly topologized A-module F, and if C is the adherence of $\mathcal{N}$, then $u(C)$ is the adherence of $u(\mathcal{N})$. (2) If each member of $\mathcal{N}$ is closed and if M is a closed submodule of E, then*

$$M + \bigcap_{N \in \mathcal{N}} N = \bigcap_{N \in \mathcal{N}} (M + N).$$

Proof. (1) For each $N \in \mathcal{N}$, $u(\overline{N}) \subseteq \overline{u(N)}$ as u is continuous. As $\overline{N}$ is linearly compact by 28.6, $u(\overline{N})$ is closed by 28.3 and 28.6, so $u(\overline{N}) \supseteq \overline{u(N)}$. Thus $u(\overline{N}) = \overline{u(N)}$ for each $N \in \mathcal{N}$. Consequently, if

$$c \in C = \bigcap_{N \in \mathcal{N}} \overline{N},$$

then

$$u(c) \in u(\bigcap_{N \in \mathcal{N}} \overline{N}) \subseteq \bigcap_{N \in \mathcal{N}} u(\overline{N}) = \bigcap_{N \in \mathcal{N}} \overline{u(N)},$$

the adherence of $u(\mathcal{N})$. Conversely, let

$$d \in \bigcap_{N \in \mathcal{N}} \overline{u(N)}.$$

Then for each open submodule V of F and each $N \in \mathcal{N}$, $(d+V) \cap u(N) \neq \emptyset$, so $u^{-1}(d+V) \cap N$ is a coset of a submodule of E; let $G_{V,N} = u^{-1}(d+V) \cap N$. As E is linearly compact, there exists c belonging to $\overline{G}_{V,N}$ for all $N \in \mathcal{N}$ and all open submodules V of F. In particular, for each open submodule V of F, since $u^{-1}(d+V)$ is open and thus closed, $c \in u^{-1}(d+V)$, that is, $u(c) - d \in V$. Therefore as F is Hausdorff, $u(c) = d$. Also for each $N \in \mathcal{N}$, $c \in \overline{G}_{F,N} \subseteq \overline{N}$, that is, $c \in C$.

(2) Let ϕ be the canonical epimorphism from E to E/M. If $N \in \mathcal{N}$, then $\phi(N)$ is closed by 28.3 and 28.6, so

$$\phi(\bigcap_{N \in \mathcal{N}} N) = \bigcap_{N \in \mathcal{N}} \phi(N)$$

by (1), whence

$$M + \bigcap_{N \in \mathcal{N}} N = \phi^{-1}(\phi(\bigcap_{N \in \mathcal{N}} N)) = \phi^{-1}(\bigcap_{N \in \mathcal{N}} \phi(N))$$
$$= \bigcap_{N \in \mathcal{N}} \phi^{-1}(\phi(N)) = \bigcap_{N \in \mathcal{N}} (N + M). \bullet$$

28.21 Theorem. *Let E be the cartesian product of a family $(E_\lambda)_{\lambda \in L}$ of nonzero A-modules. If F is a submodule of E that is linearly compact for the discrete topology and dense in E for the cartesian product topology, where each E_λ is given the discrete topology, then L is finite.*

Proof. By 28.4, F is also linearly compact for the topology induced by the cartesian product topology, so by 28.6, $F = E$. Consequently, $\bigoplus_{\lambda \in L} E_\lambda$ is a submodule of F and therefore is also linearly compact for the discrete topology by 28.6. Replacing F by $\bigoplus_{\lambda \in L} E_\lambda$ in the preceding argument, we conclude that $\bigoplus_{\lambda \in L} E_\lambda = E$, whence L is finite. •

28.22 Theorem. *If a strictly linearly compact A-module E is the direct sum of closed submodules $M_1, \ldots, M_n$, then E is the topological direct sum of $M_1, \ldots, M_n$.*

Proof. By (1) of 28.16 and 28.17, $\prod_{k=1}^n M_k$ is strictly linearly compact, so as

$$(x_1, \ldots, x_n) \to \sum_{k=1}^n x_k$$

is a continuous isomorphism from $\prod_{k=1}^n M_k$ to E, it is a topological isomorphism by 28.10. •

Exercises

28.1 Let E be an A-module. A *Zariski topology* on E is a linear topology for which all submodules are closed. (a) A Zariski topology is Hausdorff. (b) If F is a submodule of E, a Zariski topology on E induces Zariski topologies on F and E/F. (c) A linear topology on E stronger than a Zariski topology is a Zariski topology.

28.2 Let E be an A-module furnished with a Zariski topology. (a) Let f be a continuous homomorphism from a Hausdorff linearly topologized A-module D to E, and let $\hat{f}$ be the continuous extension of f to a continuous homomorphism from $\hat{D}$ to $\hat{E}$. The kernel of $\hat{f}$ is the closure $\hat{K}$ in $\hat{D}$ of the kernel K of f. [Use 7.20.] (b) (Zariski [1945]) In particular, if f is a monomorphism, so is $\hat{f}$. (c) (Zariski [1945]) If E is complete for a Zariski topology, E is complete for any stronger linear topology. (d) If $M_1, \ldots, M_n$ are submodules of E and if $\sum_{i=1}^{n} a_i = 0$ where $a_i \in \widehat{M_i}$, the closure of M_i in $\hat{E}$, for all $i \in [1, n]$, then for each $i \in [1, n]$ there is a net $(x_{i,\lambda})_{\lambda \in L}$ in M_i such that $\lim_{\lambda \in L} x_{i,\lambda} = a_i$ and $\sum_{i=1}^{n} x_{i,\lambda} = 0$ for all $\lambda \in L$. (e) If $M_1, \ldots, M_n$ are submodules of E such that $\sum_{i=1}^{n} M_i$ is the direct sum of $M_1, \ldots, M_n$, then $\sum_{i=1}^{n} \widehat{M_i}$ is the direct sum of $\widehat{M_i}, \ldots, \widehat{M_n}$. (f) If F and G are submodules of E, $\widehat{F \cap G} = \hat{F} \cap \hat{G}$. [Consider the mapping $(x, y) \to x - y$ from $F \times G$ to E.]

28.3 Let E be a linearly compact vector space. (a) If E is discrete, then E is finite-dimensional. (b) If U is an open subspace of E, E/U is finite-dimensional. (c) The topology of E is strictly linearly compact.

28.4 (Lefschetz [1942]) Let E be a linearly compact vector space. (a) If M and N are subspaces of E such that $M + N = E$, then for all $x, y \in E$, there exists $z \in E$ such that $z \equiv x \pmod{M}$ and $z \equiv y \pmod{N}$. (b) If $H_1, \ldots, H_n$ are subspaces of E of codimension 1, none of which contains the intersection of the others, and if $x_1, \ldots, x_n \in E$, there exists $z \in E$ such that $z \equiv x_i \pmod{H_i}$ for all $i \in [1, n]$. (c) There is a family $(H_\lambda)_{\lambda \in L}$ of open subspaces of E of codimension 1 such that

$$\bigcap_{i=1}^{n} H_{\lambda_i} \not\subseteq H_{\lambda_0}$$

whenever $\lambda_0, \lambda_1, \ldots, \lambda_n$ are distinct members of L and

$$\bigcap_{\lambda \in L} H_\lambda = \{0\}.$$

(d) The function $\Phi : z \to (z + H_\lambda)_{\lambda \in L}$ from E to $\prod_{\lambda \in L}(E/H_\lambda)$ is a topological isomorphism. Thus E is topologically isomorphic to the cartesian product of discrete one-dimensional vector spaces. (e) Conversely, the cartesian product of discrete one-dimensional vector spaces is a linearly compact vector space.

28.5 Let A be a ring with identity. For each left ideal J of A and each $c \in A$, we define $(J : c)$ by

$$(J : c) = \{x \in A : xc \in J\}.$$

Let $\mathcal{T}$ be a linear topology on A. For each left ideal J, let

$$J' = \{c \in A : (J : c) \text{ is open}\}.$$

Topology $\mathcal{T}$ is a *Gabriel topology* if a left ideal J is open whenever J' is open. (a) If J is a left ideal, J' is a left ideal containing J. (b) If I and J are left ideals and if $c \in A$, then $(I \cap J)' = I' \cap J'$ and $(J : c)' = (J' : c)$. (c) For each left ideal J of A, define J_n recursively for all $n \in \mathbb{N}$ by $J_0 = J$, $J_{n+1} = (J_n)'$. If I and J are left ideals and if $c \in A$, then $(J_n)' = (J')_n$, $(I \cap J)_n = I_n \cap J_n$, and $(J_n : c) = (J : c)_n$ for all $n \in \mathbb{N}$. (d) The set of all left ideals J such that J_n is open for some $n \in N$ is a fundamental system of neighborhoods of zero for the weakest Gabriel topology on A stronger than $\mathcal{T}$.

29 Linearly Compact Semisimple Rings

Here we shall describe all linearly compact semisimple rings. Basic to our description is the ring of A of all linear operators on a discrete vector space E over a discrete division ring K, furnished with the topology of pointwise convergence, that is, the weakest topology on A such that for all $x \in E$, $u \to u(x)$ is continuous from A to E.

First, we shall extend two definitions introduced on pages 206 and 210: If E is an A-module and if L is a subset of A, we shall call the *annihilator of L in E* the submodule of all $x \in E$ such that $ux = 0$ for all $u \in L$ and denote it by $\text{Ann}_E(L)$, and if F is a subset of E, we shall call the *annihilator of F in A* the left ideal of all $u \in A$ such that $ux = 0$ for all $x \in F$ and denote it by $\text{Ann}_A(F)$.

If A is a ring of linear operators on a vector space E, we shall regard E as an A-module under the scalar multiplication $(u, x) \to u(x)$. Thus E is a simple A-module if and only if A is a primitive ring of endomorphisms of the additive group E.

Let E be a topological space, F a set. The *topology of pointwise* (or *simple*) *convergence* on E^F, the set of all functions from F to E, is the weakest topology on E^F such that for all $x \in E$, $u \to u(x)$ is continuous from E^F to E. Thus the topology of pointwise convergence is simply the cartesian product topology on E^F, regarded as the cartesian product of $(E_x)_{x \in F}$ where $E_x = E$ for all $x \in F$. We shall also call the topology it induces on any subset of E^F the topology of pointwise convergence on that set, and denote it by $\mathcal{T}_s$.

In particular, let A be the ring of all linear operators on a discrete vector space E over a discrete division ring K, whence $A \subseteq E^E$. A fundamental system of neighborhoods of zero for $\mathcal{T}_s$ on A is then $\{\mathrm{Ann}_A(X): X$ is a finite subset of $E\}$, or equivalently, $\{\mathrm{Ann}_A(M): M$ is a finite-dimensional subspace of $E\}$, since clearly if M is the subspace generated by a finite subset X, $\mathrm{Ann}_A(X) = \mathrm{Ann}_A(M)$. If B is a ring of linear operators on E, clearly B is a dense ring of linear operators if and only if B is (topologically) dense in A for the topology $\mathcal{T}_s$.

29.1 Theorem. *Let A be the ring of all linear operators on a discrete vector space E over a discrete division ring K. Furnished with the topology $\mathcal{T}_s$ of pointwise convergence, A is a strictly linearly compact ring, and E is a topological A-module.*

Proof. For any subset X of E, $\mathrm{Ann}_A(X)$ is a left ideal of A, and clearly the intersection of all the annihilators of all finite subsets of E is the zero ideal. Thus $\mathcal{T}_s$ is a Hausdorff linear topology on A. To show that $\mathcal{T}_s$ is a ring topology, therefore, it suffices by 2.15 to show that for any $v \in A$, $u \to u \circ v$ from A to A is continuous at zero. But for any finite subset X of E, $v(X)$ is finite, and clearly

$$\mathrm{Ann}_A(v(X)) \circ v \subseteq \mathrm{Ann}_A(X).$$

Moreover, E is a topological A-module by 2.16, for (TM 4) holds as the image of $A \times (0)$ under scalar multiplication is (0).

For each $b \in E$, let E_b be the topological A-module E. Let B be a basis of the vector space E. Then $f : u \to (u(b))_{b \in B}$ is a continuous isomorphism from the topological A-module A to the topological A-module $\prod_{b \in B} E_b$. To show that f is also open, let M be a finite-dimensional subspace of E. Then there is a finite subset C of B such that M is contained in the subspace generated by C. Let $H_b = (0)$ if $b \in C$, $H_b = E$ if $b \in B \setminus C$; then $\prod_{b \in B} H_b$ is an open neighborhood of zero in $\prod_{b \in B} E_b$ contained in $f(\mathrm{Ann}_A(M))$. Thus by 5.18, f is a topological isomorphism from A to $\prod_{b \in B} E_b$. Since E is complete, so is $\prod_{b \in B} E_b$ by (2) of 7.8, and therefore A is also complete by 7.14.

Consequently, to show that $\mathcal{T}_s$ is strictly linearly compact, it suffices by (2) of 28.15 to show that if M is a finite-dimensional subspace of E, then $A/\mathrm{Ann}_A(M)$ is an artinian A-module. Let $\{b_1, \ldots, b_n\}$ be a basis of M. Then $u \to (u(b_1), \ldots, u(b_n))$ is an epimorphism from the A-module A to the A-module $\prod_{i=1}^{n} E_{b_i}$ whose kernel is $\mathrm{Ann}_A(M)$. Since E is a simple A-module, it is trivially artinian. Hence by 27.6, $A/\mathrm{Ann}_A(M)$ is an artinian A-module. •

29.2 Theorem. *Let A be the ring of all linear operators on a discrete vector space E over a discrete division ring K. There is no linearly compact topology on the A-module A that is strictly stronger than the topology T_s of pointwise convergence.*

Proof. Let T be a linearly compact topology on the A-module A stronger than T_s. Let L be a left ideal of A that is open for T, let B be a basis of the vector subspace $\mathrm{Ann}_E(L)$, and for each $b \in B$ let E_b be the A-module E. Then $f : u \to (u(b))_{b \in B}$ is an epimorphism from A to $\prod_{b \in B} E_b$ with kernel $\mathrm{Ann}_A(\mathrm{Ann}_E(L))$. Thus $\prod_{b \in B} E_b$ is isomorphic to $A/\mathrm{Ann}_A(\mathrm{Ann}_E(L))$, which by 28.3 is linearly compact for the discrete topology as $\mathrm{Ann}_A(\mathrm{Ann}_E(L)) \supseteq L$. By 28.21, B is finite, so $\mathrm{Ann}_A(\mathrm{Ann}_E(L))$ is open for T_s. But also, as L is open, thus closed and hence by (2) of 28.6 linearly compact for the topology induced by T, L is also linearly compact and hence closed for T_s by 28.4 and (1) of 28.6. By 25.8, L is dense in $\mathrm{Ann}_A(\mathrm{Ann}_E(L))$ for T_s. Therefore $L = \mathrm{Ann}_A(\mathrm{Ann}_E(L))$. Thus $T = T_s$. •

29.3 Theorem. *Let u be a continuous epimorphism from a [strictly] linearly compact ring A to a Hausdorff linearly topologized topological ring B. Then B is a [strictly] linearly compact ring, [u is an open mapping], and if L is a closed left ideal of A, $u(L)$ is a closed left ideal of B.*

Proof. We convert B into an A-module by defining $a.b$ to be $u(a)b$ for all $a \in A$, $b \in B$. This scalar multiplication is continuous from $A \times B$ to B since u is continuous and multiplication on B is continuous. Moreover, u is a homomorphism from the A-module A to the A-module B since for all $x, y \in A$,

$$u(x.y) = u(xy) = u(x)u(y) = x.u(y).$$

By 28.3 [28.11], B is a [strictly] linearly compact A-module. The left ideals of the ring B are precisely the submodules of the A-module B as u is surjective [and if J is a left ideal of B, the A-submodules and B-submodules of A/J coincide], so B is a [strictly] linearly compact ring. Moreover, if L is a closed left ideal of A, $u(L)$ is a closed left ideal of B by 28.3 and 28.6. •

29.4 Corollary. *If P is a closed ideal of a [strictly] linearly compact ring A, then A/P is a [strictly] linearly compact ring.*

29.5 Theorem. *The cartesian product of a family of [strictly] linearly compact rings is a [strictly] linearly compact ring.*

Proof. Let A be the cartesian product of a family $(A_\lambda)_{\lambda \in L}$ of [strictly] linearly compact rings. We convert each A_λ into an A-module by defining

$x . y_\lambda$ to be $pr_\lambda(x) y_\lambda$ for all $x \in A$, $y \in A_\lambda$, where pr_λ is the canonical epimorphism from A to A_λ. Since pr_λ is continuous and multiplication is continuous on A_λ, A_λ is a topological A-module. Since pr_λ is surjective, the left ideals of A_λ are precisely the A-submodules of A_λ [and if J_λ is a left ideal of A_λ, the A_λ-submodules and the A-submodules of A_λ/J_λ coincide]. Consequently, each A_λ is a [strictly] linearly compact A-module. By 28.7 [28.17], A is a [strictly] linearly compact A-module, that is, A is a [strictly] linearly compact ring. •

29.6 Theorem. *If M is an open regular maximal left ideal of a linearly compact ring A, the largest ideal $P(M)$ of A contained in M is closed, and $A/P(M)$ is topologically isomorphic to the ring of all linear operators on a discrete vector space over a discrete division ring, furnished with the topology of pointwise convergence.*

Proof. Since M is open, M is closed, so as $P(M) = \{a \in A : aA \subseteq M\}$ by 26.4, $P(M)$ is also closed. With the notation of (1) of 26.5, $\hat{A}$ is by 25.6 a dense of linear operators on the K-vector space A/M, where K is the division ring of all endomorphisms of A/M that commute with each member of $\hat{A}$, and there is an isomorphism g from $A/P(M)$ to $\hat{A}$ satisfying $g(a + P(M)) = \hat{a}$ for all $a \in A$. Since M is open, A/M is discrete. Let $\mathcal{T}$ be the topology on $\hat{A}$ for which g is a topological isomorphism. By 29.4, $\mathcal{T}$ is a linearly compact ring topology. For each $x \in A \setminus M$, $\hat{a} \to a(x + M)$ is continuous from $\hat{A}$, furnished with $\mathcal{T}$, to A/M by 5.11 since $a \to ax + M$ is a continuous epimorphism from the additive group A to A/M whose kernel contains $P(M)$. Therefore $\mathcal{T}$ is stronger that the topology of pointwise convergence on $\hat{A}$. Consequently by 29.3 applied to the identity mapping of $\hat{A}$, that topology is a linearly compact ring topology, hence is complete by 28.5, and thus $\hat{A}$ is closed in the ring B of all linear operators on the discrete K-space A/M for the topology $\mathcal{T}_s$ of pointwise convergence. Since $\hat{A}$ is also dense in B for that topology, $\hat{A} = B$. By 29.2, $\mathcal{T} = \mathcal{T}_s$. •

29.7 Theorem. *A topological ring A is semisimple and linearly compact if and only if A is topologically isomorphic to the cartesian product of a family of topological rings, each the ring of all linear operators on a discrete vector space over a discrete division ring, furnished with the topology of pointwise convergence. In particular, the intersection of the closed primitive ideals of a semisimple linearly compact ring is the zero ideal.*

Proof. The condition is sufficient by 29.1, 29.5, and 26.21. Necessity: Let $(P_\lambda)_{\lambda \in L}$ be the family of all the primitive ideals $P(M)$ where M is an open regular maximal left ideal of A, and for each $\lambda \in L$, let $A_\lambda = A/P_\lambda$. The canonical homomorphism Φ from A to $\prod_{\lambda \in L} A_\lambda$, defined by

$$\Phi(a) = (a + P_\lambda)_{\lambda \in L}$$

for all $a \in L$, is then continuous. By 29.6 we need only show that Φ is a topological isomorphism.

To establish the final assertion and the injectivity of Φ, we shall show that

$$\bigcap_{\lambda \in L} P_\lambda = (0).$$

Let $c \in A^*$. As A is semisimple, by 26.9 there exists $b \in A$ such that cb is not left advertible; let $J = \{x - xcb : x \in A\}$. Then J is a regular left ideal of A not containing cb. Moreover, J is a linearly compact A-module by 28.3 since it is the image of A under the continuous homomorphism $x \to x - xcb$. Therefore J is closed by (1) of 28.6, so by (3) of 3.3 there is an open left ideal I of A such that $cb \notin J + I$. As J is regular, so is $J + I$. By 26.3 there is a maximal regular left ideal M containing $J + I$, and $cb \notin M$ since otherwise $x = (x - xcb) + xcb \in J + M = M$ for all $x \in A$. As $J + I$ is open, so is M, and therefore $P(M) = P_\lambda$ for some $\lambda \in L$. Finally, $c \notin P_\lambda$, for otherwise $cb \in M$, a contradiction.

For each $\mu \in L$ let ϕ_μ be the canonical epimorphism from A to A_μ, a linearly compact ring with an identity element by 29.6 and 29.1. To show that $P_\lambda + P_\mu = A$ whenever λ and μ are distinct members of L, we may assume that $P_\lambda \not\subseteq P_\mu$. Then $\phi_\mu(P_\lambda) = A_\mu$ since $\phi_\mu(P_\lambda)$ is a nonzero closed ideal by 29.3 and is also dense by 25.10. Thus

$$P_\lambda + P_\mu = \phi_\mu^{-1}(\phi_\mu(P_\lambda)) = \phi_\mu^{-1}(A_\mu) = A.$$

Moreover, as each A/P_λ has an identity element, $(A/P_\lambda)^2 = A/P_\lambda$. Therefore by 24.11, $\Phi(A)$ is dense in $\prod_{\lambda \in L} A_\lambda$. By 29.3, $\Phi(A)$ is linearly compact, thus closed by 28.5, and hence closed in $\prod_{\lambda \in L} A_\lambda$. Therefore

$$\Phi(A) = \prod_{\lambda \in L} A_\lambda.$$

To show that Φ is open, let J be an open left ideal of A, and for each $\mu \in L$ let

$$J_\mu = pr_\mu(\Phi(J)) \subseteq A_\mu.$$

Then J is closed in A, so $\Phi(J)$ is closed in $\prod_{\lambda \in L} A_\lambda$ by 29.3, and consequently $\Phi(J) = \prod_{\lambda \in L} J_\lambda$ by 24.12, since each A_λ has an identity element. Thus A/J is isomorphic to $\prod_{\lambda \in L}(A_\lambda/J_\lambda)$ and hence to $\prod_{\lambda \in N}(A_\lambda/J_\lambda)$ where $N = \{\lambda \in L : J_\lambda \neq A_\lambda\}$. Consequently, since J is open, the A-module $\prod_{\lambda \in N}(A_\lambda/J_\lambda)$ is linearly compact for the discrete topology. By 28.21, therefore, N is finite, so $\Phi(J)$ is open in $\prod_{\lambda \in L} A_\lambda$. Thus Φ is open. •

29.8 Corollary. *If A is a nonzero semisimple linearly compact ring, then A is strictly linearly compact and has an identity element, and every nonzero closed left ideal of A contains a minimal left ideal.*

Proof. By 29.7, 29.1, and 29.5 the first two assertions hold. For the third, we may by 29.7 assume that $A = \prod_{\lambda \in L} A_\lambda$, where each A_λ is the ring of all linear operators on a vector space. Let J be a closed left ideal of A. By 24.12, $J = \prod_{\lambda \in L} J_\lambda$, where each J_λ is a left ideal of A_λ. As $J \neq (0)$, there exists $\mu \in L$ such that $J_\mu \neq (0)$, so by (1) of 25.21, J_μ contains a minimal left ideal N_μ. Then $in_\mu(N_\mu)$ is clearly a minimal left ideal of A contained in J, where in_μ is the canonical injection from A_μ to A. •

29.9 Corollary. *A topological ring A is semisimple and linearly compact and its topology is an ideal topology if and only if A is topologically isomorphic to the cartesian product of discrete rings, each the ring of all linear operators on a finite-dimensional vector space.*

Proof. The only Hausdorff ideal topology on the ring A of all linear operators on a vector space E is the discrete topology by (2) of 25.21, and the topology of pointwise convergence on A is discrete if and only if E is finite-dimensional. •

29.10 Corollary. *A topological ring A is commutative, semisimple, and linearly compact if and only if A is topologically isomorphic to the cartesian product of a family of discrete fields.*

Theorem 29.7 generalizes the Artin-Wedderburn theorem (27.14). Indeed, if A is a semisimple artinian ring, then A is (strictly) linearly compact for the discrete topology by (2) of 28.14, so A is isomorphic to the cartesian product of finitely many rings, each the ring of all linear operators on a finite-dimensional vector space by 29.7, since a cartesian product of topological rings with identity is discrete if and only if the number of such rings is finite and each is discrete, and the topology of pointwise convergence on the ring of all linear operators of a vector space is discrete only if the vector space is finite-dimensional.

29.11 Theorem. *A topological ring A is primitive [right primitive] and linearly compact if and only if A is topologically isomorphic [anti-isomorphic] to the ring of all linear operators on a discrete nonzero vector space over a discrete division ring, furnished with the topology of pointwise convergence.*

Proof. For primitivity, the condition is sufficient by 29.1 and necessary by 26.12, 29.7, and 25.11. The assertion for right primitivity therefore follows from 25.12. •

29.12 Theorem. *The radical of a linearly compact ring A is closed.*

Proof. By 26.9 and 26.13 it suffices to prove that the closure of an advertible left ideal J is an advertible left ideal. Let $a \in \overline{J}$, and let $\mathcal{L}$ be the set of all open left ideals of A. For each $L \in \mathcal{L}$, let

$$H_L = \{x \in A : x - xa \in L\},$$

an open left ideal since $x \to x - xa$ is continuous on A, let $a_L \in J$ be such that $a - a_L \in L$, and let x_L be the adverse of a_L. To show that the open cosets $\{x_L + H_L : L \in \mathcal{L}\}$ form a filter base, let $L, M \in \mathcal{L}$ be such that $M \subseteq L$. Then

$$a + x_L - x_L a = a + (x_L a_L - a_L) - x_L a = (a - a_L) - x_L(a - a_L) \in L$$

and similarly $a + x_M - x_M a \in M$, so

$$(x_L - x_M) - (x_L - x_M)a = [a + x_L - x_L a] - [a + x_M - x_M a] \in L + M = L$$

and therefore $x_L - x_M \in H_L$, whence $x_M + H_M \subseteq x_L + H_L$. Thus there exists

$$x \in \bigcap_{L \in \mathcal{L}} (x_L + H_L).$$

For each $L \in \mathcal{L}$, $x - x_L \in H_L$, so $x - x_L - (x - x_L)a \in L$, and thus

$$a + x - xa = [a + x_L - x_L a] + [(x - x_L) - (x - x_L)a] \in L + L = L.$$

Consequently, x is the left adverse of a. But $x \in \overline{J}$ since $\overline{J}$ is a left ideal, so by what we have just proved, x also has a left adverse. Consequently as $\circ$ is associative, a is advertible. •

29.13 Corollary. *The radical R of a linearly compact ring A is the intersection of its closed primitive ideals, and A/R is a strictly linearly compact, semisimple ring.*

Proof. By 29.12, 29.4, 29.8, and 26.16, A/R is a strictly linearly compact, semisimple ring. Therefore by 29.7, R is the intersection of the ideals $\phi^{-1}(P)$ where P is a closed primitive ideal of A/R and ϕ is the canonical epimorphism from A to A/R; but each such $\phi^{-1}(P)$ is closed as ϕ is continuous and is primitive since $A/\phi^{-1}(P)$ is isomorphic to $(A/R)/P$. •

29.14 Theorem. *Let A be a linearly compact ring that is not a radical ring, and let R be the radical of A. The following statements are equivalent:*

1° *A/R is artinian.*
2° *A/R is noetherian.*
3° *R is open.*
4° *A/R is topologically isomorphic to the cartesian product of finitely many discrete rings, each the ring of all linear operators on a finite-dimensional vector space.*
5° *Every ideal of A/R is closed.*

Proof. By 29.13, A/R is a linearly compact, semisimple ring. The equivalence of the assertions readily follows from 29.7 and 25.10, applied to the ideal of all linear operators of finite-dimensional range. •

29.15 Theorem. *If e is an idempotent of a [strictly] linearly compact ring A, then eAe is a [strictly] linearly compact ring.*

Proof. By 26.28, Ae and eAe are closed in A. Hence if L is a closed left ideal of eAe, $\overline{AL}$ is a closed left ideal of A contained in Ae. Clearly $L = e\overline{AL}$, for as e is the identity of eAe, $L = eAeL = eAL$, and thus $L \subseteq e\overline{AL} \subseteq \overline{eAL} = \overline{L} = L$. Let $(x_L + L)_{L \in \mathcal{L}}$ be a closed linear filter base on eAe. Then $(x_L + \overline{AL})_{L \in \mathcal{L}}$ is a closed linear filter base on A and hence its intersection contains some $c \in A$. Therefore for each $L \in \mathcal{L}$,

$$ec - x_L = e(c - x_L) \in e\overline{AL} = L.$$

Thus $ec \in \bigcap_{L \in \mathcal{L}}(x_L + L)$.

Assume that A is strictly linearly compact. As $x \to xe$ is a continuous epimorphism from the A-module A to the A-module Ae, the latter is strictly linearly compact by 28.3. The submodules of the A-module Ae are, of course, the left ideals of A contained in Ae. For each closed left ideal L of eAe, let $\mathcal{J}_L$ be the set of all A-submodules J of Ae such that $eJ \subseteq L$. Clearly the sum of two members of $\mathcal{J}_L$ is again a member of $\mathcal{J}_L$, so the union L' of all the members of $\mathcal{J}_L$ is a member of $\mathcal{J}_L$. As the closure of a member of $\mathcal{J}_L$ also belongs to $\mathcal{J}_L$, L' is a closed A-submodule of Ae such that $eL' = L$, since $\overline{AL} \in \mathcal{J}_L$. If L is open in eAe, then L' is open in Ae: indeed, there exists an open submodule J of Ae such that $J \cap eAe \subseteq L$; then as $J = Je$, $eJ = eJe \subseteq eAe$ and $eJ \subseteq J$, so $eJ \subseteq J \cap eAe \subseteq L$. Thus J is an open A-submodule of Ae belonging to $\mathcal{J}_L$. Therefore $J \subseteq L'$, so L' is open in Ae. Thus $L \to L'$ is an increasing injection from the set of all closed [open] left ideals of eAe to the set of all closed [open] submodules of the A-module Ae. In particular, if U is an open left ideal of eAe, $L \to L'$ is an increasing injection from the set of left ideals of eAe containing U to the set of A-submodules of Ae containing the open submodule U'; hence as the A-module Ae/U' is artinian by hypothesis, the eAe-submodule eAe/U is artinian. Consequently by 28.15, eAe is strictly linearly compact. •

Exercises

29.1 Let E be a Hausdorff vector space over a division ring K furnished with a ring topology, and let A be the ring of all continuous linear operators on E, furnished with the topology of pointwise convergence. (a) For each $v \in A$, $u \to u \circ v$ and $u \to v \circ u$ are continuous from A to A. (b) If K is nondiscrete, complete, and straight, if E is infinite-dimensional, and if every finite-dimensional subspace of E has a topological supplement, then $(u, v) \to u \circ v$ is not continuous at $(0, 0)$. [Cf. Exercise 25.3.]

29.2 If P is a closed primitive ideal of a linearly compact ring A, then P is the intersection of open regular maximal left ideals, and if M is any closed maximal left ideal containing P, then M is open, a regular left ideal, and $P = P(M)$. [Use 29.11 and 25.8.]

In the remaining exercises $M_n(K)$ denotes the ring of all n by n matrices over a topological ring K, furnished with the cartesian product topology.

29.3 $M_n(K)$ is a topological ring.

29.4 Let $\{b_1, \ldots, b_n\}$ be a basis of an n-dimensional Hausdorff vector space E over a Hausdorff division ring K, and let A be the ring of all linear operators on E, furnished with the topology of pointwise convergence. For each $(i, j) \in [1, n] \times [1, n]$, let e_{ij} be the linear operator satisfying $e_{ij}(a_j) = a_i$, $e_{ij}(a_k) = 0$ if $k \neq j$, and for each $\lambda \in K$ let $\hat{\lambda}$ be the linear operator satisfying $\hat{\lambda}(a_k) = \lambda a_k$ for all $k \in [1, n]$. Let M be the isomorphism from A to $M_n(K)$ determined by the basis $\{b_1, \ldots, b_n\}$; thus for each $u \in A$, $M(u) = (\lambda_{u;i,j})$ where

$$u(a_j) = \sum_{i=1}^{n} \lambda_{u;i,j} a_i$$

for all $j \in [1, n]$, and

$$M^{-1}((\lambda_{i,j})) = \sum_{i,j} \hat{\lambda}_{i,j} \circ e_{ij}.$$

(a) M^{-1} is continuous from $M_n(K)$ to A. (b) If E is a straight vector space, M is continuous from A to $M_n(K)$. [First show that $u \to \lambda_{u;r,s} b_1$ is continuous from A to E.] (c) If E is a straight vector space, A is a topological ring.

29.5 Let K be a Hausdorff commutative ring with identity. (a) If K is advertibly open, then $M_n(K)$ is advertibly open. [Use the determinant function.] (b) If K is a ring with continuous inversion, then $M_n(K)$ is a ring with continuous inversion. [Recall that if $A \in M_n(K)^\times$ and if adj(A) is the adjoint of A, then $A^{-1} = (\det A)^{-1} \text{adj}(A)$.]

29.6 (Kaplansky [1946]) Let K be a Gel'fand ring (Exercise 26.6), and let $n \geq 1$. (a) For each neighborhood U of zero contained in K^a there is a neighborhood V of zero contained in K^a such that if $c_{i,j}, d_j \in V$ for all $i, j \in [1, n]$, then there exist (unique) $x_1, \ldots, x_n \in U$ such that

(1) $$x_i - \sum_{k=1}^{n} c_{i,k} x_k = d_i$$

for all $i \in [1, n]$. [Use induction on n; circle each side of the first equation of (1) on the left by $c_{1,1}^a$ to arrive at an explicit expression for x_1; substitute it in the ith equation of (1) for each $i \in [2, n]$ to obtain $n-1$ equations of form (1).] (b) $M_n(K)$ is advertibly open.

29.7 Let A be the ring of all linear operators on a vector space E, and let $P = \{e \in A : e \text{ is a projection on a one-dimensional subspace of } E\}$. (a) $e \in P$ if and only if e is a nonzero idempotent and for every idempotent $f \in A$ such that $ef = fe$, either $ef = e$ or $ef = 0$. (b) For each finite subset J of P let $V(J) = \{u \in A : ue = 0 \text{ for all } e \in J\}$. Then $\{V(J) : J \text{ is a finite subset of } P\}$ is a fundamental system of neighborhoods of zero for the topology of pointwise convergence on A, where E is given the discrete topology. (c) If Φ is an isomorphism from A to the ring B of all linear operators on a vector space F, then Φ is a topological isomorphism when E and F are given the discrete topology, A and B the topology of pointwise convergence. (d) If Φ is an isomorphism from one linearly compact primitive ring to another, then Φ is a topological isomorphism. (e) A primitive ring admits at most one linearly compact ring topology.

29.8 (Warner [1960b], Arnautov and Ursul [1979]) Let $(A_\lambda)_{\lambda \in L}, (B_\mu)_{\mu \in M}$ be families of rings with identity, and let A and B be their respective cartesian products. For each $\lambda \in L$, we denote by in_λ the canonical injection from A_λ to A and by pr_λ the canonical epimorphism from A to A_λ (so that $pr_\lambda \circ in_\lambda$ is the identity automorphism of A_λ), and similarly for each $\mu \in M$. (a) If h is a homomorphism from A to A such that the restriction of h to $\bigoplus_{\lambda \in L} A_\lambda$ is the identity automorphism of $\bigoplus_{\lambda \in L} A_\lambda$, then h is the identity automorphism of A. (b) Assume that for each $\lambda \in L$ and each $\mu \in M$, A_λ and B_μ are indecomposable rings, that is, not the direct sum of two proper subrings. If h is an isomorphism from A to B, then there exist a bijection σ from M to L and, for each $\mu \in M$, an isomorphism h_μ from $A_{\sigma(\mu)}$ to B_μ such that

$$h((x_\lambda))_{\lambda \in L} = (h_\mu(x_{\sigma(\mu)}))_{\mu \in M}.$$

[Observe that if 1_λ is the identity element of A_λ, then $\{in_\lambda(1_\lambda) : \lambda \in L\}$ is the set of indempotents e in the center of A such that for every idempotent f in the center of A, either $ef = 0$ or $ef = e$. Then show that for every $\mu \in M$ there exists $\sigma(\mu) \in L$ such that $pr_\mu \circ h \circ in_{\sigma(\mu)}$ is an isomorphism from $A_{\sigma(\mu)}$ to B_μ. Use (a).] (c) If Φ is an isomorphism from one linearly compact semisimple ring to another, then Φ is a topological isomorphism. [Use (b) and Exercise 29.7.] (d) A semisimple ring admits at most one linearly compact ring topology.

29.9 A ring A is *von Neumann regular* if for each $a \in A$ there exists $x \in A$ such that $axa = a$. (a) The ring of all linear operators on a vector space is von Neumann regular. (b) An ideal of a von Neumann regular ring is a von Neumann regular ring. (c) The cartesian product of von Neumann regular rings is a von Neumann regular ring. (d) A von Neumann regular ring is semisimple. [Use the paragraph preceding 11.7.] (e) (Wiegandt [1965a]) A linearly compact ring is von Neumann regular if and only if it is semisimple.

29.10 Let e be an idempotent of a topological ring A. (a) If A is linearly compact, eA is a linearly compact ring. [Modify the proof of 29.15.] (b) (Ánh [1980]) If A is strictly linearly compact, eA is a strictly linearly compact ring. [Show that if H and L are left ideals of eA generating the same left ideal of A, then $H = L$.]

30 Strongly Linearly Compact Modules

30.1 Definition. *A topological module or ring E is* **strongly linearly compact** *if E is linearly topologized and the $\mathbb{Z}$-module E is linearly compact.*

Thus a linearly topologized module or ring is strongly linearly compact if and only if the intersection of any filter base of cosets of closed additive subgroups is nonempty. Consequently, a strongly linearly compact module or ring is linearly compact. By 28.16, if F is a closed submodule of a linearly topologized module E, E is strongly linearly compact if and only if F and E/F are strongly linearly compact. Our principal purpose here is to show that a strongly linearly compact module or ring is strictly linearly compact by determining the structure of discrete, linearly compact $\mathbb{Z}$-modules.

We recall that an abelian group D is *divisible* if for each $x \in D$ and each nonzero integer n there exists $y \in D$ such that $n.y = x$. Thus, D is divisible if and only if $n.D = D$ for each nonzero integer n. Clearly the sum of two divisible subgroups of an abelian group is a divisible subgroup, an epimorphic image of a divisible group is divisible, and the zero group is the only finite divisible group. Furthermore, a divisible group D contains no proper subgroups H of finite index, for if D/H had order n and $x \in D$, then for some $y \in D$, $x = n.y \in H$.

30.2 Theorem. *A divisible subgroup H of an abelian group G has an algebraic supplement.*

Proof. By Zorn's Lemma, G contains a subgroup K maximal for the ordering $\subseteq$ among all the subgroups L of G such that $H \cap L = (0)$. It suffices to show that $H + K = G$. Suppose there existed $x \notin H + K$. Then by maximality, $H \cap (K + \mathbb{Z}.x) \neq (0)$, so there would exist a smallest integer $n \geq 1$ such that for some $h \in H$ and some $k \in K$, $h = k + n.x \neq 0$, whence $n.x \in H + K$. As $n \neq 1$, there is a prime p dividing n. Let $y = (n/p).x$. Then $y \notin H + K$ but $p.y \in H + K$. As H is divisible, there would exist $h_1 \in H$ such that $p.h_1 = h$. Let $t = y - h_1$. Then $t \notin H + K$, so as before, $H \cap (K + \mathbb{Z}.t) \neq (0)$, and thus there would exist $m \geq 1$, $h_2 \in H$, and $k_2 \in K$ such that $h_2 = k_2 + m.t \neq 0$. Now $p.t = n.x - h = k \in K$, so $p \nmid m$ since otherwise $m.t \in K$ and thus

$$0 \neq h_2 = k_2 + m.t \in H \cap K = (0),$$

a contradiction. Therefore p and m would be relatively prime, so there would exist $a, b \in \mathbb{Z}$ such that $am + bp = 1$, whence

$$t = a.(m.t) + b.(p.t) = a.(h_2 - k_2) + b.(p.t) \in H + K,$$

a contradiction. Thus $G = H + K$. •

30.3 Theorem. *An abelian group G has a largest divisible subgroup D, and if K is an algebraic supplement of D, the zero subgroup is the only divisible subgroup of K.*

Proof. The set $\mathcal{D}$ of all divisible subgroups of G is nonempty as it contains the zero subgroup, and the union of a family of divisible subgroups, totally ordered by $\subseteq$, is clearly a divisible subgroup. Therefore by Zorn's Lemma, G contains a maximal divisible subgroup D. If H is any divisible subgroup of G, $D + H$ is also clearly divisible, so by the maximality of D, $H \subseteq D$. Thus D is the largest divisible subgroup of G, and in particular, an algebraic supplement of D can contain no nonzero divisible subgroup. •

30.4 Definition. *An abelian group G is a **torsion** group if for each $x \in G$ there exists an integer $m \geq 1$ such that $m.x = 0$. If p is a prime, G is a **p-primary** group if for each $x \in G$ there exists $r \in \mathbb{N}$ such that $p^r.x = 0$; G is a **primary** group if G is p-primary for some prime p.*

Let G be an abelian group. The subset of G consisting of all $x \in G$ such that $n.x = 0$ for some integer $n \geq 1$ is clearly the largest subgroup of G that is a torsion subgroup, and hence is called the *torsion subgroup* of G. If p a prime, the subset of G consisting of all $x \in G$ such that $p^r.x = 0$ for some $r \in \mathbb{N}$ is clearly the largest p-primary subgroup of G, called its *p-primary component*.

30.5 Theorem. *Let G be an abelian torsion group, $(T_p)_{p \in P}$ its primary components, where P is the set of primes. Then G is the direct sum of $(T_p)_{p \in P}$, and if f is an epimorphism from G to an abelian group H, then H is a torsion group whose primary components are $(f(T_p))_{p \in P}$.*

Proof. Let $x \in G$, let $m > 1$ be such that $m.x = 0$, and let $m = p_1^{r_1} \ldots p_n^{r_n}$ where $p_1, \ldots, p_n$ are distinct primes. For each $j \in [1, n]$ let $q_j = m/p_j^{r_j}$. Then $q_1, \ldots, q_n$ are relatively prime, so there exist $a_1, \ldots, a_n \in \mathbb{Z}$ such that $a_1 q_1 + \cdots + a_n q_n = 1$. Consequently,

$$x = a_1.(q_1.x) + \cdots + a_n.(q_n.x),$$

and clearly each $q_j.x \in T_{p_j}$.

Thus to establish the first assertion, we need only show that if $x_1 + \ldots + x_n = 0$ where $x_i \in T_{p_i}$ for each $i \in [1, n]$ and $p_1, \ldots, p_n$ are distinct primes, then each $x_i = 0$. For each $j \in [1, n]$ let $r_j \geq 0$ be such that $p_j^{r_j}.x_j = 0$, and for each $i \in [1, n]$ let

$$q_i = \prod_{j \neq i} p_j^{r_j}.$$

Then $q_i.x_j = 0$ for all $j \neq i$, so

$$q_i.x_i = q_i.(x_1 + \cdots + x_n) = 0.$$

As q_i and $p_i^{r_i}$ are relatively prime, there exist $a, b \in \mathbb{Z}$ such that $aq_i + bp_i^{r_i} = 1$, whence

$$x_i = a.(q_i.x) + b.(p_i^{r_i}.x_i) = 0.$$

If f is an epimorphism from G to H, clearly H is a torsion group. and for each prime p, $f(T_p)$ is contained in the p-primary component. But as $f(G) = H$ and as H is the direct sum of its primary components, $f(T_p)$ must therefore be the p-primary component of H. •

30.6 Theorem. *If G is a p-primary abelian group, then G is divisible if and only if $p.G = G$.*

Proof. Sufficiency: Let $x \in G$, let $m \geq 1$, and let $m = p^s q$ where $p \nmid q$ and $s \geq 0$. Let $n \geq 1$ be such that $p^n.x = 0$. As p^n and q are relatively prime, there exist $a, b \in \mathbb{Z}$ such that $ap^n + bq = 1$, so

$$x = a.(p^n.x) + q.(b.x) = q.(b.x).$$

By hypothesis there exists $y \in G$ such that $p^s.y = b.x$; therefore

$$m.y = q.(b.x) = x. \bullet$$

30.7 Definition. *Let p be a prime. A group G is a **basic divisible p-primary group** if G is the union of an increasing sequence $(H_n)_{n\geq 1}$ of cyclic subgroups such that the order of H_n is p^n for all $n \geq 1$.*

To justify the terminology, let G and $(H_n)_{n\geq 1}$ be as in Definition 30.7. We recall that a finite cyclic group of order m contains precisely one subgroup of order d for each divisor d of m. It readily follows that $p.H_{n+1} = H_n$ for all $n \geq 1$ and, more generally, that $p^s.H_{n+s} = H_n$ for all $n, s \geq 1$. Clearly G is p-primary since $p^n.x = 0$ for all $x \in H_n$. Also as $p.H_{n+1} = H_n$ for all $n \geq 1$, we conclude that $p.G = G$, whence G is a divisible group by 30.6.

If $x \in G$ is an element of order p^n, then x is a generator of H_n, for $x \in H_m$ for some $m \geq 1$, and clearly $m \geq n$, whence H_n is the unique subgroup of H_m of order p^n. It readily follows that the only nonzero proper subgroups of G are the groups H_n where $n \geq 1$. Consequently, the $\mathbb{Z}$-module G is an artinian module but not a a noetherian module.

The additive group $\mathbb{Q}_p/\mathbb{Z}_p$ is a basic p-primary group, as we need only let $H_n = p^{-n}\mathbb{Z}_p/\mathbb{Z}_p$ for each $n \geq 1$. Indeed, H_n clearly has order p^n, and if $x \in \mathbb{Q}_p$ and if $v_p(x) = -n$ where $n \geq 1$, then $p^n x \in \mathbb{Z}_p$, so $x \in p^{-n}\mathbb{Z}_p$.

Any basic divisible p-primary group G is generated by elements $(x_n)_{n\geq 1}$ where $x_1 \neq 0$, $p.x_1 = 0$, and $p.x_{n+1} = x_n$ for all $n \geq 1$. Consequently, any two basic divisible p-primary groups are isomorphic, for if $(x'_n)_{n\geq 1}$ is such a sequence for G', then there is a unique isomorphism f from G to G' such that $f(x_n) = x'_n$ for all $n \geq 1$. Indeed, as x_n and x'_n are generators of the unique cyclic subgroups H_n and H'_n of order p^n of G and G' respectively, there is a unique isomorphism f_n from H_n to H'_n such that $f_n(x_n) = x'_n$. If $n > 1$, then as $x_{n-1} = p.x_n$ and $x'_{n-1} = p.x'_n$, $f_n(x_{n-1}) = x'_{n-1}$, and therefore the restriction of f_n to H_{n-1} is f_{n-1}. The function f from G to G' whose restriction to each H_n is f_n is therefore a well-defined isomorphism from G to G' satisfying $f(x_n) = x'_n$ for all $n \geq 1$.

In sum, we have proved:

30.8 Theorem. *Let p be a prime. There exist basic divisible p-primary groups, and any two are isomorphic. If G is a basic divisible p-primary group, then for each $n \geq 1$ G contains exactly one subgroup H_n of order p^n, H_n is cyclic, $(H_n)_{n\geq 1}$ is an increasing sequence of subgroups of G whose union is G, and the H_n's are the only nonzero proper subgroups of G. In particular, G is an artinian $\mathbb{Z}$-module but not a noetherian $\mathbb{Z}$-module.*

The traditional notation for a basic divisible p-primary group is $\mathbb{Z}(p^\infty)$.

By 30.5, to describe all divisible torsion groups, it suffices to describe divisible primary groups:

30.9 Theorem. *Let p be a prime. An abelian group G is a divisible p-primary group if and only if G is the direct sum of a family of basic divisible*

p-primary subgroups.

Proof. The condition is clearly sufficient, since a sum of divisible [p-primary] groups is a divisible [p-primary] group. Necessity: We first observe that any nonzero divisible p-primary group H contains a basic divisible p-primary subgroup. Indeed, H contains a sequence $(x_n)_{n \geq 1}$ of elements such that $x_1 \neq 0$, $p.x_1 = 0$, and $p.x_{n+1} = x_n$ for all $n \geq 1$. By induction, the subgroup H_n generated by x_n is a cyclic group of order p^n for all $n \geq 1$, and $(H_n)_{n \geq 1}$ is an increasing sequence of subgroups of H whose union is therefore a basic divisible p-primary subgroup of H.

We may assume that G is a nonzero group. Let $(H_\lambda)_{\lambda \in L}$ be the family of all basic divisible p-primary subgroups of G. We have just seen that $L \neq \emptyset$. Let $\mathcal{N}$ be the set of all subsets N of L such that $\sum_{\lambda \in N} H_\lambda$ is the direct sum of $(H_\lambda)_{\lambda \in N}$. Clearly $\mathcal{N}$, ordered by inclusion, is inductive and therefore by Zorn's Lemma contains a maximal member M. We need only show that $G = \sum_{\lambda \in M} H_\lambda$. In the contrary case, G is the direct sum of $\sum_{\lambda \in M} H_\lambda$ and a nonzero subgroup K by 30.2, as $\sum_{\lambda \in M} H_\lambda$ is a divisible group. Then K is isomorphic to $G/\sum_{\lambda \in M} H_\lambda$, which is divisible as it is an epimorphic image of G. Therefore by the preceding, K contains a basic divisible p-primary group H_μ, where $\mu \in L \setminus M$. Clearly $M \cup \{\mu\} \in \mathcal{N}$, a contradiction of the maximality of M. Thus $G = \sum_{\lambda \in M} H_\lambda$. •

30.10 Theorem. *Let G be an abelian group. The following statements are equivalent:*

$1°$ *The $\mathbb{Z}$-module G is linearly compact for the discrete topology.*

$2°$ *G is the direct sum of a finite subgroup and finitely many basic divisible primary subgroups.*

$3°$ *The $\mathbb{Z}$-module G is artinian.*

Proof. Since a basic divisible primary group is an artinian $\mathbb{Z}$-module, $2°$ implies $3°$ by 27.5. Also $3°$ implies $1°$ by 28.14. To show that $1°$ implies $2°$, we first observe that the $\mathbb{Z}$-module $\mathbb{Z}$ is not linearly compact for the discrete topology. Indeed, if $F_n = \frac{3^n - 1}{2} + \mathbb{Z}.3^n$, it is easy to see that $(F_n)_{n \geq 1}$ is a decreasing sequence of cosets of subgroups whose intersection is empty. Let G satisfy $1°$. For each $a \in G$, the $\mathbb{Z}$-module $\mathbb{Z}.a$ is isomorphic to the $\mathbb{Z}$-module $\mathbb{Z}/\mathrm{Ann}_\mathbb{Z}(a)$; since $\mathbb{Z}.a$ is also linearly compact for the discrete topology, therefore, $\mathrm{Ann}_\mathbb{Z}(a) \neq (0)$ by the preceding. Thus G is a torsion group. By 30.5 and 28.21, the p-primary component of G is nonzero for only finitely many primes p; thus G is the direct sum of finitely many primary subgroups. Therefore to establish $2°$, we may assume that G is a p-primary group for some prime p.

By 28.3, $G/p.G$ is a linearly compact $\mathbb{Z}$-module for the discrete topology. Let F_p be the prime field $\mathbb{Z}/p.\mathbb{Z}$ of p elements. We may regard $G/p.G$ as a vector space over F_p; the subgroups of $G/p.G$ then coincide with the subspaces of the F_p-vector space $G/p.G$. Thus $G/p.G$ is a linearly compact F_p-vector space for the discrete topology. By 28.21 $G/p.G$ is finite-dimensional over F_p and hence is finite. Thus there exist $a_1, \ldots, a_m \in G$ such that if $H = \mathbb{Z}.a_1 + \cdots + \mathbb{Z}.a_m$, then $G = H + p.G$. As G is a p-primary group, H is a finite p-primary group and thus its order is p^r for some $r \geq 0$. Consequently,

$$p^r.G = p^r.H + p^{r+1}.G = p^{r+1}.G.$$

Therefore $p^s.G = p^r.G$ for all $s \geq r$. Let $K = p^r.G$. Then $p.K = K$, so K is a divisible group by 30.6, and hence G is the direct sum of K and a subgroup H_0 by 30.2. For each $s \in [1, r-1]$, $x \to p^s.x + p^{s+1}.G$ is an epimorphism from G to $p^s.G/p^{s+1}.G$ whose kernel contains $p.G$, so $p^s.G/p^{s+1}.G$ is an epimorphic image of $G/p.G$ and hence is finite. Therefore as $G/p.G, p.G/p^2.G, \ldots, p^{r-1}.G/p^r.G$ are all finite, so is $G/p^r.G$; as H_0 is isomorphic to $G/K = G/p^r.G$, therefore, H_0 is finite. Since K is also a linearly compact $\mathbb{Z}$-module for the discrete topology, K is the direct sum of finitely many basic divisible p-primary subgroups by 30.9 and 28.21. •

30.11 Corollary. *A strongly linearly compact module or ring is strictly linearly compact.*

Proof. If U is an open submodule of a strongly linearly compact A-module E, then E/U is a discrete, linearly compact $\mathbb{Z}$-module by (1) of 28.16, so E/U is an artinian $\mathbb{Z}$-module by 30.10 and *a fortiori* an artinian A-module. Consequently, E is strictly linearly compact by (2) of 28.15. •

Exercises

30.1 Let A be a ring. (a) The largest divisible subgroup of the additive group A is an ideal. (b) The torsion subgroup of the additive group G is an ideal, and for each prime p, the p-primary component of the additive group A is an ideal.

30.2 If A is a torsion ring (that is, if its additive group is a torsion group), then the largest divisible ideal of A is contained in $\mathrm{Ann}_A(A)$.

30.3 A topological A-module E is a compact, totally disconnected, torsion module if and only if E is strongly linearly compact and for some $m \geq 1$, $m.E = \{0\}$. [Necessity: Use 9.4.]

31 Locally Linearly Compact Semisimple Rings

Here we will determine the structure of linearly topologized semisimple rings having a linearly compact open left ideal.

31 LOCALLY LINEARLY COMPACT SEMISIMPLE RINGS 257

31.1 Theorem. *Let B be a ring furnished with an additive group topology such that for all $b \in B$, $x \to xb$ is continuous from B to B. Let A be a subring of B such that for all $a \in A$, $x \to ax$ is continuous from B to B and, with its induced topology, A is a linearly topologized A-module. Then $\overline{A}$ is a linearly topologized topological ring.*

Proof. Let L be a left ideal of A. For each $a \in A$, as $x \to ax$ is continuous, $a\overline{L} \subseteq \overline{aL} \subseteq \overline{L}$. Hence $A\overline{L} \subseteq \overline{L}$. For each $b \in \overline{L}$, as $x \to xb$ is continuous, $\overline{A}b \subseteq \overline{Ab}$, a subset of the closure of $A\overline{L}$ and hence of $\overline{L}$. Thus $\overline{A}\,\overline{L} \subseteq \overline{L}$. In particular, $\overline{A}\,\overline{A} \subseteq \overline{A}$, so $\overline{A}$ is a subring of B, and by 4.22 the topology induced on $\overline{A}$ is an $\overline{A}$-linear topology. Since for each $b \in \overline{A}$, $x \to xb$ is continuous from $\overline{A}$ to $\overline{A}$ by hypothesis, therefore, $\overline{A}$ is a topological ring by 2.15. •

31.2 Theorem. *If A is a semisimple topological ring having a nonzero open left ideal L that is a linearly compact A-module for its induced topology, then L contains a minimal left ideal of A.*

Proof. Since L is complete by 28.5, the topology of pointwise convergence on the A-module L^L of all functions from L to L is also complete by 7.8. The A-submodule $\mathrm{End}(L)$ of all endomorphisms of the additive group L is closed in L^L and hence is also complete. Let $\mathcal{U}$ be the fundamental system of neighborhoods of zero consisting of all open left ideals of A contained in L. For each finite subset X of L and each $U \in \mathcal{U}$, let

$$W(X, U) = \{u \in \mathrm{End}(L) : u(X) \subseteq U\}.$$

The set of all such $W(X, U)$ is then a fundamental system of neighborhoods of zero for the topology of pointwise convergence on $\mathrm{End}(L)$. That topology is an A-linear topology, for as U is a left ideal, $W(X, U)$ is an A-submodule.

With composition as multiplication, $\mathrm{End}(L)$ is also a ring. Moreover, for each $b \in \mathrm{End}(L)$, $u \to u \circ b$ is continuous at zero and hence everywhere, since

$$W(b(X), U) \circ b \subseteq W(X, U)$$

for any finite subset X of L and any $U \in \mathcal{U}$. For each $a \in A$, let $\hat{a} \in \mathrm{End}(L)$ be defined by $\hat{a}(x) = ax$ for all $x \in L$. Then $\phi : a \to \hat{a}$ is a homomorphism from the ring A to the ring $\mathrm{End}(L)$. For each $a \in A$ and each $u \in \mathrm{End}(L)$, $a.u = \hat{a} \circ u$ (since by definition, $a.u$ is the function $x \to au(x)$). Therefore as the topology of $\mathrm{End}(L)$ is an A-module topology, $u \to \hat{a} \circ u$ is continuous from $\mathrm{End}(L)$ to $\mathrm{End}(L)$ for each $a \in A$. Moreover, for each $U \in \mathcal{U}$, $W(X, U) \cap \phi(A)$ is a left ideal of $\phi(A)$ since $W(X, U)$ is a submodule of $\mathrm{End}(L)$. By 31.1, therefore the closure $\overline{\phi(A)}$ of $\phi(A)$ in $\mathrm{End}(L)$ is a linearly topologized topological ring.

Now $\phi(A)$ is also a submodule of the A-module $\text{End}(L)$ since $a.\hat{b} = \widehat{ab}$ for all $a, b \in A$. Therefore $\overline{\phi(A)}$ is a submodule of the A-module $\text{End}(L)$. To show that $\overline{\phi(A)}$ is a linearly compact ring, it therefore suffices to show that the closed A-submodules of $\overline{\phi(A)}$ coincide with the closed left ideals of $\overline{\phi(A)}$ and that $\overline{\phi(A)}$ is a linearly compact A-module. If J is a closed A-submodule of $\overline{\phi(A)}$, then for any $a \in A$ and any $u \in J$, $\hat{a} \circ u = a.u \in J$, so $\phi(A) \circ J \subseteq J$, whence

$$\overline{\phi(A)} \circ J = \overline{\phi(A)} \circ \overline{J} \subseteq \overline{J} = J$$

as $\overline{\phi(A)}$ is a topological ring. Conversely, if J is a left ideal of $\overline{\phi(A)}$, then J is an A-submodule of $\overline{\phi(A)}$ since $a.u = \hat{a} \circ u \in J$ for all $a \in A$, $u \in J$. To show that $\overline{\phi(A)}$ is a linearly compact A-module, we observe that if $\{x_1, \ldots, x_n\} \subseteq L$ and if $U \in \mathcal{J}$, then

$$u \to (u(x_1) + U, \ldots, u(x_n) + U)$$

is an A-module homomorphism from $\overline{\phi(A)}$ to $(L/U) \times \cdots \times (L/U)$ whose kernel is $\overline{\phi(A)} \cap W(\{x_1, \ldots, x_n\}, U)$. Since L/U is a discrete linearly compact A-module, so is $(L/U) \times \cdots \times (L/U)$ by 28.7, and hence each of its submodules is a discrete, linearly compact A-module. In particular

$$\overline{\phi(A)}/[W(\{x_1, \ldots, x_n\}, U) \cap \overline{\phi(A)}]$$

is a discrete, linearly compact A-module. As $\overline{\phi(A)}$ is closed in $\text{End}(L)$, $\overline{\phi(A)}$ is complete. Therefore $\overline{\phi(A)}$ is a linearly compact ring by (1) of 28.15.

Furthermore, ϕ is continuous from A into $\text{End}(L)$, for if X is a finite subset of L and if $U \in \mathcal{U}$, there exists $V \in \mathcal{U}$ such that $VX \subseteq U$, whence $V \subseteq \phi^{-1}(W(X, U))$. Consequently, $\phi(L)$ is a linearly compact, hence complete, and thus closed subset of $\text{End}(L)$. Hence $\phi(L)$ is a closed left ideal of $\overline{\phi(A)}$.

Let $x \in L$. Then $u \to u \circ \hat{x}$ is continuous from $\text{End}(L)$ to $\text{End}(L)$ as noted earlier, and $u \to \widehat{u(x)}$ is also continuous from $\text{End}(L)$ to $\text{End}(L)$ since it is the composite of the continuous function $u \to u(x)$ from $\text{End}(L)$ to L and ϕ. For each $a \in A$, $\hat{a} \circ \hat{x} = \widehat{ax} = \widehat{\hat{a}(x)}$. Consequently,

(1) $$u \circ \hat{x} = \widehat{u(x)}$$

for all $u \in \overline{\phi(A)}$ and all $x \in L$.

The restriction ϕ_L of ϕ to L is a monomorphism. Indeed, the kernel K of ϕ is clearly $\{a \in A : aL = (0)\}$. The kernel of ϕ_L is therefore $L \cap K$, a left ideal; but as $KL = (0)$, $(L \cap K)^2 = (0)$, so $L \cap K = (0)$ by 26.14 as A is semisimple.

To show that $\overline{\phi(A)}$ is semisimple, let R be its radical. As $\phi(L)$ is a left ideal of $\overline{\phi(A)}$, $R \circ \phi(L)$ is an advertible left ideal of $\overline{\phi(A)}$ contained in $\phi(L)$, so as ϕ_L is an isomorphism from L to $\phi(L)$ and as

$$\phi_L^{-1}(R \circ \phi(L)) = \phi^{-1}(R \circ \phi(L)) \cap L,$$

$\phi^{-1}(R \circ \phi(L)) \cap L$ is an advertible left ideal of A and thus is (0) by 26.13, that is,

(2) $$(0) = \phi_L^{-1}(R \circ \phi(L)) = \phi^{-1}(R \circ \phi(L)) \cap L.$$

Now

$$\phi^{-1}(R \circ \phi(L))L \subseteq \phi^{-1}(R \circ \phi(L))\phi^{-1}(\phi(L))$$
$$= \phi^{-1}(R \circ \phi(L) \circ \phi(L)) \subseteq \phi^{-1}(R \circ \phi(L)),$$

and $\phi^{-1}(R \circ \phi(L))L \subseteq L$. Hence by (2), $\phi^{-1}(R \circ \phi(L))L = (0)$, that is, $\phi^{-1}(R \circ \phi(L))) \subseteq K$, whence

(3) $$R \circ \phi(L) \subseteq \phi(K) = (0).$$

Let $r \in R$. To show that $r = 0$, it suffices to show that $r(x) = 0$ for each $x \in L$. By (3), $r \circ \hat{x} = 0$, that is, $\widehat{r(x)} = 0$ by (1), whence $r(x) \in K$ and thus $r(x) \in L \cap K = (0)$.

By 29.8 there is a minimal left ideal N_0 of $\overline{\phi(A)}$ contained in $\phi(L)$. Let

$$N = \phi_L^{-1}(N_0) = \phi^{-1}(N_0) \cap L,$$

a nonzero left ideal of A contained in L. To show that N is a minimal left ideal, let N_1 be a left ideal of A contained in N. If $\phi(LN_1) = (0)$, then

$$LN_1 \subseteq K \cap N_1 \subseteq K \cap L = (0)$$

so $N_1^2 \subseteq LN_1 = (0)$, and therefore $N_1 = (0)$ by 26.14, as A is semisimple. We may suppose, therefore, that $\phi(LN_1) \neq (0)$. For any $y, z \in L$, $\hat{a}(yz) = ayz = \hat{a}(y)z$ for all $a \in A$, so as $u \to u(yz)$ and $u \to u(y)z$ are continuous from $\text{End}(L)$ to L,

(4) $$u(yz) = u(y)z$$

for all $u \in \overline{\phi(A)}$. Consequently, $\phi(LN_1)$ is a left ideal of $\overline{\phi(A)}$, for if $y_1, \ldots, y_n \in L$ and if $z_1, \ldots, z_n \in N_1$, then for any $u \in \overline{\phi(A)}$,

$$u \circ (\sum_{i=1}^{n} y_i z_i)\hat{} = (u(\sum_{i=1}^{n} y_i z_i))\hat{} = (\sum_{i=1}^{n} u(y_i z_i))\hat{}$$
$$= (\sum_{i=1}^{n} u(y_i) z_i)\hat{} \in \phi(LN_1)$$

by (1) and (4). Furthermore,

$$\phi(LN_1) \subseteq \phi(N_1) \subseteq \phi(N) \subseteq N_0,$$

so by minimality, $\phi(LN_1) = N_0$. Therefore

$$N = \phi_L^{-1}(N_0) = \phi_L^{-1}(\phi(LN_1))$$
$$= \phi^{-1}(\phi(LN_1)) \cap L = (LN_1 + K) \cap L.$$

But as $LN_1 \subseteq L$,

$$(LN_1 + K) \cap L \subseteq LN_1 + (K \cap L) = LN_1 \subseteq N_1$$

as $K \cap L = (0)$. Thus $N = N_1$, so N is a minimal left ideal. •

31.3 Theorem. *Let E be a discrete vector space over a discrete division ring K, and let $A = \text{End}_K(E)$, furnished with the topology of pointwise convergence. Let M be a proper subspace of E, and let $L = \text{Ann}_A(M)$. With its induced topology, L is a linearly compact ring if and only if M is a finite set.*

Proof. For each $u \in L$, let $g(u)$ be the linear operator on the K-vector space E/M that is well defined by

$$g(u)(x + M) = u(x) + M.$$

It is easy to verify that g is a topological epimorphism from the ring L to the ring $\text{End}_K(E/M)$, furnished with the topology of pointwise convergence. Let N be the kernel of g. Then N is closed and the topological ring L/N is topologically isomorphic to $\text{End}_K(E/M)$ and hence, by 29.1, is a linearly compact ring. Therefore the L-module L/N is linearly compact as the left ideals of the ring L/N coincide with the submodules of the L-module L/N. Consequently by (1) of 28.16, L is a linearly compact ring if and only if N is a linearly compact L-module. Clearly $N = \{u \in L : u(E) \subseteq M\}$, so

$LN = (0)$, and thus N is a trivial L-module. Consequently, L is a linearly compact ring if and only if N is strongly linearly compact, or equivalently by 30.11 and (2) of 28.15, if and only if N/U is an artinian $\mathbb{Z}$-module for all subgroups U forming a fundamental system of neighborhoods of zero.

Necessity: Let $b \in E \setminus M$, let $U = \{u \in N : u(b) = 0\}$, an open L-submodule of N, and let Y be a linearly independent subset of E containing a basis of M such that $Y \cup \{b\}$ is a basis of E. Suppose that M contained an infinite linearly independent sequence $(x_i)_{i \geq 1}$. Let v_i be the linear operator satisfying $v_i(b) = x_i$, $v_i(y) = 0$ for all $y \in Y$. For each $k \geq 1$ let G_k be the additive subgroup generated by $\{v_i + U : i \geq k\}$. The linear independence of $(x_i)_{i \geq 1}$ insures that $v_k + U \notin G_{k+1}$ for all $k \geq 1$. Thus $(G_k)_{k \geq 1}$ is a strictly decreasing sequence of additive subgroups of N/U, so N/U is not artinian. Consequently, M is finite-dimensional.

Next, suppose that K is infinite but that there exists a nonzero vector $a \in M$. Case 1: K has characteristic zero. Let u be the linear operator satisfying $u(b) = a$ and $u(y) = 0$ for all $y \in Y$. Then $u \in N$, but if n and m are distinct integers, then $n.u - m.u \notin U$ since $(n.u - m.u)(b) \neq 0$. Consequently, $\{n.(u+U) : n \in \mathbb{Z}\}$ is a subgroup of N/U isomorphic to $\mathbb{Z}$, so N/U is not artinian. Case 2: K has prime characteristic p. Then K is infinite-dimensional over its prime subfield F_p. For each $\lambda \in K$ let u_λ be the linear operator satisfying $u_\lambda(b) = \lambda a$, $u_\lambda(y) = 0$ for all $y \in Y$. Then $\lambda \to u_\lambda + U$ is an isomorphism from the additive group K to an additive subgroup of N/U, for if $\lambda \neq \mu$, $(u_\lambda - u_\mu)(b) = (\lambda - \mu)a \neq 0$. Thus if N/U were artinian, K would be also. Let $(\lambda_i)_{i \geq 1}$ be a denumerable subset of a basis of K over F_p, and for each $k \geq 1$ let G_k be the additive group generated by $\{\lambda_i : i \geq k\}$. Clearly $\lambda_k \notin G_{k+1}$, so $(G_k)_{k \geq 1}$ is a strictly decreasing sequence of subgroups of K. Consequently, the additive group K is not an artinian $\mathbb{Z}$-module, so neither is N/U.

Thus, either K is finite and M is finite-dimensional, or $M = (0)$; equivalently, M is a finite set.

Sufficiency: Let $X = \{x_1, \ldots, x_n\} \subseteq E$, and let $U = N \cap \text{Ann}_A(X)$. Then $h : u \to (u(x_1), \ldots, u(x_n))$ is a homomorphism from the additive group N to the additive group $M \times \cdots \times M$ (n terms) with kernel U. Consequently, N/U is finite and hence is artinian. •

31.4 Theorem. *Let A be a topological ring, L a proper nonzero left ideal of A. The following statements are equivalent:*

1° *A is primitive and linearly topologized, L is open in A and is a linearly compact ring for its induced topology.*

2° *There is a topological isomorphism ϕ from A to the topological ring $\text{End}_K(E)$ of all linear operators on a discrete vector space E over a (discrete) finite field K, furnished with the topology of pointwise convergence, such* that $\phi(L)$ is the annihilator in $End_K(E)$ of a proper nonzero finite-dimensional subspace of E.

Proof. Assume 1°. Since L is a linearly compact ring, it is *a fortiori* a linearly compact A-module. By 31.2, A possesses a minimal left ideal. By 25.6 and 25.11, A has no nonzero nilpotent ideals. Therefore by 25.17 and 25.18, A has an idempotent f such that fAf is a division ring, Af is a minimal left ideal and is a right vector space over fAf under multiplication as scalar multiplication, and $\phi : a \to a_L$ is an isomorphism from A to a dense ring of linear operators on the right fAf-vector space Af, where $a_L(x) = ax$ for all $x \in Af$. Since L is Hausdorff and open, A is Hausdorff, so any minimal left ideal of A is discrete. Let $E = Af$, $K = fAf$, and let $B = End_K(E)$, furnished with the topology of pointwise convergence. The monomorphism ϕ from A to B is continuous since for each $x \in E$, $a \to a_L(x)$ is simply the continuous function $a \to ax$. Therefore by 29.3, $\phi(L)$ is linearly compact, hence complete, and thus closed in B. By 4.2, as $\phi(L)$ is a left ideal of $\phi(A)$, $\phi(L) = \overline{\phi(L)}$ is a left ideal of $\overline{\phi(A)} = B$.

Let $M = Ann_E(\phi(L))$. By 25.8, $\phi(L) = Ann_B(M)$, since $\phi(L)$ is a closed left ideal of B, so M is a proper nonzero subspace of E as $\phi(L)$ is a proper nonzero left ideal. Since $\phi(L)$ is a linearly compact subring, by 31.3 M is finite. Consequently, K is finite and hence, by Wedderburn's theorem, is a field, M is a finite-dimensional subspace, and $\phi(L)$ is open in B. Therefore $\phi(A)$ is also open and hence closed in B, so $\phi(A) = B$.

By 29.1 and 31.3, 2° implies 1°. •

31.5 Definition. *Let $(A_\lambda)_{\lambda \in I}$ be a family of rings, for each $\lambda \in I$ let L_λ be a subring of A_λ, and let D_λ be the largest subring of A_λ in which L_λ is an ideal (thus $D_\lambda = \{x \in A_\lambda : L_\lambda x \cup x L_\lambda \subseteq L_\lambda\}$). The **local direct sum** of $(A_\lambda)_{\lambda \in I}$ relative to $(L_\lambda)_{\lambda \in I}$ is the subring A of $\prod_{\lambda \in I} A_\lambda$ consisting of all $(x_\lambda)_{\lambda \in I}$ such that $x_\lambda \in D_\lambda$ for all but finitely many $\lambda \in I$. If each A_λ is a topological ring and if each L_λ is open in A_λ, the **local direct sum topology** of A is that for which the neighborhoods of zero in $\prod_{\lambda \in I} L_\lambda$, furnished with its cartesian product topology, form a fundamental system of neighborhoods of zero.*

To show that the local direct sum topology on A is indeed a ring topology, it suffices by 2.15 to verify (TR 5). Let $b = (b_\lambda)_{\lambda \in I} \in A$, and let M be the finite subset of I such that $b_\lambda \in D_\lambda$ for all $\lambda \in I \setminus M$. Let $V = \prod_{\lambda \in I} V_\lambda$ where each V_λ is a neighborhood of zero in L_λ (and hence also in A_λ) and $V_\lambda = L_\lambda$ for all $\lambda \in I \setminus N$ for some finite subset N of L. For each $\lambda \in M \cup N$, let W_λ be a neighborhood of zero in A_λ that is contained in L_λ and satisfies $W_\lambda b_\lambda \subseteq V_\lambda$ and $b_\lambda W_\lambda \subseteq V_\lambda$, and let $W_\lambda = L_\lambda$ for all $\lambda \in I \setminus (M \cup N)$. If

$\lambda \in I \setminus M \cup N$, then

$$b_\lambda W_\lambda \cup W_\lambda b_\lambda \subseteq D_\lambda L_\lambda \cup L_\lambda D_\lambda \subseteq L_\lambda = V_\lambda.$$

Therefore $bW \cup Wb \subseteq V$ where $W = \prod_{\lambda \in I} W_\lambda$.

31.6 Definition. Let A be the local direct sum of rings $(A_\lambda)_{\lambda \in I}$ relative to subrings $(L_\lambda)_{\lambda \in I}$. A subring B of A is an **algebraically dense subring** of A if $B \supseteq \prod_{\lambda \in I} L_\lambda$ and if, for each finite subset J of I, $pr_J(B) = \prod_{\lambda \in J} A_\lambda$, where pr_J is the canonical projection from $\prod_{\lambda \in I} A_\lambda$ to $\prod_{\lambda \in J} A_\lambda$. If each A_λ is a topological ring and each L_λ an open subring, the **local direct sum topology** on B is the topology induced by the local direct sum topology of A.

31.7 Theorem. If B is an algebraically dense subring of the local direct sum of semisimple rings $(A_\lambda)_{\lambda \in I}$ relative to subrings $(L_\lambda)_{\lambda \in I}$, then B is semisimple.

Proof. Let R be the radical of B. For each $\mu \in I$, the restriction to B of the canonical projection pr_μ from $\prod_{\lambda \in I} A_\lambda$ to A_μ is an epimorphism from B to A_μ, so by 26.15, $pr_\mu(R) = (0)$. Therefore $R = (0)$. •

31.8 Theorem. Let A be a topological ring. The following statements are equivalent:

$1°$ A is a semisimple linearly topologized topological ring possessing an open left ideal L that is a linearly compact ring for its induced topology.

$2°$ There is a topological isomorphism ϕ from A to $A_0 \times A_1 \times A_2$ where: A_0 is a discrete semisimple ring; A_1 is an algebraically dense subring of the local direct sum of a family $(A_\lambda)_{\lambda \in J}$ of topological rings relative to proper nonzero open left ideals $(L_\lambda)_{\lambda \in J}$, where for each $\lambda \in J$, A_λ is the topological ring $\operatorname{End}_{K_\lambda} E_\lambda$ of all linear operators on a discrete vector space E_λ over a finite field K_λ, furnished with the topology of pointwise convergence, $L_\lambda = \operatorname{Ann}_{A_\lambda} M_\lambda$ where M_λ is a nonzero proper finite-dimensional subspace of E_λ, the largest subring D_λ of A_λ in which L_λ is an ideal is $\{u \in A_\lambda : u(M_\lambda) \subseteq M_\lambda\}$, and the topology of A_1 is the local direct sum topology; $A_2 = \prod_{\mu \in M} A_\mu$, where for each $\mu \in M$, A_μ is the topological ring $\operatorname{End}_{K_\mu} E_\mu$ of all linear operators on a discrete nonzero right vector space E_μ over a division ring K_μ, furnished with the topology of pointwise convergence; and

$$\phi(L) = (0) \times \left(\prod_{\lambda \in J} L_\lambda\right) \times \left(\prod_{\mu \in M} A_\mu\right).$$

Proof. Statement $2°$ implies $1°$ by 31.7, 26.21, 29.1, and 29.5. Assume $1°$. We may further assume that $L \neq (0)$, since otherwise $2°$ holds where $A_0 = A$, $J = M = \emptyset$.

(a) We first show that there is an idempotent e such that $L = Ae$. Let Rad(L) be the radical of the ring L. By 29.12, 29.4, and 26.16, $L/\text{Rad}(L)$ is a semisimple linearly compact ring, and $L \neq \text{Rad}(L)$ since otherwise L would be a nonzero advertible left ideal of the semisimple ring A, in contradiction to 26.13. Consequently by 29.8 there exists $e \in L$ whose coset in $L/\text{Rad}(L)$ is the identity element of that ring. Thus for any $x \in L$, $x - xe \in \text{Rad}(L)$, an advertible ideal of L, whence $\{x - xe : x \in L\}$ is an advertible left ideal of A and so is (0) by 26.13. Consequently for all $x \in L$, $x = xe$. In particular, $e^2 = e$, and $Ae \subseteq L = Le \subseteq Ae$.

(b) Next we show that the intersection of the set $\mathcal{P}$ of all the primitive ideals $P(M)$, where M is an open regular maximal left ideal of A, is the zero ideal. Since each such M is closed and $P(M) = \{a \in A : aA \subseteq M\}$, each $P \in \mathcal{P}$ is closed.

First, if $z \in A$ and $ze \neq 0$, then $z \notin P$ for some $P \in \mathcal{P}$. Indeed, as A is semisimple, there exists $a \in A$ such that aze is not left advertible by 26.9. Let $D = \{x - xaze : x \in A\}$, a regular left ideal; clearly $aze \notin D$. To show that D is closed, it suffices by 4.11 to show that $L \cap D$ is closed in L. As $e \in L$, $L \cap D = \{x - xaze : x \in L\}$, the image of L under the continuous endomorphism $x \to x - xaze$ of the L-module L. Consequently $L \cap D$ is closed in L by 28.3 and 28.6. Therefore, since A is linearly topologized, there is an open left ideal I such that $aze + I \subseteq A \backslash D$. As D is a regular left ideal, so is $D + I$; therefore by 26.3 there is a maximal regular left ideal M containing $D + I$, and $zae \notin M$ since otherwise $x = (x - xaze) + xaze \in D + M = M$ for all $x \in A$. As I is open, M is open. If $z \in P(M)$, then $az \in P(M)$, so $aze \in M$, a contradiction. Thus $z \notin P(M)$.

Second, if $z' = x - ex$ for some $x \in A$ and if $z' \neq 0$, then $z' \notin P$ for some $P \in \mathcal{P}$. Indeed, there exists $a \in A$ such that $z'a$ is not left advertible by 26.9. Let $D' = \{x - xz'a : x \in A\}$. Then $e = e - ez'a \in D'$, so $L = Ae \subseteq D'$. Thus D' is an open regular left ideal not containing $z'a$. As in the preceding paragraph, there is by 26.3 an open regular maximal left ideal M' containing D' but not $z'a$. If $z' \in P(M')$, then $z'a \in M'$, a contradiction.

It follows that if $z - ez \neq 0$, then $z \notin P$ for some $P \in \mathcal{P}$. Indeed, in the contrary case, $z - ez$ would also belong to each $P \in \mathcal{P}$, whence $z - ez = 0$ by the preceding paragraph, a contradiction.

Finally, let x belong to each $P \in \mathcal{P}$. By what we proved first, $xe = 0$, so $x = x - xe$. By what we proved second, $x - ex = 0$, so $x = ex = e(x - xe) = ex - exe$. But for all $y, z \in A$, $(ey - eye)(ez - eze) = 0$. Thus

$$\left(\bigcap_{P \in \mathcal{P}} P\right)^2 = (0), \text{ so } \bigcap_{P \in \mathcal{P}} P = (0)$$

by 26.14.

(c) Next, Let $P_0 = \bigcap \{P \in \mathcal{P} : P \supseteq L\}$, let $(P_\lambda)_{\lambda \in I} = \{P \in \mathcal{P} : P \not\supseteq L\}$, and let $I' = I \cup \{0\}$. For each $\lambda \in I'$, let $A_\lambda = A/P_\lambda$, let ϕ_λ be the canonical epimorphism from A to $A/P_\lambda = A_\lambda$, and let $L_\lambda = \phi_\lambda(L)$. Then A_0 is a discrete semisimple ring, $L_0 = (0)$, and for each $\lambda \in I$, L_λ is a nonzero open left ideal of A_λ, and L_λ is a linearly compact ring by 29.3. Therefore by 31.4, for each $\lambda \in I$ we may regard A_λ as the ring $\mathrm{End}_{K_\lambda}(E_\lambda)$ of all linear operators on a discrete vector space E_λ over a discrete division ring K_λ, furnished with the topology of pointwise convergence, and we may regard L_λ as $\mathrm{Ann}_{A_\lambda}(M_\lambda)$ where M_λ is a proper finite subspace of E_λ (and thus $L_\lambda = A_\lambda$ if $M_\lambda = (0)$ and, in particular, if K is infinite).

(d) Let ϕ be the continuous homomorphism from A to $\prod_{\lambda \in I'} A_\lambda$, furnished with the cartesian product topology, defined by

$$\phi(x) = (\phi_\lambda(x))_{\lambda \in I'} = (x + P_\lambda)_{\lambda \in I'}.$$

By (b), ϕ is an isomorphism from A to $\phi(A)$. We shall show that for any finite subset Q of I',

$$pr_Q(\phi(A)) = \prod_{\lambda \in Q} A_\lambda,$$

where pr_Q is the canonical projection from $\prod_{\lambda \in I'} A_\lambda$ to $\prod_{\lambda \in Q} A_\lambda$. For this, we shall first show that if $\lambda, \gamma \in I'$ and if $\lambda \neq \gamma$, then $P_\lambda + P_\gamma = A$. If P_0 is one of P_λ, P_γ, say P_γ, then $P_\gamma = P_0 \not\subseteq P_\lambda$ since $L \subseteq P_0$ but $L \not\subseteq P_\lambda$. If P_0 is neither P_λ nor P_γ, we may assume that $P_\gamma \not\subseteq P_\lambda$. Then $\phi_\lambda(P_\gamma)$ is a nonzero ideal of A_λ and so is dense in $A_\lambda = \mathrm{End}_{K_\lambda}(E_\lambda)$ by 25.10. Therefore as $P_\lambda + P_\gamma = \phi_\lambda^{-1}(\phi_\lambda(P_\gamma))$, $P_\lambda + P_\gamma$ is dense in A. Thus it remains to show that $P_\lambda + P_\gamma$ is closed in A, and for this it suffices by 4.11 to show that $L \cap (P_\lambda + P_\gamma)$ is closed in L. By (a),

$$L \cap (P_\lambda + P_\gamma) = (P_\lambda + P_\gamma)e = P_\lambda e + P_\gamma e = (L \cap P_\lambda) + (L \cap P_\gamma).$$

As P_λ and P_γ are closed in A, $L \cap P_\lambda$ and $L \cap P_\gamma$ are closed submodules of the linearly compact L-module L, so $L \cap P_\lambda + L \cap P_\gamma$ is closed in L by (3) of 28.6. Therefore $P_\lambda + P_\gamma = A$ whenever $\lambda \neq \gamma$. Consequently, as A_λ has an identity element for all $\lambda \in I$, $pr_Q(\phi(A)) = \prod_{\lambda \in Q} A_\lambda$ for each finite subset Q of I' by 24.11.

(e) Next, we shall show that $\phi(L) = \prod_{\lambda \in I'} L_\lambda$. For this, we shall first show that for each finite subset Q of I',

$$pr_Q(\phi(L)) = \prod_{\lambda \in Q} L_\lambda.$$

Indeed, let

$$y = (\phi_\lambda(y_\lambda))_{\lambda \in Q} \in \prod_{\lambda \in Q} L_\lambda,$$

where $y_\lambda \in L$ for all $\lambda \in Q$. By (d) there exists $x \in A$ such that $pr_Q(\phi(x)) = y$, that is, $\phi_\lambda(x) = \phi_\lambda(y_\lambda)$ for all $\lambda \in Q$. Since $ze = z$ for all $z \in L$ by (a), for each $\lambda \in Q$,

$$\phi_\lambda(y_\lambda) = \phi_\lambda(y_\lambda e) = \phi_\lambda(y_\lambda)\phi_\lambda(e) = \phi_\lambda(x)\phi_\lambda(e) = \phi_\lambda(xe),$$

and $xe \in L$. Therefore $\phi(L)$ is dense in $\prod_{\lambda \in I'} L_\lambda$ for the cartesian product topology. By 29.3, $\phi(L)$ is a linearly compact ring, and so is complete and thus closed in $\prod_{\lambda \in I'} L_\lambda$. Therefore $\phi(L) = \prod_{\lambda \in I'} L_\lambda$.

(f) Next, we shall show that L is a strictly linearly compact ring. As noted in (a), $L/\mathrm{Rad}(L)$ is a semisimple linearly compact ring and therefore by 29.8 is a strictly linearly compact ring. It follows readily that $L/\mathrm{Rad}(L)$ is also a strictly linearly compact L-module. Consequently, by (2) of 28.16, it suffices to show that $\mathrm{Rad}(L)$ is a strictly linearly compact L-module. By 26.20, $\mathrm{Rad}(L) = \{u \in L : Lu = (0)\}$, so $\mathrm{Rad}(L)$ is a trivial L-module. Consequently, as $\mathrm{Rad}(L)$ is a linearly compact L-module, $\mathrm{Rad}(L)$ is a strongly linearly compact L-module, and therefore a strictly linearly compact L-module by 30.11.

(g) Next we shall show that $\phi(A)$ is a subring of the local direct sum of $(A_\lambda)_{\lambda \in I'}$ relative to the left ideals $(L_\lambda)_{\lambda \in I'}$. By (e) and (f), the restriction of ϕ to L is a topological isomorphism from L to $\prod_{\lambda \in L'} L_\lambda$. Let $\mathcal{T}$ be the topology on $\phi(A)$ for which ϕ is a topological isomorphism. Then $\prod_{\lambda \in I'} L_\lambda$ is open for $\mathcal{T}$ and its induced topology is the cartesian product topology. For each $\lambda \in I'$ let D_λ be the largest subring of A_λ in which L_λ is an ideal, that is, let $D_\lambda = \{u \in A_\lambda : L_\lambda u \subseteq L_\lambda\}$. Thus $D_0 = A_0$, and for each $\lambda \in I$, $D_\lambda = L_\lambda = A_\lambda$ if K_λ is infinite, $D_\lambda = \{u \in A_\lambda : u(M_\lambda) \subseteq M_\lambda\}$ otherwise. Let $z \in \phi(A)$. Since $x \to xz$ is continuous for $\mathcal{T}$, there is an open neighborhood U of zero for $\mathcal{T}$ such that $Uz \subseteq \prod_{\lambda \in I'} L_\lambda$, and we may assume that $pr_\lambda(U)$ is an open submodule of L_λ for each $\lambda \in I'$ and that $pr_\lambda(U) = L_\lambda$ for all but finitely many $\lambda \in I'$. Let $z = (z_\lambda)_{\lambda \in I'}$; then for all but finitely many $\mu \in I'$,

$$L_\mu z_\mu = pr_\mu(U) pr_\mu(z) = pr_\mu(Uz) \subseteq pr_\mu(\prod_{\lambda \in I'} L_\lambda) = L_\mu,$$

and hence $z_\mu \in D_\mu$.

By (d), (e), and (g), $\phi(A)$ is an algebraically dense subring of the local direct sum of $(A_\lambda)_{\lambda \in I'}$ relative to left ideals $(L_\lambda)_{\lambda \in I'}$, and since the restriction of ϕ to L is a topological isomorphism from L to $\prod_{\lambda \in I'} L_\lambda$, $\mathcal{T}$ is the local direct sum topology. Letting $M = \{\lambda \in I : L_\lambda = A_\lambda\}$, $J = I \setminus M$, we obtain the desired decomposition of $\phi(A)$ and $\phi(L)$ given in 2°. ●

31.9 Corollary. *A topological ring A is semisimple, linearly topologized, and possesses an open ideal L that is a linearly compact ring for its induced topology if and only if L is a semisimple linearly compact ring and A is the topological direct sum of a discrete semisimple subring and L.*

Proof. The condition is sufficient by 28.7 and 26.21. Necessity: If M is a proper nonzero subspace of a K-vector space E, $\mathrm{Ann}_E(M)$ is not an ideal of $\mathrm{End}_K(E)$, so in the terminology of 31.8, $J = \emptyset$. •

An abelian group is *torsionfree* if its torsion subgroup is the zero subgroup, that is, if $n.x \neq 0$ whenever x is a nonzero element of G and n is a nonzero integer.

31.10 Corollary. *Let A be a topological ring whose additive group is torsionfree. Then A is semisimple, linearly topologized, and possesses an open left ideal L that is a linearly compact ring for its induced topology if and only if L is a semisimple linearly compact ring and A is the topological direct sum of a discrete semisimple ring and L.*

Proof. Necessity: Once again, the hypothesis implies that $J = \emptyset$. •

32 Locally Compact Semisimple Rings

To apply the information thus far obtained to compact and locally compact semisimple rings with complete generality, we need a deep theorem concerning locally compact abelian groups, whose proof is beyond the scope of this book:

32.1 Theorem. *Let $\mathbb{T}$ be the multiplicative topological group of all complex numbers of absolute value one. If G is a locally compact abelian group, for each nonzero $a \in G$ there is a continuous homomorphism h from G to $\mathbb{T}$ such that $h(a) \neq 1$.*

A continuous homomorphism from G to $\mathbb{T}$ is called a *character* of G. From Theorem 32.1 we obtain important information about the connected component of zero in a locally compact ring:

32.2 Theorem. *Let C be the connected component of zero in a locally compact ring A. (1) If B is a left [right] bounded additive subgroup of A, then $BC = (0)$ [$CB = (0)$]. (2) If A is a locally compact left [right] bounded ring such that zero is the only element c of A satisfying $Ac = (0)$ [$cA = (0)$], then A is totally disconnected, and the open left [right] ideals of A form a fundamental system of neighborhoods of zero. (3) If A is a locally compact left [right] bounded ring that either has an identity element or is semisimple, then A is totally disconnected.*

Proof. (1) Let B be a left bounded additive subgroup of A, let H be the set of all characters of the locally compact additive subgroup A, and for each h, let $S_h = \{a \in A : h(Ba) = \{1\}\}$. Let $P = \{e^{i\theta} : |\theta| < \pi/2\}$, an open neighborhood of 1 in $\mathbb{T}$, and let V be a neighborhood of zero in A such that $BV \subseteq h^{-1}(P)$. Suppose that $h(bv) \neq 1$ for some $b \in B$, $v \in V$. Let $h(bv) = e^{i\theta}$ where $0 < |\theta| < \pi/2$, and let $n \geq 2$ be the smallest positive integer such that $n|\theta| \geq \pi/2$. Then $\pi/2 \leq |n\theta| < \pi$. Therefore as $h(n.bv) = e^{in\theta}$, $h(n.bv) \notin P$, but $n.bv = (n.b)v \in BV$, a contradiction. Consequently, $S_h \supseteq V$. Since S_h is clearly an additive subgroup, therefore, S_h is open and hence closed, so $S_h \supseteq C$. Thus

$$C \subseteq \bigcap_{h \in H} S_h,$$

so if $c \in C$ and $b \in B$, $h(bc) = 1$ for all $h \in H$, whence $bc = 0$ by 32.1. Clearly (2) follows from (1) and 12.16. Also (3) follows from (2) and (1), since $C^2 = (0)$, and hence $C = (0)$ if A is semisimple by 26.14. •

32.3 Corollary. *Let C be the connected component of zero in a compact ring A. (1) $AC = (0) = CA$. (2) If zero is the only element c of A such that $Ac = cA = (0)$, then A is totally disconnected. (3) If A either has an identity element or is semisimple, A is totally disconnected.*

This corollary enables us to complete a discussion begun in §5:

32.4 Corollary. *A topological ring A is compact and totally disconnected if and only if it is a closed subring of a compact ring with identity.*

Proof. The condition is necessary by 5.25. Sufficiency: Since a compact ring is bounded by 12.3, a compact ring with identity is totally disconnected by 32.3, and hence each of its subrings is. •

32.5 Theorem. *A compact, totally disconnected ring A is a strictly linearly compact ring, and its topology is an ideal topology.*

Proof. The topology of A is an ideal topology by 4.20, so A is linearly topologized and hence is clearly linearly compact. Let f be a continuous epimorphism from the compact A-module A to a Hausdorff, linearly topologized A-module B, and let K be the kernel of f. Then $f = g \circ \phi_K$ where ϕ_K is the canonical topological epimorphism from A to A/K and g is a continuous bijection from A/K to B; as A/K is compact, g is a homeomorphism by a theorem of topology, and hence f is a topological homomorphism. Thus A is strictly linearly compact by 28.10. •

32.6 Theorem. *A topological ring A is semisimple and compact if and only if it is topologically isomorphic to the cartesian product of a family of discrete rings, each the ring of all linear operators on a finite-dimensional vector space over a finite field.*

Proof. Necessity: The assertion follows from 32.3, 32.5 and 29.9, since the cartesian product of a family of nonempty topological spaces is compact only if each member of the family is compact, and a compact discrete space is finite. The condition is sufficient by Tikhonov's theorem and 26.21. •

32.7 Corollary. *A nonzero, compact, semisimple ring has an identity element.*

32.8 Corollary. *A commutative topological ring is compact and semisimple if and only if it is topologically isomorphic to the cartesian product of a family of finite fields.*

32.9 Theorem. *A topological ring A is left bounded, locally compact, and semisimple if and only if it is topologically isomorphic to the cartesian product of a discrete semisimple ring, an algebraically dense subring of the local direct sum of (discrete) finite rings $(A_\lambda)_{\lambda \in J}$, each the ring of all linear operators on a finite-dimensional vector space over a finite field, relative to proper nonzero left ideals $(L_\lambda)_{\lambda \in J}$, and a compact semisimple ring.*

Proof. The condition is sufficient by 31.8 and 32.6. Necessity: By 32.4 and 12.16, A is linearly topologized and hence has a compact open left ideal L, which is clearly linearly compact. With the notation of 31.8, for each $\lambda \in J \cup M$, L_λ is a continuous epimorphic image of L and hence is compact. Therefore A_2 is compact and, by 26.21, semisimple. Also, for each $\lambda \in J$, there exists $c_\lambda \in E_\lambda \setminus M_\lambda$, so $u \to u(c_\lambda)$ is a continuous surjection from L_λ to E_λ, and consequently E_λ is compact and discrete and thus finite. Thus by 31.8, the condition holds. •

32.10 Corollary. *A topological ring A is bounded, locally compact, and semisimple if and only if it is the topological direct sum of a discrete semisimple ring and a compact semisimple ring.*

The proof is similar to that of 31.9.

If K is a nondiscrete locally compact division ring and if A is the ring of all linear operators on a nonzero finite-dimensional vector space E over K, then there is a unique topology on A making A a topological K-vector space by 15.10, 18.17, and 13.8; that topology is a locally compact ring topology by 15.15 as A is a finite-dimensional algebra over the center C of K by 18.17. A natural problem is to determine conditions under which a topological ring may be described as topologically isomorphic to one of this type. By 25.23, a locally compact, connected, primitive ring with a minimal left ideal admits this description. We shall prove here that two other classes of locally compact rings may be so described: primitive rings with a minimal left ideal whose additive group is torsionfree (in §35 we shall show that the existence of a minimal left ideal is unneeded), and simple rings with a minimal left ideal (we need consider only totally disconnected rings of this class by 25.23 and 4.5).

32.11 Theorem. *If G is a nonzero, compact, totally disconnected abelian group, there is a prime p such that for some nonzero $a \in G$,*

$$\lim_{n \to \infty} p^n.a = 0.$$

Proof. Let $x \in G$, and let p be a prime. For each open subgroup U of G, let H_U be the set of all $z \in G$ such that z is an adherent point of $(p^n.x)_{n \geq 1}$ and, for some $m \geq 1$, $p^n.(x - z) \in U$ for all $n \geq m$. As U is open, G/U is compact and discrete and hence finite. Then $p^k.t \in U$ for some $k \geq 1$ if and only if $t + U$ belongs to the p-primary component T_p of G/U, in which case $p^s.t \in U$ where p^s is the order of T_p, and hence, as U is a subgroup, $p^n.t \in U$ for all $n \geq s$. As U is closed and as the function $z \to p^s.(x - z)$ is continuous, $\{z \in G : p^s.(x - z) \in U\}$ is closed. As the adherence of a sequence is closed, therefore, H_U is closed.

We show next that H_U is nonempty. As G is compact, the sequence $(p^n.x)_{n \geq 1}$ has an adherent point y. Therefore there is a sequence $(n_k)_{k \geq 1}$ of integers ≥ 1 such that $n_{k+1} > 2n_k$ for all $k \geq 1$ and $p^{n_k}.x - y \in U$ for all $k \geq 1$. Let $m_k = n_{k+1} - n_k$ for all $k \geq 1$. Then $n_{k+1} > m_k > n_k$ for all $k \geq 1$, so an adherent point z of $(p^{m_k}.x)_{k \geq 1}$ is a fortiori an adherent point of $(p^n.x)_{n \geq 1}$. Let r be so large that $p^{m_r}.x - z \in U$. Then

$$p^{n_r}.(z - x) = (p^{n_r}.z - p^{n_{r+1}}.x) + (p^{n_{r+1}}.x - y) + (y - p^{n_r}.x)$$
$$\in p^{n_r}.(z - p^{m_r}.x) + U + U \subseteq p^{n_r}U + U + U = U.$$

If V is an open subgroup contained in U, clearly $H_V \subseteq H_U$. Therefore $\{H_U : U \text{ is an open subgroup of } G\}$ is a filter base on compact G, and consequently there exists v belonging to each H_U. Thus for any open subgroup U, $p^n.(x - v) \in U$ for all but finitely many $n \geq 1$, so by 4.17,

$$\lim_{n \to \infty} p^n.(x - v) = 0.$$

By 4.17, G contains a proper open subgroup U. As G/U is a finite group, there is a prime p dividing the order of G/U, and consequently there exist

$b \in G$ and $s \geq 1$ such that the order of $b + U$ is p^s. Therefore $b \notin U$ but $p^n.b \in U$ for all $n \geq s$. Hence b is not an adherent point of $(p^n.b)_{n \geq 1}$. By the preceding, there exists an adherent point c of $(p^n.b)_{n \geq 1}$ such that $\lim_{n \to \infty} p^n.(b - c) = 0$. Thus if $a = b - c$, $a \neq 0$ and

$$\lim_{n \to \infty} p^n.a = 0. \bullet$$

For our next result, we need a simply proved fact about rings:

32.12 Theorem. *If an ideal I of a topological ring A is a ring with identity element e and if $J = \{x \in A : xe = 0\}$, then e belongs to the center of A, J is an ideal of A, A is the topological direct sum of I and J, and $J = \{y - ye : y \in A\}$.*

Proof. For any $x \in A$, $xe = exe = ex$ since xe and ex belong to I. Consequently, J is an ideal, and $J = \{y - ye : y \in A\}$ since $xe = 0$ if and only if there exists y such that $x = y - ye$. Since $x \to xe$ is a continuous projection on I whose kernel is J and since I and J are ideals such that $IJ = JI = (0)$, A is the topological direct sum of I and J by 15.4 and the discussion on page 112. •

32.13 Theorem. *A topological ring A is a locally compact primitive ring with a minimal left ideal whose additive group is torsionfree if and only if it is topologically isomorphic to the ring of all (continuous) linear operators on a finite-dimensional Hausdorff vector space over a nondiscrete locally compact division ring of characteristic zero.*

Proof. The condition is clearly sufficient. Necessity: By 25.22 there is an idempotent e such that Ae is a minimal left ideal, eAe is a division ring, and A is topologically isomorphic to a locally compact dense ring A_L of linear operators containing nonzero linear operators of finite rank on the right eAe-vector space Ae. Furthermore, eAe is locally compact as it is a topological epimorphic image of the additive group A, and has characteristic zero as A is torsionfree. Consequently by 16.2, 18.17, 13.8 and 7.7, it suffices to show that eAe is nondiscrete. We first note that if J is a nonzero closed ideal of A, then J is a locally compact primitive ring with a minimal left ideal. Indeed, J is locally compact as it is closed. Its image J_L in A_L under the topological isomorphism from A to A_L contains all linear operators in A_L of finite rank by 25.21, so by 25.10, 25.13, and 25.20, J_L is a primitive ring with a minimal left ideal, and thus J also has those properties.

If the connected component C of A is not the zero ideal, then by the preceding and 25.23, C satisfies the condition of the theorem and, in particular, has an identity element. Therefore by 32.12 and 25.11, $C = A$ and the condition holds for A.

Consequently, we may assume that A is totally disconnected. If, for some prime p, $\lim_{n\to\infty} p^n.x = 0$ for all $x \in A$, then in particular, $\lim_{n\to\infty} p^n.e = 0$, so eAe is not discrete as it has characteristic zero, and therefore the condition holds.

In general, A contains a compact open subgroup by 4.17 and hence by 32.11 there exist a prime p and a nonzero $a \in A$ such that $\lim_{n\to\infty} p^n.a = 0$. Let $J = \{x \in A : \lim_{n\to\infty} p^n.x = 0\}$. Clearly J is an ideal of A. Also, J is closed, for if $a \in \overline{J}$ and if U is an open additive subgroup, there exists $b \in J$ such that $a - b \in U$, and there exists $m \geq 1$ such that $p^n.b \in U$ for all $n \geq m$, so

$$p^n.a = p^n.(a-b) + p^n.b \in p^n.U + U = U$$

for all $n \geq m$. By the first part of the proof and the preceding paragraph, J satisfies the condition of the theorem and, in particular, has an identity element. Therefore by 32.15 and 25.11, $J = A$, so the condition holds for A. •

To show that a locally compact, totally disconnected, simple ring with a minimal left ideal is topologically isomorphic to the ring of all linear operators on a finite-dimensional vector space over a nondiscrete locally compact division ring, we need two preliminary theorems, the first purely topological:

32.14 Theorem. *Let E and F be topological spaces, and let H be a subset of F^E furnished with a topology such that $(u, x) \to u(x)$ is continuous from $H \times E$ to F. For any open subset O of F and any compact subset K of E, $T(K, O)$, defined by*

$$T(K, O) = \{u \in H : u(K) \subseteq O\},$$

is open in H.

Proof. Let $v \in T(K, O)$. Then $v(K) \subseteq O$, so for each $x \in K$ there are by hypothesis neighborhoods U_x of v in H and V_x of x in E such that $u(t) \in O$ for all $u \in U_x$ and all $t \in V_x$. As K is compact, there exist $x_1, \ldots, x_n \in K$ such that $\cup_{i=1}^n V_{x_i} \supseteq K$. Let $U = \cap_{i=1}^n U_{x_i}$. Then $u(x) \in O$ for all $u \in U$ and all $x \in K$, so $v \in U \subseteq T(K, O)$. Thus $T(K, O)$ is a neighborhood of each of is points and hence is open. •

32.15 Theorem. *If A is a totally disconnected locally compact ring of linear operators of finite rank on a vector space E such that for each $x \in E$, $u \to u(x)$ is continuous from A to E, furnished with the discrete topology, then A is discrete.*

32 LOCALLY COMPACT SEMISIMPLE RINGS 273

Proof. If E is finite-dimensional, then A is discrete, for if $\{c_1, \ldots, c_n\}$ is a basis of E,

$$\{0\} = \bigcap_{i=1}^{n} \{u \in A : u(c_i) = 0\},$$

a neighborhood of zero, since E is discrete. Therefore we shall assume that E is infinite-dimensional.

We shall first show that for each $n \geq 0$, the set F_n of all linear operators of rank $\leq n$ is closed in A. Indeed, let $w \in \overline{F}_n$, and let $x_1, \ldots, x_n$ be a sequence of $n+1$ vectors. There is a filter $\mathcal{F}$ on F_n converging to w, so by hypothesis, $\mathcal{F}(x_i)$ converges to $w(x_i)$ for each $i \in [1, n+1]$. Since E is discrete, there exists $H_i \in \mathcal{F}$ such that $u(x_i) = w(x_i)$ for all $u \in H_i$. Let $u \in \bigcap_{i=1}^{n+1} H_i$. As $u \in F_n$, there exist scalars $\lambda_1, \ldots, \lambda_{n+1}$, not all zero, such that

$$\sum_{i=1}^{n+1} \lambda_i u(x_i) = 0.$$

Hence

$$\sum_{i=1}^{n+1} \lambda_i w(x_i) = \sum_{i=1}^{n+1} \lambda_i u(x_i) = 0.$$

Therefore rank $w \leq n$.

By 9.4, A is a Baire space. Hence as $\bigcup_{i=1}^{\infty} F_n = A$, there exists $n \geq 0$ such that F_n has an interior point v, so by 4.17, there is an open additive subgroup G of A such that $v + G \subseteq F_n$. For any $w \in G$,

$$\text{rank } w \leq \text{rank}(v + w) + \text{rank}(-v) \leq n + \text{rank } v,$$

so the ranks of members of G are bounded. Let m be the largest of the ranks of members of G, and let $u \in G$ have rank m. Let $x_1, \ldots, x_m \in E$ be such that $\{u(x_1), \ldots, u(x_m)\}$ is a basis of the range M of u. As E is discrete, V, defined by

$$V = \{v \in G : v(x_i) = 0, 1 \leq i \leq m\},$$

is an open neighborhood of zero in A.

We shall show that if $v \in V$, then $v(E) \subseteq M$. If not, let $v \in V$ and $y \in E$ be such that $v(y) \notin M$. Then $u + v \in G$, so $\text{rank}(u+v) \leq m$. But $(u+v)(x_i) = u(x_i)$ if $i \in [1, m]$, and $(u+v)(y) = u(y) + v(y) \notin M$ since $v(y) \notin M$; hence $u(x_1), \ldots, u(x_m), u(y) + v(y)$ is a linearly independent sequence of $m+1$ vectors belonging to the range of $u + v$, a contradiction.

Since E is infinite-dimensional, there exist $y_1, \ldots y_m \in E$ such that $u(x_1), \ldots, u(x_m), y_1, \ldots, y_m$ is a linearly independent sequence of $2m$ vectors. As A is dense, there exists $w \in A$ such that $w(u(x_i)) = y_i$ for each

$i \in [1, m]$. As $v \to wv$ is continuous, there is a neighborhood U of zero in A such that $U \subseteq V$ and $wU \subseteq V$. To show that $U = \{0\}$, let $v \in U$, $x \in E$. Then $v(x) \in M$, so there exist scalars $\lambda_1, \ldots, \lambda_m$ such that

$$v(x) = \sum_{i=1}^{m} \lambda_i u(x_i).$$

Consequently,

$$wv(x) = \sum_{i=1}^{m} \lambda_i y_i,$$

but $wv(x) \in M$ as $wv \in V$; hence $wv(x) = 0$, so $\lambda_i = 0$ for all $i \in [1, m]$, whence $v(x) = 0$. Thus $U = \{0\}$, so the topology of A is the discrete topology. •

32.16 Theorem. *A topological ring A is a nondiscrete, locally compact, totally disconnected, simple ring with a minimal left ideal if and only if A is topologically isomorphic to the ring of all (continuous) linear operators on a finite-dimensional Hausdorff vector space over a nondiscrete, locally compact, totally disconnected division ring K, furnished with its unique topology as a topological vector space over K.*

Proof. The condition is sufficient by the discussion on page 269. Necessity: By 25.22, as A is primitive, there is an idempotent e in A such that Ae is a minimal left ideal, eA is a minimal right ideal, and eAe is a division ring, and furthermore, with their induced topologies, eAe is a locally compact ring, Ae is a straight, locally compact, right vector space over eAe, eA is a straight, locally compact, left vector space over eAe, and there is a topological [anti-]isomorphism $a \to a_L$ [$a \to a_R$] from A to a dense ring A_L [A_R] of linear operators of finite rank on the right [left] eAe-vector space Ae [eA] such that $(u, x) \to u(x)$ is continuous from $A_L \times Ae$ [$A_R \times eA$] to Ae [eA]. If eAe is not discrete, then the conclusion holds by 18.17, 13.8, and 16.2. Consequently, we shall assume that eAe is discrete and prove that A is discrete.

We shall first prove the conclusion under the additional assumption that the eAe-vector space Ae is generated by a compact neighborhood V of zero. By 25.20, e_L is a projection on a one-dimensional subspace M of Ae; let N be the kernel of e_L. As Ae is straight, M is a discrete subspace and hence is closed, so $V \cap M$ is a compact discrete subset and hence is finite. Therefore there is an open neighborhood W of zero in Ae such that $W \cap M = \{0\}$. Now $N = e_L^{-1}(W)$, for if $x \in e_L^{-1}(W)$, then $e_L(x) \in W \cap M = \{0\}$; but $e_L^{-1}(W)$ is open as e_L is continuous. Let

$$U = \{u \in A : u_L(V) \subseteq N\}.$$

By 32.14, U_L is an open neighborhood of zero in A_L. As V generates Ae,

$$U = \{u \in A : u_L(Ae) \subseteq N\},$$

so $(eU)_L = e_L \circ U_L = \{0\}$. As $eA \cap U \subseteq eU = \{0\}$, therefore, eA is discrete. Consequently by 32.15, A_R is discrete. Therefore A is discrete.

We turn, finally, to the general case. Let $E = Ae$, and let V be a compact neighborhood of zero in E. If $V = \{0\}$, then by 32.14 A is discrete; we shall assume, therefore, that V contains a nonzero vector. Let F be the subspace of E generated by V. Then F is locally compact, hence complete and thus closed, so the subring B of A_L, defined by

$$B = \{u \in A_L : u(F) \subseteq F\},$$

is a closed and hence locally compact subring of A_L. For each $u \in B$, let u_F be the restriction of u to F, and let ρ be the epimorphism from B to a ring B' of continuous linear operators of finite rank on F defined by $\rho(u) = u_F$. The kernel H of ρ then satisfies

$$H = \{u \in B : u(F) = \{0\}\}.$$

Now H is clearly closed in B, so the topological ring B/H is Hausdorff and hence locally compact by 5.2. We topologize B' so that the algebraic isomorphism from B/H to B' induced by ρ is a topological isomorphism. Thus ρ is a topological epimorphism from B to B'.

To show that B' is a dense ring of linear operators on F, let $x_1, \ldots, x_n$ be a linearly independent sequence of vectors of F, and let $y_1, \ldots, y_n \in F$. By 25.21 A_L contains a projection p on the subspace generated by $\{y_1, \ldots, y_n\}$, and there exists $u \in A_L$ such that $u(x_i) = y_i$ for all $i \in [1, n]$. Then $pu(E) \subseteq F$, so $(pu)_F \in B'$ and $(pu)_F(x_i) = y_i$ for all $i \in [1, n]$. As a subspace of E, F is straight. Moreover, $(v, x) \to v(x)$ from $B' \times F$ to F is continuous since $(u, x) \to u(x)$ is continuous from $B \times F$ to F and ρ is an open mapping from B to B'. Therefore from the first part of the proof, B' is discrete. Consequently, H is open in B. But as V generates F,

$$B \supseteq \{u \in A_L : u(V) \subseteq V\},$$

a neighborhood of zero in A_L by 32.14. Hence B is an open subset of A_L, so H is an open left ideal of B.

For each $x \in E$, let

$$H_x = \{u \in A_L : u(x) = 0\}.$$

Let z be a nonzero vector of F. Then $H_z \supseteq H$, so H_z is an open left ideal of A_L. For any $x \in E$ there exist $g \in A_L$ such that $g(z) = x$ and a neighborhood W of zero in A_L such that $Wg \subseteq H_z$; hence $W \subseteq H_x$, so H_x is open. Therefore for each $x \in E$, $u \to u(x)$ is continuous from A_L to E, furnished with the discrete topology. Consequently by 32.15, A_L is discrete. Therefore A is discrete. •

Next, we shall construct an example to show that the hypothesis of 32.13 that the additive group of A be torsionfree and that the hypothesis of 32.16 that A have a minimal left ideal may not be omitted. For this, we shall use a special example of an inductive limit of rings:

Let $(A_n)_{n \geq 1}$ be a sequence of sets, and for each $n \geq 1$ let $\phi_{n+1,n}$ be an injection from A_n to A_{n+1}. The *inductive limit* of the sets $(A_n)_{n \geq 1}$ relative to the injections $(\phi_{n+1,n})_{n \geq 1}$ is the set A of all sequences $(a_n)_{n \geq m}$ such that $a_m \in A_m$, either $m = 1$ or $m > 1$ and $a_m \notin \phi_{m,m-1}(A_{m-1})$, and for all $n \geq m$, $\phi_{n+1,n}(a_n) = a_{n+1}$. The *index* of $(a_n)_{n \geq m}$ is the integer m. For each $n \geq 1$, we shall denote the identity map of A_n by $\phi_{n,n}$, and for each $r > n$ we shall denote by $\phi_{r,n}$ the injection $\phi_{r,r-1} \circ \phi_{r-1,r-2} \circ \ldots \circ \phi_{n+1,n}$ from A_n to A_r. Thus, for any $(a_n)_{n \geq m} \in A$, $a_n = \phi_{n,m}(a_m)$ for all $n \geq m$.

For each $q \geq 1$, let A'_q be the subset of A consisting of all elements of index $\leq q$. It is easy to see that for each $a \in A_q$ there is a unique element $(a_n)_{n \geq m}$ in A'_q such that $a_q = a$; we denote that element by $\chi_q(a)$. The function χ_q is readily shown to be a bijection from A_q to A'_q, and is called the *canonical embedding* of A_q in A. If $r \geq q$, clearly

(1) $$\chi_r \circ \phi_{r,q} = \chi_q,$$

and

(2) $$A = \bigcup_{q=1}^{\infty} A'_q.$$

Assume, in addition, that each A_q is a ring and each $\phi_{q+1,q}$ a monomorphism from A_q to A_{q+1}. We may then define a ring structure on each A'_q so that χ_q is an isomorphism. If $r \geq q$, A'_q is then a subring of A'_r by (1). Consequently by (2), A has a unique ring structure such that each A'_q is a subring of A.

The example we shall use is that where A_n is the ring of all square matrices of order 2^n over a finite field F, and where $\phi_{n,n+1}$ associates to each $X \in A_n$ the matrix in A_{n+1} that, in block form, is

$$\begin{pmatrix} X & 0 \\ 0 & X \end{pmatrix}.$$

32 LOCALLY COMPACT SEMISIMPLE RINGS

For each $q \geq 1$, let B_q be the subring of A_q consisting of all matrices that are of the block form

$$\begin{pmatrix} 0 & Y \\ 0 & 0 \end{pmatrix}$$

where Y is a square matrix of order 2^{q-1}. Clearly

(3) $$B_q^2 = (0)$$

for all $q \geq 1$. If $r > q$,

(4) $$(\phi_{r,q}(A_q))B_r \cup B_r(\phi_{r,q}(A_q)) \subseteq B_r$$

since for any $X \in A_q$, $\phi_{r,q}(X)$ is a matrix of the block form

$$\begin{pmatrix} Z & 0 \\ 0 & Z \end{pmatrix}.$$

For each $n \geq 0$ and each $r \geq 1$, let

(5) $$C_{n,n+r} = \sum_{k=1}^{r} \chi_{n+k}(B_{n+k}) = \chi_{n+r}(\sum_{k=1}^{r} \phi_{n+r,n+k}(B_{n+k}))$$

by (1). We shall show by induction on r that for each $n \geq 0$, $C_{n,n+r}$ is a nilpotent ring with index of nilpotency $\leq 2^r$. As $C_{n,n+1} = \chi_{n+1}(B_{n+1})$, the assertion holds for $r = 1$ by (3). Assume the assertion is true for r. Let

$$D = \sum_{k=2}^{r+1} \phi_{n+r+1,n+k}(B_{n+k}) = \sum_{k=1}^{r} \phi_{n+1+r,n+1+k}(B_{n+1+k})$$

$$= \chi_{n+1+r}^{-1}(C_{n+1,n+1+r}).$$

Then

(6) $$C_{n+1,n+1+r} = \chi_{n+1+r}(D) \subseteq \chi_{n+1+r}(\sum_{k=1}^{r+1} \phi_{n+r+1,n+k}(B_{n+k}))$$

$$= C_{n,n+r+1}.$$

By our inductive hypothesis, $C_{n+1,n+1+r}$ is a ring, so $D^2 \subseteq D$, and therefore by (3) and (4),

$$(\sum_{k=1}^{r+1}\phi_{n+r+1,n+k}(B_{n+k}))^2 = \phi_{n+r+1,n+1}(B_{n+1})^2 + \phi_{n+r+1,n+1}(B_{n+1})D +$$

$$+ D\phi_{n+r+1,n+1}(B_{n+1}) + D^2$$

$$\subseteq (0) + \sum_{k=2}^{r+1}\phi_{n+r+1,n+k}(\phi_{n+k,n+1}(B_{n+1})B_{n+k}) +$$

$$+ \sum_{k=2}^{r+1}\phi_{n+r+1,n+k}(B_{n+k}\phi_{n+k,n+1}(B_{n+1})) + D^2$$

$$\subseteq \sum_{k=2}^{n+1}\phi_{n+r+1,n+k}(B_{n+k}) + D^2 = D + D^2 = D.$$

Consequently,

$$C_{n,n+r+1}^2 = \chi_{n+r+1}((\sum_{k=1}^{r+1}\phi_{n+r+1,n+k}(B_{n+k}))^2)$$

$$\subseteq \chi_{n+1+r}(D) = C_{n+1,n+1+r}.$$

Thus by (6),
$$C_{n,n+r+1}^2 \subseteq C_{n+1,(n+1)+r} \subseteq C_{n,n+r+1}.$$

Therefore $C_{n,n+r+1}$ is a ring, and moreover

$$C_{n,n+r+1}^{2^{r+1}} \subseteq C_{n+1,(n+1)+r}^{2^r} = (0)$$

by our inductive hypothesis.

For each $m \geq 1$, let

$$U_m = \bigcup_{r=1}^{\infty} C_{m-1,m-1+r} = \bigcup_{k=m}^{\infty} (\sum_{i=m}^{k} \chi_i(B_i)).$$

Then $(U_m)_{m \geq 1}$ is a fundamental system of neighborhoods of zero for a ring topology $\mathcal{T}$ on A. Indeed, U_m is the union of the increasing sequence of subrings and hence is a subring, so (TRN 1) of 3.5 holds; if $i > q$, then by (1) and (4),

$$\chi_q(A_q)\chi_i(B_i) \cup \chi_i(B_i)\chi_q(A_q) = \chi_i(\phi_{i,q}(A_q)B_i) \cup \chi_i(B_i\phi_{i,q}(A_q)) \subseteq \chi_i(B_i),$$

so for any $m > q$,

(7) $$A'_q U_m \cup U_m A'_q \subseteq U_m,$$

and therefore (TRN 2) of 3.5 holds by (2). In particular, if $1 \leq q < m$ then $\chi_q(B_q)U_m \cup U_m\chi_q(B_q) \subseteq U_m$ by (7), whereas if $q \geq m$, $\chi_q(B_q)U_m \cup U_m\chi_q(B_q) \subseteq U_m U_m \subseteq U_m$. Thus for each $m \geq 1$, U_m is an ideal of the ring U_1.

Suppose there were a nonzero $b \in \bigcap_{m=1}^{\infty} U_m$. As $b \in U_1$, there would exist an integer $q \geq 1$ and elements $b_1 \in B_1, \ldots, b_q \in B_q$ such that $b = \sum_{i=1}^{q} \chi_i(b_i)$, and as $b \in U_{q+1}$ there would exist for some integer $r > q$ elements $b_{q+1} \in B_{q+1}, \ldots, b_r \in B_r$ such that $b = \sum_{j=q+1}^{r} \chi_j(b_j)$ and $b_r \neq 0$. Consequently,

$$\chi_r(b_r) = \sum_{i=1}^{q} \chi_i(b_i) - \sum_{j=q+1}^{r-1} \chi_j(b_j),$$

so by (1),

$$b_r = \sum_{i=1}^{q} \chi_r^{-1}(\chi_i(b_i)) - \sum_{j=q+1}^{r-1} \chi_r^{-1}(\chi_j(b_j))$$
$$= \sum_{i=1}^{q} \phi_{r,i}(b_i) - \sum_{j=q+1}^{r-1} \phi_{r,j}(b_j)$$
$$= \phi_{r,r-1}(\sum_{i=1}^{q} \phi_{r-1,i}(b_i) - \sum_{j=q+1}^{r-1} \phi_{r-1,j}(b_j)),$$

and thus $b_r \in B_r \cap \phi_{r,r-1}(A_{r-1}) = (0)$, a contradiction. Therefore T is Hausdorff.

For any $m \geq 2$, $U_1 = \sum_{i=1}^{m-1} \chi_i(B_i) + U_m$; as $\sum_{i=1}^{m-1} \chi_i(B_i)$ is finite, so is U_1/U_m. Consequently by 5.22, there is a topological isomorphism from U_1 to a dense subring of $\varprojlim_{m \geq 2}(U_1/U_m)$, a closed subset of the cartesian product of finite discrete spaces by 5.20 and hence a compact set. Therefore by 8.4 the completion $\widehat{U}_1$ of U_1 is compact. Hence by 4.22, the completion $\widehat{A}$ of A is locally compact.

Next, we shall show that each element a of $\widehat{U}_1$ is a topological nilpotent. By 4.22, $(\widehat{U}_n)_{n \geq 1}$ is a fundamental system of neighborhoods of zero in $\widehat{A}$, and by 4.2, each $\widehat{U}_n$ is an ideal in $\widehat{U}_1$. Let $n \geq 1$. By 3.3, $\widehat{U}_1 = U_1 + \widehat{U}_n$, so there exist $b \in U_1$ and $c \in \widehat{U}_n$ such that $a = b + c$. By definition and an earlier result, U_1 is a union of nilpotent rings, and hence is a nil ring.

Therefore $b^r = 0$ for some $r \geq 1$. Hence as $\widehat{U}_n$ is an ideal of $\widehat{U}_1$, for any $m \geq r$,
$$a^m = (b+c)^m \in b^m + \widehat{U}_n = \widehat{U}_n.$$
Thus $\lim_{n \to \infty} a^n = 0$.

Since $\phi_{n+1,n}$ takes the identity matrix of A_n to the identity matrix of A_{n+1} for all $n \geq 1$, A has an identity 1, and hence $\widehat{A}$ does also by 4.4. Therefore $\widehat{A}$ is not a radical ring, so to show that it is simple, it suffices to show that the ideal generated by any nonzero $a \in \widehat{A}$ is $\widehat{A}$. There exists $n \geq 1$ such that $a \notin \widehat{U}_n$. By 3.3, $a = b + c$ where $b \in A$ and $c \in \widehat{U}_1$. By (2), there exists $q \geq n$ such that $b \in \chi_q(A_q)$. By 3.3, there exist $u \in U_1$ and $v \in \widehat{U}_{q+1}$ such that $c = u + v$. For some $r > q$ there exist $u_1 \in \chi_1(B_1), \ldots, u_r \in \chi_r(B_r)$ such that $u = \sum_{i=1}^r u_i$. Let $b' = b + \sum_{i=1}^q u_i$, $c' = v + \sum_{i=q+1}^r u_i$. Then $b' \in A'_q$, $c' \in \widehat{U}_{q+1}$, and $a = b + c = b + u + v = b' + c'$. As $a \notin \widehat{U}_n$ and hence $a \notin \widehat{U}_{q+1}$, $b' \neq 0$. As A'_q is a simple ring, there exist $x_1, \ldots, x_s, y_1, \ldots, y_s \in A'_q$ such that
$$1 = \sum_{i=1}^s x_i b' y_i.$$
By (7),
$$\widehat{A'_q \widehat{U}_{q+1}} \cup \widehat{\widehat{U}_{q+1} A'_q} \subseteq \widehat{A'_q U_{q+1}} \cup \widehat{U_{q+1} A'_q} \subseteq \widehat{U}_{q+1},$$
so
$$\sum_{i=1}^s x_i c' y_i \in \widehat{U}_{q+1} \subseteq \widehat{U}_1.$$
Let $z = -\sum_{i=1}^s x_i c' y_i$. As $z \in \widehat{U}_1$, z is a topological nilpotent, and therefore $1 - z$ is invertible by 11.16. Thus
$$\sum_{i=1}^s x_i a y_i (1-z)^{-1} = (\sum_{i=1}^s x_i b' y_i + \sum_{i=1}^s x_i c' y_i)(1-z)^{-1} = (1-z)(1-z)^{-1} = 1.$$
Consequently, A is a simple ring.

We shall finally establish that A is not isomorphic to the ring of all linear operators on a finite-dimensional vector space by showing that A has an infinite sequence $(e_n)_{n \geq 1}$ of nonzero idempotents such that $e_n e_m = 0$ if $m \neq n$. Let S_n be the square matrix of order 2^{n-1} and T_n the square matrix of order 2^n defined by
$$S_n = \begin{pmatrix} 0 & \cdots & 0 & 0 \\ \vdots & \ddots & \vdots & \\ 0 & \cdots & 0 & 0 \\ 0 & \cdots & 0 & 1 \end{pmatrix}, \quad T_n = \begin{pmatrix} S_n & 0 \\ 0 & 0 \end{pmatrix},$$

and let $e_n = \chi_n(T_n)$. If $m < n$, then $\phi_{n,m}(T_m)T_n = 0 = T_n\phi_{n,m}(T_m)$ since T_n has a nonzero entry only on the diagonal in the row numbered 2^{n-1}, whereas $\phi_{n,m}(T_m)$ has nonzero entries only on the diagonal in rows whose numbers are odd multiples of 2^{m-1}.

In sum, $\widehat{A}$ is a locally compact simple (in particular, primitive) ring with identity that is not algebraically isomorphic to the ring of all linear operators on a finite-dimensional vector space over a division ring.

Exercises

32.1 A left bounded locally compact primitive ring is discrete.

32.2 Let A be the local direct sum of $(A_n)_{n\geq 1}$ relative to open subrings $(L_n)_{n\geq 1}$ where for each $n \geq 1$ and some prime p, A_n is the field $\mathbb{Q}_p$ of p-adic numbers and L_n is the ring $\mathbb{Z}_p$ of p-adic integers. Then A is a commutative, semisimple, locally compact, metrizable ring with identity, but $x \to x^{-1}$ is not continuous on $A^\times$.

32.3 A nonzero topological ring A is a locally compact, advertibly open, semisimple ring satisfying the minimum condition on closed left ideals if and only if A is the topological direct sum of finitely many subrings, each either the ring of all linear operators on a nonzero, Hausdorff, finite-dimensional vector space over a nondiscrete locally compact division ring, furnished with its unique topology as a finite-dimensional algebra over the center of the division ring, or the discrete ring of all linear operators on a finite-dimensional vector space over a division ring. [Use Exercise 27.4.]

32.4 (Kaplansky [1947c]) Let F be a finite field, furnished with the discrete topology, and let $E = F^\mathbb{N}$, furnished with the cartesian product topology. For each $n \in \mathbb{N}$ let

$$M_n = \{(x_k)_{k\in\mathbb{N}} \in E : x_k = 0 \text{ for all } k < n\},$$

and let $A = \{u \in \text{End}(E): \text{there exists } q \in \mathbb{N} \text{ such that } u(M_n) \subseteq M_n \text{ for all } n \geq q\}$, $J = \{u \in \text{End}(E): u(M_n) \subseteq M_n \text{ for all } n \geq 0\}$. (a) A is a primitive ring of endomorphisms of E. (b) $\{u \in A : u(M_0) = (0)\}$ is a minimal left ideal of A. (c) For each $n \in \mathbb{N}$, let $V_n = \{u \in J : u(E) \subseteq M_n\}$. Then $(V_n)_{n\geq 0}$ is a fundamental system of neighborhoods of zero for a compact ring topology on J. [Observe that it is the weakest topology on J for which $u \to u(x)$ is continuous from J to E for each $x \in E$.] (d) The additive group topology $\mathcal{T}$ on A for which $(V_n)_{n\geq 0}$ is a fundamental system of neighborhoods of zero is a locally compact ring topology on A. (e) A contains an infinite sequence $(e_n)_{n\geq 0}$ of nonzero idempotents such that $e_n e_m = 0$ whenever $n \neq m$, and hence A is not isomorphic to the ring of all linear operators on a finite-dimensional vector space over a division ring. In sum, A is a locally compact primitive ring with an identity and a minimal left ideal that is not a simple ring.

32.5 In the example of a locally compact simple ring $\widehat{A}$ in the text, for each $n \geq 1$ let $e_n = \chi_n(E_n)$, where E_n is the square matrix of order 2^n having 1 in the first row and column and zeros elsewhere. Show that $(\widehat{A}e_n)_{n \geq 1}$ is a strictly decreasing sequence of closed left ideals whose intersection is (0).

CHAPTER VIII

LINEAR COMPACTNESS IN RINGS WITH RADICAL

The behavior of the powers of the radical R of a ring A is intimately connected with the existence of a strictly linearly compact topology on A. Specifically, a linearly compact ring A admits a weaker strictly linearly compact topology if and only if R is "transfinitely nilpotent." To establish this, we shall first show in §33 that if T is a linearly compact topology on a module E, of all the linearly compact topologies on E weaker than T there is a weakest T_*.

Next, in §34 we shall determine conditions under which an orthogonal family of idempotents in A/R can be "lifted" to an orthogonal family of idempotents of A. The possibility of doing so is of crucial importance for establishing certain structure theorems. In particular, if A is commutative and linearly compact, any family of orthogonal idempotents of A/R may be lifted to A, but we need additional restrictions, including the transfinite nilpotence of R, to establish the corresponding result for noncommutative linearly compact rings. Lifting idempotents is a technique employed in the proofs of most of the theorems concerning locally compact rings, given in §35.

Rings linearly compact for the radical topology, for which the powers of the radical form a fundamental sytem of neighborhoods of zero for a Hausdorff ring topology, offer a natural domain for generalizations of theorems concerning artinian rings, since artinian rings are linearly compact for the discrete topology, which is the radical topology as the radical of an artinian ring is nilpotent. Generalizations of some classical theorems about artinian rings are given in §36.

33 Linear Compactness in Rings with Radical

We first establish that of all the Hausdorff linear topologies on an A-module weaker than a given linearly compact topology, there is a weakest.

33.1 Definition. Let E be an A-module. A proper submodule M of E is **sheltered** if the set of submodules of E strictly containing M has a smallest member, called the **shelter** of M.

Thus S is the shelter of M if and only if $S \supset M$ and for any submodule N of E, if $N \supset M$, then $N \supseteq S$.

33.2 Theorem. *If N is a proper submodule of an A-module E, then N is the intersection of sheltered submodules of E.*

Proof. Let $x \in E \setminus N$. The set of submodules containing N but not x is inductive for the inclusion relation and hence contains a maximal member M by Zorn's Lemma. Each submodule properly containing M therefore contains the submodule S generated by $M \cup \{x\}$, so M is a sheltered submodule with shelter S, and M contains N but not x. •

33.3 Definition. Let $\mathcal{T}$ be a linear topology on an A-module E. The **Leptin topology** associated to $\mathcal{T}$ is the additive group topology $\mathcal{T}_*$ on E for which the finite intersections of the sheltered submodules open for $\mathcal{T}$ is a fundamental system of neighborhoods of zero, and $\mathcal{T}$ is a **Leptin topology** if $\mathcal{T} = \mathcal{T}_*$.

A sheltered submodule open for $\mathcal{T}$ is again a sheltered submodule open for $\mathcal{T}_*$ and conversely, so $(\mathcal{T}_*)_* = \mathcal{T}_*$. Thus the Leptin topology associated to a linear topology is indeed a Leptin topology.

33.4 Theorem. *If $\mathcal{T}$ is a linear topology on an A-module E, then $\mathcal{T}_*$ is a linear topology on E having the same closed submodules as $\mathcal{T}$. In particular, if $\mathcal{T}$ is Hausdorff, so is $\mathcal{T}_*$.*

Proof. If U is a submodule open for $\mathcal{T}_*$ and if $b \in E$, then as U is open for $\mathcal{T}$, there is a neighborhood T of zero in A such that $T.b \subseteq U$. Therefore by 3.6, $\mathcal{T}_*$ is a linear topology.

Let M be a proper submodule of E that is closed for $\mathcal{T}$. By 3.3, M is the intersection of the submodules $M + U$ where U is a submodule open for $\mathcal{T}$, and $M + U$ is a proper subset of E. Each such $M + U$ is the intersection of sheltered submodules by 33.2, each of which is necessarily open for $\mathcal{T}$ as $M + U$ is, and is therefore open for $\mathcal{T}_*$. Thus M is the intersection of submodules open and hence closed for $\mathcal{T}_*$, so M is closed for $\mathcal{T}_*$. Conversely, as $\mathcal{T}_*$ is weaker than $\mathcal{T}$, each submodule closed for $\mathcal{T}_*$ is closed for $\mathcal{T}$. •

33.5 Theorem. *Let $\mathcal{T}$ be a linearly compact topology on an A-module E. Every filter base of submodules of E whose adherence is (0) converges to zero for $\mathcal{T}_*$.*

Proof. Let $\mathcal{F}$ be a filter base of closed submodules of E such that $\bigcap_{F \in \mathcal{F}} F = (0)$. Let U be an open sheltered submodule, and let S be its shelter. By 28.20,
$$U = U + \bigcap_{F \in \mathcal{F}} F = \bigcap_{F \in \mathcal{F}} (U + F).$$

Consequently, there exists $F \in \mathcal{F}$ such that $U + F \not\supseteq S$, so $U + F = U$, that is, $F \subseteq U$. As each neighborhood of zero for $\mathcal{T}_*$ contains the intersection of finitely many sheltered submodules open for $\mathcal{T}$, therefore, $\mathcal{F}$ converges to zero for $\mathcal{T}_*$. •

33.6 Corollary. *A linearly compact topology $\mathcal{T}$ on an A-module E is a Leptin topology if and only if every filter base of submodules of E whose adherence is (0) converges to zero.*

Proof. The condition is necessary by 33.5. Sufficiency: Let $\mathcal{V}$ be the filter base of submodules open for $\mathcal{T}_*$. By 4.8 and 33.4, $\mathcal{V}$ is a filter base of submodules closed for $\mathcal{T}$ whose adherence is $\{0\}$. Therefore $\mathcal{V}$ converges to zero for $\mathcal{T}$, that is, $\mathcal{T}$ is weaker than and hence identical with $\mathcal{T}_*$. •

33.7 Theorem. *Let $\mathcal{T}$ be a linear compact topology on an A-module E, and let $\mathfrak{C}(\mathcal{T})$ be the set of all linear topologies on E having the same closed submodules as $\mathcal{T}$. Each member of $\mathfrak{C}(\mathcal{T})$ is linearly compact. All members of $\mathfrak{C}(\mathcal{T})$ have the same associated Leptin topology, which is the weakest topology belonging to $\mathfrak{C}(\mathcal{T})$. Every Hausdorff linear topology on E weaker than some member of $\mathfrak{C}(\mathcal{T})$ belongs to $\mathfrak{C}(\mathcal{T})$.*

Proof. The zero submodule of E is closed for each member of $\mathfrak{C}(\mathcal{T})$, so each member of $\mathfrak{C}(\mathcal{T})$ is Hausdorff. Since a Hausdorff linear topology is linearly compact if and only if every closed linear filter base has a nonempty intersection, all members of $\mathfrak{C}(\mathcal{T})$ are linearly compact as $\mathcal{T}$ is.

By 33.4, $\mathcal{T}_* \in \mathfrak{C}(\mathcal{T})$. Let $\mathcal{S}$ be a Hausdorff linear topology on E weaker than some member of $\mathfrak{C}(\mathcal{T})$, and let $\mathcal{V}$ be the filter base of submodules of E open for $\mathcal{S}$. Each $V \in \mathcal{V}$ is also closed for $\mathcal{S}$ by 4.8, hence for the member of $\mathfrak{C}(\mathcal{T})$ stronger than $\mathcal{S}$, and therefore also for $\mathcal{T}$. As $\mathcal{S}$ is Hausdorff, $\bigcap_{V \in \mathcal{V}} V = (0)$. Consequently by 33.5, $\mathcal{V}$ converges to zero for $\mathcal{T}_*$, that is, $\mathcal{T}_*$ is weaker than $\mathcal{S}$. Thus $\mathfrak{C}(\mathcal{T})$ contains a member stronger than $\mathcal{S}$ and a member weaker than $\mathcal{S}$, so $\mathcal{S} \in \mathfrak{C}(\mathcal{T})$. In particular, for any $\mathcal{S} \in \mathfrak{C}(\mathcal{T})$, $\mathcal{S}_* \in \mathfrak{C}(\mathcal{T})$ by 33.4, so $\mathcal{T}_*$ is weaker than $\mathcal{S}_*$. Interchanging $\mathcal{S}$ and $\mathcal{T}$, we conclude that $\mathcal{T}_* = \mathcal{S}_*$. •

33.8 Corollary. *The minimal members in the class of all linearly compact topologies on an A-module E, ordered by inclusion, are precisely the linearly compact Leptin topologies.*

33.9 Corollary. *Let $\mathcal{T}$ be a linearly compact topology on an A-module E, and let M be a closed submodule of E. The Leptin topology $(\mathcal{T}_M)_*$ associated to the topology $\mathcal{T}_M$ induced on M by $\mathcal{T}$ is the topology $(\mathcal{T}_*)_M$ induced on M by $\mathcal{T}_*$.*

Proof. By 33.4, $\mathfrak{C}(\mathcal{T}_M) = \mathfrak{C}((\mathcal{T}_*)_M)$. If $\mathcal{F}$ is a filter base of closed submodules of M such that $\bigcap_{F \in \mathcal{F}} F = (0)$, then $\mathcal{F}$ converges to zero for $\mathcal{T}_*$ by

33.5 and hence for $(T_*)_M$. By 33.6, therefore $(T_*)_M$ is a Leptin topology on M belonging to $\mathfrak{C}(T_M)$. Consequently by 33.7, $(T_*)_M = (T_M)_*$. •

33.10 Theorem. *If u is a continuous homomorphism from a linearly compact A-module E to a linearly compact A-module F, then u is also continuous when E and F are furnished with their associated Leptin topologies.*

Proof. Replacing F with $u(E)$ if necessary, we may by 33.9 assume that u is surjective. To establish the result in this case, it suffices to show that if S is the shelter of a submodule V of T, then $u^{-1}(S)$ is the shelter of $u^{-1}(V)$. First, $u^{-1}(V) \subset u^{-1}(S)$, for if $u^{-1}(V) = u^{-1}(S)$, then

$$V = u(u^{-1}(V)) = u(u^{-1}(S)) = S,$$

a contradiction. Let N be a submodule of E strictly containing $u^{-1}(V)$. Then $N = N + u^{-1}(0)$ as $u^{-1}(0) \subseteq u^{-1}(V) \subset N$. Consequently, $V \subset u(N)$, for if $V = u(N)$, then

$$u^{-1}(V) = u^{-1}(u(N)) = N + u^{-1}(0) = N,$$

a contradiction. Hence $S \subseteq u(N)$, so

$$u^{-1}(S) \subseteq u^{-1}(u(N)) = N + u^{-1}(0) = N. \bullet$$

33.11 Theorem. *Let A be a linearly compact ring, and let E be a linearly compact A-module. The Leptin topology of A is also a ring topology, and the Leptin topology of E makes E a topological module over A, furnished with its Leptin topology.*

Proof. Let T be the topology of A. Since T_* is a linear topology, to show that it is a ring topology we need only show that for each $c \in A$, $R_c : x \to xc$ is continuous for T_*. But R_c is an endomorphism of the A-module A that is continuous for T. Consequently, R_c is continuous for T_* by 33.10. The second assertion is similarly established. •

The following discussion of simple and semisimple modules will yield a new criterion for a linearly compact module to be strictly linearly compact.

33.12 Definition. A **simple** *module is a nonzero module whose only proper submodule is the zero module.* A **semisimple** *module is one that is generated by the union of its simple submodules.*

For example, the simple submodules of a vector space are precisely its one-dimensional subspaces, so a vector space is a semisimple module. If $(E_\lambda)_{\lambda \in L}$ is a family of simple A-modules, then $\bigoplus_{\lambda \in L} E_\lambda$ is clearly semisimple. If A is a ring, the simple submodules of the A-module A are precisely the minimal left ideals of the ring A.

33.13 Theorem. *If A is a semisimple artinian ring, then A is a semisimple A-module.*

The assertion follows from (2) of 25.21 and 27.14.

33.14 Theorem. *Let E be an A-module that is generated by a family $(M_\lambda)_{\lambda \in L}$ of simple submodules. If F is a submodule of E, there exists a subset J of L such that E is the direct sum of F and $(M_\lambda)_{\lambda \in J}$. In particular, E is the direct sum of a subfamily of $(M_\lambda)_{\lambda \in L}$.*

Proof. Let $\mathcal{C} = \{C \subseteq L :$ the submodule $F + \sum_{\lambda \in C} M_\lambda$ is the direct sum of F and $(M_\lambda)_{\lambda \in C}\}$. Trivially, $\emptyset \in \mathcal{C}$. To show that $\mathcal{C}$, ordered by inclusion, is inductive, let D be the union of a totally ordered subset of $\mathcal{C}$. If $x + \sum_{\lambda \in D} x_\lambda = 0$ where $x \in F$ and $x_\lambda \in M_\lambda$ for all $\lambda \in D$ and $x_\lambda = 0$ for all but finitely many $\lambda \in D$, then there is a member C of the totally odered subset such that $x_\lambda = 0$ for all $\lambda \in D \setminus C$, whence $x = 0$ and $x_\lambda = 0$ for all $\lambda \in D$ since $F + \sum_{\lambda \in C} M_\lambda$ is the direct sum of F and $(M_\lambda)_{\lambda \in C}$. Thus by Zorn's Lemma there is a maximal subset J of L such that $F + \sum_{\lambda \in J} M_\lambda$ is the direct sum of F and $(M_\lambda)_{\lambda \in J}$. For each $\mu \in L$, M_μ is a submodule of $F + \sum_{\lambda \in J} M_\lambda$, for otherwise $(F + \sum_{\lambda \in J} M_\lambda) \cap M_\mu = (0)$ as M_μ is simple, whence

$$F + \sum_{\lambda \in J \cup \{\mu\}} M_\lambda$$

would be the direct sum of F and $(M_\lambda)_{\lambda \in J \cup \{\mu\}}$, a contradiction of the maximality of J. •

33.15 Corollary. *If F is a submodule of a semisimple module E, then E/F and F are semisimple modules.*

Proof. By 33.14, the quotient module of any submodule of E is semisimple, and F has a supplement H. As F is isomorphic to E/H, F is also semisimple. •

33.16 Theorem. *If A is a ring with identity such that the A-module A is semisimple, then every unitary A-module E is semisimple.*

Proof. Let S be a set of generators for the A-module E (for example, let $S = E$), and for each $s \in S$ let A_s be the A-module A. Then $f : (\lambda_s)_{s \in S} \to \sum_{s \in S} \lambda_s s$ is an epimorphism from $\bigoplus_{s \in S} A_s$ to E, so E is semisimple by the remark following 33.12 and 33.15. •

33.17 Theorem. *An A-module E is artinian if and only if E is linearly compact for the discrete topology and for every proper submodule U of E, E/U contains a simple submodule.*

Proof. The condition is necessary by 28.14, 27.3, and the fact that any nonzero artinian module F contains a simple submodule, namely, a submodule minimal in the set of all nonzero submodules, ordered by inclusion.

Sufficiency: Let $(N_i)_{i \geq 1}$ be a decreasing sequence of proper submodules, and let $N = \bigcap_{i=1}^{\infty} N_i$. The hypotheses for E imply the same hypotheses for E/N, so replacing E by E/N if necessary, we may assume that $\bigcap_{i=1}^{\infty} N_i = (0)$. Let M be the submodule of E generated by the union of all the simple submodules. As M is also linearly compact for the discrete topology, M is the direct sum of finitely many simple submodules by 33.14 and 28.21. Consequently, M is artinian by 27.6. Therefore by 28.14, M is discrete for the topology induced by $\mathcal{D}_*$, the Leptin topology associated to the discrete topology $\mathcal{D}$, so there is a submodule S of E that is open for $\mathcal{D}_*$ such that $S \cap M = (0)$. Consequently, S contains no simple submodules.

Suppose that $S \neq (0)$. By Zorn's Lemma there is a submodule U of E that is maximal among all the submodules of E whose intersection with S is (0). As $S \neq (0)$, $U \neq E$, so by hypothesis there is a submodule T of E containing U such that T/U is a simple submodule of E/U. Then $T \cap S \neq (0)$ by the maximality of U. As $S \cap U = (0)$, the restriction ϕ_S to S of the canonical epimorphism from E to E/U is an isomorphism from S to $(S+U)/U$, so $\phi_S(T \cap S)$ is a nonzero submodule of T/U and hence is T/U. Thus $T \cap S$ is a simple submodule of S, in contradiction to the conclusion of the preceding paragraph.

Therefore $S = (0)$, so $\mathcal{D}_*$ is the discrete topology. As $\bigcap_{i=1}^{\infty} N_i = (0)$, $(N_i)_{i \geq 1}$ converges to zero for $\mathcal{D}_*$ by 33.5. Consequently, for some $q \geq 1$, $N_q = (0)$. •

33.18 Corollary. *A linear topology $\mathcal{T}$ on an A-module E is strictly linearly compact if and only if $\mathcal{T}$ is linearly compact and for every proper open submodule U of E, E/U contains a simple submodule.*

The assertion follows readily from (2) of 28.15 and 33.17.

33.19 Theorem. *A linearly compact module E over a strictly linearly compact ring A is strictly linearly compact.*

Proof. By 33.18 it suffices to show that if U is a proper open submodule of E, then E/U contains a simple submodule. If E/U is a trivial A-module, then E/U is a discrete, strongly linearly compact module, hence is artinian by 30.10, and in particular contains a simple submodule. In the contrary case there exists $x \in E$ such that $A.(x+U)$ is not the zero submodule of E/U. Let $L = \{a \in A : ax \in U\}$, a proper open left ideal of A. Then $a \to a.(x+U)$ is an epimorphism from the A-module A to $A.(x+U)$ with kernel L. By (2) of 28.15, A/L is an artinian A-module, so $A.(x+U)$ is also; in particular, $A.(x+U)$ contains a simple submodule, so E/U does also. •

33.20 Definition. *Let A be a topological ring, and let ξ_A be the smallest ordinal number whose corresponding cardinal number is the smallest of those strictly greater than that of the set of all subsets of A. Let J be a closed ideal of A. We define J_λ recursively for each ordinal number λ such that $1 \leq \lambda < \xi_A$ as follows: $J_1 = J$; if J_α is defined for all $\alpha < \lambda$ and if $\lambda = \mu+1$ (that is, if λ has an immediate predecessor μ), we define J_λ to be $\overline{J_\mu J}$, but if λ has no immediate predecessor, we define J_λ to be $\bigcap_{\alpha<\lambda} J_\alpha$.*

Clearly $(J_\lambda)_{\lambda<\xi_A}$ is a decreasing family of closed ideals, so as the cardinality of all ordinals $< \xi_A$ exceeds that of the set of all subsets of A, there exists $\gamma < \xi_A$ such that $J_{\gamma+1} = J_\gamma$, and it follows readily that $J_\lambda = J_\gamma$ for all $\lambda \in [\gamma, \xi_A)$; the smallest such ordinal γ we shall call the *transfinite index* of J. If γ is the transfinite index of J, we shall also denote J_γ by J_* and say that J is *transfinitely nilpotent* if $J_* = (0)$.

An inductive argument establishes that for each integer $n \geq 1$, $J_n = \overline{J^n}$. Indeed, if $J_k = \overline{J^k}$, then $J_k J \subseteq \overline{J^{k+1}}$ as $\{x \in A : xJ \subseteq \overline{J^{k+1}}\}$ is closed, so again

$$J_{k+1} = \overline{J_k J} \subseteq \overline{J^{k+1}}.$$

Conversely, $J^{k+1} \subseteq J_k J \subseteq J_{k+1}$, so $\overline{J^{k+1}} \subseteq J_{k+1}$ as J_{k+1} is closed.

33.21 Theorem. *The radical R of a strictly linearly compact ring A is transfinitely nilpotent.*

Proof. Let γ be the transfinite index of R. It suffices to show that for any open left ideal J of A, $R_* \subseteq J$. Let $L = \{x \in A : R_* x \subseteq J\}$. As R_* is an ideal, L is a left ideal. As $L \supseteq J$, L is open.

Suppose that $L \neq A$. Then A/L is a nonzero artinian A-module by 28.15, so there is a left ideal L' of A containing L such that L'/L is a simple A-module. Clearly $\{x \in L'/L : A.x = (0)\}$ is a submodule of L'/L and hence is either L'/L or (0). In the former case, $A.(L'/L) = (0)$, so in particular, $R.(L'/L) = (0)$. In the latter case, for any nonzero $x \in L'/L$, $A.x$ is a nonzero submodule of L'/L and hence $A.x = L'/L$. Therefore if for each $a \in A$, $\hat{a}$ is the endomorphism of the abelian group L'/L defined by $\hat{a}(x) = a.x$ for all $x \in L'/L$, $a \to \hat{a}$ is an epimorphism from A to a primitive ring of endomorphisms of L'/L. Its kernel is therefore a primitive ideal and hence contains R, so $R.(L'/L) = (0)$. Therefore in both cases, $RL' \subseteq L$. Consequently,

$$R_\gamma RL' \subseteq R_\gamma L = R_* L \subseteq J,$$

so

$$R_* L' = R_{\gamma+1} L' = \overline{R_\gamma R L'} \subseteq J$$

since J is closed. Therefore $L' \subseteq L$, a contradiction. Consequently, $L = A$, that is, $R_*A \subseteq J$, so $R_\gamma R \subseteq R_*A \subseteq J$, whence $R_* = R_{\gamma+1} = \overline{R_\gamma R} \subseteq J$. •

33.22 Theorem. *If A is a bounded, strictly linearly compact ring with radical R, then the filter base $(R^n)_{n \geq 1}$ converges to zero, and in particular,*

$$\bigcap_{n=1}^{\infty} \overline{R^n} = (0).$$

Proof. Let J be an open ideal. By 28.15, A/J is an artinian A-module and hence an artinian ring. Let ϕ be the canonical epimorphism from A to A/J, and let S be the radical of A/J. By 26.15, $\phi(R) \subseteq S$, so for all $n \geq 1$, $\phi(R^n) \subseteq S^n$. By 27.15 S is nilpotent, so for some $m \geq 1$, $R^n \subseteq J$ for all $n \geq m$. Thus by 12.16, $(R^n)_{n \geq 1}$ converges to zero and, in particular, its adherence $\bigcap_{n=1}^{\infty} \overline{R^n}$ is (0). •

33.23 Theorem. *If R is the radical of a linearly compact ring A, then A/R_* is strictly linearly compact for its Leptin topology.*

Proof. We shall show by transfinite induction that for all $\lambda \in [1, \xi_A)$, A/R_λ is strictly linearly compact for its Leptin topology. Since the submodules of the A-module A/R_λ and the left ideals of the ring A/R_λ coincide, the Leptin topologies of the topological A-module A/R_λ and of the topological ring A/R_λ coincide, and hence the assertion that the topological ring A/R_λ is strictly linearly compact for its Leptin topology is equivalent to the corresponding assertion about the topological A-module A/R_λ.

If $\lambda = 1$, then $A/R_\lambda = A/R$ and hence A/R_λ is a strictly linearly compact ring by 29.12 and 29.8. Assume that the A-module A/R_μ is strictly linearly compact for its Leptin topology. To show the corresponding assertion for the A-module $A/R_{\mu+1}$, it suffices by 28.16 to show that, for the subspace and quotient topologies determined by the Leptin topology of that module, $R/R_{\mu+1}$ and $(A/R_{\mu+1})/(R/R_{\mu+1})$ are strictly linearly compact A-modules. Since $R_\mu R \subseteq R_{\mu+1}$, we may regard $R/R_{\mu+1}$ as a module over A/R_μ. Neither the Leptin topology of a linearly topologized module nor the property of strict linear compactness depends on the topology of the underlying scalar ring, so by 33.11, our assumption, and 33.19, the Leptin topology of the (A/R_μ)-module $R/R_{\mu+1}$ is strictly linearly compact. Since the submodules of the (A/R_μ)-module $R/R_{\mu+1}$ and those of the A-module $R/R_{\mu+1}$ coincide, the Leptin topologies of the (A/R_μ)-module $R/R_{\mu+1}$ and of the A-module $R/R_{\mu+1}$ coincide, and therefore the Leptin topology of the A-module $R/R_{\mu+1}$ is strictly linearly compact. By 33.9, however, that topology is the subspace topology induced by the Leptin topology of the A-module $A/R_{\mu+1}$.

Furnished with the quotient topologies determined by the given topology of A, the topological A-module $(A/R_{\mu+1})/(R/R_{\mu+1})$ is topologically isomorphic to the topological A-module A/R, which is strictly linearly compact as noted above. The topology of $(A/R_{\mu+1})/(R/R_{\mu+1})$ is therefore minimal by 28.13 and hence coincides with the weaker quotient topology on $(A/R_{\mu+1})/(R/R_{\mu+1})$ induced by the Leptin topology of $A/R_{\mu+1}$. Thus the Leptin topology of the A-module $A/R_{\mu+1}$ is strictly linearly compact.

Finally, assume that λ has no immediate predecessor and that the Leptin topologies of the A-modules A/R_μ are strictly linearly compact for all $\mu < \lambda$. The function Δ from the A-module A to the A-module $\prod_{\mu<\lambda}(A/R_\mu)$, defined by

$$\Delta(x) = (x + R_\mu)_{\mu<\lambda}$$

for all $x \in A$, is a continuous homomorphism, where each A/R_μ is given its Leptin topology. By 28.3 and 28.6, $\Delta(A)$ is closed and hence, by 28.16 and 28.17, strictly linearly compact. The kernel of Δ is $\bigcap_{\mu<\lambda} R_\mu$, which by definition is R_λ, and hence Δ induces a continuous isomorphism from the A-module A/R_λ to the strictly linearly compact A-module $\Delta(A)$. Therefore there is a strictly linearly compact A-module topology on A/R_λ weaker than its quotient topology. By 28.13 and 33.7, that topology is the Leptin topology of A/R_λ. •

33.24 Theorem. *If A is a linearly compact ring, the Leptin topology of A is strictly linearly compact if and only if the radical R of A is transfinitely nilpotent.*

The assertion follows from 33.21 and 33.23.

33.25 Theorem. *If A is a linearly compact commutative ring with radical R, then the Leptin topology of A is strictly linearly compact if and only if $\bigcap_{n=1}^\infty \overline{R^n} = (0)$.*

Proof. The condition is necessary by 33.22 and sufficient by 33.24. •

For an example where the equivalent conditions of Theorem 33.25 fail to hold, let A be the valuation ring of a proper, nondiscrete, real valuation v satisfying the properties of 28.9 (a proof that such valuations exist is given in *Topological Fields*, Theorem 31.24). The maximal ideal M of A is its radical as A is local ring, and $M^2 = M$ since, if $v(x) > 0$, there exists $y \in A$ such that $v(x) > v(y) > 0$ as v is not discrete, so $x = (xy^{-1})y \in M^2$. Therefore $M_* = M$, so the Leptin topology of A is not strictly linearly compact. Since the ideals of A are totally ordered by inclusion, the valuation topology of A is the weakest Hausdorff ideal topology on A and hence is the Leptin topology of A.

Exercises

33.1 (Prüfer [1923a]) Let G be an abelian group in which the zero subgroup is sheltered. (a) No subgroup of G is the direct sum of two proper subgroups. (b) G is p-primary for some prime p. (c) If $a, b \in G$, either $\mathbb{Z}.a \subseteq \mathbb{Z}.b$ or $\mathbb{Z}.b \subseteq \mathbb{Z}.a$. [Apply to the subgroup generated by a and b the theorem that a finitely generated abelian p-primary group is the direct sum of cyclic p-primary subgroups.] (d) The subgroups of G are totally ordered by inclusion. (e) G is either a cyclic p-primary group or a basic divisible p-primary group. [If G is not a noetherian $\mathbb{Z}$-module, show that it contains a basic divisible p-primary subgroup, and apply 30.2.]

33.2 (Prüfer [1923a,b]) Let G be an abelian group. (a) A proper subgroup H of G is sheltered if and only if G/H is either a cyclic p-primary group or a basic divisible p-primary group for some prime p. [Use Exercise 33.1.] (b) If N is a proper subgroup of G, then the $\mathbb{Z}$-module G/N is artinian if and only if N is the intersection of finitely many sheltered subgroups. [Use (a) and 33.2.] (c) The completion of G for the Leptin topology associated to the discrete topology on G is a strictly linearly compact $\mathbb{Z}$-module.

33.3 Let E be a nonzero A-module. The only Hausdorff linear topology on E is the discrete topology if and only if there exist sheltered submodules $U_1, \ldots, U_n$ of E such that $U_1 \cap \ldots \cap U_n = (0)$. [Necessity: Apply 33.2 to the zero submodule.]

33.4 Let E be a nonzero A-module. Then E is artinian if and only if every proper submodule of E is the intersection of finitely many sheltered modules.

33.5 Let A be a commutative ring with identity. A proper ideal U of A is *sheltered* and S is its *shelter* if U is a sheltered submodule of the A-module A with shelter S. The only Hausdorff ideal topology on A is the discrete topology if and only if there exist sheltered ideals $U_1, \ldots, U_n$ of A such that $U_1 \cap \ldots \cap U_n = (0)$. [Use Exercise 33.3.]

33.6 Let A be a commutative ring with identity. For any ideal J of A, the *annihilator* of J, denoted by $\mathrm{Ann}(J)$ (or $\mathrm{Ann}_A(J)$) is the ideal $\{x \in A : Jx = (0)\}$, and for any ideals I and J of A, we denote by $(I : J)$ the ideal $\{x \in A : xJ \subseteq I\}$. Assume that the zero ideal of A is sheltered, let S be the shelter of (0), and let $M = \mathrm{Ann}(S)$. (a) M is a maximal ideal. [Show that A/M is a field by observing that if $a \notin M$, then $Aas = S$ for some $s \in S$. (b) M is the set of all zero-divisors in A. (c) $S = \mathrm{Ann}(M)$ and is a principal ideal. [Regard $\mathrm{Ann}(M)$ as a vector space over A/M, and observe it has a smallest nonzero subspace.]

33.7 Let A be a noetherian commutative ring with identity in which the zero ideal is sheltered, and let M be the annihilator of the shelter of (0). (a) $\bigcap_{n=1}^{\infty} M^n = (0)$. [Use 20.13.] (b) M is a nilpotent ideal. [Use Exercise

33.5.] (c) M is the only maximal ideal of A. [Use 26.14.] (d) A is artinian. [Use 27.17.]

33.8 Let A be a commutative ring with identity having sheltered ideals $U_1, \ldots, U_n$ such that

$$\bigcap_{i=1}^{n} U_i = (0), \qquad U_i \not\subseteq \bigcap_{j \neq i} U_j$$

for each $i \in [1, n]$. For each $i \in [1, n]$, let S_i be the shelter of U_i, let $M_i = (U_i : S_i)$, and let $K(i) = \{j \in [1, n] : M_j = M_i\}$. (a) Each M_i is a maximal ideal of A. [Apply Exercise 33.6 to A/U_i.] (b) For each $i \in [1, n]$,

$$\text{Ann}(M_i) = (\bigcap_{j \in K(i)} S_j) \cap (\bigcap_{j \notin K(i)} U_j).$$

[For inclusion, observe that $M_i + M_j = A$ if $j \notin K(i)$.] (c) For each $i \in [1, n]$, $\text{Ann}(M_i) \neq (0)$. [Use (b) and observe that $U_i + \bigcap_{j \neq i} U_j \supseteq S_i$.] (d) For each $i \in [1, n]$, $\text{Ann}(M_i)$ is finitely generated. [Use (b) to establish a monomorphism from $\text{Ann}(M_i)$ to $\prod_{j \in K(i)}(S_j/U_j)$, and apply Exercise 33.6]. (e) $\bigcup_{i=1}^{n} M_i$ is the set of all zero-divisors of A. (f) $M_1, \ldots, M_n$ are the only maximal ideals having a nonzero annihilator. [Use (e) and Exercise 24.4.] (g) If $b \neq 0$, then $Ab \cap S_i \neq (0)$ for each $i \in [1, n]$. (h) If $a \neq 0$, then $Aa \cap (\bigcap_{i=1}^{n} S_i) \neq (0)$. [Use (g) repeatedly.] (i) If $a \in \bigcap_{i=1}^{n} S_i$, then $Aa \cap \text{Ann}(M_j) \neq (0)$ for some $j \in [1, n]$. [Proceed by contradiction, to arrive at a nonzero element of $\bigcap_{i=1}^{n} U_i$.] (j) If $a \neq 0$, $Aa \cap \text{Ann}(M_j) \neq (0)$ for some $j \in [1, n]$. [Use (b), (h), and (i).]

33.9 Let A be a commutative ring with identity satisfying the following properties:

1° A has only finitely many maximal ideals, $M_1, \ldots, M_n$ whose annihilators are nonzero.

2° The annihilator of each maximal ideal of A is a finitely generated ideal.

3° Each nonzero principal ideal of A has a nonzero intersection with the annhilator of some maximal ideal.

Let $\mathcal{T}$ be a nondiscrete ideal topology on A, and let $\mathcal{V}$ be a fundamental system of neighborhoods of zero for $\mathcal{T}$ consisting of open ideals. (a) If $\text{Ann}(M_i)$ is considered a vector space over the field A/M_i, then there exists $V_0 \in \mathcal{V}$ such that if $d_i = \dim(V_0 \cap \text{Ann}(M_i))$ for each $i \in [1, n]$, then $d_i \leq \dim(V \cap \text{Ann}(M_i))$ for all $V \in \mathcal{V}$. (b) For some $i \in [1, n]$, $d_i > 0$. (c) $\mathcal{T}$ is not Hausdorff.

33.10 (Hochster [1968])A commutative ring with identity admits no nondiscrete, Hausdorff ideal topology if and only if 1°−3° of Exercise 33.9 hold. [Use Exercises 33.5, 33.8, and 33.9.]

33.11 (Hochster [1968]) If A is a commutative noetherian ring with identity, then A admits no nondiscrete, Hausdorff ideal topology if and only if A is artinian. [Necessity: With the terminology of Exercise 33.8, observe that every maximal ideal contains U_i for some $i \in [1,n]$ [use 24.10] and hence is M_i [use Exercise 33.7(d)]. Finally, apply 20.13 and 27.16.]

33.14 (Kurke [1967], Warner [1971], Ánh [1981b]) Let T be a linearly compact topology on an A-module E, where A is given the discrete topology. (a) The set $\mathcal{U}$ of all closed submodules U of E such that E/U is linearly compact for the discrete topology is a filter base. (b) $\mathcal{U}$ is a fundamental system of neighborhoods of zero for a linearly compact topology T^* on E that is stronger than T. [Use 7.21 and 28.15.] (c) T^* is the strongest of all the linearly compact topologies on E that are stronger than T.

33.15 If T and S are linearly compact topologies on an A-module E, then T and S have the same closed submodules if and only if $T_* \subseteq S \subseteq T^*$.

33.16 A linearly compact topology T on an A-module E is maximal if $T = T^*$ (Exercise 33.14). If T is a maximal linearly compact topology on E and if M is a closed submodule of E, then the induced topology on E/M is maximal.

33.17 If T is the topology of the valuation ring A of a complete, discrete valuation of a field, then T^* is the discrete topology. [Use the remark after 28.9 and 18.2.]

33.18 Let $\mathbb{Z}_p$ be the compact ring of p-adic integers, for each $k \geq 1$ let E_k be the discrete $\mathbb{Z}_p$-module $\mathbb{Z}_p/p^k\mathbb{Z}_p$, and let $E = \prod_{k=1}^{\infty} E_k$. Let M be the range of the monomorphism Δ, defined by

$$\Delta(x) = (x + p^k \mathbb{Z}_p)_{k \geq 1}$$

from $\mathbb{Z}_p$ to E; thus M is a compact submodule of E. Then $(T^*)_M$, the topology induced on M by T^*, is not discrete. [Observe by using Exercise 33.15 that if V is a submodule open for T^*, then

$$V + \bigoplus_{k=1}^{\infty} E_k = E,$$

and conclude that for a suitable $m \geq 1$, $p^m \cdot \Delta(1) \in M \cap V$. (b) Let T_M be the topology induced on M by T. Then $(T_M)^*$ is the discrete topology. [Use Exercise 33.17.] (c) Conclude that $(T^*)_M$ is linearly compact but not maximal.

33.19 Let T be a linearly compact topology on an A-module E. The following statements are equivalent:

1° T is maximal.

2° For every closed submodule M of E, if E/M is linearly compact for the discrete topology, then M is open.

3° If u is an open epimorphism from E to a linearly compact A-module F and if the kernel of u is closed, then u is a topological epimorphism.

4° Every continuous epimorphism from a linearly compact A module D to E is a topological epimorphism.

5° If u is a homomorphism from E to a linearly compact A-module F whose graph is closed, then u is continuous.

33.20 If E and F are linearly compact A-modules with topologies T and S respectively, and if u is a continuous homomorphism from E to F, then u is also continuous when E and F are furnished with topologies T^* and S^* respectively. [Use 5° of Exercise 33.19.]

33.21 If E is a linearly compact module whose topology is maximal and if E is the direct sum of closed submodules M and N, then E is the topological direct sum of M and N, and the topologies induced on M and N are maximal.

33.22 If T is a linearly compact ring topology on a ring A, then T^* is also a ring topology.

33.23 If $(A_\lambda)_{\lambda \in L}$ is a family of linearly compact rings with identity whose topologies are maximal, then the cartesian product topology on $\prod_{\lambda \in L} A_\lambda$ is maximal. [Use 24.12.]

33.24 (Kurke [1967]) Let A be a commutative ring with identity, and let E be a linearly compact, unitary A-module. Then E is strictly linearly compact if and only if for each proper open submodule U of E and each $x \in E$ there is a sequence $(M_i)_{1 \leq i \leq n}$ of maximal ideals of A such that $M_1 \ldots M_n x \subseteq U$. [Necessity: Consider A/J where J is the kernel of $a \to ax + U$ from A to E/U. Sufficiency: Establish the criterion of 33.18.]

34 Lifting Idempotents

If A is a complete, equicharacteristic, local ring with maximal ideal (or radical) M, then by 21.8 and 21.14, A contains an isomorphic copy of its residue field A/M, that is, the entire field A/M may be "lifted" to A. Here we shall determine conditions on a topological ring A with radical R under which idempotents of A/R and, more generally, orthogonal families of idempotents may be lifted to A. From these results we may derive some important structure theorems.

We first note that if e is an idempotent of a topological ring A, the epimorphism f from the additive group A to the additive group eAe, defined by $f(x) = exe$, is a topological epimorphism, since if U is a neighborhood of zero in A, $U \cap eAe \subseteq eUe$. Thus if $\mathcal{U}$ is a fundamental system of neighborhoods of zero in A, $e\mathcal{U}e$ is a fundamental system of neighborhoods of zero in eAe.

Here, if A is a ring with radical R and if $x \in A$, we shall often denote the element $x + R$ of A/R by $\bar{x}$.

34.1 Theorem. *Let A be a linearly compact ring with radical R, and let ϕ be the canonical epimorphism from A to A/R. If ϵ is a nonzero idempotent of A/R and if L is a closed left ideal of A containing R such that $\epsilon \in \phi(L)$, there is an idempotent $e \in L$ such that $\bar{e} = \epsilon$, and for any such e, the restriction ϕ_e of ϕ to eAe is a topological epimorphism from eAe to $\epsilon(A/R)\epsilon$ whose kernel is the radical eRe of eAe.*

Proof. Let $\mathcal{E}$ be the set of all ordered pairs (f, J) such that $f \in A$, J is a closed left ideal of A contained in R, $f + J \subseteq L$, $Jf \subseteq J$, $\bar{f} = \epsilon$, and $f^2 - f \in J$. If f is an element of L such that $\bar{f} = \epsilon$, then $(f, R) \in \mathcal{E}$ by 29.12, so $\mathcal{E} \neq \emptyset$. We define an ordering on $\mathcal{E}$ by declaring $(e, I) \leq (f, J)$ if and only if $e + I \supseteq f + J$, or equivalently, if and only if $I \supseteq J$ and $e + I = f + I$. To show that the ordered set $(\mathcal{E}, \leq)$ is inductive, let $\mathcal{C}$ be a totally ordered subset of $\mathcal{E}$, and let $\mathcal{J} = \{I : (f, I) \in \mathcal{C} \text{ for some } f \in A\}$. As A is linearly compact, $\bigcap \{f + I : (f, I) \in \mathcal{C}\} \neq \emptyset$ and hence is a coset $f_0 + I_0$ of I_0, where $I_0 = \bigcap_{I \in \mathcal{J}} I$. For any $(f, I) \in \mathcal{C}$, $I_0 f_0 \subseteq I \cdot (f + I) \subseteq If + I \subseteq I$, and

$$f_0^2 \in (f + I) \cdot (f + I) \subseteq f^2 + fI + If + I^2 \subseteq f + I = f_0 + I.$$

Therefore $I_0 f_0 \subseteq I_0$ and $f_0^2 - f_0 \in I_0$. Thus $(f_0, I_0) \in \mathcal{E}$, and clearly $(f_0, I_0) = \sup \mathcal{C}$.

By Zorn's Lemma, therefore, $\mathcal{E}$ has a maximal element (e, I), and we need only show that $e^2 = e$. Let $a = e^2 - e$, and let $f = e - 2ea + a$. As a and e commute,

$$\begin{aligned} f^2 - f &= e^2 + 4e^2 a^2 + a^2 - 4e^2 a + 2ea - 4ea^2 - e + 2ea - a \\ &= -4(e^2 - e)a + 4(e^2 - e)a^2 + a^2 + e^2 - e - a \\ &= -4a^2 + 4a^3 + a^2 = 4a^3 - 3a^2. \end{aligned}$$

Let $J = \mathbb{Z}.a^2 + Aa^2$, the left ideal generated by a^2. Since $e^2 - e = a$, $\bar{J} \subseteq I$, and $f^2 - f \in J \subseteq \bar{J}$. Also $Jf = fJ \subseteq J$, so $\bar{J}f \subseteq \bar{J}$. Clearly $\bar{f} = \bar{e} = \epsilon$. Thus $(f, \bar{J}) \in \mathcal{E}$. As $a \in I$, $f + I = e + I$. Hence $(f, \bar{J}) \geq (e, I)$, so $(f, \bar{J}) = (e, I)$. Thus $a = e^2 - e \in I = \bar{J}$. Let U be an open left ideal. By 3.3, $a \in J + U$, so there exist an integer n and an element x of A such that $a - (n.a^2 + xa^2) \in U$. As $a \in R$, $n.a + xa \in R$ and hence $n.a + xa$ has an adverse b by 26.9. Therefore

$$\begin{aligned} 0 = 0 \cdot a &= [b \circ (n.a + xa)]a = [b + n.a + xa - b(n.a + xa)]a \\ &= b[a - n.a^2 - xa^2] + n.a^2 + xa^2. \end{aligned}$$

Thus
$$n.a^2 + xa^2 = -b[a - (n.a^2 + xa^2)] \in U$$
and consequently
$$a = (a - (n.a^2 + xa^2)) + (n.a^2 + xa^2) \in U.$$
Therefore a belongs to every open left ideal; hence $a = 0$, so $e^2 = e$.

If f is the topological epimorphism from the additive group A to additive group eAe defined by $f(x) = exe$ and if $\bar{f}$ is the corresponding function from A/R to $\epsilon(A/R)\epsilon$, then $\bar{f} \circ \phi = \phi_e \circ f$, so ϕ_e is a topological epimorphism by 5.3. The kernel $R \cap eAe$ of ϕ_e is eRe, the radical of eAe by 26.17. •

34.2 Corollary. *Let A be a commutative linearly compact ring with radical R. If ϵ is a nonzero idempotent of A/R, there is a unique idempotent $e \in A$ such that $\bar{e} = \epsilon$.*

Proof. If e and f are nonzero idempotents of A such that $\bar{e} = \bar{f}$, then $e - ef$ and $f - ef$ are idempotents of A belonging to R as $\bar{e} - \overline{ef} = \bar{0} = \bar{f} - \overline{ef}$, so $e = ef = f$ since no nonzero idempotent is advertible (page 83). •

An *orthogonal family* of idempotents in a ring A is a family $(e_\lambda)_{\lambda \in L}$ of nonzero idempotents such that $e_\lambda e_\mu = 0$ whenever λ and μ are distinct elements of L.

34.3 Theorem. *Let A be a linearly compact ring with identity whose associated Leptin topology is strictly linearly compact, and let R be the radical of A. If $(e_\lambda)_{\lambda \in L}$ is a family of idempotents of E such that $(\bar{e}_\lambda)_{\lambda \in L}$ is a summable orthogonal family of idempotents of A/R whose sum is $\bar{1}$, then the function ϕ from the A-module A to the A-module $\prod_{\lambda \in L} Ae_\lambda$, defined by*
$$\phi(x) = (xe_\lambda)_{\lambda \in L}$$
is a continuous isomorphism. If, furthermore, A is strictly linearly compact, then ϕ is a topological isomorphism.

Proof. Clearly ϕ is a homomorphism from the A-module A to the A-module $\prod_{\lambda \in L} Ae_\lambda$. To show that ϕ is a monomorphism, it therefore suffices to show that if $x \in A$ and if $xe_\lambda = 0$ for all $\lambda \in L$, then $x = 0$. By 29.12, R is closed. Let γ be the transfinite index of R. To show that $x = 0$, it suffices by 33.21 to show that $x \in R_\mu$ for all ordinals $\mu \leq \gamma$. First,
$$\bar{x} = \bar{x}(\sum_{\lambda \in L} \bar{e}_\lambda) = \sum_{\lambda \in L} \bar{x}\bar{e}_\lambda = \bar{0},$$
by 10.16, so $x \in R = R_1$. Assume that $x \in R_\nu$ for all ordinals $\nu < \mu$, and assume first that μ has an immediate predecessor ν. Then $R_\mu = \overline{R_\nu R}$,

so in particular $R_\nu R \subseteq R_\mu$, and thus we may regard R_ν/R_μ as a unitary right topological module over A/R where scalar multiplication is defined by $(z + R_\mu).(a + R) = za + R_\mu$ for all $z \in R_\nu$ and all $a \in A$. Then

$$x + R_\mu = (x + R_\mu).(\sum_{\lambda \in L} \overline{e}_\lambda) = x \sum_{\lambda \in L} e_\lambda + R_\mu = \sum_{\lambda \in L} xe_\lambda + R_\mu = R_\mu.$$

Thus $x \in R_\mu$. If μ has no immediate predecessor, then $x \in \bigcap_{\nu < \mu} R_\nu = R_\mu$. Therefore $x \in R_* = \{0\}$.

To show that ϕ is surjective, we first show that by induction on n that for any nonempty finite subset F of L and any $\mu \in F$, there exists $z \in A$ such that $ze_\mu = e_\mu$ and $ze_\lambda = 0$ for all $\lambda \in F \setminus \{\mu\}$. If F has one element e_μ, we may let $z_\mu = e_\mu$. Assume the assertion is true for all subsets of L having $k - 1$ elements. Let F be a subset of L having k elements, and let $\lambda_1 \in F$. Let $F - \{\lambda_1\} = \{\lambda_2, \ldots, \lambda_k\}$. By our inductive hypothesis, for each $i \in [1, k-1]$ there exists $z_{\lambda_i} \in A$ such that $z_{\lambda_i} e_{\lambda_i} = e_{\lambda_i}$ and $z_{\lambda_i} e_{\lambda_j} = 0$ for all $j \in [1, k-1] \setminus \{i\}$. Then for each $i \in [1, k-1]$, $e_{\lambda_k} e_{\lambda_i} \in R$ as $\overline{e}_{\lambda_k} \overline{e}_{\lambda_i} = \overline{0}$. Consequently, $1 - \sum_{i=1}^{k-1} e_{\lambda_k} e_{\lambda_i} z_{\lambda_i}$ is invertible by 26.11. We define z by

$$z = z_{\lambda_1} - z_{\lambda_1} e_{\lambda_k} (1 - \sum_{i=1}^{k-1} e_{\lambda_k} e_{\lambda_i} z_{\lambda_i})^{-1} (e_{\lambda_k} - \sum_{i=1}^{k-1} e_{\lambda_k} e_{\lambda_i} z_{\lambda_i}).$$

If $j \in [1, k-1]$,

$$(e_{\lambda_k} - \sum_{i=1}^{k-1} e_{\lambda_k} e_{\lambda_i} z_{\lambda_i}) e_{\lambda_j} = e_{\lambda_k} e_{\lambda_j} - e_{\lambda_k} e_{\lambda_j} e_{\lambda_j} = 0,$$

so

$$ze_{\lambda_1} = z_{\lambda_1} e_{\lambda_1} = e_{\lambda_1}$$

and, if $j \in [2, k-1]$,

$$ze_{\lambda_j} = z_{\lambda_1} e_{\lambda_j} = 0.$$

Finally,

$$ze_{\lambda_k} = z_{\lambda_1} e_{\lambda_k} - z_{\lambda_1} e_{\lambda_k} (1 - \sum_{i=1}^{k-1} e_{\lambda_k} e_{\lambda_i} z_{\lambda_i})^{-1} (e_{\lambda_k} e_{\lambda_k} - \sum_{i=1}^{k-1} e_{\lambda_k} e_{\lambda_i} z_{\lambda_i} e_{\lambda_k})$$

$$= z_{\lambda_1} e_{\lambda_k} - z_{\lambda_1} e_{\lambda_k} (1 - \sum_{i=1}^{k-1} e_{\lambda_k} e_{\lambda_i} z_{\lambda_i})^{-1} (1 - \sum_{i=1}^{k-1} e_{\lambda_k} e_{\lambda_i} z_{\lambda_i}) e_{\lambda_k}$$

$$= z_{\lambda_1} e_{\lambda_k} - z_{\lambda_1} e_{\lambda_k} e_{\lambda_k} = 0.$$

Since ϕ is continuous, $\phi(A)$ is closed in $\prod_{\lambda \in L} Ae_\lambda$ by 28.3 and 28.6. To show that ϕ is surjective, therefore, we need only show that $\phi(A)$ is dense, and for that, it suffices to show that for any finite sequence $\lambda_1, \ldots, \lambda_k$ of distinct elements of L and for any elements $a_{\lambda_1}, \ldots, a_{\lambda_k}$ of A, there exists $z \in A$ such that $ze_{\lambda_j} = a_{\lambda_j} e_{\lambda_j}$ for all $j \in [1, k]$. We have just proved that for each $i \in [1, k]$ there exists $z_{\lambda_i} \in A$ such that $z_{\lambda_i} e_{\lambda_i} = e_{\lambda_i}$ and $z_{\lambda_i} e_{\lambda_j} = 0$ for all $j \in [1, k]$ such that $j \neq i$. Let

$$z = \sum_{i=1}^{k} a_{\lambda_i} z_{\lambda_i}.$$

Then for each $j \in [1, k]$,

$$ze_{\lambda_j} = \sum_{i=1}^{k} a_{\lambda_i} z_{\lambda_i} e_{\lambda_j} = a_{\lambda_j} e_{\lambda_j}.$$

If A is strictly linear compact, then ϕ is a topological isomorphism by 28.10. •

34.4 Theorem. *Let A be a strictly linearly compact ring with identity 1, and let R be the radical of A. If $(\epsilon_\lambda)_{\lambda \in L}$ is a summable orthogonal family of idempotents of A/R [whose sum is $\bar{1}$], there is a summable orthogonal family of idempotents $(e_\lambda)_{\lambda \in L}$ in A [whose sum is 1] such that $\bar{e}_\lambda = \epsilon_\lambda$ for all $\lambda \in L$.*

Proof. We first assume that $\sum_{\lambda \in L} \epsilon_\lambda = \bar{1}$. By 34.1 and the Axiom of Choice, there is a family $(f_\lambda)_{\lambda \in L}$ of idempotents in A such that $\bar{f}_\lambda = \epsilon_\lambda$ for all $\lambda \in L$. By 34.3, the function ϕ from the topological A-module A to the topological A-module $\prod_{\lambda \in L} Af_\lambda$, defined by $\phi(x) = (xf_\lambda)_{\lambda \in L}$ for all $x \in A$, is a topological isomorphism. For each $\mu \in L$, let in_μ be the canonical injection from Af_μ into $\prod_{\lambda \in L} Af_\lambda$, and let $e_\mu = \phi^{-1}(in_\mu(f_\mu))$.

For each $\mu \in L$,

$$\phi(e_\mu^2) = e_\mu \phi(e_\mu) = e_\mu.(e_\mu f_\lambda)_{\lambda \in L} = (e_\mu f_\lambda)_{\lambda \in L} = \phi(e_\mu),$$

so $e_\mu^2 = e_\mu$. If $\nu \in L$,

$$(e_\nu f_\lambda)_{\lambda \in L} = \phi(e_\nu) = in_\nu(f_\nu),$$

so

$$e_\nu f_\nu = f_\nu \text{ and } e_\nu f_\lambda = 0$$

for all $\lambda \neq \nu$. Consequently, if $\nu \neq \lambda$,

$$\phi(e_\nu e_\lambda) = e_\nu \phi(e_\lambda) = e_\nu . in_\lambda(f_\lambda) = in_\lambda(e_\nu f_\lambda) = 0,$$

so $e_\nu e_\lambda = 0$. Also,

$$\bar{e}_\nu = \bar{e}_\nu(\sum_{\lambda \in L} \bar{f}_\lambda) = \sum_{\lambda \in L} \bar{e}_\nu \bar{f}_\lambda = \bar{e}_\nu \bar{f}_\nu = \bar{f}_\nu = \epsilon_\nu.$$

Since $(in_\lambda(f_\lambda))_{\lambda \in L}$ is clearly summable in $\prod_{\lambda \in L} Af_\lambda$, $(e_\lambda)_{\lambda \in L}$ is summable in A. Let $e = \sum_{\lambda \in L} e_\lambda$. Then $\bar{e} = \sum_{\lambda \in L} \epsilon_\lambda = \bar{1}$. Let $\alpha \in L$, let $e'_\alpha = 1 - e + e_\alpha$, and let $e'_\lambda = e_\lambda$ for all $\lambda \in L \setminus \{\alpha\}$. Then $\bar{e}'_\alpha = \bar{1} - \bar{e} + \bar{e}_\alpha = \epsilon_\alpha$, and

$$\sum_{\lambda \in L} e'_\lambda = 1.$$

In general, if $(\epsilon_\lambda)_{\lambda \in L}$ is a summable orthogonal family of idempotents in A/R with sum $\epsilon \neq \bar{1}$, the family $(\epsilon_\lambda)_{\lambda \in L \cup \{\omega\}}$, where $e_\omega = 1 - e$, is a summable orthogonal family of idempotents whose sum is $\bar{1}$. Upon applying the preceding paragraphs to this orthogonal family of idempotents, we obtain the desired conclusion by 10.7. •

34.5 Theorem. *Let A be a strictly linearly compact ring, and let R be the radical of A. If $(\epsilon_\lambda)_{\lambda \in L}$ is a summable orthogonal family of idempotents of A/R, there is a summable orthogonal family of idempotents $(e_\lambda)_{\lambda \in L}$ in A such that $\bar{e}_\lambda = \epsilon_\lambda$ for all $\lambda \in L$.*

Proof. By 29.12, 29.4, and 26.16, A/R is a linearly compact, semisimple ring and hence by 29.8 has an identity ϵ. By 34.1 there is an idempotent e in A such that $\bar{e} = \epsilon$. Let ϕ_e be the restriction to eAe of the canonical epimorphism ϕ from A to A/R, and let ϕ_{eAe} be the canonical epimorphism from eAe to eAe/eRe. By 34.1, there is a topological isomorphism χ from eAe/eRe to A/R such that $\chi \circ \phi_{eAe} = \phi_e$.

Let $(\epsilon_\lambda)_{\lambda \in L}$ be a summable orthogonal family of idempotents of A/R. Then $(\chi^{-1}(\epsilon_\lambda))_{\lambda \in L}$ is a summable orthogonal family of idempotents of eAe/eRe. By 26.17, eRe is the radical of eAe, a strictly linearly compact ring by 29.15. Therefore by 34.4, there is a summable orthogonal family $(e_\lambda)_{\lambda \in L}$ of idempotents in eAe such that $\phi_{eRe}(e_\lambda) = \chi^{-1}(\epsilon_\lambda)$ for all $\lambda \in L$. Finally, for each $\lambda \in L$,

$$\phi(e_\lambda) = \phi_e(e_\lambda) = \chi(\phi_{eRe}(e_\lambda)) = \chi(\chi^{-1}(\epsilon_\lambda)) = \epsilon_\lambda. \bullet$$

34.6 Theorem. *Let A be a strictly linearly compact commutative ring. (1) Either A is a radical ring, or there is a nonzero idempotent $e \in A$ such that A is the topological direct sum of the strictly linearly compact ring with identity Ae and the strictly linearly compact radical ring J, where $J = \{y - ye : y \in A\}$. (2) If A has an identity, A is topologically isomorphic to the cartesian product of a family of strictly linearly compact local rings.*

Proof. Let R be the radical of A. (1) As in the proof of 34.5, there is an idempotent e in A such that $\bar{e}$ is the identity element of A/R. Then Ae is a strictly linearly compact ring by 29.3 as $x \to xe$ is a continuous epimorphism from A to Ae, and e is the identity of Ae. By 32.12, A is the topological direct sum of Ae and the ideal J, where $J = \{y - ye : y \in A\}$. By 29.4, J is strictly linearly compact as it is topologically isomorphic to A/Ae. If ϕ is the canonical epimorphism from A to A/R, then J is contained in the kernel of ϕ, that is, $J \subseteq R$, so J is a radical ring by 26.18.

(2) By 29.10, A/R has a summable orthogonal family $(\epsilon_\lambda)_{\lambda \in L}$ of idempotents such that $(A/R)\epsilon_\lambda$ is a field for each $\lambda \in L$ and $\sum_{\lambda \in L} \epsilon_\lambda = \bar{1}$. By 34.4, the unique family $(e_\lambda)_{\lambda \in L}$ of idempotents of A such that $\bar{e}_\lambda = \epsilon_\lambda$ for all $\lambda \in L$ is a summable orthogonal family of idempotents whose sum is 1. By 34.3, the function ϕ from A to $\prod_{\lambda \in L} Ae_\lambda$, defined by $\phi(x) = (xe_\lambda)_{\lambda \in L}$ for all $x \in A$, is a topological isomorphism from the A-module A to the A-module $\prod_{\lambda \in L} Ae_\lambda$. Since A is commutative, ϕ is a ring isomorphism. For each $\lambda \in L$, Ae_λ/Re_λ is isomorphic to the field $(A/R)\epsilon_\lambda$ by 34.1, and therefore as Re_λ is the radical of Ae_λ by 26.17, Ae_λ is a local ring, which is strictly linearly compact by 29.15. •

34.7 Theorem. *If A is a semisimple linearly compact ring, A is bounded if and only if every orthogonal family of idempotents of A is summable.*

Proof. By 12.16, A is bounded (if and) only if its topology is an ideal topology. Necessity: By 29.9 we may assume that A is the cartesian product of a family $(A_\mu)_{\mu \in M}$ of discrete rings, each the ring of all linear operators on a finite-dimensional vector space E_μ. Let $(e_\lambda)_{\lambda \in L}$ be an orthogonal family of idempotents, and for each $\lambda \in L$, let $e_\lambda = (e_{\lambda,\mu})_{\mu \in M}$. For each $\mu \in M$, the nonzero members of $(e_{\lambda,\mu})_{\lambda \in L}$ clearly form an orthogonal family of idempotents in A_μ, so as E_μ is finite-dimensional and $\sum_{\lambda \in L} e_{\lambda,\mu}(E_\mu)$ is the direct sum of $(e_{\lambda,\mu}(E_\mu))_{\lambda \in L}$, $e_{\lambda,\mu} = 0$ for all but finitely many $\lambda \in L$. Thus $(e_{\lambda,\mu})_{\lambda \in L}$ is summable in A_μ for each $\mu \in M$, so $(e_\lambda)_{\lambda \in L}$ is summable in A by 10.10.

Sufficiency: By 29.7 and 29.9, we need only show that if E is an infinite-dimensional discrete vector space, there exists an orthogonal family of idempotents in the ring A of all linear operators on E that is not summable for the topology of pointwise convergence. Let V be a subspace of E having a denumerable basis $(b_i)_{i\geq 0}$, and let W be a supplement of V in E. For each $n \geq 1$, let e_n be the linear operator on E satisfying

$$e_n(b_0) = e_n(b_n) = b_n \qquad e_n(b_j) = 0 \text{ for all } j \neq 0, n,$$

and $e_n(x) = 0$ for all $x \in W$. Clearly $(e_n)_{n\geq 1}$ is an orthogonal family of idempotents, but $(e_n)_{n\geq 1}$ is not summable for the topology of pointwise convergence by 10.5, since no e_n belongs to the neighborhood $\{u \in A : u(b_0) = 0\}$ of zero. •

34.8 Theorem. *If A is a bounded, strictly linearly compact ring, every orthogonal family of idempotents in A is summable.*

Proof. Let R be the radical of A, and let $(e_\lambda)_{\lambda\in L}$ be an orthogonal family of idempotents of A. As A is complete by 28.5, it suffices by 10.5 and 12.16 to show that for any open (and hence closed) ideal V of A, $e_\lambda \in V$ for all but finitely many $\lambda \in L$. Now $(\bar{e}_\lambda)_{\lambda\in L}$ is an orthogonal family of idempotents in A/R ($\bar{e}_\lambda \neq 0$, since the radical of a ring contains no nonzero idempotent) and hence is summable by 34.7, so if ϕ is the canonical epimorphism from A to A/R, $\bar{e}_\lambda \in \phi(V)$ and hence $e_\lambda \in V + R$ for all but finitely many $\lambda \in L$. Therefore it suffices to show that if e is an idempotent in $V + R$, then $e \in V$. Assume that $e \in V + \overline{R^n}$, and let $e = v + r$ where $v \in V$, $r \in \overline{R^n}$. Then

$$e = e^2 = (v+r)^2 = v^2 + vr + rv + r^2 \in V + \overline{R^n}\,\overline{R^n} \subseteq V + \overline{R_n R} \subseteq V + \overline{R^{n+1}}.$$

Consequently,

$$e \in \bigcap_{n=1}^\infty (V + \overline{R^n}) = V + \bigcap_{n=1}^\infty \overline{R^n} = V$$

by 28.19 and 33.22. •

Let A be the ring of all linear operators on a K-module E. The natural way of defining a scalar multiplication on $K \times A$, when K is commutative, that makes A a K-module and, indeed, a K-algebra, is no longer available if K is not commutative, simply because if $u \in A$ and if λ is a scalar not in the center of K, the function $x \to \lambda u(x)$ need not be a linear operator. If, however, E has a basis B, then we may define a scalar multiplication, dependent upon B, that makes A a K-module by declaring, for each $u \in A$ and each scalar λ, λu to be the unique linear operator on E taking b into $\lambda u(b)$ for each $b \in B$. Then for any scalar α, $(\lambda u)(\alpha b) = \alpha(\lambda u)(b) = (\alpha\lambda)u(b)$. (In contrast, if α is invertible, then for the scalar multiplication determined by the basis αB, $(\lambda u)(\alpha b) = \lambda u(\alpha b) = (\lambda \alpha)u(b)$.) If B is finite, the linear operators $(e_{bc})_{(b,c)\in B\times B}$ corresponding to the elementary matrices determined by B (e_{bc} is the unique linear operator satisfying $e_{bc}(c) = b$ and

$e_{bc}(a) = 0$ for all $a \in B \setminus \{c\}$) form a basis of the K-module A, and $\sum_{b \in B} e_{bb} = I_E$, the identity linear operator on E. By use of this scalar multiplication, we may find copies of the ring opposite K in A, since for each $b \in B$, the function ϕ_b from K to A, defined by $\phi_b(\lambda) = e_{bb}(\lambda I_E)e_{bb}$, is an anti-isomorphism from K to the subring $e_{bb}Ae_{bb}$. Indeed, for any $\nu \in K$, e_{bb} and νI_E are easily seen to commute, so if $\lambda, \mu \in K$, $\phi(\lambda)\phi(\mu) = e_{bb}(\lambda I_E)(\mu I_E)e_{bb} = e_{bb}(\mu\lambda)I_E e_{bb} = \phi(\mu\lambda)$. If $\lambda \neq 0$, then $[e_{bb}(\lambda I_E)e_{bb}](b) = \lambda b \neq 0$, so ϕ_b is an anti-monomorphism; and finally, for any $u \in A$, if $u(b) = \sum_{c \in B} \lambda_{bc} c$, then $e_{bb}ue_{bb} = e_{bb}(\lambda_{bb}I_E)e_{bb}$, so ϕ_b is an anti-isomorphism.

34.9 Theorem. *Let A be a strictly linearly compact ring with identity, and let R be its radical. If A/R is isomorphic to the ring of all linear operators on an n-dimensional vector space over a division ring K, then A is topologically isomorphic to the topological ring $\operatorname{End}_L(E)$ of all linear operators on an n-dimensional, strictly linearly compact module E over a strictly linearly compact ring with identity L, both of which are subrings of A, where $\operatorname{End}_L(E)$ has the topology of pointwise convergence and, if S is the radical of L, L/S is isomorphic to K.*

Proof. Let $(\epsilon_{ij})_{(i,j) \in [1,n] \times [1,n]}$ be the basis of A/R corresponding to the elementary matrices determined by a given basis of the underlying vector space. By 34.4 there is an orthogonal sequence $(e_{ii})_{i \in [1,n]}$ of idempotents in A such that $\sum_{i=1}^n e_{ii} = 1$ and $\bar{e}_{ii} = \epsilon_{ii}$ for all $i \in [1,n]$. For each $j \in [2,n]$, let f_{j1} and $f_{1j} \in A$ be such that $\bar{f}_{1j} = \epsilon_{1j}$ and $\bar{f}_{j1} = \epsilon_{j1}$, and define e_{1j} by

$$e_{1j} = e_{11} f_{1j} e_{jj}.$$

Then $\bar{e}_{1j} = \epsilon_{1j}$, and

(1) $$e_{11} e_{1j} = e_{1j} = e_{1j} e_{jj}.$$

Define r_j by

$$r_j = f_{j1} e_{1j} - e_{jj}.$$

Then by (1),

(2) $$r_j e_{jj} = r_j,$$

and as $\bar{f}_{j1} \bar{e}_{1j} = \bar{e}_{jj}$, $r_j \in R$, and thus $1 + r_j$ is invertible by 26.9. By (2),

(3) $$f_{j1} e_{1j} = e_{jj} + r_j = (1 + r_j)e_{jj}.$$

Define e_{j1} by

$$e_{j1} = e_{jj}(1 + r_j)^{-1} f_{j1} e_{11}.$$

Then

(4) $$e_{j1}e_{11} = e_{j1} = e_{jj}e_{j1}$$

and $\bar{e}_{j1} = \bar{e}_{jj}\bar{1}\bar{f}_{j1}\bar{e}_{11} = \epsilon_{j1}$. Moreover, by (1) and (3),

(5) $$\begin{aligned}e_{j1}e_{1j} &= e_{jj}(1+r_j)^{-1}f_{j1}e_{11}e_{1j} = e_{jj}(1+r_j)^{-1}f_{j1}e_{1j}e_{jj}\\ &= e_{jj}(1+r_j)^{-1}(1+r_j)e_{jj}^2 = e_{jj}.\end{aligned}$$

By (1), (4), and (5),

$$\begin{aligned}(e_{11} - e_{1j}e_{j1})^2 &= e_{11}^2 - e_{11}e_{1j}e_{j1} - e_{1j}e_{j1}e_{11} + e_{1j}e_{j1}e_{1j}e_{j1}\\ &= e_{11} - e_{1j}e_{j1} - e_{1j}e_{j1} + e_{1j}e_{jj}e_{j1} = e_{11} - e_{1j}e_{j1}.\end{aligned}$$

Therefore, as

$$\overline{e_{11} - e_{1j}e_{1j}} = \bar{e}_{11} - \bar{e}_{1j}\bar{e}_{j1} = \epsilon_{11} - \epsilon_{1j}\epsilon_{j1} = \bar{0}$$

and as R contains no nonzero idempotents,

(6) $$e_{11} = e_{1j}e_{j1}.$$

For $i \in [2,n]$ and $j \in [1,n]$ we define e_{ij} by

$$e_{ij} = e_{i1}e_{1j}.$$

By (6) and (1),

$$e_{ij}e_{jk} = e_{i1}e_{1j}e_{j1}e_{1k} = e_{i1}e_{11}e_{1k} = e_{i1}e_{1k} = e_{ik},$$

and if $r \neq s$, $e_{ij}e_{rs} = e_{i1}e_{1j}e_{r1}e_{1s} = e_{i1}e_{1j}e_{jj}e_{rr}e_{r1}e_{1s} = 0$ by (1) and (4).

With scalar multiplication the restriction to $Ae_{11} \times e_{11}Ae_{11}$ of multiplication on $A \times A$, Ae_{11} is a topological right module over $e_{11}Ae_{11}$. Let $B = \{e_{11}, e_{21}, \ldots, e_{n1}\}$. Then B is a basis: Indeed, for any $x \in Ae_{11}$,

(7) $$x = xe_{11} = \sum_{j=1}^{n} e_{j1}[e_{11}e_{1j}xe_{11}],$$

and if $\sum_{j=1}^{n} e_{j1}[e_{11}a_j e_{11}] = 0$, then for each $i \in [1,n]$

$$\begin{aligned}0 = e_{1i}\sum_{j=1}^{n} e_{j1}[e_{11}ae_{11}] &= \sum_{j=1}^{n} e_{1i}e_{j1}[e_{11}a_j e_{11}]\\ &= e_{1i}e_{i1}[e_{11}a_i e_{11}] = e_{11}^2 a_i e_{11} = e_{11}a_i e_{11}.\end{aligned}$$

Let $E = Ae_{11}$ and $L = e_{11}Ae_{11}$. By 29.15 and 28.18, as $E = \sum_{j=1}^{n} e_{j1}L$, E is a strictly linearly compact module over the strictly linearly compact ring L. For each $a \in A$, let $\hat{a}$ be the endomorphism of E defined by $\hat{a}(x) = ax$, and let $\phi : a \to \hat{a}$. Then $\hat{a}$ is a linear operator on the right L-module E since for any $x, y \in A$, $\hat{a}(x.e_{11}ye_{11}) = axe_{11}ye_{11} = \hat{a}(x).e_{11}ye_{11}$. Clearly ϕ is a homomorphism from A to $\text{End}_L(E)$. If $\hat{a} = 0$, then for each $j \in [1, n]$, $0 = \hat{a}(e_{jj}) = ae_{jj}$, so $0 = \sum_{j=1}^{n} ae_{jj} = a$. Furthermore, ϕ is surjective, for if $i, j \in [1, n]$, $\hat{e}_{ji}(e_{i1}) = e_{j1}$ and, if $k \neq i$, $\hat{e}_{ji}(e_{k1}) = 0$, so $\hat{A}$ contains the linear operators corresponding to the elementary matrices determined by the basis B. Therefore ϕ is an isomorphism from A to $\text{End}_L(E)$.

We furnish $\text{End}_L(E)$ with the topology of pointwise convergence. Given $b \in E$ and a neighborhood V of zero in E, there exists an open left ideal W of zero in A such that $W \cap Ae_{11} \subseteq We_{11} \subseteq V$, and there exists an open left ideal U of zero in A such that for any $a \in U$, $e_{jj}abe_{11} \in W$ for all $j \in [1, n]$. Then for any $a \in U$, $ab \in Ae_{11} = E$, so by (7),

$$\hat{a}(b) = ab = \sum_{j=1}^{n} e_{j1}[e_{11}e_{1j}abe_{11}] = \sum_{j=1}^{n} e_{jj}abe_{11} \in W \cap Ae_{11} \subseteq V.$$

Therefore ϕ is a continuous isomorphism from A to $\hat{A}$, furnished with the topology of pointwise convergence. Thus by 29.3, ϕ is a topological isomorphism.

By 34.1, $e_{11}Ae_{11}/e_{11}Re_{11}$ is isomorphic to $\epsilon_{11}(A/R)\epsilon_{11}$ and hence to K. But by 26.17, $e_{11}Re_{11}$ is the radical S of $L = e_{11}Ae_{11}$. Therefore L/S is isomorphic to K. •

34.10 Corollary. *If A is a bounded, strictly linearly compact ring with identity whose radical is a primitive ideal, then A is topologically isomorphic to the ring of all linear operators, furnished with the topology of pointwise convergence, on a finite-dimensional, strictly linear compact module E over a strictly linearly compact ring L with identity whose radical is a regular maximal left ideal, where both E and L are subrings of A.*

Proof. Let R be the radical of A. By 29.9 and 25.11, A/R is isomorphic to the discrete ring of all linear operators on a finite-dimensional vector space over a division ring. The conclusion therefore follows from 34.9. •

Each basis B of a vector space E determines in a natural way an orthogonal family of idempotents in the ring A of all linear operators on E, namely, the family $(e_b)_{b \in B}$ where for each $b \in B$, e_b is the unique linear operator satisfying $e_b(b) = b$ and $e_b(c) = 0$ for all $c \in B \setminus \{b\}$. If E is given the discrete topology and A the topology of pointwise convergence, $(e_b)_{b \in B}$ is clearly a summable orthogonal family of idempotents whose sum is the identity linear operator 1_E. Consequently, if L is a closed left ideal of A, $L = Ae$ for some idempotent e. Indeed, let $C = \{b \in B : Ae_b \cap L \neq (0)\}$. If $b \in C$, then $L \supseteq Ae_b$ since Ae_b is a minimal left ideal, and in particular, $e_b \in L$. By 29.1, 28.5, and 10.7, $(e_b)_{b \in C}$ has a sum e, so as L is closed and contains the sum of each finite subfamily of $(e_b)_{b \in C}$, $e \in L$ and thus $Ae \subseteq L$. But for each $x \in L$,

$$x = x \sum_{b \in B} e_b = \sum_{b \in B} xe_b = \sum_{b \in C} xe_b = x \sum_{b \in C} e_b = xe,$$

so $L = Ae$. Analogously, if L is a closed right ideal of A, there is an idempotent e such that $L = eA$.

A nonzero semisimple linearly compact ring A also has a naturally associated summable orthogonal family of idempotents: By 29.7, there is a topological isomorphism ϕ from A to the cartesian product of linearly compact primitive rings $(A_\lambda)_{\lambda \in L}$, each the ring of all linear operators on a discrete vector space, furnished with the topology of pointwise convergence; if, for each $\lambda \in L$, $e_\lambda = \phi^{-1}(1_\lambda)$, where 1_λ is the identity linear operator of A_λ, then $(e_\lambda)_{\lambda \in L}$ is clearly a summable orthogonal family of idempotents in the center of A whose sum is the identity of A such that $Ae_\lambda = A_\lambda$ for all $\lambda \in L$ (moreover, it follows readily from Exercise 29.9(b) that the set $\{e_\lambda : \lambda \in L\}$ of idempotents constructed in this way is independent of particular topological isomorphism ϕ chosen). Consequently, if L is a closed left ideal of A, $L = Af$ for some idempotent f. Indeed, for any idempotent e in the center of A, $Le = L \cap Ae$ and hence Le is closed by 26.28. Let $M = \{\lambda \in L : Le_\lambda \neq 0\}$. Then for each $\lambda \in M$, Le_λ is a nonzero closed left ideal of Ae_λ and thus by the preceding there is a nonzero idempotent $f_\lambda \in Ae_\lambda$ such that $Le_\lambda = A_\lambda f_\lambda = Af_\lambda$ (as $e_\nu f_\lambda = 0$ for all $\nu \neq \lambda$). Arguing as before, we conclude that $(f_\lambda)_{\lambda \in M}$ is a summable orthogonal family of idempotents, and $L = Af$ where f is its sum. Analogously, if L is a closed right ideal of A, there is an idempotent $f \in A$ such that $L = fA$. Thus we have proved:

34.11 Theorem. *If L is a closed left [right] ideal of a semisimple linearly compact ring A, there is an idempotent $e \in A$ such that $L = Ae$ [$L = eA$].*

To illustrate the usefulness of our theorems concerning infinite orthogonal families of idempotents, we shall investigate bounded, linearly compact rings whose closed ideals are all strictly linearly compact rings.

34.12 Theorem. *Let R be a radical ring. The following assertions are equivalent:*

$1°$ *R is a strictly linearly compact ring for the discrete topology.*

2° R is a nilpotent artinian ring.
3° R is a strongly linearly compact ring for the discrete topology.

Proof. 1° implies 2° by (2) of 28.15 and 33.22. By 30.11, 3° implies 1°. To show that 2° implies 3°, we proceed by induction on the index of nilpotency of R. Assume that all nilpotent artinian rings of index $< n$ are strongly linearly compact for the discrete topology, where $n \geq 2$, and let R be an artinian ring satisfying $R^n = \{0\}$. Then R/R^{n-1} is an artinian ring by 27.4 whose index of nilpotency is $< n$ and hence is a strongly linearly compact ring and thus a strongly linearly compact R-module for the discrete topology, which is the topology induced on R/R^{n-1} by the discrete topology on R. Since R^{n-1} is a trivial (closed) submodule of the discrete, linearly compact R-module R, R^{n-1} is a strongly linearly compact R-module. Therefore R is a strongly linearly compact R-module for the discrete topology by 28.16. •

34.13 Theorem. *A bounded, strictly linearly compact radical ring R is strongly linearly compact. In particular, a commutative, strictly linearly compact radical ring is strongly linearly compact.*

Proof. The filter base $\mathcal{U}$ of open ideals of R is a fundamental system of neighborhoods of zero by (1) of 12.16. For each $U \in \mathcal{U}$, R/U is a discrete, strictly linearly compact, radical ring by 29.4 and 26.16, so R/U is a strongly linearly compact ring and hence a strongly linearly compact R-module by 34.12. As R is complete, therefore, R is a strongly linearly compact R-module and thus a strongly linearly compact ring by (1) of 28.16. •

34.14 Lemma. *Let A be a primitive linearly compact ring, R a strongly linearly compact A-module. If e is an idempotent of A such that Ae is a minimal left ideal, then either $e.R = \{0\}$ or A is finite.*

Proof. By 29.11 we may regard A as the ring of all linear operators on a discrete vector space E over a discrete division ring K, furnished with the topology of pointwise convergence. Assume that $e.R \neq \{0\}$. Then there exists $r \in R$ such that $e.r \neq 0$. Let f be the function from Ae to R defined by $f(x) = x.r$ for all $x \in Ae$. Then f is a continuous homomorphism from the A-module Ae to the A-module R; its kernel is a left ideal of A properly contained in Ae and hence is $\{0\}$, so f is a continuous monomorphism. By 29.1, A is strictly linearly compact, so Ae is a strictly linearly compact A-module by 26.28 and 28.16. Consequently, f is a topological isomorphism from Ae to $Ae.r$, a strongly linearly compact A-submodule of R by 28.3. As A is linearly topologized and Hausdorff, the induced topology on Ae is the discrete topology. Therefore $Ae.r$ is a discrete, strongly linearly compact A-module such that either every element of $Ae.r$ has infinite additive order or every element has order p for some prime p. Consequently, by 30.10, $Ae.r$ is finite, whence Ae is also. As e is a projection of E on a one-dimensional subspace of E, both the dimension of E and the cardinality of K are finite, so A is finite. •

34.15 Corollary. *If A is a strongly linearly compact semisimple ring, then A is compact.*

Proof. We may assume that $A \neq \{0\}$. By 29.7, we may regard A as the cartesian product of a family $(A_\lambda)_{\lambda \in L}$ of primitive linearly compact rings. Each A_λ is strongly linearly compact by 28.3 applied to the canonical projection from A to A_λ. As A_λ is isomorphic to the ring of all linear operators on a vector space, it contains an idempotent e_λ such that Ae_λ is a minimal left ideal of A_λ. We may regard A_λ as a topological module over itself. Since $e \in eA_\lambda$, A_λ is finite by 34.14. Thus A is compact by Tikhonov's theorem. •

34.16 Theorem. *If A is a linearly compact ring with identity whose radical R is a nonzero strongly linearly compact ring and if A/R is a primitive ring, then A/R is finite.*

Proof. By 29.12, 28.16, and 29.11, we may regard A/R as the ring of all linear operators on a discrete, nonzero vector space E over a discrete division ring K, furnished with the topology of pointwise convergence.

Case 1: $R^2 = \{0\}$. We may regard R as a unitary topological (A/R)-module under the well defined scalar multiplication $(a + R).b = ab$ for all $a \in A$, $b \in R$. As R is closed by 29.12, R is a strongly linearly compact A-module by 28.16 and hence is a strongly linearly compact (A/R)-module. By the discussion following 34.10, A/R contains a summable, orthogonal family $(e_b)_{b \in B}$ of idempotents in A/R whose sum is 1 such that for each $b \in B$, $(A/R)e_b$ is a minimal left ideal of A/R. Therefore as R is a unitary (A/R)-module, $e_b.R \neq \{0\}$ for some $b \in B$. By 34.14, A/R is finite.

Case 2: $R^2 \neq \{0\}$. If $\overline{R^2} = R$, then, with the notation of the paragraph preceding 33.21, $R_2 = R_1$, and by transfinite induction, $R_\gamma = R_1$ where γ is the transfinite index of R. But R is a strictly linearly compact ring by 30.11, so $R_\gamma = \{0\}$ by 33.21. Therefore $\overline{R^2}$ is properly contained in R, so $A/\overline{R^2}$ has the nonzero radical $R/\overline{R^2}$ by 26.16. Clearly $(R/\overline{R^2})^2 = \{0\}$, and $(A/\overline{R^2})/(R/\overline{R^2})$ is topologically isomorphic to A/R by 5.13. By 28.16, $R/\overline{R^2}$ is a strongly linearly compact R-module and hence a strongly linearly compact ring. Therefore the conclusion follows by Case 1. •

34.17 Theorem. *If A is a linearly compact ring whose radical R is a strongly linearly compact ring, and if the ideals of A contained in R and open for its induced topology form a fundamental system of neighborhoods* of zero for the topology of R, then A is strictly linearly compact and is the topological direct sum of subrings B and C, described as follows: B is topologically isomorphic to the cartesian product of a family $(B_\mu)_{\mu \in M}$ of topological rings where each B_μ is the ring of all linear operators on a discrete vector space over an infinite division ring K_μ, furnished with the topology of pointwise convergence; C is a strictly linearly compact ring containing R such that C/R is topologically isomorphic to the cartesian product of a family $(C_\nu)_{\nu \in N}$ of topological rings where each C_ν is the ring of all linear operators on a discrete vector space over a finite field K_ν, furnished with the topology of pointwise convergence; and A/R is the topological direct sum of a ring topologically isomorphic to B and C/R.

Proof. By 30.11, R is a strictly linearly compact ring and a fortiori a strictly linearly compact A-module. Moreover, A/R is a strictly linearly compact ring by 29.13 and hence is a strictly linearly compact A-module. Thus by (2) of 28.6, the A-module A, i.e., the ring A, is strictly linearly compact. We may assume that $A \neq R$, since otherwise the subrings $\{0\}$ and R satisfy the conclusions of the theorem.

By the discussion following 34.10, A/R has a summable orthogonal family $(\epsilon_\lambda)_{\lambda \in L}$ of idempotents in its center such that $\sum_{\lambda \in L} \epsilon_\lambda$ is the identity 1 of A/R and for each $\lambda \in L$, $\epsilon_\lambda(A/R)\epsilon_\lambda$ is topologically isomorphic to the ring of all linear operators on a discrete vector space over a division ring K_λ (unique to within isomorphism by 25.7), furnished with the topology of pointwise convergence. Let $M = \{\lambda \in L : K_\lambda \text{ is infinite}\}$, $N = L \setminus M$. By 34.7, $(\epsilon_\lambda)_{\lambda \in M}$ has a sum ϵ. By 34.1 there is an idempotent $e \in A$ such that $\bar{e} = \epsilon$. By 29.15, eAe is a strictly linearly compact ring. By 34.4 there is a summable orthogonal family $(e_\lambda)_{\lambda \in M}$ of idempotents in eAe such that $\bar{e}_\lambda = \epsilon_\lambda$ for all $\lambda \in M$ and $\sum_{\lambda \in M} e_\lambda = e$.

Let $\mu \in M$. As $x \to e_\mu x e_\mu$ is a continuous, $\mathbb{Z}$-linear function from R to $e_\mu R e_\mu$, $e_\mu R e_\mu$ is a linearly compact $\mathbb{Z}$-module by 28.3 and hence a strongly linearly compact ring by 29.15. By 34.1, $e_\mu A e_\mu / e_\mu R e_\mu$ is topologically isomorphic to $\epsilon_\mu(A/R)\epsilon_\mu$, and $e_\mu R e_\mu$ is the radical of $e_\mu A e_\mu$ by 26.17. Therefore by 34.16, $e_\mu R e_\mu = (0)$. Thus $e_\mu A e_\mu$ is topologically isomorphic to $\epsilon_\mu(A/R)\epsilon_\mu$ and hence to the ring of all linear operators on a discrete K_μ-vector space, furnished with the topology of pointwise convergence.

Since $e = \sum_{\lambda \in L} e_\lambda$, to show that $eR = \{0\}$, it suffices to show that for each $\mu \in M$, $e_\mu R = \{0\}$. By the discussion following 34.10, $e_\mu A e_\mu$ contains a summable orthogonal family $(e_b)_{b \in B}$ of idempotents whose sum is e_μ such that $(e_\mu A e_\mu)e_b$ is a minimal left ideal of $e_\mu A e_\mu$ for all $b \in B$. As R is $\mathbb{Z}$-linearly compact and A-linearly topologized, R is a fortiori a strongly linearly compact $e_\mu A e_\mu$-module. By 34.14, $e_b R = \{0\}$ for all $b \in B$. Hence $e_\mu R = \{0\}$.

Similarly, to show that $Re = \{0\}$, it suffices to show that for each $\mu \in M$, $Re_\mu = \{0\}$. Assume that $Re_\mu \neq \{0\}$, let $a \in R$ be such that $ae_\mu \neq 0$. Then there exists $c \in B$ such that $ae_c \neq 0$; let $K_c = e_c A e_c$, a division ring anti-isomorphic to K_μ by 25.7 and 25.18. Our hypothesis concerning the topology of R insures that the right A-module R is linearly topologized, and a fortiori the right K_c-module R is linearly topologized. Furnished with the discrete topology, K_c is clearly a strictly linearly compact right K_c-module. Let f be the function from K_c to R defined by $f(x) = ax$ for all $x \in K_c$. As $f(e_c) \neq 0$, clearly f is a continuous momomorphism from the right K_c-module K_c to the right K_c-module R. As K_c is strictly linearly compact, f is a topological isomorphism from K_c to aK_c. Therefore by 28.3 and 28.6, aK_c is a discrete, closed additive subgroup of R, and consequently is a discrete $\mathbb{Z}$-linearly compact module. By 30.10, aK_c is finite since it is isomorphic to the additive group of a division ring, in contradiction to the fact that K_μ is infinite. Thus $Re_\mu = \{0\}$.

If λ and μ are distinct members of M, then $e_\lambda e_\mu \in R$ since $\epsilon_\lambda \epsilon_\mu = \overline{0}$. Thus

$$e_\lambda e_\mu = ee_\lambda e_\mu = 0.$$

For any $x \in eAe$ and any $\mu \in M$, $e_\mu x - x e_\mu \in R$ since ϵ_μ is a central idempotent of A/R, and therefore

$$e_\mu x - x e_\mu = e(e_\mu x - x e_\mu) = 0.$$

Thus $(e_\mu)_{\mu \in M}$ is an orthogonal family of idempotents in the center of eAe whose sum is e. By 29.15, eAe is a strictly linearly ring. Consequently, $\phi : x \to (xe_\mu)_{\mu \in M}$, which is a homomorphism from the ring eAe to the cartesian product of the rings $(e_\mu A e_\mu)_{\mu \in M}$, is a topological isomorphism by 34.3. Moreover, eAe is an ideal of A, since if $a, b \in A$, $aeb - abe \in R$ as ϵ is in the center of A/R, so $eaeb - eabe = e(aeb - abe) = 0$, and similarly $beae = ebae$. Thus by 32.12, if $B = eAe$, $B_\mu = e_\mu A e_\mu$ for each $\mu \in M$, and $C = \{y - ye : y \in A\}$, then B has the desired description and A is the topological direct sum of B and C. Clearly $\phi(C) \subseteq (A/R)(1 - \epsilon)$, so as A/R is the direct sum of $(A/R)\epsilon$ and $(A/R)(1 - \epsilon)$ and as $A = B + C$, $\phi(C) = (A/R)(1 - \epsilon)$. Moreover, $1 - \epsilon$ is the sum of $(\epsilon_\nu)_{\nu \in N}$. Thus if $C_\mu = (A/R)e_\nu$, C_ν is topologically isomorphic to the ring of all linear operators on a discrete vector space over a finite field, furnished with the topology of pointwise convergence. As $Re = \{0\}$, $R \subseteq C$, and hence R is the radical of C by 26.18. Consequently, the restriction ϕ_C of ϕ to C is a continuous epimorphism from C to C/R; moreover, as C is topologically isomorphic to the A/eAe, C is strictly linearly compact by 29.4, and hence ϕ_C is a topological epimorphism. Thus C has the desired description. ●

34.18 Theorem. *Let A be a bounded, linearly compact ring with radical R. The following statements are equivalent:*

$1°$ *R is strictly linearly compact ring.*

$2°$ *A is the topological direct sum of a semisimple linearly compact ring B that has no nonzero compact ideals and a strongly linearly compact ring C.*

$3°$ *Every closed ideal of A is a strictly linearly compact ring.*

Proof. By 12.11 and 12.15, every subring of A or A/R is bounded, and by 12.16, the open ideals of A form a fundamental system of neighborhoods of zero. Consequently, by 29.9, A/R is topologically isomorphic to the cartesian product of discrete rings, each the ring of all linear operators on a finite-dimensional vector space over a division ring.

Assume $1°$. By 34.13, R is a strongly linearly compact ring. Thus by 34.17 and 29.9, A is the topological direct sum of subrings B and C, described as follows: B is topologically isomorphic to the cartesian product of a family $(B_\mu)_{\mu \in M}$ of topological rings, where each B_μ is the discrete ring of all linear operators on a finite-dimensional vector space over an infinite division ring K_μ; and C is a strictly linearly compact ring containing R such that C/R is topologically isomorphic to the cartesian product of a family of rings, each the discrete ring of all linear operators on a finite-dimensional vector space over a finite field. Thus C/R is is compact and *a fortiori* a strongly linearly compact C-module. As R is a strongly linearly compact ring, it is *a fortiori* a strongly linearly compact C-module. Thus C is a strongly linearly compact ring by 28.16.

Let J be a nonzero, closed ideal of $\prod_{\mu \in M} B_\mu$, and let $M_J = \{\lambda \in M : pr_\lambda(J) \neq (0)\}$, where pr_λ is the canonical projection from $\prod_{\mu \in M} B_\mu$ to B_λ. For each $\lambda \in M_J$, $pr_\lambda(J) = B_\lambda$ since B_λ has no proper, nonzero ideals. By 28.6, J is a linearly compact B-module, so its projection J' on $\prod_{\mu \in M_J} B_\mu$ is linearly compact and hence closed; by 24.12, $\bigoplus_{\lambda \in M_J} B_\mu \subseteq J'$, so $J' = \prod_{\mu \in M_J} B_\mu$ and hence is a strictly linearly compact ring by 29.5. Consequently, J is not compact, for otherwise, for each $\mu \in M_J$, B_μ would be compact and discrete, hence finite, and thus K_μ would be finite, a contradiction. Thus $2°$ holds.

Assume $2°$, and let J be a closed ideal of A. To show that J is a strictly linearly compact ring, it suffices to show that each of its intersections with B and C is a strictly linearly compact ring by 28.17. We have just seen that $J \cap B$ is a strictly compact ring. Moreover, $J \cap C$ is a closed $\mathbb{Z}$-submodule of C, hence is a linearly compact $\mathbb{Z}$-module, thus a strictly linearly compact $\mathbb{Z}$-module by 30.11, and *a fortiori* a strictly linearly compact ring. Thus $3°$ holds, and clearly $3°$ implies $1°$. •

Actually, if the conditions of the theorem hold, then every closed left or right ideal of A is a strictly linearly compact ring (Exercises 34.15 and 34.18).

34.19 Corollary. *If A is an artinian ring whose radical R is an artinian ring, then A is the direct sum of finitely many infinite simple artinian rings and a ring containing R whose additive subgroups satisfy the descending chain condition, and every ideal of A is an artinian ring.*

Proof. Furnished with the discrete topology, a ring is a bounded, strictly linearly compact ring if and only if it is artinian by (2) of 28.14. Thus the assertion follows from 34.18. •

To apply these results to compact rings, we need a preliminary theorem:

34.20 Theorem. *The radical R of a compact ring A is closed, and either $A = R$ or A/R is a compact, totally disconnected, semisimple ring.*

Proof. By 32.2 and 26.14, R contains the connected component C of zero. Consequently by 26.16, the radical of A/C is R/C. By 5.16, 32.5, and 29.12, R/C is closed in A/C. If ϕ is the canonical epimorphism from A to A/C, $R = \phi^{-1}(R/C)$ and hence is closed. Thus A/R is Hausdorff and hence is compact. By (3) of 5.17, A/R is totally disconnected. By 26.16, A/R is semisimple. •

34.21 Theorem. *If R is the radical of a compact, totally disconnected ring A, then the filter base $(\overline{R^n})_{n \geq 1}$ converges to zero, and in particular,*

$$\bigcap_{n=1}^{\infty} \overline{R^n} = \{0\}.$$

Proof. The assertion follows from 32.5 and 33.22. •

34.22 Theorem. *Let R be the radical of a totally disconnected compact ring A [with identity]. (1) If ϵ is a nonzero idempotent of A/R and if L is a closed left ideal of A containing R whose image in A/R contains ϵ, there is an idempotent e in L such that $\bar{e} = \epsilon$. (2) Every orthogonal family of idempotents in A or A/R is summable. (3) If $(\epsilon_\lambda)_{\lambda \in L}$ is an orthogonal family of idempotents of A/R [whose sum is $\bar{1}$], there is in A an orthogonal family of idempotents $(e_\lambda)_{\lambda \in L}$ [whose sum is 1] such that $\bar{e}_\lambda = \epsilon_\lambda$ for all $\lambda \in L$. (4) If $A \neq R$, A has a summable, orthogonal family $(e_\lambda)_{\lambda \in L}$ of idempotents such that if $e = \sum_{\lambda \in L} e_\lambda$, then $\bar{e}$ is the identity of A/R, and for each $\mu \in L$, the radical $e_\mu R e_\mu$ of $e_\mu A e_\mu$ is open in $e_\mu A e_\mu$.*

Proof. The assertions follow from 32.5, 34.1, 34.8, 34.4, 34.5, and 32.6. •

34.23 Theorem. *Let A be a compact, totally disconnected, commutative ring.* (1) *Either A is a radical ring, or there is a nonzero idempotent e such that A is the topological direct sum of the compact ring with identity Ae and the compact radical ring J, where $J = \{y - ye : y \in A\}$.* (2) *If A has an identity, A is topologically isomorphic to the cartesian product of a family of compact local rings.*

Proof. The assertions follow readily from 32.5 and 34.6. •

34.24 Theorem. *If A is a compact ring with identity whose radical is a primitive ideal, then A is topologically isomorphic to the ring of all linear operators, furnished with the topology of pointwise convergence, on a finite-dimensional compact module E over a compact ring L with identity whose radical is a regular maximal left ideal, where both E and L are subrings of A.*

The assertion follows from 32.5 and 34.10.

Exercises

34.1 If A is a linearly compact ring with identity 1 and radical R, then A/R is a division ring if and only if 1 is the only nonzero idempotent in A.

34.2 If A is a linearly compact ring with radical R and without proper zero-divisors, either A has an identity element and A/R is a division ring, or A is a radical ring.

34.3 (a) If a ring A has no nonzero nilpotents, then every idempotent of A belongs to its center. [Show that $ex = exe = xe$.] (b) A locally compact ring that has no nonzero topological nilpotents is totally disconnected.

34.4 A *metacompact* ring is a bounded, strictly linearly compact ring, and a *locally metacompact* ring is a topological ring that has an open metacompact subring. For example, a commutative topological ring is metacompact if and only if it is strictly linearly compact, and a compact ring is metacompact if and only if it is totally disconnected. Let R be the radical of a metacompact ring A. Either R is open, or zero is a cluster point of an orthogonal family of idempotents. [Use 34.5.]

34.5 (Lucke [1968]) (a) A topological ring is a metacompact ring that has no nonzero topological nilpotents if and only if it is topologically isomorphic to the cartesian product of discrete division rings. [Use 33.22.] (b) A nondiscrete topological ring A is a locally metacompact ring that has no nonzero topological nilpotents if and only if A is the topological direct sum of a discrete ring that has no nonzero nilpotents and a ring that is topologically isomorphic to the local direct sum of discrete rings that have no nonzero nilpotents with respect to division subrings. [If B is a metacompact open subring of A, use Exercise 34.4 to show that there is a summable orthogonal family $(e_\lambda)_{\lambda \in L}$ of idempotents whose sum is the identity element e of B, and use Exercise 34.3 in considering the function $x \to (xe_\lambda)_{\lambda \in L}$ from Ae to $\prod_{\lambda \in L} Ae_\lambda$. (c) A nondiscrete topological ring A is a locally compact ring having no nonzero topological nilpotents if and only if A is the topological direct sum of a discrete ring that has no nonzero nilpotents and a ring that is topologically isomorphic to the local direct sum of discrete rings that have no nonzero nilpotents with respect to finite fields.

34.6 (Lucke [1968], Blair [1976]) A topological ring A is a *Jacobson* ring if for each $x \in A$, x is a cluster point of $\{x^n : n \geq 2\}$. (a) A discrete ring is a Jacobson ring if and only if for each $x \in A$ there exists $n(x) \geq 2$ such that $x^{n(x)} = x$. (For example, a field is a (discrete) Jacobson ring if and only if each of its nonzero elements is a root of unity. A theorem of Jacobson asserts that a discrete Jacobson ring is commutative.) (b) A Hausdorff Jacobson ring has no nonzero topological nilpotents. (c) A nondiscrete topological ring A is a locally metacompact Jacobson ring if and only if A is the topological direct sum of a discrete Jacobson ring and a ring that is topologically isomorphic to the local direct sum of a family of discrete Jacobson rings relative to Jacobson fields. [Use Jacobson's theorem and Exercise 34.5.] (c) In particular, infer from Jacobson's theorem that a locally metacompact Jacobson ring is commutative. (d) A nondiscrete topological ring A is a locally compact Jacobson ring if and only A is the topological direct sum of a discrete Jacobson ring and a ring that is topologically isomorphic to the local direct sum of a family of discrete Jacobson rings relative to finite subfields.

34.7 A topological ring A is locally metacompact if and only if A is the topological direct sum of subrings A_1 and A_2, where A_1 is the topologically isomorphic to the local direct sum of locally metacompact rings $(A_\lambda)_{\lambda \in L}$ with centers $(C_\lambda)_{\lambda \in L}$ relative to metacompact open subrings $(B_\lambda)_{\lambda \in L}$ such that for each $\lambda \in L$, $B_\lambda \cap C_\lambda$ is a local ring whose identity element is that of A_λ, and where A_2 is a locally metacompact ring that has a metacompact open subring B_2 such that $B_2 \cap C_2$ is a metacompact radical ring, where C_2 is the center of B_2. [Use 34.6.]

34.8 A ring is a *boolean* ring if each of its elements is an idempotent. (a) The following statements about a topological ring A are equivalent:

1° A is a linearly compact boolean ring.

2° A is topologically isomorphic to the cartesian product of fields, each having two elements.

3° A is a compact boolean ring.

(b) A nondiscrete topological ring A is a locally compact boolean ring if and only if A is the topological direct sum of a discrete boolean ring and a ring that is topologically isomorphic to the local direct sum of a family $(A_\lambda)_{\lambda \in L}$

of discrete boolean rings with identity relative to subfields $(B_\lambda)_{\lambda \in L}$, where each B_λ is field of two elements that contains the identity element of A_λ.

34.9 Let A be a metacompact ring, and let R be the radical of A. The following statements are equivalent:

1° A is metrizable and R is open.
2° A is ultranormable.
3° A is normable.

In particular, if A is a totally disconnected, compact ring, 1° – 3° are equivalent. [Use 33.22 and Exercise 14.5.]

34.10 (Lipkina [1964b], [1966]) If A is a metacompact ring without proper zero-divisors, then 1° – 3° of Exercise 34.9 are equivalent. In particular, if A is a compact ring with identity, 1° – 3° of Exercise 34.9 are equivalent. [Use Exercises 34.2 and 34.9.]

34.11 (Øfsti [1965]) Let A be a commutative, locally compact, metacompact ring with identity. (a) If A is a local ring, then A is either compact or discrete. [Let M be the maximal ideal of A. Show that if I and J are compact open ideals such that $J \subset I$, then card$(A/M) \leq$ card(J/I). Show that for any open ideal J, A/J is finite by considering, for each $n \geq 1$, $(M^n + J)/(M^{n+1} + J)$ as an A/M-vector space and using 28.15.] (b) A is the topological direct sum of a compact ring and a discrete artinian ring. [Use 34.6.]

34.12 (Cude [1970]) Let A be a metacompact ring with identity. If A has prime characteristic p and if A/R is a finite field, where R is the radical of A, then A contains a unique subfield K mapped onto A/R by the canonical epimorphism from A to A/R. [If $\bar{a}$ generates the multiplicative group $(A/R)^*$ of order $p^m - 1$, observe that $a^{p^{mr}} - a \in R$ for all $r \geq 1$, and conclude that $(a^{p^{mk}})_{k \geq 1}$ is a Cauchy sequence.]

34.13 (Widiger [1979]) Let A be a strictly linearly compact ring with radical R. (a) The following conditions are equivalent:

1° A has a left identity.
2° For all $a \in A$, $a \in Aa$.
3° $R = AR$.

[To establish that 3° implies 1°, show that there is an idempotent e such that the additive group A is the direct sum of eA and a right ideal J contained in R. Observe that the additive group R is the direct sum of eR and J, and use 3° to show that R is also the direct sum of eR and JR. Apply 33.21.] (b) If $\bigcap_{n=1}^{\infty} \overline{R^n} = \{0\}$, the following conditions are equivalent:

1° A has a right identity.
2° For all $a \in A$, $a \in aA$.
3° $R = RA$.

(c) (Kaplansky [1947b]) A metacompact ring A has an identity if and only if $a \in Aa \cap aA$ for all $a \in A$.

34.14 Let A be the ring of all linear operators on a discrete vector space E over a division ring K, furnished with the topology of pointwise convergence. If A has a minimal left ideal Ae that is a linearly compact ring, then E is one-dimensional if K is infinite, and E is finite otherwise. [Construct a closed additive subgroup H such that $He = H$ and $eH = \{0\}$ whose additive subgroup is isomorphic to that of K if K is infinite, and to the additive subgroup $e^{-1}(0)$ otherwise.]

34.15 (Widiger [1972]) Let A be a linearly compact ring with radical R. (a) If R is metacompact, the following statements are equivalent:

1° Every closed left ideal of A is a linearly compact ring.

2° A is the topological direct sum of a subring B, topologically isomorphic to the cartesian product of a family of discrete, infinite division rings, and a strongly linearly compact subring C.

If these conditions hold, each closed left ideal of A is a strictly linearly compact ring. [Use 34.18 and Exercise 34.14.] (b) If A is bounded, the following statement is equivalent to 1° and 2° of (a): Every closed left ideal of A is a strictly linearly compact ring.

34.16 (Kertész and Widiger [1969]) Let A be an artinian ring, R its radical. Every left ideal of A is an artinian ring if and only if A is the direct sum of an ideal isomorphic to the cartesian product of finitely many infinite division rings and an ideal whose additive groups satisfy the descending chain condition. [Apply Exercise 34.15.]

34.17 (Kertész and Widiger [1969]) Let A be an artinian ring with radical R. The following statements are equivalent:

1° R is an artinian ring.

2° A is the direct sum of finitely many rings, each isomorphic to the ring of all linear operators on a finite-dimensional vector space over an infinite division ring, and an ideal that is a $\mathbb{Z}$-artinian module.

3° Every ideal of A is an artinian ring.

[Apply 34.18.]

34.18 If A is a bounded, linearly compact ring satisfying the equivalent conditions of 34.18, then every closed right ideal of A is a strictly linearly compact ring. [Use 34.11 and Exercise 29.10.] In particular, if A is an artinian ring satisfying the equivalent conditions of Exercise 34.17, then every right ideal of A is an artinian ring.

34.19 (Dinh Van Huynh [1973]) If A is a strictly linearly compact ring, then every idempotent of A is in its center if and only if A is the topological direct sum of rings B and C, where B is topologically isomorphic to the cartesian product of of a family $(A_\lambda)_{\lambda \in L}$ of strictly linearly compact rings such that for each $\lambda \in L$, A_λ/R_λ is a division ring, where R_λ is the radical of A_λ, and B is a strictly linearly compact radical ring.

35 Locally Compact Rings

We present here some theorems concerning locally compact rings whose proofs depend either on the Pontrĭagin-van Kampen theory of locally compact commutative groups or on theorems of §34 concerning the lifting of idempotents.

35.1 Theorem. (Pontrĭagin-van Kampen) *Let G be a locally compact abelian group. There is a unique natural number n such that G is the topological direct sum of a subgroup topologically isomorphic to $\mathbb{R}^n$ and a subgroup that, for its induced topology, contains a compact open subgroup.*

35.2 Theorem. *Let A be a locally compact ring, let C be the connected component of zero, and let T be the union of all the compact additive subgroups of A. Then C and T are closed ideals of A, $CT = TC = (0)$, $C+T$ is an open ideal, $C/(T\cap C)$ is a finite-dimensional topological algebra over $\mathbb{R}$, and $T/(T \cap C)$ is totally disconnected. If, moreover, $C/(T \cap C)$ has an identity element, then A is the topological direct sum of a finite-dimensional topological $\mathbb{R}$-algebra B with identity and a locally compact subring D such that $D \cap C = T \cap C$ and the connected component of zero in D is compact.*

Proof. By 35.1, the topological additive group A is the topological direct sum of a subgroup N topologically isomorphic to $\mathbb{R}^n$ for some $n \in \mathbb{N}$ and a subgroup K that contains a compact open subgroup L. Clearly T is an additive subgroup, since the sum of two compact subgroups of A is again a compact subgroup. Since A/K is topologically isomorphic to $\mathbb{R}^n$ by 15.4, and thus contains no nonzero compact subgroups, the image of T under the canonical epimorphism from A to A/K is the zero subgroup, so $T \subseteq K$. Since $L \subseteq T$, T is an open and hence closed subgroup of K, so as A is the topological direct sum of N and K, T is closed in A by 15.4. Also, T is an ideal, for if S is a compact additive subgroup containing t, then Sa and aS are compact additive subgroups containing ta and at respectively. By 32.2, $Ca = (0) = aC$ for all $a \in T$, so $CT = TC = (0)$. Since T is open in K and since A is the topological direct sum of N and K, $N+T$ is open in A, whence as $C + T \supseteq N + T$, $C + T$ is also open in A.

The connected component C_0 of zero in K is clearly contained in $L \cap C$ and hence in $T \cap C$, so $T/(T \cap C)$ is totally disconnected by (3) of 5.17. Moreover, as A is the topological direct sum of N and K,

$$C = N + C_0 \subseteq N + (T \cap C) \subseteq C,$$

so the additive topological group C is the topological direct sum of subgroups N and $T \cap C$. In particular, by 15.4 the additive group $C/(T \cap C)$ is topologically isomorphic to N. Thus there is a topological isomorphism ϕ from $\mathbb{R}^n$ to $C/(T \cap C)$, so $C/(T \cap C)$ becomes a topological vector space over $\mathbb{R}$ under the scalar multiplication defined by $r.x = r.\phi^{-1}(x)$ for all $r \in \mathbb{R}$ and all $x \in C/(T \cap C)$. As $C/(T \cap C)$ is a topological ring, for any $a, b \in C/(T \cap C)$ the functions $r \to r.(ab)$, $r \to (r.a)b$, and $r \to a(r.b)$ are all continuous from $\mathbb{R}$ to $C/(T \cap C)$; since they agree on $\mathbb{Z}$ and hence on $\mathbb{Q}$, they therefore coincide, so $r.(ab) = (r.a)b = a(r.b)$ for all $r \in \mathbb{R}$, and all $a, b \in C/(T \cap C)$. Consequently, $C/(T \cap C)$ is a topological $\mathbb{R}$-algebra.

Suppose, finally, that $C/(T \cap C)$ has an identity element $f + (T \cap C)$. In particular, $f^2 - f \in T \cap C$, so as $f \in C$ and $f^2 - f \in T$, $f(f^2 - f) = 0$, whence $f^3 = f^2$, and consequently $f^4 = f^3 = f^2$. Let $e = f^2$. Then e is an idempotent and $e + (T \cap C)$ is the identity element of $C/(T \cap C)$. Consequently, e belongs to the center of C, since for any $x \in C$, $ex - xe \in T \cap C$, so $exe - xe = (ex - xe)e \in TC = (0)$ and hence $e(xe) = xe$, and similarly $(ex)e = ex$. Therefore, as C is an ideal, so is Ce, and consequently $Ae = Ce$. Let $B = Ae$, and let $D = \{x \in A : xe = 0\}$. By 32.12, A is the topological direct sum of ideals B and D. If $x \in T \cap C$, then $xe \in TC = (0)$, so $x \in D \cap C$. Conversely, if $x \in D \cap C$, then $xe = 0$, so $x \in T \cap C$ as $e + (T \cap C)$ is the identity of $C/(T \cap C)$.

Since $B = Ae = Ce$, by 32.12 C is the topological direct sum of B and $\{x \in C : xe = 0\}$, and the latter is $D \cap C$ and therefore $T \cap C$ by the preceding paragraph. Thus by 15.4, B is topologically isomorphic to $C/(T \cap C)$ and hence to N; therefore B is an n-dimensional $\mathbb{R}$-algebra. By 15.4, D is closed and hence locally compact, so by 35.1 there is a unique $m \in \mathbb{N}$ such that the additive group D is the topological direct sum of a subgroup topologically isomorphic to $\mathbb{R}^m$ and a subgroup M that, for its induced topology, contains a compact open subgroup. Then the additive group A would be the topological direct sum of a subgroup topologically isomorphic to $\mathbb{R}^{n+m}$ and M, so $m = 0$ by 35.1, whence $D = M$. Thus D contains a compact open subgroup, and hence its connected component of zero is compact. ∙

35.3 Theorem. *If A is a locally compact ring that has no nonzero nilpotent ideals [that is semisimple], then A is the topological direct sum of its connected component C and a locally compact, totally disconnected ring D that has no nonzero nilpotent ideals [that is semisimple], and if $C \neq (0)$, C is the topological direct sum of finitely many ideals, each the ring of all linear operators on a finite-dimensional topological vector space over either $\mathbb{R}$, $\mathbb{C}$, or $\mathbb{H}$, furnished with its unique topology as a Hausdorff finite-dimensional algebra over $\mathbb{R}$.*

Proof. Assume that $C \neq (0)$, and let T be the union of all the compact additived subgroups of A. By 35.2, T is a closed ideal, and $(C \cap T)^2 \subseteq CT = (0)$, so $C \cap T = (0)$ by hypothesis [by 26.14]. Thus by 35.2, C is a finite-dimensional topological algebra over $\mathbb{R}$. In particular, C is an artinian $\mathbb{R}$-algebra. Let R be the radical of A. Then $R \cap C$ is the radical of C by 26.18. Consequently, $R \cap C$ is nilpotent by 27.15 and hence is the zero ideal by hypothesis [by 26.14]. Thus C is a semisimple, finite-dimensional algebra over $\mathbb{R}$ and hence has an identity element by 27.14. By 32.12, A is the topological direct sum of C and an ideal B, and B is necessarily totally disconnected as $B \cap C = (0)$. As every ideal of B is an ideal of A [By 26.18], B has no nonzero nilpotent ideals [B is semisimple].

By 27.14 and 27.12, C is the direct sum of $(C_i)_{1 \leq i \leq n}$ where each C_i is a primitive ring that is an artinian $\mathbb{R}$-algebra and has an identity element e_i. Each member of the associated family $(p_i)_{1 \leq i \leq n}$ of projections is continuous, since $p_i(x) = xe_i$ for all $x \in C$. Thus C is the topological direct sum of $(C_i)_{1 \leq i \leq n}$ by 15.2. By 26.25 we may regard each C_i as a dense $\mathbb{R}$-algebra of linear operators on a vector space E_i over a division ring K_i containing $\mathbb{R}$ in its center. Therefore, as C finite-dimensional over $\mathbb{R}$, so is C_i; consequently, E_i is finite-dimensional over K_i and K_i is finite-dimensional over $\mathbb{R}$. Therefore K_i is either $\mathbb{R}$, $\mathbb{C}$, or $\mathbb{H}$, and as E_i is finite-dimensional, C_i is the ring of all linear operators on E_i, and its topology is the unique topology making it a Hausdorff finite-dimensional algebra over $\mathbb{R}$. •

35.4 Theorem. *If A is a connected, locally compact ring, then A contains a connected, compact ideal K such that $AK = KA = (0)$ and A/K is a finite-dimensional topological $\mathbb{R}$-algebra.*

Proof. By 35.1, the topological additive group A is the topological direct sum of a subgroup N topologically isomorphic to $\mathbb{R}^n$ for some $n \geq 0$ and a subgroup K that, for its induced topology, contains a compact open subgroup L. By 15.4, K is topologically isomorphic to A/N and hence is connected. Therefore $L = K$. As in the proof of 35.2, the union T of all compact additive subgroups is an ideal contained in K and hence is K as K is compact. The conclusion follows from 35.2. •

35.5 Corollary. *If A is a connected, locally compact ring such that zero is the only element c satisfying $cA = Ac = \{0\}$, then A is a finite-dimensional topological $\mathbb{R}$-algebra.*

35.6 Corollary. *A connected, locally compact ring A is advertibly open.*

Proof. As $K^2 = \{0\}$, K is an advertible ideal since for each $x \in K$, $-x$ is the adverse of x. By 35.4, A/K is a finite-dimensional topological $\mathbb{R}$-algebra and hence is a complete normed algebra by 15.11 and 16.7. Therefore A/K

is advertibly open by 11.12. Consequently, A is advertibly open by (2) of 26.26. •

35.7 Theorem. *Let A be a totally disconnected, locally compact ring. (1) Either A is advertibly open, or there is a nonzero idempotent $e \in A$ such that eAe is advertibly open. (2) If e is a nonzero idempotent in A such that eAe is advertibly open, then any proper left ideal I of A containing $\{x - xe : x \in A\}$ is contained in a closed regular maximal ideal of A.*

Proof. (1) By 4.21, A contains an open, compact subring B. If B is a radical ring, then A is advertibly open by 26.9. Otherwise, let R be the radical of B. By (4) of 34.22, there is an idempotent e of B such that eRe is open in eBe and hence in eAe, so eAe is advertibly open by 26.17 and 26.9.

(2) I is contained in a left regular maximal ideal M by 26.3. Thus M is either closed or dense in A. If M were dense, then as $\phi : x \to exe$ is continuous from A to eAe, there would exist $x \in M$ such that $e - x \in \phi^{-1}(U)$, where U is the set of advertible elements of eAe. Then as $e(e-x)e = e - exe$, $e - exe$ has an adverse y. Thus

$$0 = y \circ (e - exe) = y + e - exe - ye + yexe$$
$$= e + (y - ye) - ex + (ex - exe) + yex - (yex - yexe) \in e + M,$$

so $e \in M$ and hence $M = A$, a contradiction. Thus M is closed. •

35.8 Theorem. *If A is a locally compact ring that is not a radical ring, then A contains a closed, regular maximal left ideal.*

Proof. Let C be the connected component of zero. If A/C is a radical ring, then A/C is advertibly open by 26.9, so A is advertibly open by 35.6 and (2) of 26.26, and hence A contains a closed regular maximal left ideal by 26.27. In the contrary case, as A/C is totally disconnected by 5.6, A/C contains a closed, regular maximal left ideal M by 35.7. Thus $\phi_C^{-1}(M)$ is a closed, regular maximal ideal of A, where ϕ_C is the canonical epimorphism from A to A/C. •

35.9 Theorem. *The radical R of a locally compact ring A is closed.*

Proof. Assume that $R \subset \overline{R}$. Then $\overline{R}$ is a locally compact ring whose radical is R by 26.18. Consequently, by 35.8, $\overline{R}$ contains a closed, regular maximal ideal M. By 26.7, $R \subseteq M \subset \overline{R}$, a contradiction since M is closed and R dense in $\overline{R}$. •

35.10 Theorem. *The radical R is a locally compact ring A is either A or the intersection of the closed regular maximal left ideals of A.*

Proof. Assume $A \neq R$. By 35.9, A/R is Hausdorff and hence a locally compact ring. The assertion is therefore equivalent to the statement that the intersection of the closed regular maximal left ideals of A/R is $\{0\}$. Consequently, we shall assume that A is semisimple. By 35.3, A is the topological direct sum of a locally compact connected ring C and a totally disconnected, semisimple locally compact ring. By 35.6, C is advertibly open. Consequently, every left regular maximal ideal of C is closed by 26.27. Therefore it suffices to consider the case where A is a totally disconnected, semisimple, locally compact ring.

Let M_0 be the intersection of the closed regular maximal left ideals of A. By 4.21, A has a compact open subring B with radical S. If $B = S$, then A is advertibly open by 26.8, and hence every regular maximal left ideal of A is closed by 26.27, so $M_0 = \{0\}$. In the contrary case, by (4) of 34.22, B contains a summable orthogonal family $(e_\lambda)_{\lambda \in L}$ of idempotents such that if $e = \sum_{\lambda \in L} e_\lambda$, $\bar{e}$ is the identity of B/S and for each $\mu \in L$, $e_\mu A e_\mu$ is advertibly open.

We shall first show that if $c \in M_0$ and if $\mu \in L$, then ce_μ is left advertible. Let

$$I = \{x - xe_\mu : x \in A\} + A(e_\mu - e_\mu c).$$

Clearly I is a left ideal of A. If I were proper, then $e_\mu \notin I$, and I would be contained in a closed regular maximal left ideal M by 35.7; consequently, $c \in M$, hence $e_\mu c \in M$, and therefore as $e_\mu - e_\mu c = e_\mu(e_\mu - e_\mu c) \in I$, $e_\mu \in M$ and so $M = A$, a contradiction. Thus $I = A$, so there exist $x \in A$ and $b \in A$ such that $e_\mu = x - xe_\mu + b(e_\mu - e_\mu c)$. Let $a = e_\mu - b$. Then $e_\mu = x - xe_\mu + e_\mu - ae_\mu - e_\mu c + ae_\mu c$. Multiplying both sides of that equality on the right by e_μ, we obtain $e_\mu = e_\mu - ae_\mu - e_\mu ce_\mu + ae_\mu ce_\mu$, so $ae_\mu \circ e_\mu ce_\mu = 0$. Thus $e_\mu(ce_\mu)$ is left advertible, so by 11.5, as $ce_\mu = (ce_\mu)e_\mu$, ce_μ is left advertible. Consequently, the left ideal $M_0 e_\mu$ consists of left advertible elements of A and hence is $\{0\}$ by 26.13.

Consequently, for any $c \in M_0$, $ce = \sum_{\lambda \in L} ce_\lambda = 0$ by 10.16 and the preceding, so $M_0 e = \{0\}$.

Next, we show that if M is a regular maximal left ideal containing Ae, then M is closed. Let $f \in A$ be such that $x - xf \in M$ for all $x \in A$. Assume that M is not closed; then M is dense, and hence there exists $x \in B$ such that $f - x \in M$. As $\bar{e}$ is the identity of B/S, $x - xe \in S$, so by 26.9, there exists $s \in S$ such that $s \circ (x - xe) = 0$. As $M \supseteq Ae$, $xe \in M$, so $f - x + xe \in M$, and hence $sf - sx + sxe \in M$. Also, $s - sf \in M$, so $s - sx + sxe \in M$; but $s - sx + sxe = s - s(x - xe) = -(x - xe)$. Thus as

$xe \in M$, $x \in M$, so as $f - x \in M$, $f \in M$ and hence $M = A$, a contradiction. Thus M is closed.

If A has no identity, let A_1 be the ring obtained by adjoining an identity to A. Let $c \in M_0$, and let $K = Ae + A(1 - (c - e))$, a left regular ideal of A. If K were proper, K would be contained in a left regular maximal ideal M by 26.3, and M would be closed by the preceding and hence would contain c and also $c - ec$ and so would be A, a contradiction. Hence $K = A$, so there exist $a, b \in A$ such that $ae + b(1 - c + ec) = ec - e - c$. Multiplying each term of that equality on the right by $1 - e$ and using the equality $ce = 0$, we conclude that $b(1 - e) \circ (1 - e)c = 0$, and hence $(1 - e)c$ is left advertible in A. By 11.5, $c(1 - e)$ is left advertible in A_1. But as $ce = 0$, $c(1 - e) = c$. Thus each $c \in M_0$ is right advertible in A_1 and hence in A, as A is an ideal of A_1. Consequently by 26.13, $M_0 = \{0\}$. •

35.11 Theorem. *If A is a totally disconnected, bounded, locally compact commutative ring with radical R, then either R is open, or A is the topological direct sum of a compact ring with identity and a locally compact ring having an open radical.*

Proof. Assume that R is not open. By (1) of 12.16, A contains a compact open ideal B. Then $R \cap B$, the radical of B by 26.18, is a proper ideal of B. By 29.13, 32.7, and 34.22, there is a nonzero idempotent $e \in B$ such that if $K = \{y - ye : y \in B\}$, K is a radical ring. By 32.12, A is the topological direct sum of the compact ring Ae, which is identical with Be as B is an ideal of A, and the ideal D, where $D = \{y - ye : y \in A\}$. The topological isomorphism $x \to (xe, x - xe)$ from A to $Ae \times D$ takes B into $Ae \times K$, and hence induces a topological isomorphism from the discrete space A/B to D/K. If S is the radical of D, $S \supseteq K$ by 26.18 as K is a radical ring. The canonical epimorphism $x + K \to x + S$ from D/K to D/S is a topological epimorphism, and therefore D/S is also discrete. Thus S is open in D. •

Theorem 34.1 yields further information about locally compact primitive rings:

35.12 Theorem. *A topological ring A is a locally compact primitive ring whose additive group is torsionfree if and only if it is topologically isomorphic to the ring of all linear operators on a finite-dimensional Hausdorff vector space over a nondiscrete locally compact division ring of characteristic zero.*

Proof. The condition is clearly sufficient. Necessity: By 35.3 and 25.15, A is either connected or totally disconnected, and if A is connected, the assertion holds. Consequently, we shall assume that A is totally disconnected.

Case 1: A is advertibly open, and there is a prime p such that for each $a \in A$, $\lim_{n \to \infty} p^n.a = 0$. By (2) of 26.5 there is a regular maximal left ideal M of A such that $P(M) = (0)$. Consequently by (1) of 26.5, A is isomorphic to a dense ring $\hat{A}$ of linear operators on the right vector space A/M over the division ring $D(M)/M$, where scalar multiplication is given by $(x+D).(d+M) = xd+M$ for all $x \in A$, $d \in D(M)$. Furnished with their topologies induced by that of A, M is closed by 26.27, so A/M is locally compact; as $D(M) = \{d \in A : Md \subseteq M\}$ and as M is closed, $D(M)$ is also closed and hence locally compact, so $D(M)/M$ is locally compact. Moreover, the scalar multiplication of the right $(D(M)/M)$-vector space A/M is continuous since multiplication is continuous on $A \times A$ and the canonical epimorphisms from A and $D(M)$ to A/M and $D(M)/M$ respectively are topological epimorphisms. Since $\hat{A}$ is torsionfree, the scalar division ring $D(M)/M$ has characteristic zero. Furthermore, $\lim_{n \to \infty} p^n.\delta = 0$ for every $\delta \in D(M)/M$. Consequently, $D(M)/M$ is not discrete. Therefore A/M is finite-dimensional over $D(M)/M$ by 16.2, 18.17, and 13.8, so the conclusion holds.

Case 2: There is a prime p such that for each $a \in A$, $\lim_{n \to \infty} p^n.a = 0$. By Case 1, we may assume that A is not advertibly open, so by 35.7 there is a nonzero idempotent $e \in A$ such that eAe is advertibly open. By 25.15, eAe is a primitive ring. As eAe is closed by 26.28, eAe is locally compact. Consequently, by Case 1, eAe is isomorphic to the ring of all linear operators on a finite-dimensional vector space, and hence has a minimal left ideal. As eAe is primitive and hence has no nonzero nilpotent ideals by 26.14, there is an idempotent e_1 in eAe such that $(eAe)e_1$ is a minimal left ideal by 25.17, and hence, by (1) of 25.18, $e_1(eAe)e_1$ is a division ring. As $e_1 \in eAe$, $e_1 e = e_1 = e e_1$, so $e_1 A e_1$ is a division ring, and hence Ae_1 is a minimal left ideal of A by 26.14 and (1) of 25.18. Therefore the conclusion follows from 32.13.

To prove the theorem, let A be a locally compact, totally disconnected, primitive ring such that the additive group A is torsionfree. As in the final paragraph of the proof of 32.13, A contains a nonzero, closed ideal J such that $\lim_{n \to \infty} p^n.x = 0$ for all $x \in J$. Then J is locally compact and a primitive ring by 25.15. By Case 2, J is isomorphic to the ring of all linear operators on a finite-dimensional, locally compact vector space over a nondiscrete, locally compact division ring, and hence has an identity element e. By 32.12, A is the direct sum of J and another ideal I, so by 25.15, $J = A$. •

35.13 Theorem. *Let A be a simple, nondiscrete topological ring with identity. For any neighborhood U of zero and any $a \in A$, if $aUa = \{0\}$, then $a = 0$.*

Proof. Assume that $aUa = \{0\}$ but that $a \neq 0$. As A is simple, there exist $x_1, \ldots, x_n \in A$ and $y_1, \ldots, y_n \in A$ such that

$$1 = \sum_{i=1}^{n} x_i a y_i.$$

There is a neighborhood V of zero such that $V x_i \subseteq U$ for all $i \in [1, n]$. Consequently, for each $v \in V$,

$$av = \sum_{i=1}^{n} (avx_i a) y_i = 0.$$

There is a neighborhood W of zero such that $y_i W \subseteq V$ for all $i \in [1, n]$. Consequently, if $w \in W$,

$$w = 1 \cdot w = \sum_{i=1}^{n} x_i a(y_i w) \in \sum_{i=1}^{n} x_i aV = \{0\},$$

in contradiction to the hypothesis that A is not discrete. •

35.14 Theorem. *If A is a simple, totally disconnected, locally compact ring with identity, there is a compact, open subring S of A such that the filter base $(S^n)_{n \geq 1}$ converges to zero.*

Proof. We may assume that A is not discrete. By 4.21, A contains a compact open subring B that does not contain 1. Let R be the radical of B. By 34.20, R is compact; hence if R is open, R is the desired subring S by 34.21. Therefore we shall assume that R is not open, so B is not a radical ring, and consequently B/R is a a ring with identity ϵ by 32.7. By (1) of 34.22 there is an idempotent $e \in B$ such that $\bar{e} = \epsilon$. Thus $b - be \in R$ for all $b \in B$, and $1 - e \neq 0$. By 35.13, there exists $c \in B$ such that $(1-e)c(1-e) \neq 0$, so as A is simple, there exist $x_1, \ldots, x_n \in A$ and $y_1, \ldots, y_n \in A$ such that

$$1 = \sum_{i=1}^{n} x_i (1-e) c (1-e) y_i.$$

By 4.21 there is a compact open subring S of B such that $Sx_i \cup y_i S \subseteq B$ for all $i \in [1, n]$. To show that $S^2 \subseteq R$, let $s, t \in S$, and for each $i \in [1, n]$, let $z_i = sx_i - sx_i e$, an element of R as $sx_i \in B$. Then

$$st = \sum_{i=1}^{n} sx_i(1-e)c(1-e)y_i t$$

$$= \sum_{i=1}^{n} sx_i e(1-e)c(1-e)y_i t + z_i(1-e)c(1-e)y_i t \in z_i B \subseteq R,$$

as $(1-e)c(1-e) = c - ec - ce + ece \in B$. Consequently, as $(R^n)_{n \geq 1}$ converges to zero, so does $(S^n)_{n \geq 1}$. •

35.15 Corollary. *A simple, totally disconnected, locally compact ring with identity is advertibly open.*

Proof. Each element of S is a topological nilpotent and hence, by 11.16, is advertible. •

35.16 Theorem. *Let A be a simple, locally compact ring with identity. The following statements are equivalent:*

1° *The left ideal generated by each neighborhood of zero is A.*
2° *A has no proper open left ideals.*
3° *A is topologically isomorphic to the ring of all linear operators on a finite-dimensional Hausdorff vector space over a nondiscrete locally compact division ring.*

Proof. By 4.9, 1° and 2° are equivalent. Assume 3°. Then A is a Hausdorff, finite-dimensional algebra over the center F of the underlying scalar division ring, and the topology of F is given by a proper, complete absolute value by 18.17. If L were a proper open left ideal, A/L would be a nonzero discrete vector space over F, in contradiction to 13.8. Thus 2° holds.

Assume 1°. By 35.3, A is either connected or totally disconnected, and 3° holds if A is connected. Therefore, we shall assume that A is totally disconnected.

We shall first show that if U and V are compact open subrings of A and if F is a closed subset such that $VF \subseteq U$, then F is compact. By 1°, there exist $a_1, \ldots, a_n \in A$ and $v_1, \ldots, v_n \in V$ such that

$$1 = \sum_{i=1}^{n} a_i v_i.$$

For each $i \in [1, n]$, $\overline{v_i F}$ is a closed and hence compact subset of compact U. For each $x \in F$,

$$x = \sum_{i=1}^{n} a_i v_i x \in \sum_{i=1}^{n} a_i \overline{v_i F},$$

a compact set. Hence as F is closed, F is compact.

By 35.14, A has a compact, open subring S such that the filter base $(S^n)_{n \geq 1}$ converges to zero. The filter base $\mathcal{U}$ of compact open subrings of A contained in S is a fundamental system of neighborhoods of A by 4.21. For each $U \in \mathcal{U}$, let

$$(U : S) = \{a \in A : Sa \subseteq U\}.$$

Let $U \in \mathcal{U}$. As U is closed, clearly $(U : S)$ is closed. As $S(U : S) \subseteq U$, $(U : S)$ is compact by the preceding paragraph. If J is a nonzero left ideal of A such that $J \cap S \subseteq U$, then $(U : S) \setminus S \neq \emptyset$: Indeed, there exists $y \in J \setminus S$; otherwise, J would be contained in S, therefore each element of J would be a topological nilpotent and hence advertible by 11.16; thus by 26.14, J would be a subset of the radical of A, the zero ideal by hypothesis, a contradiction. Let $S^0 = A$. There is a largest $k \in \mathbb{N}$ such that $S^k y \not\subseteq S$, since there is a neighborhood V of zero such that $Vy \subseteq S$ and there exists $m \geq 1$ such that $S^n \subseteq V$ for all $n \geq m$. Let $t \in S^k$ be such that $ty \notin S$; as

$$Sty \subseteq J \cap S^{k+1}y \subseteq J \cap S \subseteq U,$$

$ty \in (U : S) \setminus S$.

There exists $W \in \mathcal{U}$ such that for every nonzero left ideal I of A, $I \cap S \not\subseteq W$. Indeed, suppose the contrary. Then for each $U \in \mathcal{U}$, there would exist a nonzero left ideal I_U of A such that $I_U \cap S \subseteq U$; by the preceding, therefore, $(U : S) \setminus S$ would be nonempty. As S is open and $(U : S)$ compact, $\{(U : S) \setminus S : U \in \mathcal{U}\}$ would be a filter base of nonempty compact subsets of A, so there would exist

$$a \in \bigcap_{U \in \mathcal{U}} ((U : S) \setminus S).$$

Thus

$$Sa \subseteq \bigcap_{U \in \mathcal{U}} U = \{0\},$$

and $a \notin S$, so $a \neq 0$ and $aSa = \{0\}$, in contradiction to 35.13.

By the preceding and 4.20, there is an open ideal D of S such that for every nonzero left ideal I, $I \cap S \not\subseteq D$. We shall show that every nonzero closed left ideal J contains a minimal closed left ideal, that is, a left ideal maximal in the set of all nonzero closed left ideals, ordered by $\supseteq$. Let $\mathcal{I}$ be a totally ordered subset of the set $\mathcal{J}$ of all nonzero closed left ideals contained in J, and let $I_0 = \bigcap_{I \in \mathcal{I}} I$. Then $\{I \cap (S \setminus D) : I \in \mathcal{I}\}$ is a filter base of nonempty closed subsets of compact S, and hence there exists

$$c \in \bigcap_{I \in \mathcal{I}} (I \cap (S \setminus D)).$$

Thus c is a nonzero element of I_0, so I_0 is the supremum of $\mathcal{I}$ for the ordering $\supseteq$. By Zorn's Lemma, therefore, each nonzero closed left ideal J of A contains a closed minimal left ideal.

Consequently, by 35.15 and 26.29, every nonzero closed left ideal of A contains a minimal left ideal. Therefore 3° holds by 32.16. •

Exercises

35.1 Let A be the cartesian product of $\mathbb{R}$, furnished with the discrete topology, and $\mathbb{R}$, furnished with its usual topology, let addition be defined componentwise on A and multiplication by $(x,y)(z,w) = (0, xz)$. Show that A is a locally compact ring such that the connected component C of zero is the smallest nonzero closed ideal of A. In particular, A is not the topological direct sum of C and another ideal.

35.2 (Kaplansky [1947c]) If A is a totally disconnected, locally compact ring, either A is advertibly open, or zero is a cluster point of an orthogonal family of idempotents. [Use Exercise 34.3.]

35.3 (Kaplansky [1947c]) A locally compact ring A is advertibly open under any of the following conditions: (1) The set of left [right] advertible elements is a neighborhood of zero. (2) A has no proper zero-divisors. (3) A satisfies the minimum condition on closed left ideals. [In the totally disconnected case, use Exercise 35.2; in general, use 26.26.]

35.4 An idempotent of a ring A is *central* if it belongs to the center of A. Let E be the set of central idempotents of A. (a) If $e, f \in E$ and if $Ae = Af$, then $e = f$. (b) The relation $\leq$ on E satisfying $e \leq f$ if and only if $Ae \subseteq Af$ is an ordering on E. A *minimal* central idempotent is a minimal member of $E \setminus \{0\}$ for the induced ordering. (c) If $e, f \in E$, then $e \leq f$ if and only if $ef = e$. (d) Any set of minimal central idempotents is orthogonal.

35.6 A ring A is *biregular* if for each $x \in A$ there is a central idempotent $e \in A$ such that x and e generate the same ideal. (a) If $x \in A$ and if e is a central idempotent of A, then x and e generate the same ideal of A if and only if $x = xe$ and there exist $a_1, \ldots, a_n, b_1, \ldots, b_n \in A$ such that $e = \sum_{k=1}^n a_k x b_k$. (b) A biregular ring is semisimple. (c) If A is the ring of all linear operators on a vector space E, then A is biregular if and only if E is finite-dimensional. [Use 25.21.] (d) An epimorphic image of a biregular ring is biregular. (e) A simple ring is biregular if and only if it has an identity element. (f) A biregular ring with identity whose center is a local ring is simple. (g) If J is an ideal of a biregular ring A, then the ring J is biregular. (h) The cartesian product of finitely many biregular rings is biregular.

35.7 A subring B of a biregular ring A is a *strictly biregular subring* if for each $x \in B$ there is a central idempotent e of A belonging to B such that the ideals of B generated by x and e are identical (and hence the ideals of A generated by x and e are identical). (a) If a biregular ring A is the local direct sum of $(A_\lambda)_{\lambda \in L}$ relative to subrings $(B_\lambda)_{\lambda \in L}$, then A_λ is biregular for all $\lambda \in L$, and B_λ is a strictly biregular subring of A_λ for all but finitely many $\lambda \in L$. [Let $M = \{\lambda \in L : B_\lambda$ is not a strictly biregular subring of $A_\lambda\}$, and for each $\lambda \in M$, let $x_\lambda \in B_\lambda$ be such that there is no central idempotent of A_λ belonging to B_λ that generates the same ideal of B_λ as x_λ, and let $x_\mu = 0$ if $\mu \in L \setminus M$.] (b) If a ring A is the local direct sum of biregular rings $(A_\lambda)_{\lambda \in L}$ relative to strictly biregular subrings $(B_\lambda)_{\lambda \in L}$ and if each B_λ is isomorphic to the ring of all linear operators on an n_λ-dimensional vector space, then A is biregular if and only if $\{n_\lambda : \lambda \in L\}$ is bounded. [Observe that if u is a linear operator of rank 1, $\sum_{i=1}^m a_i u b_i$ is a linear operator of rank at most m.]

35.8. A topological ring A is *locally without central idempotents* if there is a neighborhood of zero that contains no nonzero central idempotents. A topological ring A is a totally disconnected, locally compact, biregular ring if and only if A is the topological direct sum of a locally compact biregular ring that is locally without central idempotents and a ring that is topologically isomorphic to the local direct sum of a family $(A_\lambda)_{\lambda \in L}$ of discrete, biregular rings with identity relative to subrings $(B_\lambda)_{\lambda \in L}$, where for each $\lambda \in L$, B_λ contains the identity of A_λ and is isomorphic to the (finite) ring of all linear operators on an n_λ-dimensional vector space over a finite field, and where $\{n_\lambda : \lambda \in L\}$ is bounded. [Use 32.6 and Exercises 35.7, 34.7, and 35.6.]

35.9 A topological ring A is a compact biregular ring if and only if A is topologically isomorphic to $\prod_{\lambda \in L} A_\lambda$, where for some $N > 0$, each A_λ is the discrete ring of all linear operators on a vector space of dimension not exceeding N over a finite field. [Use Exercises 35.6(b) and 35.7(b).]

35.10 (Skorniakov [1962]) Let A be a locally compact, totally disconnected biregular ring. (a) Let B be a compact open subring of A. If $(e_k)_{k \geq 1}$ is a sequence of orthogonal central idempotents and if $x_k \in Ae_k \cap B$ for all $k \geq 1$, then $(x_k)_{k \geq 1}$ is summable. [Recall that in a compact space, a sequence that has a unique adherent point converges to that point. To show that zero is the only adherent point c of $(x_k)_{k \geq 1}$, let e be the central idempotent generating the same ideal as c, first show that $ce_k = 0$ for all $k \geq 1$, then show that $x_k e = 0$ for all $k \geq 1$.] (b) A central idempotent of a topological ring is *discrete* or *nondiscrete* according as the ideal it generates is discrete or nondiscrete. If e is a central idempotent of a topological ring, then e is discrete if and only if there is a neighborhood U of zero such that $eU = \{0\}$. (c) If A is locally without central idempotents, then any orthogonal family of nondiscrete central idempotents in A is finite. [In the contrary case, apply (a); if $s = \sum_{j=1}^\infty x_j$ and if e is the central idempotent generating the same ideal as s, first show that $se_k = x_k$ for all $k \geq 1$, then show that for all sufficiently large n, $ee_n \in B$.] (d) If all the central idempotents of A are discrete, then A is discrete. [In the contrary case, let B be a compact open subring of A, and show that there are sequences $(x_k)_{k \geq 1}$ and $(e_k)_{k \geq 1}$ of nonzero elements satisfying the hypotheses of (a); for this, if $x_1, \ldots, x_n$

and $e_1, \ldots, , e_n$ are chosen, let

$$x_{n+1} = y - \sum_{k=1}^{n} y e_k$$

where $y \in B \setminus A(e_1 + \cdots + e_n)$ and is sufficiently small. Apply (a), and observe that if e is as in (c), then $x_n e = 0$ for all but finitely many $n \geq 1$.]

35.11 (Skorniakov [1977]) A nondiscrete central idempotent e of a topological ring is *topologically minimal* if there do not exist orthogonal, nondiscrete central idempotents e_1 and e_2 such that $e = e_1 + e_2$. A topological ring A is *conditionally simple* if there is a discrete ideal H such that A/H is a simple ring. Let A be a locally compact, totally disconnected, biregular ring that is locally without central idempotents. (a) If e is a nondisdcrete central idempotent of A, there is a topologically minimal central idempotent e_1 such that $e_1 \leq e$. [Use Exercise 35.10(c).] (b) A is the topological direct sum of a discrete biregular ring and finitely many locally compact biregular rings, each having an identity element that is a topologically minimal central idempotent. [Use (a) and Exercise 35.10(c), (d).] (c) If A has an identity element that is a topologically minimal central idempotent, then A is conditionally simple. [If A has a minimal central idempotent e that is nondiscrete, let $H = A(1 - e)$; in the contrary case, let H be the ideal generated by all the discrete central idempotents of A, use Exercises 35.5(g) and 35.10(b), (d) to show that H is discrete, and show that $A = Ae + H$ whenever e is a nondiscrete central idempotent.]

35.12 (Skorniakov [1977]) A topological ring A is a locally compact biregular ring if and only if A is the topological direct sum of subrings A_0, A_1, A_2, and A_3, described as follows: (a) A_0 is a finite-dimensional semisimple algebra over $\mathbb{R}$; (b) there is an integer $N > 0$ such that A_1 is topologically isomorphic to the local direct sum of a family $(A_\lambda)_{\lambda \in L}$ of discrete biregular rings with identity relative to subrings $(B_\lambda)_{\lambda \in L}$, where for each $\lambda \in L$, the identity element of A_λ is the only nonzero central idempotent of A_λ belonging to B_λ, and B_λ is isomorphic to the ring of all linear operators on a vector space of dimension not exceeding N over a finite field; (c) A_2 is the topological direct sum of finitely many locally compact, totally disconected, nondiscrete, biregular, conditionally simple rings with identity; (d) A_3 is a discrete biregular ring. [Use 35.5 and Exercises 35.7-8, 10-11.]

35.13 A ring is *strongly regular* if for each $a \in A$ there exists $x \in A$ such that $a^2 x = a$. (a) A strongly regular ring has no nonzero nilpotents. (b) A strongly regular ring is semisimple. [Use 26.11.] (c) An epimorphic image of a strongly regular ring is strongly regular. (d) The cartesian product of strongly regular rings is strongly regular. (e) A primitive strongly regular ring is a division ring.

35.14 A topological ring A is a compact, strongly regular ring if and only if A is topologically isomorphic to the cartesian product of finite fields.

35.15 Let A be a strongly regular ring. (a) If $a, x \in A$ satisfy $a^2 x = a$, then ax is a central idempotent. [Use Exercises 35.13(b), (e) to show that ax is an idempotent; then use Exercises 35.13(a) and 34.3.] (b) In particular, A is biregular. (c) If $a^2 x = a$ and $e = ax$, then $e = a^n x^n$ for all $n \geq 1$.

35.16 (a) If a ring A is the local direct sum of rings $(A_\lambda)_{\lambda \in L}$ relative to subrings $(B_\lambda)_{\lambda \in L}$, then A is strongly regular if and only if A_λ is strongly regular for all $\lambda \in L$ and B_λ is strongly regular for all but finitely many $\lambda \in L$. [Argue as in Exercise 35.7(a).] (b) A topological ring A is totally disconnected, locally compact, and strongly regular if and only if A is the topological direct sum of a locally compact, strongly regular ring that is locally without central idempotents (Exercise 35.8) and a ring that is topologically isomorphic to the local direct sum of a family $(A_\lambda)_{\lambda \in L}$ of discrete, strongly regular rings with identity relative to subrings $(B_\lambda)_{\lambda \in L}$, where for each $\lambda \in L$, B_λ contains the identity element of A_λ and is a finite field. [Use Exercise 35.13(e) in arguing as in Exercise 35.8.]

35.17 (Skorniakov [1977]) Let M be a discrete maximal ideal of a nondiscrete, locally compact, totally disconnected, strongly regular ring A with identity. (a) A/M is a nondiscrete locally compact division ring. [Use Exercise 35.13(e).] (b) There is a topological nilpotent $w \in A \setminus M$. [Show that there is a compact open subring V such that the restriction to V of the canonical epimorphism ϕ from A to A/M is injective and therefore a homeomorphism from V to $\phi(V)$, and use 18.17.] (c) Let $e = wx$, where $w^2 x = w$. Then A is the topological direct sum of Ae and M. [To show that $Ae + M = A$, use (a) and Exercise 35.15. If $h \in Ae \cap M$, let $d = hy$ where $h^2 y = h$, observe that $d = de$, and use Exercises 35.15(c) and 35.10(b) to show that $d = 0$.]

35.18 (Kaplansky [1949b]) A topological ring A is locally compact and strongly regular if and only if A is the topological direct sum of A_1, A_2, and A_3, described as follows: A_1 is the direct sum of finitely many nondiscrete locally compact division rings; A_2 is topologically isomorphic to the local direct sum of a family $(A_\lambda)_{\lambda \in L}$ of discrete, strongly regular rings with identity relative to finite subfields $(B_\lambda)_{\lambda \in L}$, where for each $\lambda \in L$, B_λ contains the identity of A_λ; and A_3 is a discrete strongly regular ring. [Use Exercises 35.12, 35.16(b), and 35.17.]

36 The Radical Topology

If A is a ring with radical R, the *radical topology* of A is the topology for which the powers $(R^n)_{n \geq 1}$ of R form a fundamental system of neighborhoods of zero. Thus, the radical topology is Hausdorff if and only if $\bigcap_{n=1}^\infty R^n =$

$\{0\}$.

Here we shall show that much of the theory of artinian rings has a natural extension to the theory of rings that are linearly compact for the discrete topology and Hausdorff for the radical topology. But first, we need more information about artinian rings. The artinian rings are precisely the rings strictly linearly compact for the discrete topology, which is a bounded topology, by (2) of 28.14.

We shall call a ring or ideal a torsion [torsionfree, divisible, primary] ring or ideal if its underlying additive group is a torsion [torsionfree, divisible, primary] group.

36.1 Theorem. *If L is a left ideal of an artinian ring A, then the additive group L is the direct sum of a divisible left ideal and a subgroup M satisfying $m.M = \{0\}$ for some $m \geq 1$.*

Proof. For each $q \in \mathbb{Z}$, $q.L$ is a left ideal. Consequently, there is an integer $m > 0$ such that $m.L$ is minimal in the set $\{q.L : q > 0\}$ of left ideals of A, ordered by inclusion. For any nonzero integer n, $nm.L \subseteq m.L$ and hence $n.(m.L) = nm.L = m.L$. Therefore $m.L$ is a divisible left ideal. By 30.2 there is an additive subgroup M of L such that the additive group L is the direct sum of $m.L$ and M. Then M is isomorphic to the additive group $L/m.L$, and hence $m.M = \{0\}$. •

36.2 Theorem. *If A is a nonzero, torsionfree artinian ring, then every left ideal of A is a divisible left ideal, and A has a left identity.*

Proof. Every left ideal of A is a divisible left ideal by 36.1, since A contains no nonzero torsion subgroups.

A nilpotent artinian ring is a discrete, linearly compact $\mathbb{Z}$-module by 34.12 and hence is a torsion ring by 30.10. Therefore the radical R of A is a proper ideal by 27.15, and hence A/R has an identity element by 26.16 and 27.14. By 34.1 there is an idempotent $e \in A$ such that $e + R$ is the identity of A/R, and hence $x - ex \in R$ for each $x \in A$. To show that e is a left identity of A, let $a \in A$, and let $b = a - ea$. We have just seen that the left ideal J_b generated by b is then divisible, so there exists $c \in J_b$ such that $2.c = b$. Let $n \in \mathbb{Z}$ and $d \in A$ be such that $c = n.b + db$. Then

$$(2n - 1).c = -db = -d(2.c) = (-2.d)c.$$

By the preceding, A is divisible, so there exists $h \in A$ such that $(2n - 1).h = -2.d$. Thus $(2n - 1).c = (2n - 1).hc$, so as A is torsionfree, $c = hc$, and hence $b = hb$. Consequently, $ehb = eb = e(a - ea) = 0$. Therefore $(h - eh)b = hb = b$, so by iteration $(h - eh)^k b = b$ for all $k \geq 1$. As $h - eh \in R$, a nilpotent ideal by 27.15, we conclude that $b = 0$ and hence $a = ea$. •

The torsion subgroup T of the additive group of a ring A is an ideal, since if $a, t \in A$ and $n.t = 0$, then $n.at = a(n.t) = 0$ and similarly $n.ta = 0$. Moreover, A/T is a torsionfree ring, for if $n.(a+T) = T$ where $n > 0$, then $n.a \in T$, so $mn.a = m.(n.a) = 0$ for some $m > 0$, whence $a \in T$. Also the largest divisible subgroup D of the additive group A is an ideal, since if $a \in A$, $d \in D$ and $n > 0$, there exists $b \in D$ such that $n.b = d$, whence $n.ab = a(n.b) = ad$ and similarly $n.ba = da$. We shall call D the *largest divisible ideal* of A.

If S is a subset of a ring A, the *left [right] annihilator* of S is the set of all $x \in A$ such that $xS = \{0\}$ [$Sx = \{0\}$], and the *annihilator* of S is the set of all $s \in A$ such that $xS = Sx = \{0\}$. The [left, right] annihilator of a set in a topological ring is clearly closed.

36.3 Theorem. *If D is the largest divisible ideal and T the torsion ideal of an artinian ring A, then D is contained in the annihilator of T, and $D \cap T$ is contained in the annihilator of A.*

Proof. If $d \in D$ and $t \in T$, then there exists $n > 0$ such that $n.t = 0$ and there exists $h \in D$ such that $n.h = d$, so $dt = (n.h)t = h(n.t) = 0$ and similarly $td = 0$. By 36.1 applied to A, $D + T = A$, so

$$(D \cap T)A = (D \cap T)(D + T) \subseteq TD + DT = \{0\}$$

and similarly $A(D \cap T) = \{0\}$. •

36.4 Theorem. *Let T be the torsion ideal and D the largest divisible ideal of an artinian ring A. There is a unique ideal S of A such that $S + T = A$, S has a left identity, and $S \subseteq D$. Moreover, the ring A is the direct sum of S and T, and S is a divisible ideal.*

Proof. By 36.1 applied to A, A contains a divisible subgroup D_0 such that A/D_0 is a torsion group. Since D_0 is contained in the largest divisible subgroup D, A/D is an epimorphic image of A/D_0 and hence is a torsion group. By 30.2, D has a supplement T_0; as T_0 is isomorphic to A/D, T_0 is a torsion subgroup. Now $D \cap T$ is a divisible ideal, since if $m.d = 0$ where $m > 0$ and if $b \in D$ satisfies $n.b = d$, then $nm.b = 0$, so $b \in D \cap T$. Consequently, by 30.2, the additive group D is the direct sum of an additive group B and $D \cap T$. Clearly $T = (D \cap T) + T_0$. Therefore the additive group A is the direct sum of B and T. As $B \subseteq D$, $BT = TB = \{0\}$ by 36.3. As A/T is a torsionfree artinian ring by 27.4, it has a left identity element by 36.2. Thus, B contains an element f such that $x - fx \in T$ for all $x \in A$.

Let $S = fB$. As D is an ideal, $S \subseteq D$. To show that S is a subring, let $b, c \in B$, and let $bfc = d + t$ where $d \in B$, $t \in T$. Then $fbfc = fd + ft = $

$fd \in S$ as $BT = \{0\}$. To show that $S + T = A$, let $a \in A$, and let $a = b + t$ where $b \in B$ and $t \in T$. Then $a - fa \in T$ and $ft = 0$, so

$$a = fa + (a - fa) = f(b+t) + (a - fa) = fb + (a - fa) \in S + T.$$

Consequently,

$$SA = S(S+T) \subseteq SS + fBT \subseteq S,$$
$$AS = (S+T)S \subseteq SS + TfB \subseteq S + TB = S$$

as T is an ideal, so S is an ideal. Moreover, $S \cap T = \{0\}$, for if $b \in B$ and $fb \in T$, then $b - fb \in T$, whence $b \in T$, and thus $b \in B \cap T = \{0\}$. Therefore the ring A is the direct sum of the ideals S and T. Consequently, the ring S is isomorphic to A/T and hence is torsionfree. Thus by 36.2, S has a left identity element e and is a divisible ideal.

Let S' be an ideal such that $S' + T = A$, S' has a left identity e', and $S' \subseteq D$. Let $s \in S$, and let $s = s' + t$ where $s' \in S'$ and $t \in T$. Then $t = s - s' \in D$, so $t \in D \cap T$, whence $et \in DT = \{0\}$. Therefore $s = es = es' + et = es' \in S'$. Thus $S \subseteq S'$, and similarly $S' \subseteq S$. •

We shall call the unique ideal S of A simply *the ideal supplement* of T. Thus the ideal supplement of T is a divisible, torsionfree ideal.

36.5 Theorem. *If g is an epimorphism from an artinian ring A to an artinian ring A', if T and T' are respectively the torsion ideals of A and A' and if S and S' are their ideal supplements, then $g(S) = S'$ and $g(T) = T'$.*

Proof. Let D and D' be respectively the largest divisible subgroups of A and A'. Clearly $g(T) \subseteq T'$ and $g(D) \subseteq D'$ so

$$g(S) + T' \supseteq g(S) + g(T) = g(S+T) = g(A) = A'$$

and $g(S) \subseteq D'$. Also, if e is the left identity of S, $g(e)$ is the left identity of $g(S)$. Hence by 36.4, $g(S) = S'$. As $S' + g(T) = A'$, $g(T) \subseteq T'$, and as A' is the direct sum of S' and T', we conclude that $g(T) = T'$. •

The subgroup [ideal] T of all elements of a Hausdorff commutative group [ring] A contained in some compact additive subgroup of A is closed if A is locally compact by 35.2, but need not be closed in general. For example, in the group $\mathbb{Q}(\sqrt{2})/\mathbb{Z}$, T is the dense subgroup $\mathbb{Q}/\mathbb{Z}$, the subgroup of all elements of finite order, since a countable compact group is discrete by Baire's theorem and hence finite. Much can be established about T, however, if A is complete and the open additive subgroups form a fundamental system of neighborhoods of zero:

36.6 Theorem. *If c is an element of a complete, Hausdorff, commutative group A whose open subgroups form a fundamental system of neighborhoods of zero, then the closure $[c]$ of $\mathbb{Z}.c$, the cyclic group generated by c, is either an infinite discrete group or a compact group. The set T of all elements c such that $[c]$ is compact is a closed subgroup of A. If A is the additive group of a topological ring, then T is an ideal.*

Proof. If $\mathbb{Z}.c$ is an infinite discrete group, then it is closed by 4.13 and hence is $[c]$.

Assume that $\mathbb{Z}.c$ is not discrete, and let $\mathcal{U}$ be the filter base of all open subgroups. Then for each $U \in \mathcal{U}$, $U \cap \mathbb{Z}.c$ is a nonzero subgroup of $\mathbb{Z}.c$ and hence $\mathbb{Z}.c/(\mathbb{Z}.c \cap U)$ is finite. By 5.2, the canonical homomorphism g from $\mathbb{Z}.c$ to $\varprojlim_{U \in \mathcal{U}}(\mathbb{Z}.c/(\mathbb{Z}.c \cap U))$ is a topological isomorphism from $\mathbb{Z}.c$ to a dense subgroup. Consequently by 8.4, $[c]$ is topologically isomorphic to $\varprojlim_{U \in \mathcal{U}}(\mathbb{Z}.c/(\mathbb{Z}.c \cap U))$, which is compact by 5.20 and Tikhonov's theorem.

As mentioned in the proof of 35.2, T is a subgroup and, if A is a ring, an ideal. Let $b \in \overline{T}$, and let $U \in \mathcal{U}$. Then there exists $c \in T$ such that $c - b \in U$, and there exists $n > 0$ such that $n.c \in U$. As U is a subgroup, $n.(c - b) \in U$ and therefore $n.b = n.c - n.(c - b) \in U$. Thus $\mathbb{Z}.b$ is not an infinite discrete group, and hence $b \in T$. •

36.7 Definition. *Let A be a complete, Hausdorff, abelian group [ring] whose open [additive] subgroups form a fundamental system of neighborhoods of zero. The **topological torsion subgroup [ideal]** of A is the group [ideal] T of all elements $c \in A$ such that $[c]$ is compact, and A is a **topological torsion group [ring]** if $A = T$. For each prime p, the **topological p-primary component** of T is the set T_p of all elements $c \in A$ such that*

$$\lim_{n \to \infty} p^n.c = 0,$$

*and A is a **topological p-primary group [ring]** if $A = T_p$.*

Clearly T_p is indeed a subset of T and is a subgroup [an ideal] of A. To describe the relation between the topological torsion group [ideal] and its primary components, we need the following definition:

36.8 Definition. *Let $(A_\lambda)_{\lambda \in L}$ be a family of subgroups [subrings] of a Hausdorff, abelian topological group [ring] A, and let $\mathcal{U}$ be the filter of neighborhoods of zero. We define $\mathfrak{S}_{\lambda \in L} A_\lambda$ to be the subgroup [subring] of $\prod_{\lambda \in L} A_\lambda$ consisting of all $(x_\lambda)_{\lambda \in L}$ such that for every $U \in \mathcal{U}$, $x_\lambda \in U$ for all but finitely many $\lambda \in L$. The **uniform** topology on $\mathfrak{S}_{\lambda \in L} A_\lambda$ is that for which $\{\mathfrak{S}_{\lambda \in L} A_\lambda \cap \prod_{\lambda \in L}(U \cap A_\lambda) : U \in \mathcal{U}\}$ is a fundamental system of neighborhoods of zero.*

It is easy to verify that the indicated fundamental system of neighborhoods of zero satisfies the conditions (TGB 1), (TBG 2) [and (TRN 1), (TRN 2)] on pages 20-21. If $A = A_\lambda = \mathbb{R}$ for all $\lambda \in L$, then $\mathfrak{S}_{\lambda \in L} A_\lambda$ is simply the group [ring] of all real-valued functions on L that "vanish at infinity," furnished with the uniform topology.

For each $\mu \in L$, the restriction σ_μ to $\mathfrak{S}_{\lambda \in L} A_\lambda$ of the canonical projection pr_μ from $\prod_{\lambda \in L} A_\lambda$ to A_μ is a topological epimorphism. Indeed, clearly $in_\mu(A_\mu) \subseteq \mathfrak{S}_{\lambda \in L} A_\lambda$ where in_μ is the canonical injection from A_μ to $\prod_{\lambda \in L} A_\lambda$, so $\sigma_\mu(\mathfrak{S}_{\lambda \in L} A_\lambda) = A_\mu$. The uniform topology on $\mathfrak{S}_{\lambda \in L} A_\lambda$ is stronger than that induced by the cartesian product topology on $\prod_{\lambda \in L} A_\lambda$, so σ_μ is continuous. Therefore the identity

$$pr_\mu(\mathfrak{S}_{\lambda \in L} A_\lambda \cap \prod_{\lambda \in L}(U \cap A_\lambda)) = U \cap A_\mu$$

for each $U \in \mathcal{U}$ establishes that σ_μ is a topological epimorphism.

It is easy to see that if, for each $\lambda \in L$, B_λ is a subgroup [subring] of A_λ, then $\mathfrak{S}_{\lambda \in L} B_\lambda$ is a topological subgroup [subring] of $\mathfrak{S}_{\lambda \in L} A_\lambda$, that is, the uniform topology on $\mathfrak{S}_{\lambda \in L} B_\lambda$ is the topology induced on $\mathfrak{S}_{\lambda \in L} B_\lambda$ by the uniform topology of $\mathfrak{S}_{\lambda \in L} A_\lambda$.

If A is a complete, Hausdorff, abelian group for which the open subgroups form a fundamental system of neighborhoods of zero, then by 10.5, $\mathfrak{S}_{\lambda \in L} A_\lambda$ is the set of summable families $(x_\lambda)_{\lambda \in L}$ such that $x_\lambda \in A_\lambda$ for all $\lambda \in L$.

36.9 Theorem. *Let A be a complete, Hausdorff, abelian group [ring] for which the open [additive] subgroups form a fundamental system of neighborhoods of zero. Let T be the topological torsion subgroup [ideal] of A, let P be the set of prime integers, and for each $p \in P$ let T_p be the topological p-primary component of T. The function S from $\mathfrak{S}_{p \in P} T_p$ to T, defined by*

$$S((c_p)_{p \in P}) = \sum_{p \in P} c_p,$$

is a topological isomorphism.

Proof. Let $\mathcal{U}$ be the filter base of open additive subgroups of the ideal T. Let $(c_p)_{p \in P} \in \mathfrak{S}_{p \in P} T_p$. As $\bigcup_{p \in P} T_p \subseteq T$, and as T is a subgroup, $\sum_{p \in F} c_p \in T$ for any finite subset F of P; hence as T is closed, $\sum_{p \in P} c_p \in T$. Thus the range of S is indeed contained in T. Clearly S is an additive homomorphism.

Let $c \in T$. As $[c]$ is compact, for each $U \in \mathcal{U}$ the image $([c] + U)/U$ of $[c]$ under the canonical epimorphism from T to T/U is a finite group, and hence for each prime p there exists $c_{p,U} \in T$ such that $c_{p,U} + U$ is the p-primary component of $([c]+U)/U$. Then $c_{p,U} \in U$ for all but finitely many $U \in \mathcal{U}$, and

(1) $$c + U = \sum_{p \in P} c_{p,U} + U.$$

If $U, V \in \mathcal{U}$ are such that $V \subseteq U$, the restriction to $([c]+V)/V$ of the canonical epimorphism $\phi_{U,V}$ from T/V to T/U is an epimorphism taking the pth component of $c+V$ to the p-th component of $c+U$ for each $p \in P$; in short, $c_{p,V} + U = c_{p,U} + U$, that is, $c_{p,V} + V \subseteq c_{p,U} + U$ whenever $V \subseteq U$. Thus for each $p \in P$, $\{c_{p,U} + U : U \in \mathcal{U}\}$ is a Cauchy filter base of closed sets (as $c_{p,U} + U$ is a U-small subset) and hence converges to an element $c_p \in T$. For each $U \in \mathcal{U}$, $c_p \in c_{p,U} + U$ and hence $c_p - c_{p,U} \in U$. For each $U \in \mathcal{U}$, as $c_{p,U} + U$ belongs to the pth component of $([c]+U)/U$, there exists $n \geq 1$ such that $p^n.c_{p,U} \in U$. Consequently,

(2) $$p^n.c_p = p^n.(c_p - c_{p,U}) + p^n.c_{p,U} \in U,$$

so $p^m.c_p \in U$ for all $m \geq n$ as U is a subgroup. Hence $\lim_{n \to \infty} p^n.c_p = 0$, that is, $c_p \in T_p$. For each $U \in \mathcal{U}$, the set P_U of primes such that $c_{p,U} \notin U$ is finite, and $c - \sum_{p \in P_U} c_{p,U} \in U$. Consequently, by (1) and (2), for any finite subset J of P containing P_U,

$$c - \sum_{p \in J} c_p = c - \sum_{p \in J} c_{p,U} + \sum_{p \in J}(c_{p,U} - c_p) \in U + U = U.$$

Therefore $(c_p)_{p \in P}$ is summable, and its sum is c. Thus by 10.5, $(c_p) \in \mathfrak{S}_{p \in P} T_p$. Consequently, S is an additive epimorphism.

Let $(c_p)_{p \in P} \in \mathfrak{S}_{p \in P} T_p$, and let $c = \sum_{p \in P} c_p$. Then for each $U \in \mathcal{U}$, $c_p + U$ is clearly the pth component of $c + U$ in the torsion group T/U. Therefore, if $c = 0$, then for each $U \in \mathcal{U}$, $c_p \in U$ for all $p \in P$ by 30.5, and consequently $c_p = 0$ for all $p \in P$. Thus S is injective and hence an additive isomorphism.

Let $U \in \mathcal{U}$, and let $U' = \prod_{p \in P}(U \cap T_p)$. If $(c_p)_{p \in P} \in U'$, then $c_p \in U$ for all $p \in P$ and hence $\sum_{p \in P} c_p \in U$ as U is a closed subgroup. Conversely, let $c \in U$, and let $c = \sum_{p \in P} c_p$. As $c + U$ is the zero element of T/U, each each of its components $c_p + U$ is also the zero element by 30.5, and hence $(c_p)_{p \in P} \in U'$. Thus $S(U') = U$, so S is an additive topological isomorphism.

Finally, assume that A is a topological ring. If p and q are distinct primes, then $T_p T_q = \{0\}$. Indeed, if $a \in T_p$ and if $b \in T_q$, then for any $n \geq 1$ there exist integers r_n and s_n such that $r_n p^n + s_n q^n = 1$ and hence $ab = r_n.(p^n.a)b + s_n.a(q^n.b)$. As each $U \in \mathcal{U}$ is an additive group,

$$\lim_{n \to \infty} r_n.(p^n.a)b = 0 = \lim_{n \to \infty} s_n.a(q^n.b).$$

Consequently, $ab = 0$. Therefore by 10.16, for any $(a_p)_{p\in P}$, $(b_p)_{p\in P} \in \mathfrak{S}_{p\in P}T_p$,

$$\sum_{p\in P} a_p \sum_{p\in P} b_p = \sum_{p\in P} a_p b_p.$$

Thus S is a topological isomorphism from the topological ring $\mathfrak{S}_{p\in P}T_p$ to T. •

The discrete case of Theorem 36.9 yields the ring extension of Theorem 30.5:

36.10 Corollary. *A torsion ring T is the direct sum of its primary components $(T_p)_{p\in P}$.*

Proof. Indeed, if T is given the discrete topology, for any prime p the topological p-primary component of T is its p-primary component, and $\mathfrak{S}_{p\in P}T_p$ is the discrete ring $\bigoplus_{p\in P} T_p$. •

36.11 Theorem. *Let $(A_\lambda)_{\lambda\in L}$ be a family of subgroups [subrings] of a Hausdorff, abelian topological group [ring] A. (1) If for each $\mu \in L$, A_μ is complete, then $\mathfrak{S}_{\lambda\in L}A_\lambda$ is complete. (2) If for each $\mu \in L$, A_μ is a [strictly] linearly compact module or ring, then $\mathfrak{S}_{\lambda\in L}A_\lambda$ is [strictly] linearly compact if and only if $\mathfrak{S}_{\lambda\in L}A_\lambda = \prod_{\lambda\in L} A_\lambda$, or equivalently, for each neighborhood U of zero in A, $A_\lambda \subseteq U$ for all but finitely many $\lambda \in L$.*

Proof. For each $\mu \in L$, let pr_μ be the canonical projection from $\mathfrak{S}_{\lambda\in L}A_\lambda$ to A_μ, let $\mathcal{U}$ be a fundamental system of symmetric neighborhoods of zero, and for each $U \in \mathcal{U}$, let $\tilde{U} = \mathfrak{S}_{\lambda\in L}A_\lambda \cap \prod_{\lambda\in L}(A_\lambda \cap U)$. A subset F of $\mathfrak{S}_{\lambda\in L}A_\lambda$ is $\tilde{U}$-small if and only if for all $\mu \in L$, $pr_\mu(F)$ is $(A_\mu \cap U)$-small.

(1) Let $\mathcal{F}$ be a Cauchy filter base on $\mathfrak{S}_{\lambda\in L}A_\lambda$. Then for each $\mu \in L$, $pr_\mu(\mathcal{F})$ is a Cauchy filter base on A_μ and hence converges to some $a_\mu \in A_\mu$. To show that $(a_\mu)_{\mu\in L} \in \mathfrak{S}_{\lambda\in L}A_\lambda$, let $U \in \mathcal{U}$, let $V \in \mathcal{U}$ be such that $V + V \subseteq U$, and let $F \in \mathcal{F}$ be $\tilde{V}$-small. Let $(b_\lambda)_{\lambda\in L} \in F$. For each $\mu \in L$, $a_\mu \in \overline{pr_\mu(F)} \subseteq pr_\mu(F) + V$, so there exists $(c_\lambda)_{\lambda\in L} \in F$ such that $a_\mu \in c_\mu + V$, and consequently

$$a_\mu - b_\mu = (a_\mu - c_\mu) + (c_\mu - b_\mu) \in V + V \subseteq U.$$

Therefore as $b_\mu \in U$ for all but finitely many $\mu \in L$, $a_\mu \in U$ for all but finitely many $\mu \in L$. Thus $(a_\mu)_{\mu\in L} \in \mathfrak{S}_{\lambda\in L}A_\lambda$. Moreover, for any $(x_\lambda)_{\lambda\in L} \in F$ and any $\mu \in L$, $x_\mu - a_\mu = (x_\mu - b_\mu) + (b_\mu - a_\mu) \in V + V \subseteq U$, so $(x_\lambda)_{\lambda\in L} \in (a_\lambda) + \tilde{U}$; thus $F \subseteq (a_\lambda)_{\lambda\in L} + \tilde{U}$. Consequently, $\mathcal{F}$ converges to $(a_\lambda)_{\lambda\in L}$.

(2) We may assume that each $U \in \mathcal{U}$ is an additive subgroup. If $U \in \mathcal{U}$, $(\mathfrak{S}_{\lambda\in L}A_\lambda)/\tilde{U}$ is isomorphic to the discrete module $\bigoplus_{\lambda\in L}(A_\lambda/(A_\lambda \cap U))$,

and thus by 28.21 and 28.7 [27.6] is linearly compact [artinian] if and only if for all but finitely many $\lambda \in L$, $A_\lambda/(A_\lambda \cap U) = \{0\}$, that is, $A_\lambda \subseteq U$. Consequently by (1) and 28.15, the assertion holds. •

For each prime p, the additive group $\mathbb{Q}_p$ is an example of a divisible, topological torsion group that is torsionfree, and its quotient group $\mathbb{Q}_p/\mathbb{Z}_p$, also denoted by $\mathbb{Z}(p^\infty)$, is a basic divisible p-primary group, as noted on page 253. By (3) of 18.10, the closed, nonzero, proper subgroups of $\mathbb{Q}_p$ are the groups $p^n\mathbb{Z}_p$ where $n \in \mathbb{Z}$. The topological automorphism $x \to p^n x$ of the additive group $\mathbb{Q}_p$ induces a topological isomorphism from $\mathbb{Q}_p/\mathbb{Z}_p$ to $\mathbb{Q}_p/p^n\mathbb{Z}_p$ for each $n \in \mathbb{Z}$. Thus, for each $n \in \mathbb{Z}$, $\mathbb{Q}_p/p^n\mathbb{Z}_p$ is a basic p-primary group. Consequently, the $\mathbb{Z}$-module $\mathbb{Q}_p$ is a strictly linearly compact $\mathbb{Z}$-module. The absence of subgroups topologically isomorphic to $\mathbb{Q}_p$ or $\mathbb{Q}_p/\mathbb{Z}_p$ in the additive group of a linearly compact ring, furnished with its radical topology, is both necessary and sufficient for several attractive statements. Consequently, we shall say that an additive subgroup of a topological ring is *pathological* if it is topologically isomorphic either to the additive topological group $\mathbb{Q}_p$ or to a basic divisible p-primary group for some prime p. One reason for calling these topological groups pathological is apparent from the following theorem:

36.12 Theorem. *Let A is a bounded, strictly linearly compact ring. If D is the largest divisible ideal of A and T its topological torsion ideal, then $D \cap T$ is contained in the annihilator of A. In particular, a pathological subgroup of A is contained in its annihilator.*

Proof. Let U be an open ideal of A. Then A/U is an artinian ring, and the image of $D \cap T$ under the canonical epimorphism from A to A/U is a divisible, torsion ideal. Thus by 36.3, $A(D \cap T) \subseteq U$ and $(D \cap T)A \subseteq U$. Therefore $A(D \cap T) = \{0\} = (D \cap T)A$. •

The following theorem gives a useful characterization of pathological groups:

36.13 Theorem. *Let p be a prime, and let E be a complete, Hausdorff, nonzero abelian group whose open subgroups form a fundamental system of neighborhoods of zero. If E contains a dense subgroup H generated by a family $(a_i)_{i\in\mathbb{Z}}$ of elements satisfying $p.a_{i+1} = a_i$ for all $i \in \mathbb{Z}$ and $\lim_{i\to\infty} a_{-i} = 0$, then E is topologically isomorphic either to the additive group $\mathbb{Q}_p$ or a basic divisible p-primary group.*

Proof. Let $A_p = \bigcup_{i\geq 0} p^{-i}\mathbb{Z}$, the additive subgroup of $\mathbb{Q}_p$ generated by $\{p^{-i} : i \in \mathbb{N}\}$. It is easy to verify that there is a unique epimorphism g from A_p to H satisfying $g(p^{-i}) = a_i$ for all $i \in \mathbb{Z}$ and that g is continuous. Its continuous extension $\widehat{g}$ from $\mathbb{Q}_p$ to E is a topological epimorphism since $\mathbb{Q}_p$

is a strictly linearly compact $\mathbb{Z}$-module. Thus E is topologically isomorphic either to $\mathbb{Q}_p$ or to the basic divisible p-primary group $\mathbb{Q}_p/p^n \mathbb{Z}_p$ for some $n \in \mathbb{Z}$. •

The following two lemmas enable us to infer the existence of a pathological group in a topological ring from the existence of one in a quotient ring:

36.14 Lemma. *Let ϕ be an epimorphism from an artinian ring A to an artinian ring A', let p be a prime, and let $s \in \mathbb{N}$. If $(a'_j)_{j \in \mathbb{Z}}$ is a family of elements of A' satisfying $p.a'_{j+1} = a'_j$ for all $j \in \mathbb{Z}$ and, for some $r \leq 0$, $a'_r \neq 0$ but $a'_{r-1} = 0$, then there is a family $(a_j)_{j \in \mathbb{Z}}$ of elements of A satisfying $p.a_{j+1} = a_j$ for all $j \in \mathbb{Z}$, $a_q \neq 0$ but $a_{q-1} = 0$ for some $q \leq r$, and $\phi(a_j) = a'_j$ for all $j \leq s$.*

Proof. Let T and T' be the torsion ideals of A and A' respectively, T_p and T'_p their p-primary components. By 36.5, $\phi(T) = T'$ and hence by 30.5, $\phi(T_p) = T'_p$. The additive group T_p is the direct sum of its largest divisible subgroup D_p and a subgroup B_p satisfying $m.B_p = \{0\}$ for some $m \geq 1$. Let p^k be the largest power of p dividing m. If $a \in T_p$ and if $p^n.a = 0$ where $n \geq k$, then p^k is the greatest common divisor of m and p^n, so there exist integers r and s such that $rp^n + sm = p^k$, whence $p^k.a = r.(p^n.a) + s.(m.a) = 0$. Thus $p^k.B_p = \{0\}$. Let $d \in D_p$ and $b \in B_p$ be such that $\phi(b+d) = a'_{k+s}$. Let $a_s = p^t.d$. As $a_s \in D_p$, there exists a sequence $(a_j)_{j>s}$ in D_p such that $p.a_{j+1} = a_j$ for all $j \geq s$. Let $a_j = p^{s-j}.a_s$ for all $j < s$. Clearly $(a_j)_{j \in \mathbb{Z}}$ has the desired properties. •

36.15 Lemma. *Let A be a bounded, metrizable, strictly linearly compact ring, and let U be an open ideal of A. If $c + U$ is a nonzero element of a pathological subgroup of A/U, then A contains a pathological subgroup whose image in A/U contains $c + U$.*

Proof. By 12.16 there is a decreasing sequence $(U_n)_{n \geq 0}$ of open ideals that forms a fundamental system of neighborhoods of zero such that $U_0 = U$. By hypothesis, for some prime p there is a family $(a_{0,j})_{j \in \mathbb{Z}}$ in A such that $p.a_{0,j+1} - a_{0,j} \in U_0$ for all $j \in \mathbb{Z}$, $a_{0,0} \notin U_0$, but $a_{0,-1} \in U_0$, and $c = a_{0,r}$ for some $r \geq 0$. An inductive application of 36.14 to the canonical epimorphism from A/U_i to A/U_{i-1} yields, for each $i > 0$, a family $(a_{i,j})_{j \in \mathbb{Z}}$ such that $p.a_{i,j+1} - a_{i,j} \in U_i$ for all $j \in \mathbb{Z}$, $a_{i,j} - a_{i-1,j} \in U_{i-1}$ for all $j \leq i + r$, and $a_{i,q(i)} \in U_i$ for some $q(i) \leq 0$. Given $j \in \mathbb{Z}$, the sequence $(a_{i,j})_{i \geq 0}$ is easily seen to be a Cauchy sequence and hence converges to some $b_j \in A$. Clearly $p.b_{j+1} = b_j$ for all $j \in \mathbb{Z}$, and as $a_{s,r} - a_{0,r} \in U_0$ for all $s \geq 0$, $b_r - c = b_r - a_{0,r} \in U$ as U is closed. An easy argument establishes that $\lim_{k \to \infty} b_{-k} = \lim_{k \to \infty} p^k.b_0 = 0$, so the proof is complete by 36.13. •

The following theorem generalizes the statement that a semisimple linearly compact ring is strictly linearly compact.

36.16 Theorem. *Let A be a topological ring with radical R. (1) A is linearly compact and the filter base $(R^n)_{n \geq 1}$ converges to zero if and only if A is strictly linearly compact and $\bigcap_{n=1}^{\infty} \overline{R^n} = (0)$. (2) If the radical topology of A is linearly compact, then it is strictly linearly compact.*

Proof. (1) We shall first establish that for each $n \geq 1$, $\overline{R^n}/\overline{R^{n+1}}$ is a strictly linearly compact A-module. Since $\{y \in A : Ry \subseteq \overline{R^{n+1}}\}$ is closed and contains R^n, it contains $\overline{R^n}$. Thus as $R\overline{R^n} \subseteq \overline{R^{n+1}}$, we may regard $\overline{R^n}/\overline{R^{n+1}}$ as a module over A/R that has the same submodules as the A-module $\overline{R^n}/\overline{R^{n+1}}$. By (1) of 28.16, $A/\overline{R^{n+1}}$ is a linearly compact A-module, so again by (1) of 28.16, $\overline{R^n}/\overline{R^{n+1}}$ is a linearly compact A-module and hence a linearly compact (A/R)-module. By 29.13, A/R is a strictly linearly compact ring. Therefore by 33.19, $\overline{R^n}/\overline{R^{n+1}}$ is a strictly linearly compact A/R-module and hence a strictly linearly compact A-module.

An inductive argument now establishes that $A/\overline{R^m}$ is a strictly linearly compact A-module for all $m \geq 1$. Indeed, if $A/\overline{R^n}$ is a strictly linearly compact A-module, then $(A/\overline{R^{n+1}})/(\overline{R^n}/\overline{R^{n+1}})$ is a strictly linearly compact A-module as it is topologically isomorphic to $A/\overline{R^n}$ by 5.13, so as the A-module $\overline{R^n}/\overline{R^{n+1}}$ is strictly linearly compact, the A-module $A/\overline{R^{n+1}}$ is strictly linearly compact by (2) of 28.16.

Necessity: Let U be an open left ideal of A. By hypothesis, there exists $n \geq 1$ such that $R^n \subseteq U$ and hence $\overline{R^n} \subseteq U$ as U is closed. As $\phi : x + \overline{R^n} \to x + U$ is a continuous A-linear transformation from $A/\overline{R^n}$ onto A/U, A/U is a discrete, strictly linearly compact A-module by the preceding and 28.11. Therefore A/U is an artinian A-module by (2) of 28.14. Consequently, A is strictly linearly compact by (2) of 28.15.

Sufficiency: By 33.8 and 28.13, a strictly linearly compact topology is a Leptin topology. Therefore by 33.5, as $\bigcap_{n=1}^{\infty} \overline{R^n} = (0)$, the filter base $(R^n)_{n \geq 1}$ converges to zero. Clearly (2) follows from (1). •

These considerations yield a generalization of Corollary 34.15:

36.17 Theorem. *Let A be a topological ring. (1) If A is compact and totally disconnected, then A is a bounded, strongly linearly compact ring that has no pathological subgroups. (2) If A is metrizable, then A is compact and totally disconnected if and only if A is a bounded, strongly linearly compact ring that has no pathological subgroups. (3) If A has an identity, then A is compact if and only if A is a bounded and strongly linearly compact.*

Proof. A compact ring with identity is totally disconnected by 32.3, and a totally disconnected compact ring is ideally topologized by 4.20 and hence strongly linearly compact. A compact ring is also bounded by 12.3. By 12.6, a bounded linearly compact ring is ideally topologized. (1) A pathological subgroup is complete (indeed, locally compact) but not compact and hence is not contained in a compact group.

(2) and (3): Sufficiency: Let U be a proper open ideal. Then A/U is a discrete, strongly linearly compact A-module by 28.16, hence is a strictly linearly compact A-module by 30.11, and therefore is a strictly linearly compact ring. By the condition of (2) and 36.15, A/U contains no pathological subgroups and therefore is finite by 36.10; by the hypothesis of (3) and 36.12, the same conclusion holds. If $\mathcal{U}$ is the filter base of all open ideals, A is topologically isomorphic to $\varprojlim_{U \in \mathcal{U}}(A/U)$ by 8.5, and therefore, A is compact by Tikhonov's theorem. •

The following three theorems prepare for the proof of Theorem 36.21, which characterizes those rings linearly compact for the radical topology that lack pathological subgroups.

36.18 Theorem. *Let E be an A-module, J an ideal of A. If F is a submodule of E that is closed for the J-topology of E and if $E = F + JE$, then $E = F$.*

Proof. If $E = F + J^k E$, then $JE \subseteq JF + J^{k+1}E \subseteq F + J^{k+1}E$, so $E = F + JE = F + J^{k+1}E$. Thus

$$E = \bigcap_{n=1}^{\infty}(F + J^n E) = F$$

by 3.3, as F is closed. •

36.19 Theorem. *If F is a unitary module over a semisimple artinian ring A and if F is linearly compact for the discrete topology, then F is finitely generated.*

Proof. By 33.13 and 33.16, F is a semisimple A-module and hence, by 33.14, is the direct sum of a family $(M_\lambda)_{\lambda \in L}$ of simple submodules. Thus F is isomorphic to $\bigoplus_{\lambda \in L} M_\lambda$. By 28.21, L is finite. For each $\lambda \in L$, let x_λ be a nonzero element of L_λ. Then $A = \sum_{\lambda \in L} Ax_\lambda$. •

36.20 Theorem. *If a topological ring A, furnished with the radical topology, is the topological direct sum of ideals B and C, then the induced topologies on B and C are their radical topologies.*

Proof. We shall prove, for example, that the topology induced on B is its radical topology. The radicals of B and C are $R \cap B$ and $R \cap C$ respectively by

26.18. Thus by 26.21, $R = (R \cap B) + (R \cap C)$, and consequently $(R+C)/C = ((R \cap B) + C)/C$. The restriction ϕ of the canonical epimorphism from A to A/C to B is a topological isomorphism. Consequently, the radical of A/C is $((R \cap B) + C)/C$. Thus $(R + C)/C$ is the radical of A/C. As $((R+C)/C)^n = (R^n + C)/C$ for all $n \geq 1$, the quotient topology of A/C is its radical topology. As B is topologically isomorphic to A/C, therefore, the topology induced on B is its radical topology. •

We note next that if I and J are ideals of a ring A that are finitely generated left ideals, then IJ is a finitely generated left ideal. Indeed, if $I = \sum_{i=1}^{m}(\mathbb{Z}a_i + Aa_i)$ and $J = \sum_{j=1}^{n}(\mathbb{Z}b_j + Ab_j)$, then

$$IJ = I(\sum_{j=1}^{n}\mathbb{Z}b_j) + I(\sum_{j=1}^{n}Ab_j) = \sum_{j=1}^{n}Ib_j + \sum_{j=1}^{n}IAb_j$$
$$= \sum_{j=1}^{n}Ib_j = \sum_{i=1}^{m}\sum_{j=1}^{n}(\mathbb{Z}a_ib_j + Aa_ib_j).$$

Our principal result concerning the absence of pathological groups is the following:

36.21 Theorem. *Let A be a ring linearly compact for its radical topology, and let R be the radical of A. The following statements are equivalent:*

1° *A contains no pathological subgroups.*

2° *The annihilator of A is compact.*

3° *A contains an idempotent e whose right annihilator is compact and contained in R.*

4° *R is a finitely generated left ideal.*

If these conditions hold, then for each $k \geq 1$, A/R^k is both an artinian and a noetherian A-module (and hence an artinian and a noetherian ring); if, in addition, A/R is finite, then the radical topology is compact.

Proof. If $A \neq R$, then as R is open, A/R is a nonzero, semisimple artinian ring R by 36.16 and 28.15, and consequently A/R has an identity element by 27.14. For each $n \geq 1$, R^n/R^{n+1} is an (A/R)-module; we let M_{n+1} and N_{n+1} be respectively the unitary and trivial submodules of the A/R-module R^n/R^{n+1}. If $A = R$, then R^n/R^{n+1} is a trivial (A/R)-module, so we let M_{n+1} be the zero submodule of R^n/R^{n+1} and $N_{n+1} = R^n/R^{n+1}$. Clearly R^n/R^{n+1} is the direct sum of its submodules M_{n+1} and N_{n+1}.

First we shall show that if 1° holds, then N_{n+1} is finite. Indeed, N_{n+1} is a discrete, trivial module and hence is discrete and strongly linearly compact.

By 36.15, N_{n+1} contains no pathological subgroups, so by 30.10, N_{n+1} is finite.

To show that 1° implies 4°, let M and N be the left ideals of A containing R^2 such that $M/R^2 = M_2$ and $N/R^2 = N_2$. Either M/R^2 is the zero module or M/R^2 is a discrete, linearly compact, unitary (A/R)-module and hence is finitely generated by 36.19. By the preceding paragraph, N/R^2 is finite. Thus there exist $x_1, \ldots, x_n \in M$ such that $M = Ax_1 + \ldots + Ax_n + R^2$, there exist $y_1, \ldots, y_m \in N$ such that $N = \mathbb{Z}.y_1 + \ldots + \mathbb{Z}.y_m$, and there exists $q > 0$ such that $q.y \in R^2$ for all $y \in N$. Consequently, there exist $z_1, \ldots, z_r \in R$ such that $q.y_i \in Rz_1 + \ldots + Rz_r$ for all $i \in [1, m]$. Let

$$J = Ax_1 + \ldots + Ax_n + Ay_1 + \ldots + Ay_m + Az_1 + \ldots + Az_r,$$

a closed left ideal by 28.18. Let

$$I = J + \mathbb{Z}.y_1 + \ldots + \mathbb{Z}.y_m.$$

Clearly I/J is a surjective image of $\mathbb{Z}^m/(\mathbb{Z}q)^m$ and hence is finite. Thus I is the union of finitely many cosets of J, each of which is closed, and therefore I is closed. Furthermore, $R = M + N + R^2 = I + R^2$. Therefore $R = I$ by 36.18 and hence R is a finitely generated left ideal.

To show that 4° implies 1°, suppose that A contains a pathological subgroup G. By 36.12, G is contained in the annihilator of A and hence in R by 26.14. Let n be the largest integer such that $G \subseteq R^n$. The image G' of G in R^n/R^{n+1} is then a basic primary subgroup. As R is a finitely generated left ideal, all powers of R are finitely generated left ideals, and hence R^n/R^{n+1} is a finitely generated (A/R)-module. As N_{n+1} is a direct summand of the (A/R)-module R^n/R^{n+1}, N_{n+1} is also finitely generated. As G is contained in the annihilator of A, $G' \subseteq N_{n+1}$. As N_{n+1} is discrete and strongly linearly compact, N_{n+1} is a torsion module by 30.10. Hence as N_{n+1} is finitely generated, there exists $m > 0$ such that $m.N_{n+1} = \{0\}$. Consequently, N_{n+1} contains no basic divisible group, a contradiction.

Next, we shall show that 1° and 4° imply the final statements. For any $k \geq 1$, A/R^k is an artinian A-module and hence an artinian ring by 36.16 and 28.15. Since M_{n+1} is a direct summand of the (A/R)-module R^n/R^{n+1}, M_{n+1} is also finitely generated. By 29.14, either $A = R$ or A/R is a noetherian ring. In either case, $(R^n/R^{n+1})/M_{n+1}$ is isomorphic to N_{n+1}, a finite (A/R)-module and hence a noetherian A/R-module. If $A = R$, then $M_{n+1} = \{0\}$, so R^n/R^{n+1} is a noetherian A-module. Otherwise, by 29.14, A/R is a noetherian ring with identity, so M_{n+1} is a noetherian A-module by 20.8, and consequently R^n/R^{n+1} is a noetherian (A/R)-module by 20.3. By 27.9 applied to $E = A/R^k$ and its submodules $(R^n/R^k)_{0 \leq n \leq k}$, we conclude that A/R^k is a noetherian A-module and hence a noetherian ring.

Suppose further that A/R is finite. Then M_{n+1} is also finite, so R^n/R^{n+1} is finite. An inductive argument now establishes that A/R^k is finite for all $k \geq 1$, for if A/R^n is finite, then as $(A/R^{n+1})/(R^n/R^{n+1})$ is isomorphic to A/R^n and as R^n/R^{n+1} is finite, A/R^{n+1} is also finite. The canonical mapping g from A to $\varprojlim_{n \geq 1}(A/R^n)$, defined by $g(x) = (x + R^n)_{n \geq 1}$, is a topological isomorphism by 8.5. By 5.20 and Tikhonov's theorem, $\varprojlim_{n \geq 1}(A/R^n)$ is compact.

Clearly 3° implies 2°, and 2° implies 1° by 36.12 as a pathological group is complete and noncompact.

Assume 1°. To prove 3°, we may, by the preceding, assume that $A \neq R$. By 29.8 and 34.1, A has an idempotent e such that $e + R$ is the identity of A/R. Consequently the right annihilator H of e is contained in R. We shall show by induction that for each $i \geq 1$, the right annihilator H_i of $e + R^i$ in A/R^i is finite. Clearly $H_1 = \{0\}$. Suppose that H_n is finite, and let let ϕ be the additive homomorphism from H_{n+1} to A/R^n defined by $\phi(x + R^{n+1}) = x + R^n$ for all $x \in H_{n+1}$. Clearly $\phi(H_{n+1}) \subseteq H_n$ and hence $\phi(H_{n+1})$ is finite. The kernel of ϕ, $H_{n+1} \cap (R^n/R^{n+1})$, is simply the trivial submodule N_{n+1} of the (A/R)-module R^n/R^{n+1}. Indeed, if $x + R^{n+1} \in N_{n+1}$, then in particular,

$$(1) \quad (e + R^{n+1})(x + R^{n+1}) = ex + R^{n+1} = (e + R)(x + R^{n+1}) = R^{n+1},$$

so $x + R^{n+1} \in H_{n+1}$. Conversely, if $x + R^{n+1} \in H_{n+1}$, then (1) holds, so for any $a \in A$ and any $x \in R^n$,

$$(a + R)(x + R^{n+1}) = ax + R^{n+1} = aex + (a - ae)x + R^{n+1} = R^{n+1},$$

since $aex \in R^{n+1}$ by (1) and $(a - ae)x \in RR^n = R^{n+1}$. We saw above that N_{n+1} is finite. Therefore, as the range and kernel of ϕ are finite, so is its domain H_{n+1}. Clearly H is closed. Thus $g(H)$ is a closed subset of $(\prod_{i=1}^{\infty} H_i) \cap \varprojlim_{n \geq 1}(A/R^n)$ and hence is compact, so H is compact. •

Our next theorems extend Theorems 36.2-36.5:

36.22 Theorem. *If A is a nonzero ring that is linearly compact and topologically torsionfree for its radical topology, then A has a left identity.*

Proof. A satisfies 1° of 36.21 as each element of a pathological subgroup has finite order. By 3° of 36.21, A contains an idempotent e whose right annihilator is compact and therefore $\{0\}$. Hence e is a left identity. •

36.23 Theorem. *Let A be linearly compact for the radical topology, let T be its topological torsion ideal. There is a closed ideal S of A satisfying the following properties:*

1° A is the topological direct sum of S and T.
2° S is a divisible ideal.
3° S is topologically torsionfree.
4° S has a left identity.

If L is a left ideal of A, then $L = (L \cap S) + (L \cap T)$. Finally, S is the only ideal supplement of T.

Proof. For each $n \geq 1$, let ϕ_n be the canonical epimorphism from A to A/R^n, an artinian A-module and hence an artinian ring by (2) of 36.16 and (2) of 28.15. For each $m \geq n$, let $\phi_{n,m}$ be the canonical epimorphism from A/R^m to A/R^n. Thus $\phi_{n,m} \circ \phi_m = \phi_n$ for all $m \geq n$. For each $n \geq 1$, let S_n be the ideal supplement of the torsion ideal T_n of A/R^n, and let

$$S = \bigcap_{n=1}^{\infty} \phi_n^{-1}(S_n).$$

Thus S is a closed ideal of A. By 36.5, $\phi_{k,n}(S_n) = S_k$ and $\phi_{k,n}(T_n) = T_k$ whenever $k \leq n$. We shall show that $\phi_n(S) = S_n$ for all $n \geq 1$. Let $s_n + R^n \in S_n$. By 36.5, applied to $\phi_{n,n+1}$ there exists $s_{n+1} \in A$ such that $s_{n+1} + R^{n+1} \in S_{n+1}$ and $s_{n+1} - s_n \in R^n$. Similarly, by induction, there is a sequence $(s_k)_{k \geq n}$ in A such that for all $k \geq n$, $s_{k+1} + R^{k+1} \in S_{k+1}$ and $s_{k+1} - s_k \in R^k$. Consequently, $\{s_k + R^k : n \geq k\}$ is a Cauchy filter base of closed sets and hence converges to some $s \in A$. For each $k \geq n$, $s \in s_k + R^k \in S_k$, so $s \in \phi_k^{-1}(S_k)$; if $k \in [1, n-1]$, $\phi_k(s) = \phi_{k,n}(\phi_n(s)) \in \phi_{k,n}(S_n) = S_k$, so $s \in \phi_k^{-1}(S_k)$; thus $s \in S$. In particular, $\phi_n(s) = s_n + R^n$. An entirely similar argument shows that $\phi_n(T) = T_n$ for all $n \geq 1$. Thus $\phi_n(S+T) = S_n + T_n = A/R_n$, so $A = S + T + R^n$ for all $n \geq 1$. As S and T are closed in A, $S + T$ is closed by (3) of 28.6, and therefore $A = S + T$ by 3.3.

To show that S is topologically torsionfree, let s be a nonzero element of S. Let $n \geq 1$ be such that $s \notin R^n$. As $\phi_n(s) \in S_n$, $\mathbb{Z}.\phi_n(s)$ is infinite and discrete. Consequently $\mathbb{Z}.s$ is also. In particular, $S \cap T = (0)$, so A is the topological direct sum of the ideals S and T by 36.16 and 28.22.

By 36.20, the topology induced on S is its radical topology. Consequently by 36.22, S has a left identity element e.

To show that S is divisible, let $s \in S$ and let $q > 0$. As S_n is divisible and torsionfree, there is a unique $t_n + R^n \in S_n$ such that $q.(t_n + R^n) = s + R^n$, and we may assume that $t_n \in S$ as $\phi_n(S) = S_n$. If $m \geq n$, then $q.t_m - q.t_n =$

$(q.t_m - s) - (q.t_n - s) \in R^n$, so by uniqueness, $t_m + R^n = t_n + R^n$. Thus $\{t_n + R^n : n \geq 1\}$ is a Cauchy filter base of closed sets and hence converges to some $t \in A$. Thus $t \in t_n + R^n \subseteq S + R^n$ for all $n \geq 1$, so $t \in S$ by 3.3 as S is closed. Furthermore, $q.t \in q.(t_n + R^n) = s + R^n$, hence $q.t - s \in R^n$ for all $n \geq 1$, and therefore $q.t = s$. Thus S is divisible.

Let L be a left ideal, and let $x \in L$. Then $ex \in S \cap L$. Let $x - ex = s + t$ where $s \in S$ and $t \in T$. Then $0 = e(x - ex) = es + et = s + et$. As T is an ideal, $et \in T$ and therefore $s = et = 0$. Thus $x - ex = t \in T$, so $x = ex + (x - ex) \in (L \cap S) + (L \cap T)$. In particular, if S' is an ideal supplement of T, then $S' = S' \cap S + S' \cap T = S' \cap S$. Hence $S' \subseteq S$, so $S' = S$ as $S' + T = A$. •

Consequently, we shall call S the ideal topological supplement of T. Thus the ideal topological supplement of T is a divisible, topologically torsionfree ideal.

In the following discussion, for any strongly linearly compact module H we define $D(H)$ by

$$D(H) = \bigcap_{n=1}^{\infty} n.H.$$

Since $x \to n.x$ is a continuous homomorphism, $n.H$ is closed by 28.3 and 28.6, and therefore $D(H)$ is closed; for the same reason, $n.D(H)$ is also closed. Clearly $D(H)$ contains the largest divisible subgroup of H, and if H is the additive group of an ideal of a topological ring, $D(H)$ is an ideal.

36.24 Theorem. *If H is a strongly linearly compact module, then $D(H)$ is closed, $H/D(H)$ is compact, and $D(H)$ is the largest divisible subgroup of H.*

Proof. We have already seen that $D(H)$ is closed. Let $K = H/D(H)$, a strongly linearly compact module by 28.6. Then $D(K) = \{0\}$. Indeed, let $x + D(H) \in D(K)$ and let $n \geq 1$. Then there exists $y \in H$ such that $x - n.y \in D(H)$, and thus there exists $z \in H$ such that $x - n.y = n.z$; consequently, $x = n.(y + z) \in n.H$. Therefore $x \in D(H)$.

Let $\mathcal{V}$ be the filter base of all open subgroups of K, and let $\mathcal{U}$ be the collection of all $U \in \mathcal{V}$ such that K/U is finite. Then $\mathcal{U}$ is a filter base, for if U and V are subgroups such that K/U and K/V are finite, then $x \to (x + U, x + V)$ is a homomorphism from K to $K/U \times K/V$ with kernel $U \cap V$, so $K/(U \cap V)$ is isomorphic to a subgroup of $K/U \times K/V$ and hence is finite. For any $V \in \mathcal{V}$ and any $n \geq 1$, $n.K + V \in \mathcal{U}$. Indeed, by 30.10, K/V is the direct sum of a divisible subgroup D_V and a finite subgroup F_V. Then for any $n \geq 1$, $D_V = n.D_V \subseteq n.(K/V)$, so $(K/V)/n.(K/V)$ is an epimorphic image of F_V and hence is finite. As $n.(K/V) = (n.K + V)/V$

and as $(K/V)/((n.K+V)/V)$ is isomorphic to $K/(n.K+V)$, the latter is finite, so $n.K+V \in \mathcal{U}$. Consequently,

$$\bigcap_{U \in \mathcal{U}} U \subseteq \bigcap_{n=1}^{\infty}(\bigcap_{V \in \mathcal{V}}(n.K+V) = \bigcap_{n=1}^{\infty} n.K = D(K) = \{0\}$$

by 3.3, as each $n.K$ is closed. By 30.11, K is a strictly linearly compact $\mathbb{Z}$-module. Its topology is therefore minimal among all Hausdorff $\mathbb{Z}$-linear topologies by 28.13, so $\mathcal{U}$ is a fundamental system of neighborhoods of zero. Consequently, as K is complete, K is topologically isomorphic to $\varprojlim_{U \in \mathcal{U}}(K/U)$, a closed and hence compact subset of $\prod_{U \in \mathcal{U}}(K/U)$ by 8.5 and 5.20.

To complete the proof, we need only show that $D(H)$ is divisible. Let $\mathcal{W}$ be the filter base of all open subgroups of H, and let $W \in \mathcal{W}$. By 30.10, there exist $B, F \in \mathcal{W}$ such that B and F both contain W, B/W is divisible, F/W is finite, and H/W is the direct sum of B/W and F/W. Consequently, as H/B is isomorphic to $(H/W)/(B/W)$ and hence to F/W, H/B is finite, and thus there exists $m \geq 1$ such that $m.H \subseteq B$. Consequently, if $a \in H \setminus B$, then no $h \in H$ would satisfy $m.h = a$; thus $D(H) \subseteq B$, whence $D(H) + W \subseteq B$. Now $(D(H)+W)/D(H)$ is an open subgroup of compact $H/D(H)$, so $(H/D(H))/((D(H)+W)/D(H))$ is compact and discrete and hence finite. Thus $H/(D(H)+W)$ is finite, so its subset $B/(D(H)+W)$ is also finite. As $B/(D(H)+W)$ is an epimorphic image of the divisible group B/W, $B/(D(H)+W)$ is divisible. The only finite divisible group is the zero group, however, so $D(H)+W = B$. For any $n \geq 1$, $n.(B/W) = B/W$, so $n.B + W = B$. Thus

$$n.D(H) + W = n.(D(H)+W) + W = n.B + W = B = D(H) + W.$$

As $n.D(H)$ and $D(H)$ are closed, therefore,

$$n.D(H) = \bigcap_{W \in \mathcal{W}}(n.D(H)+W) = \bigcap_{W \in \mathcal{W}}(D(H)+W) = D(H)$$

by 3.3. Thus $D(H)$ is a divisible group and so is the largest divisible subgroup of H. •

36.25 Theorem. *Let A be a ring linearly compact for its radical topology, let D be the largest divisible ideal, let T be its topological torsion ideal, let P be the set of primes and $(T_p)_{p \in P}$ the topological p-primary components of T, and let H be the annihilator of A. (1) D is closed, $H \subseteq T$,* and $H/(D \cap T)$ is compact. (2) $\mathfrak{S}_{p \in P} T_p = \prod_{p \in P} T_p$, that is, for every neighborhood U of zero, $T_p \subseteq U$ for all but finitely many $p \in P$.

Proof. Let S be the ideal supplement of T. (1) $H \cap S = \{0\}$ by 4° of 36.23, so by the final assertion of that theorem, $H = (H \cap S) + (H \cap T) = H \cap T$, and hence $H \subseteq T$. As H is a closed, trivial A-module, H is strongly linearly compact, so by 36.24, $D(H)$ is closed and the largest divisible ideal of H, and $H/D(H)$ is compact. Also by 36.23, $D = S + (D \cap T)$, and thus D is the direct sum of S and $D \cap T$. Consequently $D \cap T$ is an epimorphic image of D and hence is a divisible group. As $D \cap T \subseteq H$ by 36.12, therefore, $D \cap T \subseteq D(H)$, the largest divisible subgroup of H. But as $D(H)$ is a divisible group, $D(H) \subseteq D$ and thus $D(H) \subseteq D \cap T$. Therefore $D \cap T = D(H)$ and hence $D \cap T$ is closed. Consequently as $D = S + (D \cap T)$, D is closed, and $H/(D \cap T) = H/D(H)$, a compact ring.

(2) Since T has a topological direct summand, T is a linearly compact ring, and hence each T_p is also linearly compact by 29.3 and the remark following 36.8. Consequently, $\mathfrak{S}_{p \in P} T_p = \prod_{p \in P} T_p$ by 36.11. •

36.26 Theorem. *Let A be a ring with radical R, and let T be the topological torsion ideal of A for the radical topology. Then A is linearly compact for the radical topology and contains no pathological subgroups if and only if the radical topology is Hausdorff and complete, A/R is an artinian ring, R is a finitely generated left ideal, and A/T is a ring with left identity.*

Proof. The condition is necessary by 36.21, 36.23, and (2) of 28.15. Sufficiency: It suffices by (2) of 28.15 to show that for each $n \geq 1$, A/R^n is an artinian A-module. By 27.14, either A/R is the zero ring or A/R is a ring with identity. Assume that A/R^n is an artinian A-module. To show that A/R^{n+1} is artinian, it suffices, by 27.3, to show that R^n/R^{n+1} is an artinian A-module, or equivalently, an artinian A/R-module. As R is finitely generated, all the powers of R are finitely generated left ideals, and therefore R^n/R^{n+1} is a finitely generated A/R-module. Therefore as R^n/R^{n+1} is the direct sum of its unitary submodule M and its trivial submodule N, each of them is finitely generated. If $A = R$, $M = \{0\}$; otherwise, by 27.8, M is an artinian (A/R)-module. Let $e \in A$ be such that $ex - x \in T$ for all $x \in A$. For any $x + R^{n+1} \in N$, $ex \in R^{n+1}$, so if $x_1 = x - ex$, $x_1 + R^{n+1} = x + R^{n+1}$, and $x_1 \in T$. Thus the closure $[x_1]$ of $\mathbb{Z}.x_1$ is compact, so its image in R^n/R^{n+1} is compact and discrete and therefore finite. Thus each member of N is contained in a finite subgroup, so as N is finitely generated and trivial, N is finite and hence artinian. •

For the radical topology on a ring with identity A to be linearly compact, it suffices that there exist a linearly compact topology such that $\bigcap_{n \geq 1} \overline{R^n} =$

$\{0\}$ and that $\overline{R^2}$ be open, where R is the radical of A:

36.27 Theorem. *Let A be a linearly compact ring with identity such that $\bigcap_{n \geq 1} \overline{R^n} = \{0\}$, where R is the radical of A. The following statements are equivalent:*

1° *The topology of A is stronger than the radical topology.*
2° $\overline{R^2}$ *is open.*
3° A/R *is an artinian ring, and R is a finitely generated left ideal.*

If these statements hold, then the radical topology of A is a linearly compact topology.

Proof. Clearly 1° implies 2°. Assume 2°. Since R is closed by 29.12, $\overline{R^2} \subseteq R$ and hence R is open. Therefore A/R is an artinian A-module and hence an artinian ring by 29.14. Moreover, $R/\overline{R^2}$ is a discrete, linearly compact, unitary module over A/R and hence is finitely generated by 36.19. Thus there is a finitely generated left ideal M of A such that $R = M + \overline{R^2}$. As A has an identity element, M is closed by 28.18. If N is a left ideal such that $R = M + \overline{N}$, then $R = M + \overline{RN}$. Indeed,

$$R^2 = RM + R\overline{N} \subseteq M + \overline{RN},$$

a closed left ideal by (3) of 28.6, so $\overline{R^2} \subseteq M + \overline{RN}$, and thus $R = M + \overline{R^2} = M + \overline{RN}$. By induction, therefore, $R = M + \overline{R^n}$ for all $n \geq 1$. Thus

$$R = \bigcap_{n=1}^{\infty} (M + \overline{R^n}) = M + \bigcap_{n=1}^{\infty} \overline{R^n} = M$$

by (2) of 28.20. Therefore R is a finitely generated left ideal, and 3° holds.

Assume 3°. By 27.10, A/R^n is an artinian A-module for all $n \geq 1$. Moreover each R^n is finitely generated and hence closed by 28.18. Therefore A/R^n is a linearly compact, artinian A-module and hence its topology is discrete by 28.14. Thus R^n is open for all $n \geq 1$, so the topology of A is stronger than the radical topology.

The hypothesis implies that the radical topology is Hausdorff. Hence 1° implies that the radical topology is linearly compact by 28.4. •

If the radical topology of a ring A is linearly compact, it is strictly linearly compact by 36.16 and hence minimal in the set of all linearly compact topologies on A. When is it the weakest of all linearly compact topologies on A? For rings with identity, if the radical topology is linearly compact, then it is indeed the weakest of all linearly compact topologies:

36.28 Theorem. *Let A be a ring with identity, and let R be its radical. The radical topology of A is linearly compact if and only if it is Hausdorff and complete, A/R is an artinian ring, and R is a finitely generated left ideal. If the radical topology is linearly compact, then every linearly compact topology on A is stronger than the radical topology.*

Proof. The first assertion follows from 36.26. Assume that the radical topology on A is linearly compact. Then $\bigcap_{n \geq 1} R^n = \{0\}$, A/R is artinian, and R is a finitely generated left ideal. Consequently, for all $n \geq 1$, R^n is a finitely generated left ideal and hence is closed for any linearly compact topology $\mathcal{T}$ on A by 28.18, as A has an identity. Therefore $\mathcal{T}$ is stronger than the radical topology by 36.27. •

A natural problem is to describe those rings that are linearly compact for the discrete topology. Our considerations here are limited to the case where the radical topology is Hausdorff, a condition implying that the Leptin topology associated to the discrete topology is strictly linearly compact by 33.24 and equivalent to that statement for commutative rings by 33.25.

36.29 Theorem. *Let A be a ring, R its radical. The following statements are equivalent:*

1° *A is linearly compact for the discrete topology, and $\bigcap_{n \geq 1} R^n = \{0\}$.*

2° *A admits a bounded, strictly linearly compact topology for which every left ideal is closed.*

3° *The radical topology is a linearly compact topology for which every left ideal of A is closed.*

If these conditions hold, then every Hausdorff linear topology on A is linearly compact, and the radical topology is the weakest Hausdorff linear topology on A.

Proof. 1° and 3° are equivalent by 28.4 and 28.19, and clearly 3° implies 2°. Finally, 2° implies 1° by 33.22 and 28.19.

If the conditions hold, then every Hausdorff linear topology on A is linearly compact by 28.4, and there is a weakest Hausdorff linear topology by 33.7. By 36.16, the radical topology is strictly linearly compact and hence, by 28.13, minimal in the set of all linear Hausdorff topologies. Therefore it is the weakest Hausdorff linear topology on A. •

More interesting chacterizations may be made of those rings that, in addition, contain no basic primary subgroups. To obtain them, we need some preliminary results:

36.30 Theorem. *A linearly compact A-module E is a Baire space.*

Proof. Let $(G_n)_{n \geq 1}$ be a sequence of open dense subsets of E; we shall show that $\bigcap_{n \geq 1} G_n$ is dense, that is, that $(\bigcap_{n \geq 1} G_n) \cap P \neq \emptyset$ for any nonempty open subset P. We may assume that $P = a + M$ where M is an open submodule of E. We shall construct a sequence of points $(b_n)_{n \geq 0}$ and a decreasing sequence $(M_n)_{n \geq 0}$ of open submodules such that $b_0 = a$, $M_0 = M$, and

$$b_n + M_n \subseteq \bigcap_{k=0}^{n-1} [(b_k + M_k) \cap G_{k+1}]$$

for all $n \geq 1$. Indeed, assume that b_k and M_k satisfy those conditions for all $k \leq n$. As G_{n+1} is dense, there exists $b_{n+1} \in (b_n + M_n) \cap G_{n+1}$; as that set is open, there exists an open submodule M_{n+1} such that

$$b_{n+1} + M_{n+1} \subseteq (b_n + M_n) \cap G_{n+1} = \bigcap_{k=0}^{n} [(b_k + M_k) \cap G_{k+1}].$$

Thus $(b_n + M_n)_{n \geq 0}$ is a decreasing sequence of open and hence closed cosets of submodules, so as E is linearly compact, there exists $b \in \bigcap_{k \geq 0} (b_k + M_k)$. Consequently, $b \in (\bigcap_{n \geq 1} G_n) \cap P$. •

36.31 Theorem. *Let A be a ring with radical R such that $\bigcap_{n \geq 1} R^n = \{0\}$, R is a finitely generated left ideal, and A/R is an artinian ring. If A admits a linearly compact topology for which every left ideal of A is closed, then A is a noetherian ring.*

Proof. By 28.19 and 36.29, the radical topology is a linearly compact topology for which every left ideal is closed, so we may replace, if necessary, the given topology by the radical topology. Let $(M_n)_{n \geq 1}$ be an increasing sequence of left ideals, and let M be their union. Then M is a closed and hence linearly compact A-module, so M is a Baire space by 36.30. As each M_n is also closed, there exists $r \geq 1$ such that M_r is open in M by 4.9. As the topology is the radical topology, there exists $t \geq 1$ such that $R^t \cap M \subseteq M_r$. As $M/(R^t \cap M)$ is isomorphic to the A-submodule $(M + R^t)/R^t$ of A/R^t, a noetherian A-module by 36.21, $M/(R^t \cap M)$ is a noetherian A-module by 20.3. As M/M_r is isomorphic to the A-module $(M/(R^t \cap M))/(M_r/(R^t \cap M))$, M/M_r is also a noetherian A-module by 20.3. Consequently there exists $q \geq r$ such that $M_q/M_r = M/M_r$, and therefore $M_q = M$. •

36.32 Theorem. *If A is a linearly compact noetherian ring, then every left ideal of A is closed (or equivalently, for each $c \in A$, there exists $m \geq 1$ such that $m.c \in Ac$).*

Proof. By (3) of 28.6, it suffices to show that for any $c \in A$, the left ideal $Ac + \mathbb{Z}.c$ generated by c is closed. By 28.18, Ac is closed, so by

28.16, $\overline{Ac + \mathbb{Z}.c}/Ac$ is a linearly compact A-module. It is, however, a trivial A-module, for as $A(Ac + \mathbb{Z}.c) \subseteq Ac$, $A(\overline{Ac + \mathbb{Z}.c}) \subseteq Ac$. Consequently, $\overline{Ac + \mathbb{Z}.c}/Ac$ is strongly linearly compact and also, by 20.3, a noetherian $\mathbb{Z}$-module.

If $\overline{Ac + \mathbb{Z}.c}/Ac$ were not discrete, it would be uncountable by 36.30, and hence would contain a strictly increasing sequence of additive subgroups, in contradiction to the fact that it is a noetherian $\mathbb{Z}$-module. Thus it is discrete, so $(Ac + \mathbb{Z}.c)/Ac$ is a discrete, strongly linearly compact module. If $Ac \cap \mathbb{Z}.c = \{0\}$, then $(Ac + \mathbb{Z}.c)/Ac$ would be isomorphic to the $\mathbb{Z}$-module $\mathbb{Z}$, in contradiction to 30.10. Therefore $(Ac + \mathbb{Z}.c)/Ac$ is isomorphic to the $\mathbb{Z}$-module $\mathbb{Z}/\mathbb{Z}.m$ (i.e., $m.c \in Ac$) for some $m \geq 1$, and consequently is finite. Thus $Ac + \mathbb{Z}.c$ is the union of finitely many cosets of Ac and hence is closed. •

36.33 Theorem. *Let A be a ring, R its radical. The following statements are equivalent:*

1° *A is linearly compact for the discrete topology, $\bigcap_{n \geq 1} R^n = \{0\}$, and A contains no basic divisible primary subgroup.*

2° *A admits a bounded, strictly linearly compact topology for which every left ideal is closed, and A contains no basic divisible primary subgroup.*

3° *Furnished with the radical topology, A is linearly compact, A contains no pathological subgroups, and every left ideal of A is closed.*

4° *A/R is artinian, R is a finitely generated left ideal, the radical topology is Hausdorff and complete, A/T has a left identity where T is the topological torsion ideal for the radical topology, and every left ideal of A is closed.*

5° *A is noetherian and linearly compact for the radical topology.*

6° *A is noetherian and admits a bounded, strictly linearly compact topology.*

7° *A is noetherian, and A admits a linearly compact topology for which $\bigcap_{n \geq 1} \overline{R^n} = \{0\}$.*

8° *A is noetherian, A/R is artinian, the radical topology is Hausdorff and complete, and for each $c \in A$, there exists $m \geq 1$ such that $m.c \in Ac$.*

If these conditions hold, then the radical topology is the weakest Hausdorff linear topology on A, and for that topology A has only finitely many nonzero topological primary components; if, in addition, A/R is finite, then the radical topology is compact.

Proof. By 36.29, 1° and 2° are equivalent. Assume 1°. To establish 3°, we need only show, by 36.29, that A, furnished with the radical topology (for which every left ideal is closed), contains no subgroup G topologically isomorphic to the additive group $\mathbb{Q}_p$ for some prime p. By 36.12, G would be contained in the annihilator of A, and hence every additive subgroup of

G would be a (closed) left ideal and hence a linearly compact $\mathbb{Z}$-module. In particular, G would contain a subgroup Z algebraically isomorphic to $\mathbb{Z}$ that is a linearly compact $\mathbb{Z}$-module. Consequently, Z would be discrete by 36.30. But Z is not a discrete, linearly compact $\mathbb{Z}$-module by 30.10, since it is not an artinian $\mathbb{Z}$-module.

By 28.19, 3° implies 1°. By 36.26, 3° and 4° are equivalent. Thus 1°–4° are all equivalent, and by 36.31, they imply 5°. By 36.16, 5° implies 6°; by 33.22, 6° implies 7°; and by 36.9, each of them implies that A has only finitely many nonzero topological primary components for the radical topology.

Assume 7°. By 36.32 and 28.19, A is linearly compact for the discrete topology and $\bigcap_{n \geq 1} R^n = \{0\}$, so by 28.4 A is linearly compact for the radical topology. Suppose that A contained a basic divisible primary subgroup G. By 36.12, G is contained in the annihilator of A. Therefore G is a trivial noetherian A-module by 20.3, and thus is a noetherian $\mathbb{Z}$-module. But by definition, G is the union of a strictly increasing sequence of subgroups, a contradiction. Thus A contains no basic divisible primary subgroup, and hence 1° holds.

By 36.32, 4° and 5° imply 8°. Assume 8°. To show 5°, it suffices by 28.15 to show that for each $k \geq 1$, A/R^k is an artinian A-module. For this, it suffices by our hypothesis, 27.3, and induction to show that for each $n \geq 1$, R^n/R^{n+1} is an artinian A-module, or equivalently, an artinian (A/R)-module. Since A is noetherian, R^n/R^{n+1} is a noetherian (A/R)-module and hence its unitary submodule M_{n+1} is finitely generated. If $A = R$, M_{n+1} is the zero submodule and hence is artinian; otherwise, M_{n+1} is artinian by 27.8. Let $c_1, \ldots, c_q \in R^n$ be be such that $c_1 + R^{n+1}, \ldots, c_q + R^{n+1}$ are generators of the A/R-module R^n/R^{n+1}. By hypothesis, there exists $m \geq 1$ such that $m.c_j \in Ac_j$ for all $j \in [1, q]$. Therefore $(r_1, \ldots, r_q) \to r_1.c_1 + \ldots + r_q.c_q + M_{n+1}$ is an additive epimorphism from $\mathbb{Z}^q$ to $(R^n/R^{n+1})/M_{n+1}$ whose kernel contains $(m\mathbb{Z})^q$, so $(R^n/R^{n+1})/M_{n+1}$ is finite as $\mathbb{Z}^q/(m\mathbb{Z})^q$ has mq elements. Thus $(R^n/R^{n+1})/M_{n+1}$ is an artinian A/R-module, so R^n/R^{n+1} is also an artinian (A/R)-module by 27.3. Finally, if these conditions hold and A/R is finite, the radical topology is compact by 36.21. •

The commutative rings with identity that satisfy the equivalent conditions of 36.33 have a simple description: they are precisely the complete semilocal noetherian rings of §24. Moreover, they may be described by a property implied by those of 36.33, but not equivalent to them for the class of rings with identity (Exercise 36.4).

36.34 Theorem. *If A is a commutative ring with identity whose radical R is finitely generated and whose radical topology is linearly compact, then*

A, furnished with that topology, is the topological direct sum of finitely many complete local noetherian rings.

Proof. A/R is a discrete, semisimple, linearly compact ring and hence is isomorphic to the cartesian product of finitely many fields by 29.10. Therefore A has only finitely many maximal ideals, i.e., A is a semilocal ring. By (2) of 36.16 and (2) of 34.6, A is topologically isomorphic to the cartesian product of a family $(A_\lambda)_{\lambda \in L}$ of strictly linearly compact local rings. Therefore as A is semilocal, L is finite, and thus A is the topological direct sum of finitely many strictly linearly compact local rings $A_1, \ldots, A_n$. By 26.21, R is the direct sum of the maximal ideals $M_1, \ldots, M_n$, where each M_i is the maximal ideal of A_i; thus each M_i is also finitely generated. Moreover, the topology induced on each A_i is its radical (or natural) topology by 36.20. As A is complete by 28.5, so is each A_i by 7.8. Thus each A_i is noetherian by (2) of 24.17. •

36.35 Corollary. *If A is a commutative ring with identity, the following statement is equivalent to those of Theorem 36.33:*

9° *A is linearly compact for the radical topology, and R is a finitely generated ideal.*

Moreover, the commutative rings with identity satisfying 8° are precisely the complete semilocal noetherian rings.

Proof. Clearly 5° of 36.33 implies 9°. Assume 9°. Then A is strictly linearly compact by 36.16, so by 36.34 A is a semilocal ring that is Hausdorff and complete for its radical topology. By (1) of 24.16, A/R has only finitely many ideals; hence A/R is artinian. Therefore by 36.28, the radical topology is linearly compact, so 5° of 36.33 and the final assertion hold. •

To describe the class of commutative rings (including those not having an identity) that satisfy the conditions of 36.29, we begin with a preliminary result:

36.36 Theorem. *If A is a compact noetherian ring, then A is totally disconnected, and its annihilator H is finite.*

Proof. Every additive subgroup of H is an ideal. If H were uncountable, then there would exist a strictly increasing sequence $(G_n)_{n \geq 1}$ of subgroups of H, a contradiction. If H were countably infinite, then as H is compact and hence a Baire space, H would be discrete, in contradiction to our hypothesis that A is compact. Thus H is finite. Consequently, the connected component C of zero is finite by 32.3, and thus is the zero ideal. •

36.37 Theorem. *A commutative topological ring A is a strictly linearly compact noetherian ring if and only if A is the topological direct sum of finitely many complete local noetherian rings and a commutative, compact, noetherian, radical ring J.*

Proof. Necessity: By (1) of 34.6, A is the topological direct sum of strictly linearly compact rings B and J, where B is either the zero ring or a strictly linearly compact ring with identity, and J is a strictly linearly compact radical ring. Both B and J are noetherian rings by 20.4, as they are isomorphic respectively to A/J and A/B. By 36.34 B is the topological direct sum of finitely many complete local noetherian rings. By the final assertion of 36.33, the given topology of A is the radical topology, since a strictly linearly compact topology is minimal by 28.13. Therefore the topologies induced on B and J are their radical topologies by 36.20. Also J satisfies 6° of 36.33. Thus by the concluding statements of that theorem, J is compact as it is a radical ring.

Sufficiency: By 36.36, J is totally disconnected and hence, by 32.5, a strictly linearly compact ring. Consequently, A is strictly linearly compact by 29.5. •

Any finite, nilpotent ring is a noetherian radical ring; a nondiscrete example of a compact, noetherian radical ring is the maximal ideal (or, more generally, any proper nonzero ideal) of the ring $\mathbb{Z}_p$ of p-adic integers (Exercise 36.3).

36.38 Theorem. *Let A is a commutative noetherian ring. (1) The radical topology of A is Hausdorff. (2) A ring topology $\mathcal{T}$ on A is a linearly compact topology if and only if it is an ideal topology stronger than the radical topology and the radical topology is linearly compact. (3) If the radical topology is linearly compact, then A, furnished with the radical topology, is the topological direct sum of finitely many complete local noetherian rings and a commutative, compact, noetherian, radical ring.*

Proof. (1) If A has an identity, the radical topology is Hausdorff by 24.14. In the contrary case, let A_1 be the (commutative) ring obtained by adjoining an identity element to A. Thus A is an ideal of A_1, and A/A_1 is isomorphic to the commutative noetherian ring $\mathbb{Z}$. Both A and A_1/A are noetherian A_1-modules since they are noetherian rings, so by 20.3, A_1 is a noetherian A_1-module, that is, a noetherian ring. Consequently by 24.14, $\bigcap_{n \geq 1} R_1^n = \{0\}$, where R_1 is the radical of A_1. But by 26.19, R_1 is the radical of A. (2) and (3) are immediate consequences of 36.33 and 36.37. •

36.39 Theorem. *Let A be a compact ring with identity. The following statements are equivalent:*

1° The topology of A is the radical topology.
2° $\overline{R^2}$ is open.
3° A/R is finite, and R is a finitely generated left ideal.

Proof. By 32.3, A is totally disconnected and hence, by 32.5, a bounded, strictly linearly compact ring. By 34.21, $\bigcap_{n \geq 1} R^n = \{0\}$. Since a compact topology is a minimal Hausdorff topology, 1° and 2° are equivalent by 36.27. Also 1° implies that A/R is compact and discrete, and hence finite. Therefore 1° and 3° are equivalent by 36.27. •

36.40 Theorem. *If A is a noetherian ring with radical R, then A is compact if and only if the radical topology of A is Hausdorff and complete, the topology of A is the radical topology, A/R is finite, and for each $c \in A$ there exists $m \geq 1$ such that $m.c \in Ac$.*

Proof. Necessity: By 36.36 and 32.5, A satisfies 6° of 36.33, so by that theorem, as a compact topology is a minimal Hausdorff topology, we need only verify that A/R is finite. But as R is open and A compact, A/R is a compact, discrete ring and hence is finite.

Sufficiency: A satisfies 8° of 36.33. Consequently by the final statement of 36.33, A is compact. •

36.41 Theorem. *A commutative topological ring A is a compact noetherian ring if and only if A is the topological direct sum of finitely many compact local noetherian rings and a commutative, compact, noetherian radical ring.*

Proof. The assertion is an immediate consequence of 36.37. •

By 36.40, a complete local noetherian ring A with maximal ideal M is compact if and only if the residue field A/M is finite.

Exercises

36.1 Which of the properties listed in 4° and 8° of 36.33 fail to hold if: (a) A is the trivial ring whose additive group is $\mathbb{Z}$? (b) A is the trival ring whose additive group is $\mathbb{Z}(p^\infty)$, where p is a prime? (c) $A = \mathbb{Z}[[X]]$?

36.2 If A is a linearly compact ring with identity, and if A satisfies the Ascending Chain Condition on closed left ideals (if $(J_n)_{n \geq 1}$ is an increasing sequence of closed left ideals, then there exists $m \geq 1$ such that $J_n = J_m$ for all $n \geq m$), then A is noetherian.

36.3 Let A be the valuation ring of a complete, discrete valuation v of a field whose value group is $\mathbb{Z}$, and let M be its maximal ideal. (a) M is a radical ring with no proper zero-divisors. (b) The following statements are equivalent:

1° M is compact.
2° A/M is finite.
3° M is a noetherian ring.
4° M is a linearly compact ring.

[Use 18.7. To show that 2° implies 3°, show that if I is a nonzero ideal of M and if $n = \inf\{v(x) : x \in I\}$, then $M_{n+1} \subseteq I$, and observe that M^n/M^{n+1} is finite. If 3° holds, to show 2°, first show that A/M is a finitely generated $\mathbb{Z}$-module, and conclude that the characteristic of A/M is a prime. To show that 4° implies 2°, use 34.13, and observe that M/M_2 is a vector space over A/M.] (c) In particular, if M is the maximal ideal of the ring of $\mathbb{Z}_p$ of p-adic integers, where p is a prime, then M has the properties of (a) and (b).

36.4 (Warner [1971]) Let S be the semigroup consisting of all n-tuples where $n \in \mathbb{N}$ ($\emptyset$ is considered the 0-tuple) whose entries are either 0 or 1, with multiplication defined by juxtaposition:

$$(a_1, \ldots, a_n)(b_1, \ldots, b_m) = (a_1, \ldots, a_n, b_1, \ldots, b_m).$$

Let $\leq$ be the well ordering of S satisfying $(a_1, \ldots, a_n) < (b_1, \ldots, b_m)$ if and only if either $n < m$ or $n = m$ and, for some $j \in [1, n]$, $a_i = b_i$ for all $i < j$, $a_j = 0$, and $b_j = 1$. Thus $\emptyset$ is the identity element of S and also is the smallest element of S, and if $s \leq s'$ and $t \leq t'$, the $st \leq s't'$. Let K be a finite field, let $K_s = K$ for all $s \in S$, and let $A = \prod_{s \in S} K_s$, where addition is defined componentwise and multiplication is defined by

$$(a_s)_{s \in S}(b_s)_{s \in S} = (\sum_{qr=s} a_q b_r)_{s \in S}.$$

(a) A is a ring, and the radical R of A consists of all $(a_s)_{s \in S}$ such that $a_\emptyset = 0$. (b) If each K_s is given the discrete topology and A the cartesian product topology, then A is a compact ring with identity, and the topology of A is the radical topology. (c) Let $s_k \in S$ be the $(k+2)$-tuple whose first and last entries are 0 and whose remaining entries are 1, and let $e_k = (\delta_{s.s_k})_{s \in S}$, where $\delta_{s.s_k}$ is 1 or 0 according as $s = s_k$ or $s \neq s_k$. Show that the left ideal generated by $\{e_k : k \geq 1\}$ is not finitely generated. (d) Conclude that A satisfies 9° of 36.35 but not the conditions of 36.33.

36.5 Let A be a ring, R its radical, T is torsion ideal. Then A is an artinian ring that contains no basic, divisible primary subgroups if and only if R is a nilpotent, finitely generated left ideal, A/R is artinian, and A/T is a ring with left identity. [Use 36.26.]

36.6 Let A be an artinian ring, R its radical. The following statements are equivalent:

1° A contains no basic, divisible, primary subgroup.

2° The annihilator of A is finite.

3° A contains an idempotent e whose right annihilator is finite and contained in R.

4° R is a finitely generated left ideal.

5° A is noetherian.

[Use 36.21.] In particular: (Szele and Fuchs [1955]) A contains no basic, divisible, primary subgroup if and only if A is noetherian; (Szele and Fuchs [1955]) If the annihilator of A is $\{0\}$, then A is noetherian; (Hopkins [1938]) If A has a left identity, then A is noetherian.

36.7 (Kovács [1954]) If A is an infinite ring such that every proper left ideal of A is finite, then either A is a division ring or A is a trivial ring whose additive group is a p-primary group for some prime p. [Use Exercise 34.17, 30.10, and 36.3.]

36.8 (a) A commutative, strictly linearly compact ring with identity is a semilocal ring if and only if it is advertibly open. (b) If A is a commutative topological ring with identity, then A is an advertibly open, strictly linear compact ring whose radical is finitely generated if and only if A is a complete semilocal noetherian ring whose topology is its natural topology.

CHAPTER IX

COMPLETE LOCAL NOETHERIAN RINGS

In this chapter we refine the description of complete local noetherian rings given in Theorem 23.6 for those that either are equicharacteristic or are nonequicharacteristic but in which $p.1$ is not a zero-divisor, where p is the characteristic of the residue field. Most of this chapter will be devoted to topics in commutative algebra. The Principal Ideal Theorem, a cornerstone of commutative algebra, is the subject of §37. In §38 we introduce Krull dimension and discuss regular local rings. In §39 the description is given with special attention to complete regular local rings. In §40 we show that complete local noetherian domains have the Japanese property, a fact needed in Chapter 10.

In this chapter, by a "noetherian ring" is meant a "commutative noetherian ring with identity."

37 The Principal Ideal Theorem

We begin with some results on modules that are both noetherian and artinian.

37.1 Definition. *Let E be a module. A* **Jordan-Hölder sequence** *of submodules of E is a finite, strictly decreasing sequence $(M_i)_{0 \leq i \leq n}$ such that $M_0 = E$, $M_n = \{0\}$, and for each $i \in [1,n]$, M_{i-1}/M_i is a simple module. The* **length** *of a strictly decreasing sequence of submodules is defined to be $+\infty$ if the sequence is infinite, otherwise n where $n+1$ is the number of terms in the sequence. E has* **finite length** *if E has a Jordan-Hölder sequence.*

37.2 Theorem. *A module E has finite length if and only if E is noetherian and artinian.*

Proof. Necessity: A simple module is clearly noetherian and artinian, so the assertion follows from 27.9. Sufficiency: The set of submodules of finite length is nonempty, since it contains the zero submodule and hence contains a maximal member N, as E is noetherian. Suppose that $N \neq E$. The set of submodules of E properly containing N is then nonempty and hence contains a minimal member M, as E is artinian. By the minimality of M,

M/N is a simple module. Therefore as N has a Jordan-Hölder sequence, so does M, a contradiction of the maximality of N. Thus $N = E$. •

37.3 Corollary. *If M is a submodule of a module E, then E has finite length if and only if M and E/M have finite length.*

Proof. The assertion follows from 37.2, 20.3, and 27.3. •

37.4 Theorem. *If a module E has a Jordan-Hölder sequence, then the length of any strictly decreasing sequence of submodules of E is at most that of the Jordan-Hölder sequence.*

Proof. Let $S = \{n \in \mathbb{N} :$ for every submodule M of E that has a Jordan-Hölder-sequence of length n, every strictly decreasing sequence of submodules of M has length at most $n\}$. Clearly $0 \in S$, since the zero submodule is the only submodule of E having a Jordan-Hölder sequence of length 0. Also, $1 \in S$, for simple submodules of E are the only ones having a Jordan-Hölder sequence of length 1. Assume that $r \geq 2$ and that S contains all natural numbers $< r$, where r does not exceed the length of the Jordan-Hölder sequence of E. Let F be a submodule of E that has a Jordan-Hölder sequence $(M_i)_{0 \leq i \leq r}$ of length r. By 37.2, a strictly decreasing sequence of submodules of F is finite; let $(N_j)_{0 \leq j \leq s}$ be such a sequence, where by adding terms if necessary, we may assume that $N_0 = F$ and $N_s = \{0\}$. Clearly M_1 has a Jordan-Hölder sequence of length $r - 1$, so any strictly decreasing sequence of submodules of M_1 has length at most $r - 1$. In particular, if $N_1 \subseteq M_1$, then $(N_j)_{1 \leq j \leq s}$ is a strictly decreasing sequence of submodules of M_1, so $s - 1 \leq r - 1$ and hence $s \leq r$.

Consequently, we may assume that $N_1 \not\subseteq M_1$. As F/M_1 is simple and as $N_1 \neq F$, N_1 does not properly contain M_1, so by our assumption $N_1 \not\supseteq M_1$ and thus $M_1 \cap N_1 \subset M_1$. By 37.3, $M_1 \cap N_1$ has a Jordan-Hölder sequence of length, say, t. Adjoining M_1 at the beginning of the sequence, we obtain a strictly decreasing sequence of submodules of M_1 of length $t+1$, so $t + 1 \leq r - 1$ and therefore $t \leq r - 2$. Since $N_1 \not\subseteq M_1$ and since F/M_1 is simple, $M_1 + N_1 = F$, so $N_1/(M_1 \cap N_1)$ is simple as it is isomorphic to $(M_1 + N_1)/M_1 = F/M_1$. Consequently, adjoining N_1 at the beginning of a Jordan-Hölder sequence for $M_1 \cap N_1$, we obtain a Jordan-Hölder sequence for N_1 of length $t + 1 \leq r - 1$. As $r - 1 \in S$, therefore, and as $(N_j)_{1 \leq j \leq s}$ is a strictly decreasing sequence of submodules of N_1, $s - 1 \leq r - 1$ and thus $s \leq r$. Consequently, by induction S contains q, the length of the given Jordan-Hölder sequence of E, so the conclusion holds. •

37.5 Corollary. *Any two Jordan-Hölder sequences of a module have the same length.*

Consequently, we may define the *length* of a module of finite length to be the length of all its Jordan-Hölder sequences. The following theorem is easy to prove:

37.6 Theorem. *Let E be a module of finite length. If M is a submodule of E, then*

$$\text{length}(E) = \text{length}(M) + \text{length}(E/M).$$

If M is a proper submodule of E, then $\text{length}(M) < \text{length}(E)$. If E is the direct sum of submodules M and N, then

$$\text{length}(E) = \text{length}(M) + \text{length}(N).$$

To begin our investigation of complete local noetherian rings, we gather some equivalent conditions for a commutative ring with identity to be artinian:

37.7 Theorem. *Let A be a commutative ring with identity. The following statements are equivalent:*

$1°$ *A is artinian.*

$2°$ *A is the direct sum of finitely many local noetherian rings whose maximal ideals are nilpotent.*

$3°$ *A is a semilocal noetherian ring whose radical is nilpotent.*

$4°$ *A is a semilocal ring whose radical is finitely generated and nilpotent.*

$5°$ *A is a semilocal ring whose maximal ideals are finitely generated and nilpotent.*

Proof. By 27.17, $1°$ and $2°$ are equivalent. Each of $2°$–$5°$ implies that the radical topology on A is discrete and hence complete, and that is also implied by $1°$ by 27.15. Therefore by 24.17, $3°$ and $4°$ are equivalent. The equivalence of $1°$ and $9°$ of 36.33 and 36.35 establish the equivalence of $4°$ and $1°$, for the existence of an identity element prevents the existence of a basic divisible primary subgroup by 36.3, and an artinian ring is linearly compact for the discrete topology by 28.14. Clearly $2°$–$4°$ imply $5°$. By (2) of 24.17, $5°$ implies $2°$. •

An ideal P of a commutative ring with identity is prime if and only if for any two ideals I and J of A, if $IJ \subseteq P$, then either $I \subseteq P$ or $J \subseteq P$. Indeed, the special case where I and J are principal ideals is the definition of a prime ideal. Conversely, suppose that $IJ \subseteq P$ but neither I nor J is contained in P. Then there exist $a \in I$ and $b \in J$ such that neither a nor b belongs to P, yet $ab \in IJ \subseteq P$. Thus P is not prime.

To obtain another criterion for a commutative ring with identity to be noetherian, we need the following theorem:

37.8 Theorem. *If A is a commutative noetherian ring with identity, every proper ideal of A either is a prime ideal or contains a product of nonzero prime ideals.*

Proof. If not, the set $\mathcal{A}$ of proper ideals that neither are prime nor contain a product of nonzero prime ideals contains a maximal member M. In particular, M is not prime, so there exist ideals I, J such that $I \not\subseteq M$, $J \not\subseteq M$, and $IJ \subseteq M$. Both $M + I$ and $M + J$ properly contain M, and $(M+I)(M+J) \subseteq M + IJ = M$. Moreover, $M+I$ and $M+J$ are proper ideals; for example, if $M + I = A$, then

$$(M+I)(M+J) = A(M+J) = M + J \not\subseteq M,$$

a contradiction. Consequently, $M+I$ and $M+J$ are nonzero proper ideals of A that do not belong to $\mathcal{A}$ as they properly contain M, so each contains a product of nonzero prime ideals, whence M does also, a contradiction. •

37.9 Theorem. *If A is a commutative ring with identity, then A is artinian if and only if A is noetherian and each prime ideal of A is maximal.*

Proof. Necessity: By 37.7 we need only establish that a prime ideal P of an artinian ring A is maximal. For this, it suffices by 27.4 to show that an artinian integral domain D is a field. Let x be a nonzero element of D. Then $(Dx^n)_{n \geq 1}$ is a decreasing sequence of ideals, so there exists $q \geq 1$ such that $x^q = dx^{q+1}$ for some $d \in A$, whence $1 = dx$.

Sufficiency: By 37.8 applied to the zero ideal and our hypothesis, either the zero ideal is maximal, or there exist distinct maximal ideals $M_1, \ldots, M_n$ such that if $N = M_1 M_2 \ldots M_n$, then $N^q = \{0\}$ for some $q \geq 1$. If $\{0\}$ is a maximal ideal, then A is a field and hence is artinian, so we need only consider the second possibility. First, if M is any maximal ideal, then as M is prime and contains N^q, M contains some M_k and hence $M = M_k$. Thus A is semilocal. By 24.11, $N = M_1 \cap \ldots \cap M_n$, the radical of A, so the radical of A is nilpotent. Therefore by 37.7, A is artinian. •

37.10 Definition. *Let A be a commutative ring with identity. A **minimal prime ideal** of A is a prime ideal properly containing no other prime ideal. If J is a proper ideal of A, a **minimal prime ideal over** J is a prime ideal P containing J such that there are no prime ideals containing J that are properly contained in P.*

For example, the zero ideal is the only minimal prime ideal of an integral domain. Clearly P is a minimal prime ideal over J if and only if $P \supseteq J$ and P/J is a minimal prime ideal of A/J.

37.11 Theorem. *If A is a commutative ring with identity, every prime ideal P of A contains a minimal prime ideal.*

Proof. The set $\mathcal{P}$ of all prime ideals contained in P is nonempty, as $P \in \mathcal{P}$. Ordered by $\supseteq$, $\mathcal{P}$ is inductive, for if $\mathcal{C}$ is a chain in $\mathcal{P}$, clearly $\bigcap_{Q \in \mathcal{C}} Q$ is a prime ideal. Thus by Zorn's Lemma, $\mathcal{P}$ contains a member P_0 maximal for $\supseteq$, that is, P_0 is minimal among all prime ideals contained in P. Consequently, P_0 is a minimal prime ideal. •

37.12 Corollary. *If A is a commutative ring with identity and if J is a proper ideal of A, then every prime ideal of A containing J contains a minimal prime ideal over J, and in particular, there exist minimal prime ideals over J.*

Proof. Applying 37.11 to A/J yields the first assertion, and the second follows from it since J is contained in a maximal ideal. •

In commutative algebra, the importance of considering the subrings $S^{-1}A$ of the total quotient ring of a commutative ring with identity A, where S is a multiplicative subset of cancellable elements of A, arises from the fact that the ordered set of prime ideals of $S^{-1}A$ is, in a natural way, isomorphic to the ordered set of prime ideals of A not meeting S:

37.13 Theorem. *Let A be a commutative ring with identity, and let S be a multiplicative set of cancellable elements of A. If J is an ideal of $S^{-1}A$, then $J = (S^{-1}A)(J \cap A)$. Moreover, $Q \to (S^{-1}A)Q$ is an order-preserving bijection from the set of all prime ideals of A not meeting S to the set of all prime ideals of $S^{-1}A$, and its inverse is $N \to N \cap A$.*

Proof. Clearly $J \supseteq (S^{-1}A)(J \cap A)$. If $x \in J$, then $sx \in A$ for some $s \in S$, so $sx \in J \cap A$, and thus $x = s^{-1}(sx) \in (S^{-1}A)(J \cap A)$.

Let Q be a prime ideal of A not meeting S. Then $(S^{-1}A)Q$ is a proper ideal of $S^{-1}A$, for if $1 \in (S^{-1}A)Q$, then for some $s \in S$, $s = s \cdot 1 \in Q$, a contradiction. To show that $(S^{-1}A)Q$ is a prime ideal of $S^{-1}A$, let $z, w \in S^{-1}A$ be such that $zw \in (S^{-1}A)Q$. There exist $r, s, t \in S$ such that $rz \in A$, $sw \in A$, and $tzw \in Q$, whence $t(rz)(sw) = rs(tzw) \in Q$. As $t \notin Q$, either $rz \in Q$ or $sw \in Q$, that is, either $z = r^{-1}(rz) \in (S^{-1}A)Q$ or $w = s^{-1}(sw) \in (S^{-1}A)Q$. Also, if N is a prime ideal of $S^{-1}A$, clearly $N \cap A$ is a prime ideal of A.

Therefore we need only show that if Q is a prime ideal of A not meeting S, then $(S^{-1}A)Q \cap A = Q$, since we have already seen that for any ideal J of $S^{-1}A$, $J = (S^{-1}A)(J \cap A)$. Clearly $Q \subseteq (S^{-1}A)Q \cap A$. If $x \in (S^{-1}A)Q \cap A$, then $sx \in Q$ for some $s \in S$, so $x \in Q$ as $s \notin Q$. •

Let P be a prime ideal of an integral domain A. Then $A \setminus P$ is a multiplicative set; the integral domain $(A \setminus P)^{-1}A$ is usually denoted by A_P and is called the *localization of A at P*, for by 37.13, A_P is a local ring whose maximal ideal is $A_P P$.

37.14 Theorem. *Let A be a commutative ring with identity. If S is a multiplicative subset of A and if J is an ideal of A such that $J \cap S = \emptyset$, then there is a prime ideal containing J such that $P \cap S = \emptyset$.*

Proof. The set $\mathcal{J}$ of all ideals I of A such that $I \supseteq J$ and $I \cap S = \emptyset$ is nonempty and is clearly inductive for inclusion. Consequently by Zorn's Lemma, $\mathcal{J}$ contains a maximal member P. Suppose that there exist $a, b \in A \setminus P$ such that $ab \in P$. By the maximality of P, neither $P + Aa$ nor $P + Ab$ would belong to $\mathcal{J}$, so there would exist $x, y \in P$ and $c, d \in A$ such that $x + ca \in S$ and $y + db \in S$. Therefore $(x + ca)(y + db) \in S$, but

$$(x + ca)(y + db) \in P + Aab = P,$$

a contradiction. Thus P is a prime ideal. •

37.15 Definition. *Let A be a commutative ring with identity, J an ideal of A. The **radical** of J, denoted by $\mathrm{rad}(J)$, is defined by*

$$\mathrm{rad}(J) = \{x \in A : x^n \in J \text{ for some } n \geq 1\};$$

*J is a **radical ideal** if $J = \mathrm{rad}(J)$.*

Clearly $\mathrm{rad}(J)$ is an ideal, for if $x^n, y^m \in J$, then $(x+y)^{n+m} \in J$ by the Binomial Theorem, and $(ax)^n \in J$ for any $a \in A$. Moreover, $\mathrm{rad}(\mathrm{rad}(J)) = \mathrm{rad}(J)$, so the radical of any ideal is a radical ideal. For example, $\mathrm{rad}(\{0\})$ is the ideal of all nilpotent elements of A.

37.16 Theorem. *Let A be a commutative ring with identity. If J is a proper ideal of A, then $\mathrm{rad}(J)$ is the intersection of all the prime ideals of A containing J and hence of all the minimal prime ideals over J.*

Proof. If P is a prime ideal of A containing J and if $x \in \mathrm{rad}(J)$, then $x^n \in J \subseteq P$ for some $n \geq 1$, so $x \in P$. Conversely, let $x \in A \setminus \mathrm{rad}(J)$, and let $S = \{x^n : n \in \mathbb{N}\}$, a multiplicative subset of A. Then $J \cap S = \emptyset$, so by 37.14 there is a prime ideal P of A containing J such that $P \cap S = \emptyset$, whence $x \in A \setminus P$. Thus $\mathrm{rad}(J)$ is the intersection of all the prime ideals of A containing J. The final assertion is therefore a consequence of 37.12. •

37.17 Theorem. *A proper radical ideal of a noetherian ring A is the intersection of finitely many prime ideals.*

Proof. In the contrary case, the set $\mathcal{J}$ of all proper radical ideals that are not the intersections of finitely many prime ideals is nonempty and hence

has a maximal member J. Clearly J itself is not a prime ideal, so there exist $b, c \in A \setminus J$ such that $bc \in J$. Then $\mathrm{rad}(J + Ab)$ and $\mathrm{rad}(J + Ac)$ are radical ideals not belonging to $\mathcal{J}$, so each is the intersection of finitely many prime ideals. Therefore we shall obtain a contradiction by showing that
$$\mathrm{rad}(J + Ab) \cap \mathrm{rad}(J + Ac) = J.$$
Let $x \in \mathrm{rad}(J + Ab) \cap \mathrm{rad}(J + Ac)$. Then there exist $n, m \in \mathbb{N}$ such that $x^n \in J + Ab$ and $x^m \in J + Ac$, so
$$x^{n+m} \in (J + Ab)(J + Ac) \subseteq J + Abc = J,$$
whence $x \in J$. •

37.18 Theorem. *If P is a prime ideal of a commutative ring A with identity and if $J_1, \ldots, J_n$ are ideals of A such that $P \supseteq \bigcap_{k=1}^n J_k$, then $P \supseteq J_i$ for some $i \in [1, n]$.*

Proof. In the contrary case, there would exist $x_k \in J_k \setminus P$ for each $k \in [1, n]$. But then
$$x_1 x_2 \ldots x_n \in (\bigcap_{k=1}^n J_k) \setminus P$$
since P is a prime ideal, a contradiction. •

37.19 Theorem. *If J is a proper ideal of a noetherian ring A, there are only finitely many minimal prime ideals over J.*

Proof. There are finitely many prime ideals $Q_1, \ldots, Q_n$ whose intersection is the intersection of all the prime ideals of A containing J by 37.16 and 37.17. By 37.12 there exist minimal prime ideals $P_1, \ldots, P_n$ over J such that $P_i \subseteq Q_i$ for all $i \in [1, n]$, so $\bigcap_{i=1}^n Q_i = \bigcap_{i=1}^n P_i$. Let P be a minimal prime ideal over J. Then
$$P \supseteq \bigcap_{i=1}^n Q_i = \bigcap_{i=1}^n P_i.$$
By 37.18, $P \supseteq P_k$ for some $k \in [1, n]$, so $P = P_k$ as P is minimal over J. •

37.20 Theorem. *If M is the maximal ideal of a local noetherian ring A and if J is a proper ideal of A, then M is a minimal prime ideal over J if and only if $J \supseteq M^t$ for some $t \geq 1$.*

Proof. Necessity: A/J is a local noetherian ring by 20.4 whose only prime ideal is the maximal ideal M/J. By 37.9 and 27.15, $(M/J)^t = \{0\}$ for some $t \geq 1$, that is, $M^t \subseteq J$. Sufficiency: If P is a prime ideal containing J, then P contains M^t for some $t \geq 1$, and hence $P = M$. •

37.21 Theorem. *Let A be a noetherian integral domain. If S is a multiplicative subset of A, then $S^{-1}A$ is noetherian.*

Proof. Let J be an ideal of $S^{-1}A$. By 37.13, $J = (S^{-1}A)(J \cap A)$, and by hypothesis there exist $x_1, \ldots, x_n \in J \cap A$ such that $J \cap A = Ax_1 + \ldots + Ax_n$. Consequently, $J = (S^{-1}A)x_1 + \ldots + (S^{-1}A)x_n$. Thus $S^{-1}A$ is noetherian. •

37.22 Theorem. *Let A be a noetherian ring. The set Z of zero-divisors of A is the union of finitely many prime ideals, and every minimal prime ideal of A is contained in Z.*

Proof. For each $c \in A$, let $\mathrm{Ann}(c)$ be the annihilator of c, and let $\mathcal{A} = \{\mathrm{Ann}(c) : c \in A^*\}$. Clearly $Z = \bigcup_{B \in \mathcal{A}} B$. Let $\mathcal{A}_0$ be the set of all maximal members of $\mathcal{A}$, ordered by inclusion. As A is noetherian, each member of $\mathcal{A}$ is contained in a member of $\mathcal{A}_0$, so $Z = \bigcup_{B \in \mathcal{A}_0} B$. To show that Z is a union of prime ideals, therefore, it suffices to show that each member of $\mathcal{A}_0$ is a prime ideal.

Let $J = \mathrm{Ann}(c) \in \mathcal{A}_0$, and let $a, b \in A$ be such that $ab \in J$ but $b \notin J$. Then $bc \neq 0$. Clearly $\mathrm{Ann}(bc) \supseteq \mathrm{Ann}(c)$, so by the maximality of J, $\mathrm{Ann}(bc) = \mathrm{Ann}(c)$. As $a \in \mathrm{Ann}(bc)$, therefore, $a \in \mathrm{Ann}(c) = J$.

Let $C = \{c \in A^* : \mathrm{Ann}(c) \in \mathcal{A}_0\}$, and let I be the ideal generated by C. As A is noetherian, there exist $c_1, \ldots, c_n \subset C$ such that $I = Ac_1 + \ldots + Ac_n$. To show that $\mathcal{A}_0 = \{\mathrm{Ann}(c_i) : 1 \leq i \leq n\}$, let $c \in C$. Then there exist $x_1, \ldots, x_n \in A$ such that $c = x_1 c_1 + \ldots + x_n c_n$, so

$$\bigcap_{i=1}^{n} \mathrm{Ann}(c_i) \subseteq \mathrm{Ann}(c),$$

and thus $\mathrm{Ann}(c_1)\mathrm{Ann}(c_2)\ldots\mathrm{Ann}(c_n) \subseteq \mathrm{Ann}(c)$. By the preceding, $\mathrm{Ann}(c)$ is a prime ideal, so $\mathrm{Ann}(c_k) \subseteq \mathrm{Ann}(c)$ for some $k \in [1, n]$, and therefore $\mathrm{Ann}(c) = \mathrm{Ann}(c_k)$ by the maximality of $\mathrm{Ann}(c_k)$. Thus $\mathcal{A}_0$ is finite.

Finally, we wish to show that if Q is a minimal prime ideal of A, then $Q \subseteq Z$. Let $S = \{ab \in A : a \in A \setminus Q, b \in A \setminus Z\}$. Clearly S is a multiplicative subset of A, so by 37.14 there is a prime ideal P such that $P \cap S = \{0\}$. Now $P \subseteq Z \cap Q$, for if z is a nonzero element of P, then $z \notin S$ but $z = 1 \cdot z = z \cdot 1$, so as $1 \notin Q$ and $1 \notin Z$, $z \in Z \cap Q$. Thus by the minimality of Q, $Q = P \subseteq Z$. •

37.23 Lemma. *Let A be an integral domain, and let $u, y \in A^*$. (1) The A-modules $(Au + Ay)/Au$ and $(Au^2 + Auy)/Au^2$ are isomorphic. (2) If $au^2 \in Ay$ implies that $au \in Ay$ for all $a \in A$, then the A-modules Au/Au^2 and $(Au^2 + Ay)/(Au^2 + Auy)$ are isomorphic.*

Proof. The function $x \to ux$ is an A-linear isomorphism from $Au + Ay$ to $Au^2 + Auy$ and takes Au to Au^2, so (1) holds. (2) Clearly the functions f and

g from A to Au/Au^2 and to $(Au^2+Ay)/(Au^2+Auy)$ respectively, defined by $f(x) = xu + Au^2$ and $g(x) = xy + Au^2 + Auy$, are A-epimorphisms. Thus to establish (2), it suffices to show that they have the same kernel. The kernel of f is clearly Au. If $x \in Au$, then $xy \in Auy$, so $g(x) = 0$. Conversely, suppose that $g(x) = 0$, that is, that $xy = au^2 + buy$ where $a, b \in A$. Then $au^2 \in Ay$, so by hypothesis $au = cy$ for some $c \in A$, whence $xy = cuy + buy$, therefore $x = (c+b)u \in Au$, and finally $f(x) = 0$. •

37.24 Theorem. (*Principal Ideal Theorem*) *If A is a local noetherian ring and if the maximal ideal M of A is a minimal prime ideal over Ax for some $x \in A$, then every prime ideal of A other than M is a minimal prime ideal.*

Proof. In the contrary case there would exist prime ideals P and Q such that $M \supset P \supset Q$. Then A/Q would be a local integral domain with maximal ideal M/Q, a minimal prime ideal over the principal ideal generated by $x + Q$, and P/Q would be a nonzero, nonmaximal prime ideal. Replacing A by A/Q if necessary, we may therefore assume that A is a local noetherian domain whose maximal ideal M is a minimal prime ideal over Ax and that A contains a nonzero, nonmaximal prime ideal P. To obtain a contradiction, let y be a nonzero element of P, and for each $k \geq 1$, let $J_k = \{a \in A : ax^k \in Ay\}$. As $(J_k)_{k \geq 1}$ is an increasing sequence of ideals of A, there exists $n \geq 1$ such that $J_k = J_n$ for all $k \geq n$. Then $ax^{2n} \in Ay$ implies that $ax^n \in Ay$. Let $u = x^n$; then $au^2 \in Ay$ implies that $au \in Ay$.

Clearly M is a minimal prime ideal over both Au and Au^2. Therefore A/Au^2 is a noetherian ring having precisely one prime ideal, so A/Au^2 is an artinian ring by 37.9. Therefore every finitely generated, unitary (A/Au^2)-module has finite length by 27.8, 20.8, and 37.2. Thus, as $(Au+Ay)/Au^2$ and $(Au^2+Ay)/Au^2$ are finitely generated, unitary (A/Au^2)-modules, they have finite length. As $[(Au+Ay)/Au^2]/[Au/Au^2]$ is isomorphic to $(Au+Ay)/Au$, we conclude from 37.6 that

$$\text{length}[(Au + Ay)/Au^2] = \text{length}[Au/Au^2] + \text{length}[(Au + Ay)/Au].$$

Similarly,

$$\text{length}[(Au^2 + Ay)/Au^2] = \text{length}[(Au^2 + Auy)/Au^2] + \\ + \text{length}[(Au^2 + Ay)/Au^2 + Auy)].$$

By 37.23,

$$\text{length}[(Au + Ay)/Au] = \text{length}[(Au^2 + Auy)/Au^2],$$
$$\text{length}[Au/Au^2] = \text{length}[(Au^2 + Ay)/(Au^2 + Auy)].$$

Therefore
$$\text{length}[(Au + Ay)/Au^2] = \text{length}[(Au^2 + Ay)/Au^2],$$
so again by 37.6, $(Au + Ay)/Au^2 = (Au^2 + Ay)/Au^2$. Thus $Au + Ay = Au^2 + Ay$, so $u = cu^2 + dy$ for some $c, d \in A$. Since M is a minimal prime ideal over Au, $u \in M$, and therefore $1 - cu$ is a unit of A. Consequently as $dy = (1 - cu)u$, we conclude that $u \in Ay \subseteq P$, a contradiction of our assumption that M is a minimal prime ideal over Au. •

Exercises

37.1 Let A be an integral domain. An element p of A is a *principal prime* if $p \neq 0$ and Ap is a prime ideal. (a) If $p \in A^*$, then p is a principal prime if and only if for all $a, b \in A$, if $p|ab$, then either $p|a$ or $p|b$. (b) If p is a principal prime and if $p|a_1 \ldots a_n$, then for some $i \in [1, n]$, $p|a_i$. (c) If p and q are principal primes and if $p|q$, then $q|p$, that is, p and q are associates. (d) If every noninvertible element of A^* is a product of principal primes, then A is a unique factorization domain, that is, the hypothesis holds and for any finite sequences $(p_i)_{1 \leq i \leq n}$ and $(q_j)_{1 \leq j \leq m}$ of principal primes, if $\prod_{i=1}^n p_i = \prod_{j=1}^m q_j$, then $m = n$ and there is a permutation σ of $[1, n]$ such that for each $i \in [1, n]$, p_i and $q_{\sigma(i)}$ are associates. [Use (b) and (c) and induction on n.]

37.2 Let A be an integral domain. The set S consisting of all invertible elements of A and all products of finite sequences of principal primes is a multiplicative set such that for all $a, b \in A$, if $ab \in S$, then $a \in S$ and $b \in S$. [Proceed by induction on the number of principal primes whose product is ab.]

37.3 Let A be an integral domain. (a) If A is a unique factorization domain, then every nonzero prime ideal of A contains a principal prime. (b) Conversely, if every nonzero prime ideal of A contains a principal prime, then A is a unique factorization domain. [With S defined as in Exercise 37.2, show that if $c \in A^* \setminus S$, then $Ac \cap S = (0)$, and apply 37.14.]

37.4 If A is a principal ideal domain, then $A[[X]]$ is a unique factorization domain, and the principal primes of $A[[X]]$ are the associates of X and the principal primes of A. [Apply (b) of Exercise 37.3. If P is a prime ideal not containing X, show that $P \cap A = Ap$ where p is a prime of A, and that P is the principal ideal of $A[[X]]$ generated by p. For this, argue as in the proof of 23.2.]

38 Krull Dimension and Regular Local Rings

38.1 Definition. The **height** of a prime ideal P of a noetherian ring A, denoted by $\text{ht}(P)$, is the supremum of the lengths of all the strictly decreasing sequences $(P_k)_{0 \leq k \leq m}$ of prime ideals of A such that $P_0 = P$.

For example, the height of a minimal prime ideal of A is zero, a special case of the following theorem:

38.2 Theorem. *If A is a noetherian ring and if J is a proper ideal of A generated by n elements, then for any prime ideal P of A that is minimal over J, $\mathrm{ht}(P) \leq n$.*

Proof. We shall proceed by induction on n, the assertion being true by the definition of a minimal prime ideal if $n = 0$, that is, if $J = \{0\}$. Assume, therefore, that $n \geq 1$ and that the assertion holds for any prime ideal in a noetherian ring that is minimal over an ideal generated by $n-1$ elements. We shall obtain a contradiction from the supposition that P is a prime ideal that is minimal over an ideal J generated by n elements, and that there is a strictly decreasing sequence $(P_k)_{0 \leq k \leq m}$ of prime ideals of length $m > n$, where $P_0 = P$.

We make several reductions: First, by replacing A with A/P_m, J with $(J + P_m)/P_m$, and each P_k with P_k/P_m, we may assume that A is an integral domain. Second, by replacing the integral domain A by A_P, J by $A_P J$, and each P_k by $A_P P_k$, we may assume, by 37.13 and 37.21, that A is a local noetherian domain whose maximal ideal is P. Third, we may assume that there are no prime ideals strictly between P and P_1; indeed, in the contrary case, there is an ideal P_1' maximal in the set of all prime ideals N such that $P \supset N \supset P_1$; replacing $(P_k)_{0 \leq k \leq m}$ by $(P_k')_{0 \leq k \leq m+1}$ where $P_0' = P_0$ and $P_{k+1}' = P_k$ for all $k \in [1, m]$, we obtain a sequence of length $m + 1 > n$ such that there are no prime ideals strictly between P and P_1'. Consequently, we assume that A is a local noetherian domain whose maximal ideal is P and that there are no prime ideals strictly between P and P_1.

Let S be a set of n elements generating J. Since P is a minimal prime ideal over J, $J \not\subseteq P_1$, so there exists $a_1 \in S$ not belonging to P_1; let $a_2, \ldots, a_n$ be the remaining members of S. Then $Aa_1 + P_1$ contains P_1 properly, so P is a minimal prime ideal over $Aa_1 + P_1$. By 37.20, $Aa_1 + P_1 \supseteq P^t \supseteq J^t$ for some $t \geq 1$. In particular, for each $k \in [2, n]$ there exists $c_k \in A$ and $b_k \in P_1$ such that $a_k^t = c_k a_1 + b_k$. Let $I = Ab_2 + \ldots + Ab_n$. Since $m > n$, $\mathrm{ht}(P_1) > n - 1$, so by our inductive hypothesis applied to I, P_1 is not a minimal prime ideal over I. Consequently, by 37.12, there exists a prime ideal Q of A such that $P_1 \supset Q \supseteq I$. Since

$$P \supseteq Aa_1 + Q \supseteq Aa_1 + I \supseteq Aa_1 + Aa_2^t + \ldots + Aa_n^t \supseteq J^{nt}$$

and since P is a minimal prime ideal over J, P is the only prime ideal of A containing $Aa_1 + Q$. Let the image under the canonical epimorphism from A to A/Q of an element x be denoted by x^* and that of an ideal H by H^*. Then in A/Q, P^* is a minimal prime ideal over $(A/Q)a_1^*$, but, as

$P \supset P_1 \supset Q$, $P^* \supset P_1^* \supset \{0\}$, in contradiction to 37.24, since $\{0\}$ is the only minimal prime ideal of A/Q. •

Thus, if P is a prime ideal in a noetherian ring A, $\operatorname{ht}(P)$ is finite, since $\operatorname{ht}(P) \leq n$ where P is generated by n elements. If Q is a prime ideal properly contained in P, clearly $\operatorname{ht}(P) \geq 1 + \operatorname{ht}(Q)$, that is, $\operatorname{ht}(P) > \operatorname{ht}(Q)$.

38.3 Theorem. *Let $P_1, \ldots, P_n$ be ideals of a commutative ring A with identity, all but at most two of which are prime ideals. If B is a subring of A such that*

$$B \subseteq \bigcup_{j=1}^{n} P_j,$$

then $B \subseteq P_k$ for some $k \in [1, n]$.

Proof. We first consider the case $n = 2$. Here, neither ideal need be prime. If $B \subseteq P_1 \cup P_2$ but $B \not\subseteq P_1$ and $B \not\subseteq P_2$, then there would exist $x \in B \setminus P_1$ and $y \in B \setminus P_2$, whence $x \in P_2$ and $y \in P_1$ as $B \subseteq P_1 \cup P_2$; but then $x + y \in B \subseteq P_1 \cup P_2$, whereas $x + y \notin P_1$ as $y \in P_1$ and $x \notin P_1$, and $x + y \notin P_2$ as $x \in P_2$ and $y \notin P_2$, a contradiction.

Assume next that $n \geq 3$ and that the assertion holds for $n - 1$ ideals. If $B \cap P_j \subseteq \bigcup_{k \neq j} P_k$ for some $j \in [1, n]$, then

$$B = \bigcup_{k=1}^{n} (B \cap P_k) \subseteq \bigcup_{k \neq j} P_k,$$

so by our inductive hypothesis, $B \subseteq P_k$ for some $k \in [1, n]$. In the contrary case, for each $j \in [1, n]$ there exists $x_j \in (B \cap P_j) \setminus \bigcup_{k \neq j} P_k$, and as $n \geq 3$, P_r is prime for some $r \in [1, n]$. Let $x = x_r + y_r$, where

$$y_r = \prod_{k \neq r} x_k.$$

Then $y_r \notin P_r$ as P_r is prime, so $x \notin P_r$ as $x_r \in P_r$; also, for any $i \neq r$, $y_r \in P_i$, so $x \notin P_i$ as $x_r \notin P_i$. Hence

$$x \in B \setminus \bigcup_{j=1}^{n} P_j = \emptyset,$$

a contradiction. •

We need an extension of 38.2:

38.4 Theorem. *Let A be a noetherian ring, and let P be a prime ideal of A. If J is an ideal of A generated by n elements that is contained in P, then P/J is a prime ideal of A/J, and*

$$\mathrm{ht}(P) \leq n + \mathrm{ht}(P/J).$$

Proof. We proceed by induction on $\mathrm{ht}(P/J)$, the assertion holding if $\mathrm{ht}(P/J) = 0$ by 38.2. Let $k > 0$, assume that the inequality holds whenever J is an ideal of A generated by n elements and contained in P such that $\mathrm{ht}(P/J) < k$, and let I be an ideal generated by m elements and contained in P such that $\mathrm{ht}(P/I) = k$. By 37.19, there are finitely many prime ideals $Q_1, \ldots, Q_r$ minimal over I, and P is not among them as $k > 0$. If $P \subseteq \bigcup_{i=1}^r Q_i$, then $P \subseteq Q_j$ for some $j \in [1, r]$ by 38.3, so $P = Q_j$ by the minimality of Q_j, a contradiction. Therefore there exists

$$c \in P \setminus \bigcup_{i=1}^r Q_i.$$

Let $J = I + Ac$, an ideal generated by $m+1$ elements. If $(P_k)_{0 \leq k \leq s}$ is a strictly decreasing sequence of prime ideals such that $P_0 = P$ and $P_s \supseteq J \supset I$, then $P_s \supseteq Q_t$ for some $t \in [1, r]$ by 37.12, whence $P_s \supset Q_t \supseteq I$ as $c \in P_s \setminus Q_t$. Thus $\mathrm{ht}(P/J) + 1 \leq \mathrm{ht}(P/I)$. Consequently, $\mathrm{ht}(P/J) < k$, so by our inductive hypothesis,

$$\mathrm{ht}(P) \leq (m+1) + \mathrm{ht}(P/J) \leq m + 1 + (\mathrm{ht}(P/I) - 1) = m + \mathrm{ht}(P/I). \bullet$$

38.5 Definition. *Let A be a noetherian ring. The **height** of a proper ideal J of A, denoted by $\mathrm{ht}(J)$, is the minimum of the heights of the minimal prime ideals over J.*

By 38.2, if J is a proper ideal generated by n elements, $\mathrm{ht}(J) \leq n$. If I is an ideal contained in J, then $\mathrm{ht}(J) \geq \mathrm{ht}(I)$. Indeed, if P is any minimal prime ideal over J, then P contains a minimal prime ideal Q over I by 37.12, so $\mathrm{ht}(P) \geq \mathrm{ht}(Q) \geq \mathrm{ht}(I)$, and therefore $\mathrm{ht}(J) \geq \mathrm{ht}(I)$.

38.6 Definition. *The **Krull dimension**, or simply the **dimension** of a local noetherian ring A, denoted by $\dim(A)$, is the height of its maximal ideal.*

For example, by 37.9 the local noetherian rings of dimension zero are precisely the local artinian rings, that is, by 37.7, the local noetherian rings whose maximal ideal is nilpotent.

If A is a local noetherian ring of dimension d, the height of each nonmaximal prime ideal of A is strictly less than d, so the maximal ideal M of A is a minimal prime ideal over a proper ideal J if and only if $\mathrm{ht}(J) = d$.

38.7 Theorem. *Let A be a local noetherian ring of dimension d whose maximal ideal M is generated by n elements. If J is a proper ideal of A such that $\operatorname{ht}(J) \geq s$, where $0 \leq s \leq d$, there is a sequence $u_1, \ldots, u_n$ of elements of A generating M such that M is a minimal prime ideal over $J + Au_{s+1} + \ldots + Au_d$.*

Proof. It suffices to prove by induction that for each $i \in [s, d]$ there is a sequence $u_1, \ldots, u_n$ of elements of A generating M such that $\operatorname{ht}(J + Au_{s+1} + \ldots + Au_i) \geq i$, for applying this result to $i = d$ yields an ideal of height d, over which M is therefore a minimal prime ideal.

The statement is trivially true if $i = s$. Assume that it is true if $s \leq i < d$, and let $u_i, \ldots, u_n$ be generators of M such that $\operatorname{ht}(J + Au_{s+1} + \ldots + Au_i) \geq i$. If $\operatorname{ht}(J + Au_{s+1} + \ldots + Au_i) \geq i + 1$, then also

$$\operatorname{ht}(J + Au_{s+1} + \ldots + Au_i + Au_{i+1}) \geq i + 1,$$

so the same sequence of generators serves for $i+1$. By 37.12 and 37.19, the set $\mathcal{P}$ of minimal prime ideals over $J + Au_{s+1} + \ldots + Au_i$ is nonempty and finite, and as noted above, $M \notin \mathcal{P}$. Let $P_1 \in \mathcal{P}$; as $P_1 \neq M$, there exists $m \in [1, n]$ such that $u_m \notin P_1$. Let $P_2, \ldots, P_k$ be the remaining members of $\mathcal{P}$ not containing u_m, and let $P_{k+1}, \ldots, P_h$ be the members of $\mathcal{P}$ containing u_m. For each $j \in [k+1, h]$ there exists $m(j) \in [1, n]$ such that $u_{m(j)} \notin P_j$, since $P_j \neq P$, whence $m(j) \neq m$. By 37.13 there exists

$$c_j \in (\bigcap_{i \neq j} P_i) \setminus P_j$$

for each $j \in [k+1, h]$. Let

$$u'_m = u_m + \sum_{j=k+1}^{h} c_j u_{m(j)}.$$

As $m(j) \neq m$ for all $j \in [k+1, h]$,

$$u_m \in Au_1 + \ldots + Au_{m-1} + Au'_m + Au_{m+1} + \ldots + Au_n.$$

Therefore $u_1, \ldots, u_{m-1}, u'_m, u_{m+1}, \ldots, u_n$ generate M. Clearly u'_m belongs to no member of $\mathcal{P}$; therefore each minimal prime ideal over $J + Au_{s+1} + \ldots + Au_i + Au'_m$ does not belong to $\mathcal{P}$ and hence strictly contains some member of $\mathcal{P}$ by 37.12. Consequently,

$$\operatorname{ht}(J + Au_{s+1} + \ldots + Au_i + Au'_m) \geq i + 1.$$

If $m \neq i+1$, interchanging u_{i+1} and u'_m in the sequence $u_1, \ldots, u_{m-1}, u'_m, u_{m+1}, \ldots, u_n$ yields the desired sequence of generators of M for $i+1$. ∎

From 38.2 and 38.7 applied to the zero ideal, we obtain:

38.8 Theorem. *Let A be a local noetherian ring of dimension d, and let M be its maximal ideal. There is an ideal J generated by d elements such that M is a minimal prime ideal over J, and any set of generators of an ideal over which M is a minimal prime ideal contains at least d elements.*

If A is a local noetherian ring with maximal ideal M, then M/M^2 is a finitely generated (A/M)-vector space.

38.9 Definition. *Let A be a local noetherian ring, M its maximal ideal. The **vector dimension** of A, denoted by $\mathrm{vdim}(A)$, is the dimension of the (A/M)-vector space M/M^2.*

38.10 Theorem. *Let A be a local noetherian ring, M its maximal ideal. Then $x_1, \ldots, x_r$ generate M if and only if the M^2-cosets of $x_1, \ldots, x_r$ generate the (A/M)-vector space M/M^2. In particular, there is a set of generators of M containing $\mathrm{vdim}(A)$ elements, and every set of generators of M contains at least $\mathrm{vdim}(A)$ elements.*

Proof. The condition is clearly necessary. Sufficiency: Let
$$F = Ax_1 + \ldots + Ax_r.$$
Then F is closed in M for the M-topology of the A-module M by 24.14, and by hypothesis, $M = F + M^2$. Consequently by 36.18, $M = F$. •

By 38.8 and 38.10, $\dim(A) \leq \mathrm{vdim}(A)$.

38.11 Definition. *A commutative ring with identity A is a **regular local ring** if A is a local noetherian ring such that $\dim(A) = \mathrm{vdim}(A)$.*

By 38.10, a local noetherian ring A of dimension d is a regular local ring if and only if there are d elements generating its maximal ideal.

To characterize regular local rings of dimension 1, we need the following theorem:

38.12 Theorem. *If A is a local noetherian ring that is not an integral domain, and if the principal ideal Ac is a prime ideal, then Ac is a minimal prime ideal.*

Proof. In the contrary case, there is a prime ideal Q properly contained in Ac. Let $a \in Q$. Then $a = xc$ for some $x \in A$. Suppose that $a = yc^n$ for some $y \in A$. Then $yc^n \in Q$ but $c^n \notin Q$, so $y \in Q \subset Ac$. Therefore $y = zc$ for some $z \in A$, and thus $a = zc^{n+1}$. Consequently, by induction,
$$a \in \bigcap_{n=1}^{\infty} Ac^n \subseteq \bigcap_{n=1}^{\infty} M^n = \{0\}$$
by 20.16, where M is the maximal ideal of A. Hence $Q = \{0\}$, so A is an integral domain, a contradiction. •

38.13 Theorem. *Let A be a local ring. (1) A is a regular local ring of dimension 0 if and only if A is a field. (2) A is a regular local ring of dimension 1 if and only if A is the valuation ring of a discrete valuation of a field.*

Proof. (1) Any field is clearly a regular local ring of dimension zero. Conversely, if A is a regular local ring of dimension zero with maximal ideal M, then $M/M^2 = \{0\}$, so $M = M^2$ and hence $M = \bigcap_{n \geq 1} M^n = \{0\}$ by 20.16, that is, A is a field.

(2) A discrete valuation ring A is a regular local ring of dimension 1, for its maximal ideal M is a nonzero principal ideal, whence $\mathrm{vdim}(A) = 1$, and $\dim(A) = 1$ since by 18.2, M and $\{0\}$ are its only prime ideals.

Conversely, let A be a regular local ring of dimension 1. The maximal ideal M of A is then a principal ideal. By 38.12, A is an integral domain, for otherwise M would be a minimal prime ideal, so the dimension of A would be zero. By 20.17, A is the valuation ring of a discrete valuation. •

38.14 Theorem. *Let A be a local noetherian ring with maximal ideal M. If $c \in M \setminus M^2$, then $\mathrm{vdim}(A/Ac) = \mathrm{vdim}(A) - 1$.*

Proof. We denote the image under the canonical epimorphism from A to A/Ac of an element x of A by x^*. Let $r = \mathrm{vdim}(A/Ac)$, and let $y_1, \ldots, y_r \in M$ be such that the $(M/Ac)^2$-cosets of $y_1^*, \ldots, y_r^*$ generate the $(A/Ac)/(M/Ac))$-vector space $(M/Ac)/(M/Ac)^2$. Then

$$M/Ac = (A/Ac)y_1^* + \ldots + (A/Ac)y_r^*$$

by 38.10, so

$$M = Ac + Ay_1 + \ldots + Ay_r.$$

Consequently, the M^2-cosets of $c, y_1, \ldots, y_r$ generate M/M^2, so we need only show that if

$$tc + \sum_{i=1}^{r} t_i y_i \in M^2,$$

where $t, t_1, \ldots, t_r \in A$, then $t, t_1, \ldots, t_r$ all belong to M. Since $c^* = 0$,

$$\sum_{i=1}^{r} t_i^* y_i^* \in (M^2 + Ac)/Ac = (M/Ac)^2,$$

whence each $t_i^* \in M/Ac$ by the definition of r, and therefore each $t_i \in M$. Consequently,

$$\sum_{i=1}^{r} t_i y_i \in M^2,$$

so $tc \in M^2$, and therefore $t \in M$ since otherwise t would be invertible, whence $c \in M^2$, a contradiction. •

38.15 Theorem. *If A is a regular local ring of dimension d with maximal ideal M and if $c \in M \setminus M^2$, then A/Ac is a regular local ring of dimension $d-1$.*

Proof. We have

$$d - 1 = \operatorname{vdim}(A) - 1 = \operatorname{vdim}(A/Ac) \geq \dim(A/Ac)$$
$$= \operatorname{ht}(M/Ac) \geq \operatorname{ht}(M) - 1 = d - 1,$$

the equalities and inequalities holding respectively by hypothesis, 38.14, the remark preceding 38.11, the definition of $\dim(A/Ac)$, 38.4, and the definition of $\dim(A)$. Thus $\operatorname{vdim}(A/Ac) = \dim(A/Ac) = d - 1$. •

38.16 Theorem. *A regular local ring is an integral domain.*

Proof. We proceed by induction on the dimension d of a regular local ring, the assertion holding if d is either 0 or 1 by 38.13 and the discussion preceding it. Let $d > 0$, assume that every regular local ring of dimension $< d$ is an integral domain, and let A be a regular local ring of dimension d, M its maximal ideal. For any $x \in M \setminus M^2$, A/Ax is also a regular local ring of dimension $d - 1$ by 38.15, so A/Ax is an integral domain by our inductive hypothesis, that is, Ax is a prime ideal. Suppose that A were not an integral domain, and let $P_1, \ldots, P_s$ be its minimal prime ideals (which are finite in number by 37.19). By 38.12, for each $x \in M \setminus M^2$, Ax is a minimal prime ideal, so

$$M \setminus M^2 \subseteq \bigcup_{i=1}^{s} P_i,$$

whence

$$M \subseteq M^2 \cup \left(\bigcup_{i=1}^{s} P_i\right).$$

Since $d > 0$, $M^2 \subset M$; therefore by 38.3, $M \subseteq P_i$ for some $i \in [1, s]$, so M is a minimal prime ideal and thus $d = 0$, a contradiction. Hence A is an integral domain.•

38.17 Theorem. *If A is a regular local ring of dimension d, then $A[[X]]$ is a regular local ring of dimension $d + 1$.*

Proof. For each ideal J of A, let

$$J' = \{\sum_{k=0}^{\infty} c_k X^k : c_0 \in J, c_k \in A \text{ for all } k \geq 1\}.$$

Thus J' is an ideal of $A[[X]]$ satisfying $J' \cap A = J$. Clearly if M is a maximal ideal of A, M' is the maximal ideal of $A[[X]]$, and if P is a prime ideal of A, P' is a prime ideal of $A[[X]]$. Consequently, if $(P_k)_{0 \leq k \leq d}$ is a sequence of prime ideals of A such that

$$M = P_0 \supset P_1 \supset \ldots \supset P_d = (0),$$

then

$$M' = P_0' \supset P_1' \supset \ldots \supset P_d' = (X) \supset \{0\}$$

is a sequence of ideals of $A[[X]]$ of length $d+1$. Thus $\dim(A[[X]]) \geq d+1$. If $x_1, \ldots, x_d$ generate M, clearly $x_1, \ldots, x_d, X$ generate the maximal ideal M' of $A[[X]]$. Thus $\mathrm{vdim}(A[[X]]) \leq d+1$. Consequently, as $\dim(A[[X]]) \leq \mathrm{vdim}(A[[X]])$, we conclude that $\dim(A[[X]]) = \mathrm{vdim}(A[[X]]) = d+1$. •

38.18 Corollary. *If K is a field, $K[[X_1, \ldots, X_d]]$ is a regular local ring of dimension d. If C is a discrete valuation ring, $C[[X_1, \ldots, X_{d-1}]]$ is a regular local ring of dimension d.*

Proof. The assertion follows by induction from 38.17 and 38.13. •

Exercises

38.1 Let A be a topological ring with identity, $Q(A)$ its total quotient ring, each of whose elements is of the form b/c where $b \in A$ and c is a cancellable element belonging to the center of A. The *topological quotient ring of A*, denoted by $Q_{\mathrm{top}}(A)$, is the subring $S^{-1}A$ of $Q(A)$ where S is the multiplicative set of all cancellable elements c of A belonging to its center such that $x \to cx$ is an open mapping from A to A. If B is a subring of $Q(A)$ containing A, then the neighborhoods of zero in A form a fundamental system of neighborhoods of zero for a ring topology on B if and only if $B \subseteq Q_{\mathrm{top}}(A)$.

38.2 Let A be a local noetherian ring, furnished with its natural topology. (a) $Q(A) = A$ if and only if $\dim(A) = 0$. (b) The following statements are equivalent:

1° $A \subset Q_{\mathrm{top}}(A)$.
2° $Q(A) = Q_{\mathrm{top}}(A) \neq A$.
3° $\dim(A) = 1$.

[Use 37.20, 37.22, and 37.24.]

38.3 Let A be a semilocal noetherian ring that is the direct sum of finitely many local noetherian rings $A_1, \ldots, A_n$, furnished with its natural topology, and let R be the radical of A. The following statements are equivalent:

$1°$ Each maximal ideal of A has height ≤ 1, and R contains a cancellable element.
$2°$ $Q_{\text{top}}(A) = Q(A)$, and $Q_{\text{top}}(A)$ has no proper open ideal.
$3°$ $Q_{\text{top}}(A)$ contains an invertible topological nilpotent.
$4°$ $\dim(A_i) = 1$ for each $i \in [1, n]$.

39 Complete Regular Local Rings

Here we shall show that a complete local noetherian ring A that is either an equicharacteristic local ring or a nonequicharacteristic local ring in which $p.1$ is not a zero-divisor, where p is the characteristic of its residue field, is a finitely generated module over a subring A_0 that is a power series ring over a Cohen ring. To do so, we need information concerning integral extensions of a ring. In this section, by a subring of a ring with identity is meant either the zero subring or a subring containing the identity, and all modules are assumed to be unitary.

39.1 Definition. *Let A be a subring of a commutative ring with identity B. An element x of B is **integral** over A if x is a root of a monic polynomial in $A[X]$.*

Thus x is integral over A if and only if there exist $a_0, \ldots, a_{n-1} \in A$ such that $x^n + a_{n-1}x^{n-1} + \ldots + a_1 x + a_0 = 0$.

39.2 Theorem. *Let A be a subring of a commutative ring with identity B, and let $x \in B$. The following statements are equivalent:*

$1°$ *x is integral over A.*
$2°$ *$A[x]$ is a finitely generated A-module.*
$3°$ *x belongs to a subring C of B that is a finitely generated A-module.*
$4°$ *There is a finitely generated submodule M of the A-module B such that $xM \subseteq M$ and for any $y \in A[x]$, $yM = (0)$ only if $y = 0$.*

Proof. To show that $1°$ implies $2°$, assume that

$$x^n + a_{n-1}x^{n-1} + \ldots + a_1 x + a_0 = 0,$$

where $a_0, a_1, \ldots, a_{n-1} \in A$, and for each $q \geq 0$ let M_q be the A-submodule generated by $1, x, \ldots, x^{n+q-1}$. For each $q \geq 0$,

$$x^{n+q} = -a_{n-1}x^{n+q-1} - \ldots - a_0 x^q \in M_q,$$

so $M_{q+1} = M_q$. Consequently, $M_q = M_0$ for all $q \geq 0$ by induction. As

$$A[x] = \bigcup_{q=0}^{\infty} M_q,$$

therefore, $A[x] = M_0$, a finitely generated A-module. Clearly 2° implies 3°, and 3° implies 4° as we may take $M = C$, which contains 1.

Finally, assume 4°, and let $M = Au_1 + \ldots + Au_n$. For each $i \in [1, n]$ there exist $a_{i1}, \ldots, a_{in} \in A$ such that

$$xu_i = \sum_{j=1}^{n} a_{ij} u_j.$$

Thus for each $i \in [1, n]$,

$$\sum_{j=1}^{n} (a_{ij} - \delta_{ij} x) u_j = 0,$$

where $\delta_{ij} = 1$ if $i = j$, $\delta_{ij} = 0$ if $i \neq j$. Consequently, if

$$N = \begin{bmatrix} a_{11} - x & a_{12} & a_{13} & \cdots & a_{1n} \\ a_{21} & a_{22} - x & a_{23} & \cdots & a_{2n} \\ a_{31} & a_{32} & a_{33} - x & \cdots & a_{3n} \\ \vdots & \vdots & \vdots & \ddots & \vdots \\ a_{n1} & a_{n2} & a_{n3} & \cdots & a_{nn} - x \end{bmatrix},$$

then

$$N \cdot \begin{bmatrix} u_1 \\ \vdots \\ u_n \end{bmatrix} = \begin{bmatrix} 0 \\ \vdots \\ 0 \end{bmatrix}.$$

Multiplying this equation on the left by the adjoint of N and recalling that $(\operatorname{adj} N)N = (\det N)I_n$ where $\det N$ is the determinant of N and I_n is the identity matrix of order n, we conclude that $(\det N)u_i = 0$ for all $i \in [1, n]$. Thus $(\det N)u = 0$ for all $u \in M$, so by hypothesis, $\det N = 0$. Consequently, x is a root of the monic polynial $(-1)^n \det(a_{ij} - \delta_{ij} X)$, whose coefficients belong to A. •

39.3 Theorem. *Let A be a subring of a commutative ring with identity B. If $x_1, \ldots, x_n$ are elements of B that are integral over A, then $A[x_1, \ldots, x_n]$ is a finitely generated A-module.*

Proof. For each $k \in [1, n]$, let $A_k = A[x_1, \ldots, x_k]$. By 39.2, A_1 is a finitely generated A-module. If $k > 1$, x_k is clearly integral over A_{k-1}, and thus A_k, which is $A_{k-1}[x_k]$, is a finitely generated A_{k-1}-module. An inductive argument therefore establishes that A_k is a finitely generated A-module for all $k \in [1, n]$. •

39.4 Theorem. *If A is a subring of a commutative ring with identity B, then the elements of B integral over A form a subring of B.*

Proof. If x and y are integral over A, then $-x$, $x+y$, and xy all belong to $A[x,y]$, a finitely generated A-module by 39.3. Consequently, as $A[x,y]$ is a subring of B, $-x$, $x+y$, and xy are integral over A by 39.2. •

39.5 Definition. *Let A be a subring of a commutative ring with identity B. The **integral closure** of A in B is the subring A' of B consisting of all elements of B integral over A; A is **integrally closed** in B if $A' = A$. B is **integral over** A if $A' = B$.*

39.6 Theorem. *Let A and B be subrings of a commutative ring with identity C such that $A \subseteq B$. If $x \in C$ is integral over B and if B is integral over A, then x is integral over A.*

Proof. There exist $b_0, \ldots, b_{n-1} \in B$ such that $x^n + b_{n-1}x^{n-1} + \ldots + b_0 = 0$, so x is integral over $A[b_0, \ldots, b_{n-1}]$. Consequently, $A[b_0, \ldots, b_{n-1}, x]$ is a ring that is a finitely generated $A[b_0, \ldots, b_{n-1}]$-module; but $A[b_0, \ldots, b_{n-1}]$ is a finitely generated A-module by 39.3; hence $A[b_0, \ldots, b_{n-1}, x]$ is a finitely generated A-module, so x is integral over A by 39.2. •

39.7 Corollary. *If A is a subring of a commutative ring with identity B, the integral closure of A in B is an integrally closed subring of B.*

39.8 Theorem. *Let B be a commutative ring with identity that is integral over a subring A. If P is a prime ideal of A and Q' an ideal of B such that $Q' \cap A \subseteq P$, then there is a prime ideal P' of B such that $P' \cap A = P$ and $Q' \subseteq P'$.*

Proof. Let $\mathcal{J}$ be the set of all ideals J' of B such that $J' \cap A \subseteq P$ and $Q' \subseteq J'$. Then $Q' \in \mathcal{J}$, so $\mathcal{J} \neq \emptyset$. Ordered by inclusion, $\mathcal{J}$ is clearly inductive and therefore contains a maximal member P'. Clearly $P' \supseteq Q'$. We shall show that $P' \cap A = P$ and that P' is a prime ideal of B.

Suppose that there exists $x \in P \setminus P'$. Then $(P' + Bx) \cap A \not\subseteq P$ by the maximality of P', so there exist $p \in P'$ and $b \in B$ such that $p + bx \in A \setminus P$; let $y = p + bx$. As b is integral over A, there exist $a_0, \ldots, a_{n-1} \in A$ such that $b^n + a_{n-1}b^{n-1} + \ldots + a_0 = 0$, whence $(bx)^n + a_{n-1}x(bx)^{n-1} + \ldots + a_0 x^n = 0$. As $y \equiv bx \pmod{P'}$, $y^n + a_{n-1}xy^{n-1} + \ldots + a_0 x^n$ belongs to P' and thus to $P' \cap A \subseteq P$, since $x, y \in A$. Therefore as $x \in P$, a prime ideal, we conclude that $y^n \in P$ and hence $y \in P$, a contradiction, Thus $P' \cap A = P$.

To show that P' is prime, let J'_1 and J'_2 be ideals of B containing P' such that $J'_1 J'_2 \subseteq P'$. Let $J_1 = J'_1 \cap A$, $J_2 = J'_2 \cap A$. Then $J_1 J_2 \subseteq P$, so, as P is prime, either J_1 or J_2 is contained in P, say J_1. But then $J'_1 \in \mathcal{J}$ as $J'_1 \supseteq P' \supseteq Q'$, so by the maximality of P', $J'_1 = P'$. Thus P' is a prime ideal of B. •

Applying 39.8 to the case $Q = \{0\}$, we obtain:

39.9 Corollary. *Let B be a commutative ring with identity that is integral over a subring A. If P is a prime ideal of A, there is a prime ideal P' of B such that $P' \cap A = P$.*

39.10 Theorem. *Let B be a commutative ring with identity that is integral over a subring A. If P and Q are prime ideals of B such that $P \subset Q$, then $P \cap A \subset Q \cap A$.*

Proof. Let $x \in Q \setminus P$. As x is integral over A, there exists a monic polynomial $X^n + a_{n-1}X^{n-1} + \ldots + a_0 \in A[X]$ of lowest possible degree such that $x^n + a_{n-1}x^{n-1} + \ldots + a_0 \in P$. As
$$x(x^{n-1} + a_{n-1}x^{n-2} + \ldots + a_1) \equiv -a_0 \pmod{P},$$
and $x \notin P$, we conclude that $a_0 \notin P$, for otherwise $x^{n-1} + a_{n-1}x^{n-2} + \ldots + a_1 \in P$, in contradiction to the definition of n. Thus $a_0 \notin P \cap A$, but $a_0 \in A \cap Q$ since $x^n + a_{n-1}x^{n-1} + \ldots + a_0 \in P \subseteq Q$ and $x \in Q$. •

39.11 Theorem. *Let B be a commutative ring with identity that is integral over a subring A, and let P be a prime ideal of B. Then P is a maximal ideal of B if and only if $P \cap A$ is a maximal ideal of A.*

Proof. As $P \cap A$ is a proper ideal of A, it is contained in a maximal ideal M. By 39.8 there is a prime ideal M' of B such that $M' \supseteq P$ and $M' \cap A = M$. Consequently, if P is maximal, then $M' = P$, so $M = P \cap A$, that is, $P \cap A$ is maximal.

Similarly, P is contained in a maximal ideal M' of B. Consequently, $M' \cap A$ is a proper ideal of A containing $P \cap A$, so if $P \cap A$ is maximal, then $M' \cap A = P \cap A$, whence $M' = P$ by 39.10, that is, P is maximal. •

39.12 Theorem. *Let f be an epimorphism from a local noetherian ring A to a local noetherian ring B. (1) If A and B are furnished with their natural topologies, f is a topological epimorphism. (2) If A is a complete local noetherian ring, so is B. (3) If A is an integral domain and if $\dim(A) = \dim(B)$, then f is an isomorphism.*

Proof. Let M be the maximal ideal of A. Then $f(M)$ is the maximal ideal of B. (1) Since $f(M^n) = f(M)^n$ and $M^n \subseteq f^{-1}(f(M)^n)$, f is a toopological epimorphism. (2) Let K be the kernel of f. Since f induces a topological isomorphism from A/K to B by (1), B is complete by 7.14. (3) Assume that $K \neq (0)$, let $d = \dim(B)$, and let $(P_k)_{0 \leq k \leq d}$ be a strictly decreasing sequence of prime ideals of B. Then as $f^{-1}(P_d) \supseteq K \supset (0)$,
$$f^{-1}(P_0) \supset \ldots \supset f^{-1}(P_d) \supset (0),$$
and therefore $\dim(A) \geq d+1$ as (0) and each $f^{-1}(P_k)$ are prime ideals of A, a contradiction. Thus $K = (0)$, so f is an isomorphism. •

39.13 Theorem. *Let A be a commutative ring with identity, and let A_0 be a semilocal subring of A containing the identity of A such that A is a finitely generated A_0-module. (1) A is a semilocal ring, and its natural topologies as a ring or A_0-module are identical. (2) If A_0 is a noetherian ring, then A is a semilocal noetherian ring and the natural topology of A induces on A_0 its natural topology; if, further, A_0 is open for a ring topology T on A inducing on A_0 its natural topology, then T is the natural topology of A. (3) If A and A_0 are both local noetherian rings, then $\dim(A) = \dim(A_0)$. (4) If A_0 is a complete semilocal noetherian ring, then A is a semilocal noetherian ring that is complete for its natural topology, which induces on A_0 its natural topology.*

Proof. Let $M_1, \ldots, M_s$ be the maximal ideals of A_0, and let R_0 be the radical of A_0. By 24.11,

$$R_0 = \bigcap_{i=1}^{s} M_i = M_1 M_2 \ldots M_s.$$

For each $i \in [1, s]$, $A/M_i A$ is a finitely generated (A_0/M_i)-vector space and hence is an artinian (A_0/M_i)-algebra with identity by 27.8. Therefore $A/M_i A$ is an artinian ring, so there are only finitely many maximal ideals $N_{i,1}, \ldots, N_{i,s(i)}$ of A containing $M_i A$ by 27.17. Let

$$R_i = \bigcap_{j=1}^{s(i)} N_{i,j} = N_{i,1} N_{i,2} \ldots N_{i,s(i)}$$

by 24.11. As $R_i/M_i A$ is the radical of $A/M_i A$, there exists $t(i) \geq 1$ such that $R_i^{t(i)} \subseteq M_i A$ by 27.15. By 39.11, the radical R of A is the intersection of the ideals $N_{i,j}$, where $i \in [1, s]$ and for each such i, $j \in [1, s(i)]$. Thus by 24.11, $R = R_1 \ldots R_s$, so if $t = \sup\{t(i) : i \in [1, s]\}$,

$$R^t \subseteq R_1^{t(1)} \ldots R_s^{t(s)} \subseteq (M_1 A) \ldots (M_s A) = (M_1 \ldots M_s) A = R_0 A,$$

and furthermore,

$$R = \bigcap_{i=1}^{s} R_i \supseteq \bigcap_{i=1}^{s} M_i A \supseteq R_0 A.$$

Therefore the natural topology of the semilocal ring A and the natural topology of the A_0-module A are the same, since for any $n \geq 1$, $R^{nt} \subseteq (R_0 A)^n \subseteq R^n$.

Assume further that A_0 is noetherian. By 20.8, A is a noetherian A_0-module and *a fortiori* is a noetherian ring. By 24.3 and (1), the topology induced on A_0 by the natural topology of A is the natural topology of A_0. Suppose that A_0 is open for a ring topology T on A that induces on A_0 its natural topology, and for each ideal J_0 of A_0, let

$$(A_0 : J_0) = \{x \in A : J_0 x \subseteq A_0\},$$

an A_0-submodule of A. Then

$$A = \bigcup_{n=1}^{\infty} (A_0 : R_0^n)$$

since A_0 is open and $(R_0^n)_{n \geq 1}$ is a fundamental system of neighborhoods of zero for T. Thus as A is a noetherian A_0-module, $A = (A_0 : R_0^q)$ for some $q \geq 1$, that is, $R_0^q A \subseteq A_0$. Hence for each $k \geq 1$,

$$R_0^{t(q+k)} \subseteq R^{t(q+k)} \subseteq (R_0 A)^{q+k} = R_0^{q+k} A = R_0^k R_0^q A \subseteq R_0^k A_0 = R_0^k.$$

Therefore T is the natural topology of the semilocal noetherian ring A.

(3) By 39.10, $\dim(A_0) \geq \dim(A)$. If $(P_k)_{0 \leq k \leq d}$ is a strictly decreasing sequence of prime ideals of A_0, then by 39.9 there is a prime ideal P'_d of A such that $P'_d \cap A_0 = P_d$, and by 39.8, there is a (strictly) decreasing sequence $(P'_k)_{0 \leq k \leq d}$ of prime ideals of A such that $P'_k \cap A_0 = P_k$ for all $k \in [0, d]$. Thus $\dim(A) \geq \dim(A_0)$.

(4) By 39.11, $N_{i,j} \cap A_0 = M_i$ for all $i \in [1, s]$, $j \in [1, s(i)]$. Therefore $R \cap A_0 = R_0$. Hence $R_0^n = (R \cap A_0)^n \subseteq R^n \cap A_0$, so the topology induced on A_0 by the natural topology of A, which is Hausdorff by (4) of 24.16, is weaker than the natural topology of A_0. Consequently, the two topologies are the same by 36.35 and 36.33. In particular, A is a topological A_0-module when both A and A_0 are furnished with their natural topologies. Consequently, as A is a finitely generated A_0-module, A is complete by 36.35 and 28.5. •

In the following discussion, we shall use the notational abbreviations for elements of a power series ring or an epimorphic image thereof, introduced on pages 195-6. Thus if $y_1, \ldots, y_m$ is a sequence of elements of a ring A indexed by $[1, m]$ and if $n = (n_1, \ldots, n_m) \in \mathbb{N}^m$, then $y_1^{n_1} y_2^{n_2} \ldots y_m^{n_m}$ is abbreviated to y^n.

39.14 Lemma. Let A be a complete equicharacteristic local noetherian ring, let $d = \dim(A)$, let K be a Cohen subfield of A, and let $x_1, \ldots, x_d$ generate an ideal J over which the maximal ideal M of A is a minimal prime ideal. For any family $(c_n)_{n \in \mathbb{N}^d}$ of elements of K indexed by $\mathbb{N}^d$, the family $(c_n x_1^{n_1} \ldots x_d^{n_d})_{(n_1, \ldots, n_d) \in \mathbb{N}^d}$ is summable in A, and the function S_0 from $K[[X_1, \ldots, X_d]]$ to A, defined by

$$S_0\Big(\sum_{n \in \mathbb{N}^d} c_n X_1^{n_1} \ldots X_d^{n_d}\Big) = \sum_{n \in \mathbb{N}^d} c_n x_1^{n_1} \ldots x_d^{n_d},$$

is an isomorphism from $K[[X_1, \ldots, X_d]]$ to a subring A_0 of A, A is a finitely generated A_0-module, and the natural topology of A induces on A_0 its natural topology.

Proof. By 23.5 and 23.6, A_0 is a complete local noetherian ring and hence a linearly compact ring by 36.35. Since $X_1, \ldots, X_d$ generate the maximal ideal of $K[[X_1, \ldots, X_d]]$ by 23.4, $x_1, \ldots, x_d$ generate the maximal ideal M_0 of A_0, and thus
$$M_0 A = Ax_1 + \ldots + Ax_d.$$
Let $J = Ax_1 + \ldots + Ax_d$. By 37.20 there exists $t \geq 1$ such that $M^t \subseteq J$. Let $y_1, \ldots, y_m$ generate M. By 23.5, if $z \in A$, there is a family $(c_r)_{r \in \mathbb{N}^m}$ of elements in K such that
$$z = \sum_{r \in \mathbb{N}^m} c_r y^r = \sum_{|r|<t} c_r y^r + \sum_{|r| \geq t} c_r y^r$$
by 10.8 (where, if $r = (r_1, \ldots, r_m)$, $|r| = r_1 + \ldots + r_m$). Let
$$F = \sum_{|r|<t} A_0 y^r.$$
Since $y^r \in J$ whenever $|r| \geq t$,
$$\sum_{|r| \geq t} c_r y^r \in J$$
as J is a closed ideal by 24.14. Thus as K is contained in A_0,
$$A = F + J = F + M_0 A.$$
For any $n \geq 1$, $J^n = (M_0 A)^n = M_0^n A$, and therefore, as $M^t \subseteq J \subseteq M$, $M^{tn} \subseteq M_0^n A \subseteq M^n$, so the M_0-topology of the A_0-module A is the natural topology of A. Thus A is a topological A_0-module when each is furnished with its natural topology. As A_0 is a linearly compact ring, F is a closed submodule of the A_0-module A by 28.18. Therefore by 36.18, $A = F$, that is, A is a finitely generated A_0-module. Consequently by 39.13, $\dim(A_0) = \dim(A) = d$ and the topology induced on A_0 by the natural topology of A is the natural topology of A_0. Furthermore, by (3) of 39.12, 38.18, and (b) of 23.4, S_0 is an isomorphism. •

39.15 Lemma. *Let A be a complete nonequicharacteristic local noetherian ring of characteristic zero whose residue field has prime characteristic p, let $d = \dim(A)$, let C be a Cohen subring of A, and let $x_1, \ldots, x_{d-1}$, $p.1$ generate an ideal J over which the maximal ideal M of A is a minimal prime ideal. For any family $(c_n)_{n \in \mathbb{N}^{d-1}}$ of elements of C indexed by $\mathbb{N}^{d-1}$, the family $(c_n x_1^{n_1} \cdots x_{d-1}^{n_{d-1}})_{(n_1,\ldots,n_{d-1}) \in \mathbb{N}^{d-1}}$ is summable in A, and the function S_0 from $C[[X_1, \ldots, X_{d-1}]]$ to A, defined by*

$$S_0(\sum_{n \in \mathbb{N}^{d-1}} c_n X_1^{n_1} \cdots X_{d-1}^{n_{d-1}} = \sum_{n \in \mathbb{N}^{d-1}} c_n x_1^{n_1} \cdots x_{d-1}^{n_{d-1}},$$

is an isomorphism from $C[[X_1, \ldots, X_{d-1}]]$ to a subring A_0 of A, A is a finitely generated A_0-module, and the natural topology of A induces on A_0 its natural topology.

The proof is exactly like that of 39.14.

39.16 Theorem. *Let A be a complete equicharacteristic local noetherian ring of dimension d, and let k be the residue field of A. (1) A is a regular local ring if and only if A is isomorphic to $k[[X_1, \ldots, X_d]]$. (2) A contains a complete equicharacteristic regular local ring A_0 such that A is a finitely generated A_0-module and the topology induced on A_0 by the natural topology of A is the natural topology of A_0.*

Proof. By 21.8 and 21.14, we may identify k with a Cohen subfield K of A. Let M be the maximal ideal of A. (1) The condition is sufficient by 38.18. Necessity: M is generated by d elements $x_1, \ldots, x_d$, so the homomorphism S of 23.5 is an epimorphism. Consequently, in the terminology of 39.14, $A = A_0$, which is isomorphic to $K[[X_1, \ldots, X_d]]$. (2) By 38.8, M is a minimal prime ideal over an ideal J generated by d elements $x_1, \ldots, x_d$. The conclusion therefore follows from 39.14. •

39.17 Definition. *A nonequicharacteristic local ring A is **unramified** if $p.1 \notin M^2$, where M is its maximal ideal, p the characteristic of its residue field, and A is **ramified** if $p.1 \in M^2$.*

39.18 Theorem. *Let A be a complete, nonequicharacteric, local noetherian ring of characteristic zero and dimension d, and let p be the characteristic of its residue field. (1) A is an unramified regular local ring if and only if A is isomorphic to $C[[X_1, \ldots, X_{d-1}]]$ where C is the Cohen ring of characteristic zero whose residue field is isomorphic to that of A. (2) If $p.1$ is not a zero-divisor of A and if the maximal ideal M of A is generated by n elements, then A contains a complete unramified regular local ring A_0 such that A is a finitely generated A_0-module, A_0 contains a Cohen subring*

C of A, the topology induced on A_0 by the natural topology of A is the natural topology of A_0, there exist generators $u_1, \ldots, u_n$ of M such that $p.1, u_2, \ldots, , u_d$ generate the maximal ideal M_0 of A_0, and $M = M_0 A$.

Proof. By 21.20 we may identify C with a Cohen subring of A. (1) The condition is sufficient by 38.18. Necessity: As $p.1 \in M \setminus M^2$ and as $\mathrm{vdim}(A) = d$, there exist $x_1, \ldots, x_d \in M$ whose M^2-cosets generate the (A/M)-vector space M/M^2 such that $x_d = p.1$. Consequently, $x_1, \ldots, x_d$ generate M by 38.10. We first observe that the homomorphism S of 23.5 from $C[[X_1, \ldots, X_{d-1}]]$ to A determined by the sequence $x_1, \ldots, x_{d-1}$ is an epimorphism. Indeed, if $z \in A$, by 23.5 applied to the generators $x_1, \ldots, x_{d-1}, x_d = p.1$ of M there is a family $(c_{r,s})$ of elements of C indexed by $\mathbb{N}^{d-1} \times \mathbb{N}$ such that $(c_{r,s} x^r (p.1)^s)_{(r,s) \in \mathbb{N}^{d-1} \times \mathbb{N}}$ is summable and

$$z = \sum_{(r,s) \in \mathbb{N}^{d-1} \times \mathbb{N}} c_{r,s} x^r (p.1)^s.$$

For each $r \in \mathbb{N}^{d-1}$, $(c_{r,s}(p.1)^s)_{s \geq 0}$ is summable by 10.5 as $c_{r,s}(p,1)^s \in M^n$ whenever $s \geq n$; let

$$b_r = \sum_{s=0}^{\infty} c_{r,s}(p.1)^s \in A.$$

For each $r \in \mathbb{N}^{d-1}$,

$$b_r x^r = \sum_{s=0}^{\infty} c_{r,s}(p.1)^s x^r$$

by 10.16. Hence by 10.8,

$$z = \sum_{r \in \mathbb{N}^{d-1}} (\sum_{s=0}^{\infty} c_{r,s}(p.1)^s) x^r = \sum_{r \in \mathbb{N}^{d-1}} b_r x^r = S(\sum_{r \in \mathbb{N}^{d-1}} b_r X^r).$$

Consequently, in the terminology of 39.15, $A = A_0$, which is isomorphic to $C[[X_1, \ldots, X_{d-1}]]$.

(2) By 37.22 and 37.12, $\mathrm{ht}(p.A) \geq 1$ (and hence $\mathrm{ht}(p.A) = 1$ by 38.2). By 38.7 there exist generators $u_1, \ldots, u_n$ of M such that $p.1, u_2, \ldots, u_d$ generate an ideal J over which M is a minimal prime ideal. Let $x_i = u_{i+1}$ for each $i \in [1, d-1]$, $x_d = p.1$. By 39.15, the conclusion follows. ∎

From (2) of 39.18 and 38.16, a complete nonequicharacteristic regular local ring is finitely generated over a complete nonequicharacteristic unramified regular local ring, but a more precise description, analogous to that given in 22.7 of complete discrete valuations whose valuation rings are nonequicharacteristic, is available.

39.19 Definition. *If A is a commutative ring with identity and if A_0 is a local subdomain of A, A is an **Eisenstein extension** of A_0 if there exists $u \in A$ such that $A = A_0[u]$ and u is a root of an Eisenstein polynomial over A_0.*

39.20 Theorem. *If A is a complete nonequicharacteristic regular local ring, A is an Eisenstein extension of a subring A_0 that is a complete unramified regular local ring, and the topology induced on A_0 by the natural topology of A is the natural topology of A_0.*

Proof. Let $d = \dim(A)$. By 38.16, $p.1$ is not a zero divisor. By the remark following 38.11 and by (2) of 39.18, and with the terminology of that theorem, there are generators $u_1, \ldots, u_d$ of the maximal ideal M of A such that $p.1, u_2, \ldots, u_d$ generate the maximal ideal M_0 of A_0, and there exists $t \geq 1$ lsuch that $M^t \subseteq M_0 A$. Thus there is a smallest natural number s such that $u_1^s \in M_0 A$. To show that $A = A_0[u_1]$, let $z \in A$. By 23.5 applied to the sequence $u_1, x_1 = u_2, \ldots, x_{d-1} = u_d$, there is a family $(c_{k,r})_{(k,r) \in \mathbb{N} \times \mathbb{N}^{d-1}}$ such that

$$z = \sum_{(k,r) \in \mathbb{N} \times \mathbb{N}^{d-1}} c_{k,r} u_1^k x^r.$$

Now

$$\sum_{k \geq s,\, r \in \mathbb{N}^{d-1}} c_{k,r} u_1^k x^r \in J = M_0 A$$

since $u_1^s \in M_0 A$ and $M_0 A$ is closed by 24.14. Thus by 10.8 and 10.16,

$$z = \sum_{k=0}^{s-1} \Big(\sum_{r \in \mathbb{N}^{d-1}} c_{k,r} x^r \Big) u_1^k + \sum_{k \geq s,\, r \in \mathbb{N}^{d-1}} c_{k,r} u_1^k x_r \in \sum_{k=0}^{s-1} A_0 u_1^k + M_0 A.$$

As A is a topological A_0-module and as A_0 is linearly compact by 36.35, $\sum_{k=0}^{s-1} A_0 u_1^k$ is closed by 28.18. Therefore

$$A = \sum_{k=0}^{s-1} A_0 u_1^k$$

by 36.18. In particular, there exist $a_0, \ldots, a_{s-1} \in A_0$ such that $u_1^s = \sum_{k=0}^{s-1} a_k u_1^k$, so $f(u_1) = 0$ where

$$f(X) = X^s - \sum_{k=0}^{s-1} a_k X^k.$$

Suppose there exist integers $i \in [0, s-1]$ such that $a_i \notin M_0$, and let h be the smallest such i. Then

$$u_1^h(u_1^{s-h} - \sum_{k=0}^{s-h-1} a_{h+k}u_1^k) = u_1^s - \sum_{k=h}^{s-1} a_k u_1^k = \sum_{k=0}^{h-1} a_k u_1^k \in M_0 A.$$

As $u_1^{s-h} - \sum_{k=1}^{s-h-1} a_{h+k} u_k \in M$ but $a_h \notin M_0$ and hence $a_h \notin M$, $u_1^{s-h} - \sum_{k=0}^{s-h-1} a_{h+k} u_1^k \notin M$ and hence is a unit of A, so $u_1^h \in M_0 A$, a contradiction of the definition of s. Therefore $a_i \in M_0$ for all $i \in [0, s-1]$.

Let $Q = Au_2 + \ldots + Au_d$. The maximal ideal M/Q of A/Q is then generated by $u_1 + Q$, so $\mathrm{ht}((M/Q)) \leq 1$ by 38.2. If $\mathrm{ht}(M/Q) = 0$, M would be a minimal prime ideal over A, an ideal generated by $d - 1$ elements, in contradiction to 38.8. Thus $\mathrm{ht}(M/Q) = 1$, so as M/Q is a principal ideal, A/Q is a regular local ring. In particular, A/Q is an integral domain by 38.16, so Q is a prime ideal of A. Suppose that $a_0 \in M_0^2$. Then as each $a_i \in M_0$, we conclude that

$$u_1^s = \sum_{k=0}^{s-1} a_k u_l^k \in M_0 M \subseteq (p.A + Au_2 + \ldots + Au_d)M \subseteq p.M + Q.$$

Thus there exists $b \in M$ such that $u_1^s - p.b \in Q$. As $b \in M = Au_1 + Q$, there exists $c \in A$ such that $b - cu_1 \in Q$. Hence

$$u_1(u_1^{s-1} - p.c) = u_1^s - p.cu_1 = (u_1^s - p.b) + p.(b - cu_1) \in Q.$$

Now $u_1 \notin Q$ since otherwise Q would be M, whereas $\mathrm{ht}(M/Q) = 1$. Therefore as Q is a prime ideal, $u_1^{s-1} - p.c \in Q$, whence $u_1^{s-1} \in p.A + Q = M_0 A$, a contradiction of the definition of s. Thus $a_0 \notin M_0^2$, so f is an Eisenstein polynomial •

To complete the description of complete nonequicharacteristic regular local rings, we need to establish the converse of 39.20: An Eisenstein extension of a complete unramified regular local ring is a complete regular local ring. To do so, and to establish another property of complete local noetherian domains needed in Chapter 10, we wish to show that if C is a field or principal ideal domain, $C[[X]]$ is integrally closed.

39.21 Definition. *An integral domain is **integrally closed** if it is integrally closed in its quotient field.*

An integral domain A may be integrally closed without $A[[X]]$ being integrally closed. Consequently, we shall consider another property of integral domains that implies integral closure such that if A has it, then $A[[X]]$ has it.

39.22 Definition. *Let A be an integral domain, K its quotient field. An element x of K is **almost integral** over A if there exists $d \in A^*$ such that $dx^n \in A$ for all $n \in \mathbb{N}$. A is **completely integrally closed** if every element of K that is almost integral over A belongs to A.*

39.23 Theorem. *Let A be an integral domain, K its quotient field, $x \in K$. (1) If x is integral over A, then x is almost integral over A. (2) If A is noetherian, then x is integral over K if and only if x is almost integral over K. (3) If A is completely integrally closed, then A is integrally closed. (4) If A is noetherian, A is completely integrally closed if and only if A is integrally closed. (5) If A is a unique factorization domain (in particular, if A is a principal ideal domain), then A is completely integrally closed.*

Proof. (1) By 39.2, there exist $c_1, \ldots, c_m \in K$ such that $A[x] = Ac_1 + \ldots + Ac_m$; let $c_i = a_i b_i^{-1}$ where $a_i \in A$, $b_i \in A^*$ for all $i \in [1, m]$, and let $d = b_1 b_2 \ldots b_m$. Then $dA[x] \subseteq Aa_1 + \ldots + Aa_m \subseteq A$, so $dx^n \in A$ for all $n \in \mathbb{N}$.

(2) Sufficiency: Let $d \in A^*$ be such that $dx^n \in A$ for all $n \in \mathbb{N}$. Then $A[x] \subseteq Ad^{-1}$. As A is noetherian, so is the finitely generated A-module Ad^{-1}, so its submodule $A[x]$ is also a finitely generated A-module. By 39.2, therefore, x is integral over A.

(5) Suppose that $d \in A^*$ and that $dx^n \in A$ for all $n \in \mathbb{N}$ but that $x \notin A$. Then there exist an irreducible element p of A and elements $a, b \in A^*$ such that $x = a/pb$ where p does not divide a. Let $d = p^m c$ where p does not divide c. Then

$$dx^{m+1} = (ca^{m+1})/(pb^{m+1})$$

where p does not divide ca^{m+1}, so $dx^{m+1} \notin A$, a contradiction. •

39.24 Theorem. *If A is a completely integrally closed integral domain, so is $A[[X]]$.*

Proof. Let K be the quotient field of A. Then the field $K((X))$ contains the quotient field L of $A[[X]]$. Let $f \in L$ be almost integral over $A[[X]]$. Then f is almost integral over $K[[X]]$, a principal ideal domain by 18.2, since it is the valuation ring of the discrete valuation ord on $K((X))$, as noted on page 148. Therefore $f \in K[[X]]$ by (5) of 39.23. Let $g \in A[[X]]$ be such that $g \neq 0$ and $gf^n \in A[[X]]$ for all $n \in \mathbb{N}$, and let

$$f = \sum_{k=0}^{\infty} a_k X^k, \qquad g = \sum_{k=0}^{\infty} b_k X^k,$$

where for each $k \in \mathbb{N}$, $a_k \in K$ and $b_k \in A$. If $a_k \notin A$ for some $k \geq 0$, let i

be the smallest such integer, and let

$$f_1 = \sum_{k=0}^{i-1} a_k X^k \in A[X].$$

Clearly $g(f - f_1)^n \in A[[X]]$ for all $n \in \mathbb{N}$. Let j be the smallest of the integers k such that $b_k \neq 0$. The coefficient of X^{j+mi} in $g(f - f_1)^m$ is then $b_j a_i^m$ for all $m \in \mathbb{N}$. Consequently, $b_j a_i^m \in A$ for all $m \in \mathbb{N}$, so $a_i \in A$ by hypothesis, a contradiction. Therefore $f \in A[[X]]$. •

39.25 Corollary. *If A is a completely integrally closed integral domain, then $A[[X_1, \ldots, X_n]]$ is completely integrally closed for each $n \geq 1$.*

39.26 Corollary. *Complete equicharacteristic regular local rings and complete nonequicharacteristic unramified regular local rings are completely integrally closed.*

Proof. The assertion follows from 39.16, 39.18, (5) of 39.23, and 39.25.•

39.27 Theorem. *Let A be an integrally closed integral domain, and let K be its quotient field. (1) If g and h are monic polynomials over K such that $gh \in A[X]$, then both g and h belong to $A[X]$. (2) If f is a monic irreducible polynomial in $A[X]$, then f is a prime polynomial in $K[X]$. (3) If u is a root of a monic irreducible polynomial $f \in A[X]$, then $A[u]$ is an integral domain.*

Proof. (1) Let $f = gh$, and let L be a splitting field of f over K. Then there exist $a_1, \ldots, a_n, b_1, \ldots, b_m \in L$ such that

$$g = \prod_{i=1}^{n}(X - a_i), \qquad h = \prod_{j=1}^{m}(X - b_j).$$

For all $i \in [1, m]$, $f(a_i) = g(a_i)h(a_i) = 0$, so a_i is integral over A. The coefficients of g are sums of products of the a_i's and hence are integral over A by 39.4. Therefore as the coefficients of g also belong to K, $g \in A[X]$ since A is integrally closed. Similarly, $h \in A[X]$.

(2) Let $f = gh$ where g and h are monic polynomials over K. By (1), g and h belong to $A[X]$. Consequently, as f is irreducible in $A[X]$, either g or h is the constant polynomial 1, so f is a prime polynomial in $K[X]$.

(3) $A[u] \subseteq K[u]$, a field since f is a prime polynomial over K by (2). •

39.28 Theorem. *An Eisenstein polynomial f over a local noetherian domain A is irreducible in $A[X]$.*

Proof. In the contrary case, $f = gh$ where g and h are not units of $A[X]$. Since the leading coefficients of g and h are units of A (as f is monic), we may assume that g and h are monic polynomials whose respective degrees, r and s, are strictly less than the degree n of f. Let M be the maximal ideal of A, and for any polynomial $q \in A[X]$, let $\bar{q}$ be the image of q in $(A/M)[X]$ under the epimorphism from $A[X]$ to $(A/M)[X]$ induced by the canonical epimorphism from A to A/M. As f is an Eisenstein polynomial, $\bar{f} = X^n$. Consequently, $\bar{g} = X^r$ and $\bar{h} = X^s$ as X is a prime in the principal ideal domain $(A/M)[X]$. Thus all the nonleading coefficients of g and h belong to M. In particular, as $r > 0$ and $s > 0$, the constant coefficients of g and h belong to M, Consequently, the constant coefficient of f belongs to M^2, a contradiction. •

39.29 Theorem. *If A_0 is a complete, nonequicharacteristic, unramified regular local ring of dimension d and if $A = A_0[u]$ where u is a root of an Eisenstein polynomial f over A_0, then A is a complete regular local ring of dimension d whose residue field is canonically isomorphic to that of A_0.*

Proof. Since A is a finitely generated A_0-module, A is a semilocal noetherian ring that is complete for its natural topology by (4) of 39.13. By 39.28, 39.26, and (3) of 39.27, A is an integral domain. Therefore A is a complete local noetherian domain. By 39.13, $\dim(A) = d$. Consequently, to show that A is regular, it suffices to show that $\mathrm{vdim}(A) \leq d$ by the remark following 38.10.

Let M and M_0 be the maximal ideals of A and A_0 respectively, and let

$$f = X^n + a_{n-1}X^{n-1} + \ldots + a_1 X + a_0.$$

By 39.11, $M \cap A_0 = M_0$, and therefore $M \supseteq M_0 A$. As $a_0 \in M_0 \setminus M_0^2$, there exist $x_2, \ldots, x_d \in M_0$ such that the M_0^2-cosets of $a_0, x_2, \ldots, x_d$ generate the (A_0/M_0)-vector space M_0/M_0^2. Consequently, $a_0, x_2, \ldots, x_d$ generate M_0 by 38.10. As

$$u^n = -a_{n-1}u^{n-1} - \ldots - a_1 u - a_0 \in M_0 A \subseteq M,$$

$u \in M$, and $a_0 \in Au$. We shall show that $u, x_2, \ldots, x_d$ generate M. Let $c \in M$. There exist $c_0, \ldots, c_{n-1} \in A_0$ such that $c = c_0 + c_1 u + \ldots + c_{n-1}u^{n-1}$. Consequently,

$$c_0 \in M \cap A_0 = M_0 = A_0 a_0 + A_0 x_2 + \ldots + A_0 x_n.$$

Therefore as $a_0 \in Au$, $c \in Au + Ax_2 + \ldots + Ax_n$. The final assertion results from the fact that $c \equiv c_0 \pmod{M}$. •

Exercises

39.1 (I. S. Cohen [1945]) If A is a regular local ring of dimension d and if $x_1, \ldots, x_d$ generate the maximal ideal of A, then for each $k \in [1, d]$, $A/(Ax_1 + \ldots + Ax_k)$ is a regular local ring of dimension $d - k$, and hence $Ax_1 + \ldots + Ax_k$ is a prime ideal of A. [Use 38.14 and 38.15 and induction.]

39.2 (I. S. Cohen [1945]) Let p be an odd prime and let $A_0 = \mathbb{Z}_p[[X]]$, the ring of power series in one variable over the p-adic integers. Let $A = A_0[u]$, where u is a root of the Eisenstein polynomial $Y^2 - X^2 - p$. (a) A is a regular local ring whose maximal ideal M is generated by X and u. (b) M is also generated by $X - u$ and $X + u$. (c) A does not contain a discrete valuation ring B such that A is isomorphic to $B[[Y]]$. [Use Exercise 39.1 to show that $X - u$ and $X + u$ are principal primes, and apply Exercise 37.5.]

40 The Japanese Property

Here we shall show that a complete local noetherian domain is Japanese in the following sense:

40.1 Definition. *An integral domain A is **Japanese** if for every finite-dimensional extension field L of the quotient field K of A, the integral closure of A in L is a finitely generated A-module.*

Actually, if A is any noetherian, integrally closed integral domain, its integral closure in any *separable* finite-dimensional extension L of its quotient field K is a finitely generated A-module. This, in turn, depends ultimately on the fact that the trace linear form (or functional) on the K-vector space L is not the zero linear form, a fact from field theory whose proof, for completeness, is given below:

40.2 Theorem. *Let L and Ω be fields. The set of all monomorphisms from L to Ω is linearly independent in the Ω-vector space Ω^L of all functions from L to Ω.*

Proof. We proceed by induction. Assume that any n distinct monomorphisms are linearly independent, let $\sigma_1, \ldots, \sigma_{n+1}$ be $n+1$ distinct monomorphisms, and let $\lambda_1, \ldots, \lambda_{n+1} \in \Omega$ satisfy $\sum_{k=1}^{n+1} \lambda_k \sigma_k = 0$. As $\sigma_{n+1} \neq \sigma_1$, there exists $a \in L$ such that $\sigma_{n+1}(a) \neq \sigma_1(a)$. For each $x \in L$,

$$0 = \sum_{k=1}^{n+1} \lambda_k \sigma_k(ax) = \sum_{k=1}^{n+1} \lambda_k \sigma_k(a) \sigma_k(x)$$

and also

$$0 = \sigma_{n+1}(a) \sum_{k=1}^{n+1} \lambda_k \sigma_k(x) = \sum_{k=1}^{n+1} \lambda_k \sigma_{n+1}(a) \sigma_k(x).$$

Subtracting, we obtain

$$0 = \sum_{k=1}^{n} \lambda_k[\sigma_k(a) - \sigma_{n+1}(a)]\sigma_k(x)$$

for all $x \in L$, so by our inductive hypothesis, $\lambda_k[\sigma_k(a) - \sigma_{n+1}(a)] = 0$ for all $k \in [1,n]$, and in particular, $\lambda_1 = 0$ as $\sigma_1(a) \neq \sigma_{n+1}(a)$. Thus $\sum_{k=2}^{n+1} \lambda_k \sigma_k = 0$, so by our inductive hypothesis, $\lambda_2 = \ldots = \lambda_{n+1} = 0$. •

40.3 Definition. Let L be a finite-dimensional separable extension of a field K, and let $\sigma_1, \ldots, \sigma_n$ be the K-monomorphisms from L into an algebraic closure Ω of L. For each $a \in A$, the **trace** of a over K is the element $\text{Tr}_{L/K}(a)$ defined by

$$\text{Tr}_{L/K}(a) = \sum_{k=1}^{n} \sigma_k(a).$$

By the definition of separability, K is the fixed field of $\{\sigma, \ldots, \sigma_n\}$, so $\text{Tr}_{L/K}(a) \in K$ for all $a \in L$. Moreover, if K is the quotient field of a subdomain A and if a is integral over A, then so is $\text{Tr}_{L/K}(a)$. Indeed, if f is a monic polynomial over A such that $f(a) = 0$, then $f(\sigma_k(a)) = \sigma_k(f(a)) = \sigma_k(0) = 0$ for all $k \in [1,n]$, so $\text{Tr}_{L/K}(a)$ is integral over A by 39.4. In particular, if A is integrally closed, $\text{Tr}_{L/K}(a) \in A$ for all $a \in A$.

If K has characteristic zero and if $[L:K] = n$, then $\text{Tr}_{L/K}(1) = n.1 \neq 0$. More generally:

40.4 Theorem. *If L is a finite-dimensional separable extension of a field K, there exists $a \in L$ such that $\text{Tr}_{L/K}(a) \neq 0$.*

Proof. Let $\{a_1, \ldots, a_n\}$ be a basis of L over K, and let $\sigma_1, \ldots, \sigma_n$ be the K-monomorphisms from L into an algebraic closure Ω of L. If $(x_1, \ldots, x_n) \in \Omega^n$ satisfies

(1) $$\sum_{j=1}^{n} x_j \sigma_j(a_i) = 0$$

for all $i \in [1,n]$, then clearly

$$\sum_{j=1}^{n} x_j \sigma_j = 0,$$

so $x_1 = \ldots = x_n = 0$ by 40.2. Therefore if $(x_1, \ldots, x_n) = (1, 1, \ldots, 1)$, (1) is incorrect for some $i \in [1,n]$, that is, $\text{Tr}_{L/K}(a_i) \neq 0$. •

40.5 Theorem. *Let A be an integral domain, K its quotient field, L a finite-dimensional extension field of K, and let A' be the integral closure of A in L. (1) L is the quotient field of A'. (2) If A is integrally closed and if L is a separable extension of K, there is a basis $\{b_1, \ldots, b_n\}$ of the K-vector space L such that $A' \subseteq \sum_{k=1}^{n} Ab_k$.*

Proof. If $X^m + a_{m-1}X^{m-1} + \ldots + a_1 X + a_0$ is the minimal polynomial of $x \in L$ over K, there exists $s \in A^*$ such that $sa_k \in A$ for all $k \in [0, m-1]$, and the equality

$$(sx)^m + sa_{m-1}(sx)^{m-1} + \ldots + s^{m-1}a_1(sx) + s^m a_0 = 0$$

establishes that $sx \in A'$. In particular, L is the quotient field of A'. Consequently, there is a basis $\{a_1, \ldots, a_n\}$ of the K-vector space L consisting of elements of A'.

For each $y \in L$, $y' : x \to \text{Tr}_{L/K}(xy)$ is a linear form on the K-vector space L. Moreover, $T : y \to y'$ is a linear transformation from L to the K-vector space L^* of all linear forms on L. If $y \neq 0$, then $y' \neq 0$, for by 40.4 there exists $c \in L$ such that $\text{Tr}_{L/K}(c) \neq 0$, so $y'(cy^{-1}) \neq 0$. Therefore T is injective and hence is an isomorphism from L to L^* as both are n-dimensional over K. Consequently, there is a basis $\{b_1, \ldots, b_n\}$ of L such that for each $j \in [1, n]$, $b'_j(a_j) = 1$ and $b'_j(a_i) = 0$ if $i \neq j$.

Let $x \in A'$, and let $x = \sum_{j=1}^{n} \lambda_j b_j$ where $\lambda_j \in K$ for all $j \in [1, n]$. For each $i \in [1, n]$,

$$\text{Tr}_{L/K}(a_i x) = \sum_{j=1}^{n} \lambda_j \text{Tr}_{L/K}(a_i b_j) = \sum_{j=1}^{n} \lambda_j b'_j(a_i) = \lambda_i.$$

By 39.4, $a_j x \in A'$, so $\text{Tr}_{L/K}(a_j x) \in A' \cap K$, which is A by hypothesis. Thus

$$x = \sum_{j=1}^{n} \text{Tr}_{L/K}(a_j x) b_j \in \sum_{j=1}^{n} Ab_j. \bullet$$

40.6 Corollary. *If A is an integrally closed noetherian integral domain and if A' is the integral closure of A in a finite-dimensional separable extension L of its quotient field K, then A' is a finitely generated A-module.*

Proof. By 40.5 and 20.8, A' is a submodule of a noetherian A-module and hence is itself finitely generated by 20.3. $\bullet$

40.7 Theorem. (1) *An integrally closed, noetherian integral domain of characteristic zero is Japanese.* (2) *If A is a noetherian integral domain of prime characteristic p, then A is Japanese if and only if for each purely inseparable finite-dimensional extension field E of the quotient field K of A, the integral closure of A in E is a finitely generated A-module.*

Proof. (1) follows from 40.6 and the fact that a field of characteristic zero is perfect, that is, each of its finite-dimensional extension fields is a separable extension.

(2) Sufficiency: Let L be a finite-dimensional extension field of K, and let N be the smallest normal extension of K containing L that is contained in an algebraic closure Ω of L (if $L = K[c_1, \ldots, c_n]$ and if f_j is the minimal polynomial of c_j over K for each $j \in [1, n]$, N is the splitting field of $f_1 f_2 \ldots f_n$ in Ω over L). Then $[N : K] = [N : L][L : K] < \infty$. Let E be the fixed field of the group of all K-automorphisms of N. Then N is a separable extension of E and E is a purely inseparable extension of K. Let B be the integral closure of A in E. By 39.7, B is an integrally closed subring of E and hence, by (1) of 40.5, is an integrally closed integral domain. By hypothesis, B is a finitely generated A-module. Consequently, B is a noetherian A-module and a fortiori a noetherian integral domain by 20.8. Let C be the integral closure of B in N. By 40.6, C is a finitely generated B-module, therefore a finitely generated A-module, and consequently a noetherian A-module by 20.8. By 39.6, the integral closure A' of A in L is the A-submodule $C \cap L$ of C. Thus A' is a noetherian A-module by 20.3 and, in particular, is a finitely generated A-module. •

40.8 Theorem. *If A is an integrally closed, complete local noetherian domain and if A contains a principal prime ideal P such that A/P is Japanese, then A is Japanese.*

Proof. By 40.7 we may assume that A has prime characteristic p, and we need only prove that if E is a purely inseparable finite-dimensional extension field of the quotient field K of A, then the integral closure B of A in E is a finitely generated A-module. As $[E : K] < +\infty$, there exists $n \geq 1$ such that $x^{p^n} \in K$ for all $x \in E$. Let $q = p^n$. Then $B = \{x \in E : x^q \in A\}$, for if $x \in B$, then $x^q \in B \cap K = A$ as A is integrally closed, and if $x^q \in A$, then x is integral over A and hence belongs to B.

Let $c \in A$ be such that $P = Ac$. If c does not have a qth root in E, let $E_0 = E[a]$ where a is a root of $X^q - c$. Then $x^q \in K$ for all $x \in E_0$ as $\{x \in E_0 : x^q \in K\}$ is a subfield of E_0 containing E and a. Moreover, $[E_0 : K] = [E_0 : E][E : K] < +\infty$. If the integral closure B_0 of A in E_0 is a finitely generated A-module, then it is a noetherian A-module by 20.8, so its submodule B is also finitely generated. Thus, by replacing E with E_0, if necessary, we may assume that E contains an element a such that $a^q = c$.

By 39.9 there is a prime ideal Q of B such that $Q \cap A = P$. Then $Ba = Q = \{x \in E : x^q \in P\}$. Indeed, $a^q = c \in P \subseteq Q$, so $Ba \subseteq Q$. If $x \in Q$, then $x^q \in Q \cap A = P$. Finally, if $x^q \in P$, then as $P = Ac = Aa^q$, $(x/a)^q \in A$, and therefore $x/a \in B$, that is, $x \in Ba$.

Since $Q \cap A = P$, we may regard B/Q as a module over A/P. We shall show next that B/Q is a finitely generated (A/P)-module. Let A_P and B_Q be the localizations of A and B at P and Q respectively. The maximal ideal $A_P P$ of A_P is $A_P c$ since $A_P P = A_P Ac = A_P c$, and similarly the maximal ideal of B_Q is $B_Q a$. Moreover, $B_Q \cap K = A_P$. Indeed, if $z \in B_Q \cap K$, then there exists $s \in B \setminus Q$ such that $sz \in B$; then $s^q \in A \setminus P$ and $s^q z \in B \cap K = A$ as A is integrally closed, so $z \in A_P$. Consequently by 37.21 and 20.17, B_Q is the valuation ring of a discrete valuation w of L whose restriction v to K is a valuation with valuation ring A_P. Let $\phi_{w,v}$ be the canonical monomorphism, defined on page 156, from the residue field $k_v = A_P/A_P P$ of v to the residue field $k_w = B_Q/B_Q Q$ of w. Let $\psi_v : x + P \to x + A_P P$ and $\psi_w : x + Q \to x + B_Q Q$ be the canonical monomorphisms from A/P to k_v and from B/Q to k_w respectively, and let $i : x + P \to x + Q$ be the canonical injection from A/P to B/Q. Clearly

$$\phi_{w,v} \circ \psi_v = \psi_w \circ i.$$

Let $B' = \psi_w(B/Q)$ and $A' = \psi_w(i(A/P))$. Since B is integral over A, clearly B/Q is integral over $i(A/P)$, and therefore B' is integral over A'. Also, k_w and $\phi_{w,v}(k_v)$ are the quotient fields of B' and A' respectively. Moreover,

$$[k_w : \phi_{w,v}(k_v)] = f(w/v) \leq [E:K]$$

by 19.8. By hypothesis, A/P is Japanese, so A' is also. Therefore the integral closure C' of A' in k_w is a finitely generated A'-module. As A is noetherian, so is A/P by 20.4; hence A' is also noetherian, and therefore C' is a noetherian A'-module by 20.8. Consequently, the A'-submodule B' of C' is noetherian, so B/Q is a noetherian module over $i(A/P)$, that is, B/Q is a noetherian (A/P)-module.

As $Q = Ba$, for any $k \geq 1$, $Q^k = Ba^k$. For any $n \geq 1$, $A \cap Q^{qn} = P^n$, and in particular, $A \cap Q^q = P$. Indeed, if $x \in A \cap Q^{qn}$, then $x = ba^{qn}$ for some $b \in B$, so $x = bc^n$; consequently, $b = xc^{-n} \in K$, so $b \in K \cap B = A$, and therefore $x = bc^n \in P^n$. Thus the Q-topology of B induces on A the P-topology. Moreover, the Q-topology of B is Hausdorff, for if $x \in \bigcap_{n \geq 1} Q^n$, then

$$x^q \in \bigcap_{n=1}^{\infty} Q^{nq} \cap A = \bigcap_{n=1}^{\infty} P^n = (0)$$

by 20.16, so also $x = 0$. Since, by the preceding, $A \cap Q^q = P$, for each $k \in [1, q]$,

$$P = A \cap Q^q \subseteq A \cap Q^k \subseteq A \cap Q = P.$$

Thus $A \cap Q^k = P$, so we may regard B/Q^k as a module over A/P. If $k \in [1, q-1]$, $t \to ta^k + Q^{k+1}$ is an A-module epimorphism from B to Q^k/Q^{k+1} whose kernel is Q. Therefore the (A/P)-module Q^k/Q^{k+1} is a noetherian (A/P)-module by 20.3. By 27.10 applied to the (A/P)-submodules Q^n/Q^k of B/Q^k, where $n \in [0, k-1]$, B/Q^k is a noetherian (A/P)-module for all $k \geq 1$. In particular, B/Q^q is a noetherian (A/P)-module and hence a noetherian A-module, so as $Q^q = Ba^q = Bc = PB$, there is a finitely generated submodule F of the A-module B such that $B = F + PB$.

By the preceding, B, furnished with the Q-topology, is a linearly topologized, Hausdorff topological module over A, furnished with the P-topology. By hypothesis, 36.35, and 36.38, A is linearly compact for the P-topology. Therefore F is closed in B by 28.18. Consequently, $B = F$ and hence B is a finitely generated A-module by 36.18. •

40.9 Theorem. *A complete local noetherian integral domain is Japanese.*

Proof. A field is certainly Japanese, and a complete discrete valuation ring of characteristic zero is also Japanese by 20.17, (5) and (3) of 39.23, and (1) of 40.7. Let C be a field or a complete discrete valuation ring of characteristic zero, and for each $n \in \mathbb{N}$, let $B_n = C[[X_1, \ldots, X_n]]$. We have just observed that B_0 is Japanese, and an inductive argument establishes that B_n is Japanese for all $n \in \mathbb{N}$. Indeed, assume that B_{n-1} is Japanese. By 23.4, B_n is a complete local noetherian integral domain, and by 39.25 and (3) and (5) of 39.23, B_n is integrally closed. Consequently, as X_n is a prime of B_n and as $B/B_n X_n$ is isomorphic to B_{n-1}, B_n is Japanese by 40.8. Therefore by (1) of 39.16 and (1) of 39.18, a complete regular local ring that is either equicharacteristic or unramified is Japanese.

Let A be a complete local noetherian integral domain. By (2) of 39.16 and (2) of 39.18 and the preceding, A contains a Japanese subdomain A_0 and elements $c_1, \ldots, c_n$ such that $A = A_0 c_1 + \ldots + A_0 c_n$. Let K and K_0 be the quotient fields of A and A_0 respectively. Each c_i is integral over A_0 by 39.2 and hence is algebraic over K_0, so $K_0[c_1, \ldots, c_n]$ is a finite-dimensional field extension of K_0. Clearly $K = K_0[c_1, \ldots, c_n]$. If L is a finite-dimensional field extension of K, A and A_0 have the same integral closure C in L by 39.6, so as $[L : K_0] = [L : K][K : K_0] < +\infty$, C is a finitely generated A_0-module and *a fortiori* a finitely generated A-module. •

Exercises

40.1 If $(A_\lambda)_{\lambda \in L}$ is a family of subrings of a commutative ring with identity B that are integrally closed in B, then $\bigcap_{\lambda \in L} A_\lambda$ is integrally closed in B.

40.2 If A is the valuation ring of a real valuation of a field K, then A is integrally closed in K.

40.3 (Nagata [1962]) Let K be a field of prime characteristic p such that $[K : K^p] = +\infty$, where K^p is the range of the monomorphism $x \to x^p$ from K to K (for example, let $K = F_p(X_1, X_2, \dots)$ where F_p is the prime field of p elements, for then $K^p = F_p(X_1^p, X_2^p, \dots)$). Let $B = K[[X]]$, and let A be the union of all the subrings $F[[X]]$ of $K[[X]]$ where F is a subfield of K containing K^p and $[F : K^p] < +\infty$. (a) A is a discrete valuation ring whose maximal ideal is AX. [Observe that $AX \subseteq BX$ in establishing 5° of 20.17.] Let Q be the quotient field of A in $K((X))$. (b) There exists $c \in B \setminus A$, and for any such c, $[Q(c) : Q] = p$ and $Q(c)$ is the quotient field of $A[c]$. (c) Let D be the integral closure of $A[c]$ in $Q(c)$. Then D is the integral closure of A in $Q(c)$, and $D = B \cap Q(c)$. [Apply Exercises 40.1 and 40.2 to B.] (d) $BX \cap D = DX$, and consequently D is a discrete valuation ring whose maximal ideal is DX. [Observe that B in integral over A and that if $gX \in D$, then $g \in Q(c)$.] (e) $D = K + DX$, and hence $D = A + DX$. [Use (d).] (f) D is not a finitely generated A-module, and hence A is not Japanese. [In the contrary case, apply 36.18 and 24.14 to the A-module D.]

CHAPTER X

LOCALLY CENTRALLY LINEARLY COMPACT RINGS

Topological rings formed from finite-dimensional algebras over discretely valued fields are the subject of this chapter. In §41, complete discretely valued fields are characterized as those nondiscrete topological fields having an open, strictly linearly compact subring, and complete discretely valued division rings finite-dimensional over their centers are similarly characterized as those nondiscrete topological division rings that are locally centrally linearly compact, that is, that have a linearly topologized open subring that is a strictly linearly compact module over its center. An immediate consequence is Jacobson's theorem that the topology of a nondiscrete, totally disconnected, locally compact division ring is given by a discrete valuation.

In §42 topological rings with identity that are indecomposable finite-dimensional algebras over complete, discretely valued fields are characterized as locally centrally linearly compact rings of zero or prime characteristic whose centers are local rings having no proper open ideals, and in §43 locally centrally linearly compact rings whose centers have no proper open ideals are described, and applications of these results to locally compact rings are given.

41 Complete Discretely Valued Fields and Division Rings

Basic to our characterization of complete, discretely valued division rings finite-dimensional over their centers are the following definitions:

41.1 Definition. *A topological ring is* **locally strictly linearly compact** *if it contains an open, strictly linearly compact subring. A topological ring is* **centrally linearly compact** *if it is a linearly topologized ring that is a strictly linearly compact module over its center. A topological ring is* **locally centrally linearly compact** *if it contains an open, centrally linearly compact subring.*

For example, a totally disconnected, locally compact ring is locally centrally linearly compact by 4.21 and the remark following Definition 28.10.

41 COMPLETE DISCRETELY VALUED FIELDS AND DIVISION RINGS

41.2 Theorem. *If f is a continuous epimorphism from a centrally linearly compact ring B to a Hausdorff, linearly topologized ring B', then B' is centrally linearly compact.*

Proof. Let C and C' be the centers of B and B'. Since the kernel of f is an ideal, B' becomes a C-module under the well defined scalar multiplication

$$c.f(b) = f(cb)$$

for all $c \in C$, $b \in B$. Under this definition, f is C-linear, so B' is a strictly linear compact C-module by 28.11. Since $f(C) \subseteq C'$, every C'-submodule of B' is a C-submodule, so B' is a fortiori a strictly linearly compact C'-module. •

41.3 Theorem. *The cartesian product B of a family $(B_\lambda)_{\lambda \in L}$ of centrally linearly compact rings is centrally linearly compact.*

Proof. For each $\mu \in L$ let C_μ be the center of B_μ. Then center C of B is then the cartesian product of $(C_\lambda)_{\lambda \in L}$. Each B_μ becomes a C-module under the scalar multiplication defined by

$$(c_\lambda)_{\lambda \in L}.b = c_\mu b$$

for all $(c_\lambda)_{\lambda \in L} \in C$ and all $b \in B_\mu$. As the C_μ-submodules and the C-submodules of B_μ coincide, B_μ is a strictly linearly compact C-module. Consequently by 28.17, B is a strictly linearly compact C-module. •

41.4 Theorem. *Let A be a nondiscrete, locally centrally linearly compact ring with identity 1. If either A is a division ring or the center C of A is a topological ring having no proper open ideals, then there is an open, centrally linearly compact subring B of A that contains 1.*

Proof. By hypothesis, A contains an open, centrally linearly compact subring B'. Let B be the subring of A generated by B' and 1. For each left ideal J of B', let $(J:J) = \{b \in A : bJ \subseteq J\}$. Then for each left ideal J of B',

$$B \subseteq (J:J)$$

and hence J is also a left ideal of B, for $(J:J)$ is a subring of A that contains B' and 1 and hence B. Thus every open left ideal of B' is also an open left ideal of B, so B is linearly topologized.

We next show that B is a strictly linearly compact module over the center, $C_{B'}$, of B'. Case 1: A is a division ring. Since A is nondiscrete, B' contains a nonzero element b. Then $B = Bbb^{-1} \subseteq BB'b^{-1}$, so $B \subseteq B'b^{-1}$, for as we have just seen, B' is a left ideal of B. Since $x \to xb^{-1}$ is a topological isomorphism from the $C_{B'}$-module B' to the $C_{B'}$-module $B'b^{-1}$, $B'b^{-1}$ is a strictly linearly compact $C_{B'}$-submodule. As B is an open and thus closed submodule of the $C_{B'}$-module $B'b^{-1}$, therefore, B is a strictly linearly compact $C_{B'}$-module by 28.16.

Case 2: C has no proper open ideals. The ideal of C generated by $B' \cap C$ is thus C, that is, $(B' \cap C)C = C$. In particular, there exist $x_1, \ldots, x_n \in B' \cap C$ and $c_1, \ldots, c_n \in C$ such that $1 = x_1 c_1 + \ldots + x_n c_n$. Let $B'' = B'c_1 + \ldots + B'c_n$. As we saw above, $B \subseteq (B' : B')$, so

$$b = (bx_1)c_1 + \ldots + (bx_n)c_n \in B''$$

for each $b \in B$. In particular, $B' \subseteq B''$, so B'' is a linearly topologized $C_{B'}$-module. By 28.18 B'' is a strictly linearly compact $C_{B'}$-module. As B is an open and thus closed submodule of B'', therefore, B is also a strictly linearly compact $C_{B'}$-module.

Now $C_{B'}$ is contained in the center C_B of B, for $\{x \in A : xc = cx \text{ for all } c \in C_{B'}\}$ is a subring of A containing B' and 1 and hence B. Therefore B is a fortiori a strictly linearly compact C_B-module. •

The valuation ring V of a complete discretely valued field is a complete local noetherian ring by 20.17, and hence its natural topology is linearly compact by 36.35 and thus strictly linearly compact by 36.16. Consequently, a complete discretely valued field is locally strictly linearly compact. To establish the converse, we need the following description of discrete valuation rings:

41.5 Theorem. *An integral domain A is a discrete valuation ring if and only if A is an integrally closed, local noetherian domain such that $\dim(A) = 1$.*

Proof. Necessity: As A is a local principal ideal domain by 18.2, A is integrally closed by (3) and (5) of 39.23. Moreover, $\dim(A) = 1$ by (2) of 38.13.

Sufficiency: Let K be the quotient field of A, and let M be its maximal ideal. By 20.16, $\bigcap_{n \geq 1} M^n = \{0\}$. By 38.8 there exists $c \in A^*$ such that M is a minimal prime ideal over Ac. Thus $M^t \subseteq Ac$ for some $t \geq 1$ by 37.20. Therefore there is a smallest natural number r such that $M^r \subseteq Ac$, and $r \geq 1$ since $c \in M$. Consequently, there exists $b \in M^{r-1} \setminus Ac$, so $Mb \subseteq Ac$ but $b \notin Ac$. Thus $M(b/c) \subseteq A$, so $M(b/c)$ is an ideal of A. If $M(b/c) \subseteq M$, then b/c would be integral over A by 39.2, so b/c would belong to A by hypothesis, and therefore $b \in Ac$, a contradiction. Thus $M(b/c) = A$, so $M = A(c/b)$, and in particular, $c/b \in M$. Therefore M is a principal ideal, so A is a discrete valuation ring by 20.17. •

41 COMPLETE DISCRETELY VALUED FIELDS AND DIVISION RINGS

41.6 Theorem. *A field K, furnished with a ring topology, is complete and discretely valued if and only if it is nondiscrete and locally strictly linearly compact. Moreover, if B is an open, strictly linearly compact subring of K containing 1, then the valuation ring V of of the valuation defining its topology is a finitely generated B-module.*

Proof. We have just seen that the condition is necessary. Sufficiency: By 41.4, K contain an open, strictly linearly compact subring B that contains 1. As $y \to yx$ is a homeomorphism from K to K for each $x \in K^*$, Bx is open for every $x \in K^*$, and hence every nonzero ideal of B is open and thus closed. Let R be the radical of B. By 33.22, $\bigcap_{n \geq 1} R^n = (0)$, so by 36.34, B is the direct sum of finitely many complete local noetherian subrings. As B has no proper zero-divisors, therefore, B is a complete local noetherian integral domain. By 28.13 and 36.38, the topology induced on B by that of K is its natural topology.

As K is not discrete and as the maximal ideal M of B is open, M contains a nonzero element c. Consequently, $\dim(B) > 0$ by 37.22. As the topology of B is its natural topology, $\lim_{n \to \infty} c^n = 0$. As Bc is open, $Bc \supseteq M^t$ for some $t \geq 1$, so M is a minimal prime ideal over Bc by 37.20. Consequently, $\dim(B) = 1$ by 37.24. The quotient field of B is K, for if $x \in K$, then $\lim_{n \to \infty} c^n x = 0$, so $c^m x \in B$ for some $m \geq 1$, and thus x is the quotient $c^{-m}(c^m x)$ of elements of B. By 40.9, the integral closure V of B in K is a finitely generated B-module. By (2) and (4) of 39.13, V is a semilocal noetherian ring complete for its natural topology, which is the topology it inherits from K. By 24.19, V is a complete local noetherian integral domain. Moreover, $\dim(V) = 1$ by (3) of 39.13. Therefore V is the valuation ring of a discrete valuation v of K by 39.7 and 41.5. Since $V \supseteq B$, V is open. Therefore the given topology of K is that defined by v. Moreover, K is complete by 7.6. •

Before establishing a generalization of 41.6, we need several preliminary results.

41.7 Theorem. *A Hausdorff, finite-dimensional algebra A with identity over a complete, discretely valued field K is a locally centrally linearly compact ring.*

Proof. Let v be a valuation with value group $\mathbb{Z}$ that defines the topology of K, let V be the valuation ring of v, and for each $m \geq 0$, let $V_m = \{\lambda \in K : v(\lambda) \geq m\}$. Let $\{e_1, \ldots, e_n\}$ be a basis of the K-vector space A such that $e_1 = 1$, and let

$$e_i e_j = \sum_{k=1}^{n} \alpha_{ijk} e_k$$

for all $i, j \in [2, n]$, $k \in [1, n]$. Let $\lambda \in K^*$ be such that $v(\lambda) \geq 0$ and
$$v(\lambda) \geq \sup\{-v(\alpha_{ijk}) : i, j \in [2, n], k \in [1, n]\}.$$
Let $g_1 = 1$, $g_j = \lambda e_j$ for all $j \in [2, n]$, and let
$$B = Vg_1 + \ldots + Vg_n, \qquad B_m = V_m g_1 + \ldots + V_m g_n$$
for all $m \geq 0$. Easy calulations establish that B is a subring of A and that B_m is an ideal of B for all $m \geq 0$. Since
$$(\lambda_1, \ldots, \lambda_n) \to \sum_{k=1}^n \lambda_k g_k$$
is a topological isomorphism from the K-vector space K^n to A by 15.10 and 13.8, B is an open subring of A and $(B_m)_{m \geq 0}$ is a fundamental system of neighborhoods of zero. As we noted above, V is strictly linearly compact, so by 28.18, B is a strictly linearly compact V-module and a fortiori a centrally linearly compact ring, as we may identify V with $V.1$, a subring of the center of B. •

41.8 Theorem. *If B is an open subring of a topological ring A with identity and if either A has an invertible topological nilpotent or the center C of A is a topological ring having no proper open ideals, then the center of B is $B \cap C$.*

Proof. Clearly $B \cap C$ is contained in the center C_B of B. To show that $C_B \subseteq B \cap C$, let $c \in C_B$ and $a \in A$; we wish to show that $ca = ac$.

Case 1: A has an invertible topological nilpotent b. As
$$\lim_{n \to \infty} b^n = 0 = \lim_{n \to \infty} ab^n,$$
there exists $m \geq 0$ such that $b^m \in B$ and $ab^m \in B$. Hence
$$(ca)b^m = c(ab^m) = (ab^m)c = a(b^m c) = a(cb^m) = (ac)b^m.$$
Thus $ca = ac$.

Case 2: C has no proper open ideals. Let $V = \{x \in A : xa \in B\}$. As B is open, V is a neighborhood of zero in A. Let $W = \{x \in C : x(ca - ac) = 0\}$. Then $V \cap C \subseteq W$, for if $x \in V \cap C$, then
$$x(ca) = (xc)a = (cx)a = c(xa) = (xa)c$$
since $x \in C$, $c \in C_B$, and $xa \in B$. Thus W is an open ideal of C, so $W = C$ by hypothesis. Therefore $1 \in W$, so $ca = ac$. •

41.9 Theorem. Let K be a Hausdorff division ring, let B be an open subring of K containing 1, and let R be the radical of B. If B is strictly linearly compact, then B is a noetherian ring, B/R is a division ring, the induced topology of B is its R-topology, and R is the set of all topological nilpotents of B.

Proof. As B is open and as $y \to xy$ is a homeomorphism from K to K for each $x \in K^*$, Bx is open for every $x \in K^*$, and hence every nonzero left ideal of B is open. Let $S = \bigcap_{n \geq 1} R^n$. Assume that $S \neq (0)$. Then S is open, so B/S is an artinian B-module by 28.15 and hence is an artinian ring. Consequently, its radical, which is R/S by 26.16, is nilpotent by 27.15. Thus there exists $n \geq 1$ such that $R^n \subseteq S$. Therefore $(0) \neq R^n = R^{n+1} = \ldots$, in contradiction to 33.21. Therefore $\bigcap_{n \geq 1} R^n = (0)$.

Since every nonzero left ideal of B is open and hence closed, B is linearly compact for the discrete topology by 28.19, so by 36.29 B is linearly compact for the radical topology, the weakest Hausdorff linear topology on B. By 28.13, therefore, the given topology of B is the radical topology. As B contains an identity element, B contains no pathological subgroups by 36.12. Consequently by 36.33, B is a noetherian ring and B/R is artinian ring. By 34.1, every idempotent of B/R is the R-coset of an idempotent of B. But as K is a division ring, B has no idempotents except 0 and 1. Thus by 26.16, B/R is an artinian semisimple ring whose only idempotents are 0 and 1, so B/R is a division ring by 27.14. In particular, if $x \in B \setminus R$, then $x + R$ is not a nilpotent of B/R, so as B/R is discrete, x is not a topological nilpotent of B. Thus R is the set of topological nilpotents of B. •

41.10 Theorem. A division ring K furnished with a ring topology is complete, discretely valued, and finite-dimensional over its center C if and only if K is nondiscrete and locally centrally linearly compact, in which case the topology of C is also given by a complete, discrete valuation.

Proof. Necessity: As K is discretely valued, there exists $a \in K^*$ such that $\lim_{n \to \infty} a^n = 0$. Consequently, as $[K : C] < +\infty$, $C[a]$ is a finite-dimensional field extension of C whose topology is not discrete and thus is given by a discrete valuation. By 18.6, the topology of C is also given by a discrete valuation, and as C is closed in K, C is complete. By 41.7, K is a locally centrally linearly compact ring.

Sufficiency: By 41.4, there is an open, centrally linearly compact subring B of K that contains 1. As B is linearly topologized and is strictly linearly compact over its center C_B, B is a strictly linearly compact ring. Let R be the radical of B. By 41.9, R is the set of topological nilpotents of B, and the (nondiscrete) topology of B is the radical topology. Thus B contains a nonzero topological nilpotent a. Let K_0 be the closed subfield generated by C and a, let $B_0 = B \cap K_0$, and let R_0 be the radical of B_0. As $a \in B_0$, the induced topology of B_0 is not discrete. Since the open left ideals of B form a fundamental system of neighborhoods of zero for B, the open ideals of B_0 form a fundamental system of neighborhoods of zero for B_0. By 41.8, the center C_B of B is $B \cap C$. Thus $C_B \subseteq K_0 \cap B = B_0$, so B_0 is a closed C_B-submodule of B, hence is a strictly linearly compact C_B-module by 28.16, and a fortiori is a strictly linearly compact ring. By 41.6, therefore, the induced topology of K_0 is defined by a complete discrete valuation. By 41.9, the topology of B_0 is its R_0-topology, B_0/R_0 is a field, and R_0 is the set of topological nilpotents of B_0. Thus $R_0 = R \cap B_0$.

As $C_B \subseteq B_0$, B is also a strictly linearly compact B_0-module. The topology of the B_0-module B is its R_0-topology. Indeed, as K is a division ring, every nonzero right ideal of B is open, and in particular, as a^n is a nonzero element of R_0^n, $R_0^n B$ is open for all $n \geq 1$. On the other hand, $R_0^n B \subseteq R^n$ for all $n \geq 1$.

As the topology of B is the R-topology, B/R is a discrete, strictly linearly compact B_0-module by 28.16 and hence, as $R_0 = B_0 \cap R$, a discrete, strictly linearly compact B_0/R_0-vector space. By (2) of 28.14, B/R is an artinian vector space and hence is finite-dimensional. Therefore there is a finitely generated B_0-submodule F of B such that $B = F + R$, or equivalently, $B = F + RB$ as B has an identity. As F is finitely generated, F is closed by 28.18. Therefore by 36.18, $B = F$.

Let $B = B_0 x_1 + \ldots + B_0 x_n$. Then $K = K_0 x_1 + \ldots + K_0 x_n$, for if $z \in K$, there exists $t \geq 1$ such that $a^t z \in B$; thus there exist $b_1, \ldots, b_n \in B_0$ such that $a^t z = b_1 x_1 + \ldots + b_n x_n$, so

$$z = (a^{-t} b_1) x_1 + \ldots + (a^{-t} b_n) x_n \in K_0 x_1 + \ldots + K_0 x_n.$$

Let K_0' be the division ring of K consisting of all elements of K commuting with each element of K_0. By 18.15, $\dim_C K_0' \leq \dim_{K_0} K \leq n$. But as K_0 is commutative, K_0' contains K_0, so $\dim_C K_0 \leq n$. As the topology of K_0 is given by a complete discrete valuation and as C is closed, therefore, the topology of C is also given by a complete discrete valuation by 18.6. Furthermore,

$$[K : C] \leq [K : K_0][K_0 : C] \leq n^2,$$

so the topology of K is the only Hausdorff topology making K a topological vector space over C by 15.10 and 13.8, and that topology is defined by a discrete valuation by 17.13. •

As noted after Definition 41.1, a totally disconnected, locally compact ring is locally centrally linearly compact. Therefore from 41.10 we recover Jacobson's theorem for nondiscrete, totally disconnected, locally compact division rings, from which the structure of such division rings is readily ascertained: The topology of the center C of a nondiscrete, totally disconnected, locally compact division ring K is given by a complete, discrete valuation, and K is finite-dimensional over C (Theorems 18.11 and 18.16).

Exercises

41.1 We extend the definition of a prime ideal to arbitrary rings: An ideal P of a ring with identity A is *prime* if P is a proper ideal and for all ideals I, J of A, if $IJ \subseteq P$, then either $I \subseteq P$ or $J \subseteq P$. Extend Theorem 37.8 as follows: If A is a ring such that every nonempty set of ideals, ordered by inclusion, contains a maximal member, then every proper ideal of A is either a prime ideal or contains a product of prime ideals.

41.2 Let A be a linearly compact ring whose radical R satisfies $\bigcap_{n \geq 1} R^n = \{0\}$ such that every ideal is closed for the radical topology. (a) Every nonempty set of ideals of A, ordered by inclusion, contains a maximal member. [Modify the proof of 36.31.] (b) If R is the radical of A, A/R is artinian.

41.3 Let A be a ring, R its radical. (a) If A is strictly linearly compact and every nonzero ideal of A is open, then the topology of A is the radical topology. [Argue as in 41.9.] (b) (Ánh [1977a]) If A is strictly linearly compact and has an identity, then every nonzero ideal of A is open if and only if $\bigcap_{n \geq 1} R^n = \{0\}$, A/R is artinian, R is a finitely generated left ideal, and every nonzero prime ideal of A is a maximal ideal. [Use 36.27 and Exercises 41.1 and 41.2.] (c) (Warner [1961]) If A is compact, then every ideal is open if and only if every ideal of A is closed and every nonzero proper prime ideal of A is a regular maximal ideal.

41.4 Let A be a nondiscrete, strictly linearly compact commutative ring with identity such that the ideals of A are totally ordered by inclusion. (a) Every nonzero ideal is open. (b) The topology of A is the radical topology. [Use Exercise 41.3.] (c) Every ideal is a principal ideal. [Use 36.33.] (d) R is a maximal ideal. [Consider A/R.] (e) Every element of $A \setminus R$ is invertible. [Use 11.16.] (f) A is an integral domain. [Use (a) and (b).] (g) The quotient field K of A, topologized by declaring the neighborhoods of zero in A a fundamental system of neighborhoods of zero, is a topological ring. (h) Let $c \in A$ be such that $Ac = R$, and let $v : K \to \mathbb{Z} \cup \{+\infty\}$ be defined by

$$v(x) = n \in \mathbb{Z} \text{ if } x \in Ac^n \setminus Ac^{n+1},$$
$$v(0) = +\infty.$$

Then v is a complete, discrete valuation of K whose valuation ring is A, and the topology on A defined by v is its given topology. (i) Conversely, the valuation ring of a complete, discrete valuation of a field K is a nondiscrete, strictly linearly compact ring with identity whose ideals are totally ordered by inclusion.

42 Finite-dimensional Algebras

Here we shall characterize finite-dimensional, indecomposable Hausdorff algebras with identity over complete, discretely valued fields.

42.1 Theorem. *If A is a topological integral domain that has no proper open ideals and contains an open subring B that is a complete local noetherian domain whose induced topology is its natural topology, then A is a field whose topology is given by a discrete valuation.*

Proof. Let K be the quotient field of A. To show that K is also the quotient field of B, it suffices to show that each $a \in A$ is a quotient of elements of B. By hypothesis, the maximal ideal of B contains a nonzero element t; as $\lim_{n \to \infty} t^n = 0 = \lim_{n \to \infty} at^n$, there exists $m \geq 1$ such that t^m and at^m belong to B, so $a = (at^m)t^{-m}$ is a quotient of elements of B.

By 40.9, the integral closure C' of B in K is a finitely generated B-module and hence a noetherian B-module by 20.8. Therefore the integral closure C of B in A is a submodule of the noetherian B-module C' and hence is itself a finitely generated B-module. Consequently by 39.4 and 39.13, C is a complete semilocal noetherian ring whose induced topology is its natural topology. By 24.19, C is the direct sum of finitely many complete local noetherian rings, so as A has no proper zero-divisors, C is actually a complete local noetherian domain.

Thus $C \subset A$ since A has no proper open ideals. For each ideal J of C, let $(C:J) = \{x \in A : xJ \subseteq C\}$, and let M be the maximal ideal of of C. As C is open and as $(M^n)_{n \geq 1}$ is a fundamental system of neighborhoods of zero, $A = \bigcup_{n \geq 1}(C:M^n)$. If I and J are ideals of C such that $C = (C:I) = (C:J)$, then $C = (C:IJ)$. Indeed, if $x \in A$ and $xIJ \subseteq C$, then for any $y \in I$, $xyJ \subseteq C$, so $xy \in (C:J) = C$; thus $xI \subseteq C$, so $x \in (C:I) = C$. Therefore if $C = (C:M)$, then $C = (C:M^n)$ for all $n \geq 1$, so $C = \bigcup_{n \geq 1}(C:M^n) = A$, a contradiction. Consequently, there exists $a \in (C:M) \setminus C$. Thus $aM \subseteq C$. If $aM \subseteq M$, then a would be integral over C by 39.2 and hence $a \in C$, a contradiction. Thus as aM is an ideal of the local ring C, $aM = C$. Consequently, $M \neq (0)$, and there exists $b \in M$ such that $ab = 1$. As b is invertible, $x \to xb$ is a homeomorphism from A to A, so there exists $n \geq 1$ such that $M^n \subseteq Cb \subseteq M$. Therefore by 37.20 and 37.24, $\dim(C) = 1$, that is, M and (0) are the only prime ideals of C.

Let x be a nonzero element of M. Then M is the only prime ideal of C containing Cx, so by 37.20 there exists $t \geq 1$ such that $M^t \subseteq Cx$, whence

$M^{n+t} \subseteq M^n x$ for all $n \geq 1$. Consequently, $z \to zx$ is an open mapping from A to A. In particular, Ax is open, so as A has no proper open ideals, $Ax = A$, that is, x is invertible in A. Thus every element of C^* is invertible in A. Finally, if $y \in A^*$, then $yb^n \in C^*$ for some $n \geq 1$ as $\lim_{n \to \infty} b^n = 0$, so yb^n is invertible in A, whence y is also. Therefore A is a field. •

42.2 Theorem. *Let A be a commutative topological ring with identity 1 that contains no proper open ideal. If A contains an open semilocal subring B such that $1 \in B$ and B is strictly linearly compact for its induced topology, then B is a complete semilocal noetherian ring furnished with its natural topology, A is an artinian ring all of whose ideals are closed, and A contains an invertible topological nilpotent.*

Proof. (a) The radical R of B is open. Indeed, R is closed by 29.12, so the topology induced on B/R is a Hausdorff ideal topology. By (1) of 24.16, B/R is artinian, so the induced topology is discrete by (1) of 28.14, that is, R is open.

(b) R is a finitely generated ideal. By (a), the ideal AR of A generated by R is open and hence is A, so there exist $a_1, \ldots, a_m \in A$ and $b_1, \ldots, b_m \in R$ such that $\sum_{i=1}^m a_i b_i = 1$. Let $J = Bb_1 + \ldots + Bb_m$. Then J is open, for if $I = \{x \in B : xa_i \in B \text{ for all } i \in [1, n]\}$, then I is an open ideal, and $I \subseteq J$ since if $x \in I$, then $x = \sum_{i=1}^m (xa_i) b_i \in J$. Consequently by (2) of 28.15, B/J is an artinian B-module and hence an artinian ring, so B/J is a noetherian ring by 37.7 and hence a noetherian B-module. In particular, as $R \supseteq J$, R/J is a finitely generated B-module, so there exist $c_1, \ldots, c_s \in R$ such that $R = Bc_1 + \ldots + Bc_s + J$. Consequently, R is finitely generated as J is.

(c) B is a complete semilocal noetherian ring furnished with its natural topology. For all $n \geq 1$, R^n is a finitely generated ideal by (b) and hence is closed by 28.18. Moreover, the radical topology is stronger than the topology induced on B by 33.22. Consequently by 7.21, the radical topology is complete. Therefore B is a complete semilocal noetherian ring by 24.17. By 36.35 and 36.33, the radical topology is the weakest Hausdorff linear topology on B and hence is the topology induced on B.

(d) A is a noetherian ring all of whose ideals are closed. Indeed, let J be an ideal of A. Then $J \cap B$ is closed in B by (c), 36.35, and 36.33, so J is closed by 4.11. To show that $J = A(J \cap B)$, let $c \in J$. Since $J \cap B$ is an open subset of J, $A(J \cap B)$ is an open submodule of the topological A-module J. As $R_c : x \to xc$ is a continuous A-module homomorphism from A to J, therefore, $R_c^{-1}(A(J \cap B))$ is an open ideal of A and hence is A. Therefore $1 \in R_c^{-1}(A(J \cap B))$, that is, $c \in A(J \cap B)$. By (c) there exist $c_1, \ldots, c_m \in J \cap B$ such that $J \cap B = Bc_1 + \ldots + Bc_m$, so $J = A(J \cap B) = Ac_1 + \ldots + Ac_m$. Thus A is noetherian.

(e) A is artinian. By (d) and 37.9, it suffices to show that each prime ideal P of A is maximal. By (d), P is closed, so A/P is Hausdorff. Let ϕ be the canonical epimorphism from A to A/P. Clearly A/P has no proper open ideals, and as ϕ is a topological epimorphism, $\phi(B)$ is ideally topologized and hence is a strictly linearly compact ring by 29.3. By (c) applied to A/P and its subring $\phi(B)$, $\phi(B)$ is a complete semilocal noetherian ring whose induced topology is its natural topology. Consequently, $\phi(B)$ is the direct sum of complete local rings by 24.19, so $\phi(B)$ is a complete local domain as A/P has no proper zero-divisors. By 42.1, A/P is a field, that is, P is maximal. Thus A is artinian.

(f) A contains an invertible topological nilpotent. By (e) and 27.17, A is semilocal. An inductive argument establishes that if the union of finitely many closed subsets of a topological space contains an interior point, then one of the sets contains an interior point. Let $M_1, \ldots, M_r$ be its maximal ideals, and let

$$G = A \setminus \bigcup_{i=1}^{r} M_i,$$

the set of invertible elements of A. By (d), each M_i is closed, so G is open. Moreover, G is dense, for if $\bigcup_{i=1}^{r} M_i$ contained an interior point, some M_i would also, and thus M_i would be open by 4.9, a contradiction of our hypothesis. Therefore as R is open by (a), there exists $b \in G \cap R$, and again by (c), $\lim_{n \to \infty} b^n = 0$. •

42.3 Theorem. *If E is a Hausdorff topological vector space over a nondiscrete, complete, straight division ring K, then no proper subspace of E is open.*

Proof. If M were a proper open subspace, E/M would be a nonzero discrete topological vector space over K, in contradiction to 13.1. •

42.4 Definition. *A ring [algebra] is **indecomposable** if A is not the direct sum of two proper subrings [subalgebras].*

42.5 Definition. *A commutative algebra A over a field K is a **Cohen algebra** if A is local and if K is canonically isomorphic to its residue field, that is, if $\lambda \to \lambda.1 + M$ is an isomorphism from K to A/M, where M is the maximal ideal of A.*

42.6 Theorem. *Let A be a commutative topological ring with identity. The following statements are equivalent:*

$1°$ *A is a locally strictly linearly compact local ring whose characteristic is either zero or a prime, and A has no proper open ideals.*

$2°$ A is a Hausdorff, finite-dimensional Cohen algebra over a complete, discretely valued field.

$3°$ A is a Hausdorff, indecomposable, finite-dimensional algebra over a complete, discretely valued field.

Proof. Clearly $2°$ implies $3°$, since a local ring contains no idempotents other than 0 and 1. Assume $3°$. Then A is an artinian ring as it is a finite-dimensional algebra with identity, so A is local by 37.7. By 42.3, A contains no proper open ideals. Consequently, $1°$ holds by 41.7.

Assume $1°$. By 41.4, A contains an open strictly linearly compact subring B that contains 1. By 34.6, B is topologically isomorphic to the cartesian product of local rings. Consequently, as A is local and thus has no idempotents other than 0 and 1, B itself is a local ring. By 42.2, B is a complete local noetherian ring whose induced topology is its natural topology, and A is a local artinian ring all of whose ideals are closed. Thus by 27.15, the maximal ideal Q of A is nilpotent; let $r \geq 1$ be such that $Q^r = \{0\}$, and let ϕ be the canonical epimorphism from A to A/Q. As Q is closed but not open, A/Q is Hausdorff but not discrete. Since B is local, open, and ideally topologized, so is $\phi(B)$; therefore by 29.3, $\phi(B)$ is strictly linearly compact. Consequently by 41.6, the topology of A/Q is given by a complete discrete valuation v, and the valuation ring V of v is a finitely generated $\phi(B)$-module.

Let $c_1, \ldots, c_n \in A$ be such that $c_1 = 1$ and $V = \phi(B)[\phi(c_1), \ldots, \phi(c_n)]$, and let $B' = B[c_1, \ldots, c_n]$. Clearly $\phi(B') = V$. For each $i \in [1, n]$, $\phi(c_i)$ is integral over $\phi(B)$ by 39.2, so there is a monic polynomial $f \in B[X]$ such that $\phi(f(c_i)) = 0$, that is, $f(c_i) \in Q$; thus $f(c_i)^r = 0$, and consequently c_i is integral over B. By 39.3, B' is a finitely generated B-module. By 39.13, B' is a complete semilocal noetherian ring whose induced topology is its natural topology. By 24.19, B' is the direct sum of finitely many complete local noetherian rings. Consequently as A is local and thus has no idempotents other than 0 and 1, B' is a complete local noetherian ring. Thus, by replacing B with B' if necessary, we may assume that $\phi(B) = V$.

Let M be the maximal ideal of B. The restriction ϕ_B of ϕ to B is an epimorphism from B to V and hence induces an isomorphism Φ_B from the residue field B/M of B to the residue field $V/\phi(M)$ of V such that $\Phi_B \circ \sigma = \rho \circ \phi_B$, where σ and ρ are respectively the canonical epimorphisms from B to B/M and from V to $V/\phi(M)$. Since Q is nilpotent, the characteristic of A/Q is zero if the characteristic of A is. Therefore B and V have the same characteristic. We shall show that A contains a subfield K such that the restriction to K of ϕ is a topological isomorphism from K to A/Q.

Case 1: V is equicharacteristic (and therefore B is also equicharacteristic). By 21.20, B contains a subfield k mapped onto B/M by σ. Thus

$\Phi_B \circ \sigma$ maps k isomorphically onto the residue field $V/\phi(M)$ of V, so as $\Phi_B \circ \sigma = \rho \circ \phi_B$, the restriction ϕ_k of ϕ to k is an isomorphism from k onto a Cohen subfield of V. By 22.1, there is a topological isomorphism F from the ring $k[[X]]$ of formal power series over k, furnished with its natural topology, to V that extends ϕ_k. Let $x \in B$ be such that $F(X) = \phi(x)$. Then $x \in M$, so by 23.5,

$$S : \sum_{k=0}^{\infty} c_k X^k \to \sum_{k=0}^{\infty} c_k x^k$$

is an epimorphism from $k[[X]]$ to a k-subalgebra $k[[x]]$ of B. The induced topology of $k[[x]]$ is a Hausdorff ideal topology that is not discrete, since x is a topological nilpotent. Therefore as the ideals of $k[[x]]$ are totally ordered by inclusion (as those of $k[[X]]$ are), the nonzero ideals of $k[[x]]$ form a fundamental system of neighborhoods of zero for the induced topology. Consequently, S is a topological epimorphism from $k[[X]]$ to $k[[x]]$. As F, S, and the restriction $\phi_{k[[x]]}$ of ϕ to $k[[x]]$ are continuous and since $\phi_{k[[x]]} \circ S$ and F agree on k and at X, we conclude that $\phi_{k[[x]]} \circ S = F$. Therefore S is injective and hence is a topological isomorphism, so as F is also a topological isomorphism, $\psi_{k[[x]]}$ is a topological isomorphism from $k[[x]]$ to V. In particular, as $\phi_{k[[x]]}$ is injective, $k[[x]] \cap Q = (0)$, so $k[[x]]$ has a quotient field K in A. Since $\phi(k[[x]]) = V$, clearly $\phi(K) = A/Q$. To show that the restriction of ϕ to K is a topological isomorphism from K to A/Q, therefore, it suffices to show that $k[[x]]$ is open in K. Since $k[[x]]$ is a discrete valuation ring, $k[[x]]$ is maximal in the set of proper subrings of K by 17.14. But $B \cap K$ is a proper subring of K containing $k[[x]]$ since x is not invertible in B; therefore $B \cap K = k[[x]]$, so $k[[x]]$ is open in K.

Case 2: V is not equicharacteristic, that is, A has characteristic zero and the residue field of B has prime characteristic p. As ϕ_B is an epimorphism from B to V with kernel $B \cap Q$, $\dim(B/(B \cap Q)) = \dim(V) = 1$. As Q is nilpotent, every prime ideal of B contains $B \cap Q$, so

$$\dim(B) = \dim(B/(B \cap Q)) = 1.$$

As A is equicharacteristic, $p.1$ is invertible in A and hence is not a zero-divisor of B. Therefore by 39.18, B contains a Cohen ring B_0 such that B is a finitely generated B_0-module and the topology induced on B_0 is its natural topology. As no nonzero element of B_0 is nilpotent, $B_0 \cap Q = (0)$, so B_0 has a quotient field K_0 in A. If B contained K_0, then K_0 would be integral over B_0 and hence by 39.13, $0 = \dim(K_0) = \dim(B_0)$, that is, B_0 would be a field, a contradiction. Thus $B \cap K_0$ is a proper subring of K_0 containing B_0. But as in Case 1, the discrete valuation ring B_0 is maximal in the set of proper subrings of K_0 by 17.14, so $B \cap K_0 = B_0$. Thus B_0 is open for the topology induced on K_0, so the induced topology of K_0 is defined by a complete discrete valuation whose valuation ring is B_0.

Let $a_1, \ldots, a_n \in B$ be such that $B = B_0[a_1, \ldots, a_n]$, and let $L_0 = \phi(K_0)$. Then

$$V = \phi(B) = \phi(B_0)[\phi(a_1), \ldots, \phi(a_n)] \subseteq L_0[\phi(a_1), \ldots, \phi(a_n)].$$

As each a_i is integral over B_0, each $\phi(a_i)$ is integral over $\phi(B_0)$ and hence is algebraic over L_0. Consequently, $L_0[\phi(a_1), \ldots, \phi(a_n)]$ is a subfield of A/Q containing V and thus is A/Q, and moreover, $[A/Q : L_0] < +\infty$. By 21.10, a field K that is maximal in the set of all subfields of A containing K_0 is a Cohen subfield of A, that is, $\phi(K) = A/Q$. Consequently, there exist $b_1, \ldots, b_m \in K$ such that $L_0\phi(b_1) + \ldots + L_0\phi(b_m) = A/Q$. Therefore $K = K_0 b_1 + \ldots + K_0 b_m$, for if $z \in K$, there exist $t_1, \ldots, t_m \in K_0$ such that

$$\phi(z) = \sum_{i=1}^{m} \phi(t_i)\phi(b_i),$$

whence

$$z - \sum_{i=1}^{m} t_i b_i \in K \cap Q = (0).$$

Thus $[K : K_0] < +\infty$. Therefore by 15.10 and 13.6, the induced topology on K is the only topology making K a Hausdorff K_0-vector space, and that topology is complete and discretely valued by 16.8. Furthermore, as K is straight, its topology is minimal in the set of all Hausdorff ring topologies on K by 13.2, so as the restriction ϕ_K of ϕ to K is a continuous isomorphism from K to A/Q, ϕ_K is a topological isomorphism.

We have left to show that A is a finite-dimensional K-vector space. Let $x_1, \ldots, x_m \in Q$ be such that $Q = Ax_1 + \ldots + Ax_m$. By 23.5 (with the notation of the proof of that theorem),

$$A = \sum_{|s|<r} Kx^s$$

since, as $Q^r = (0)$, $x^s = 0$ whenever $|s| \geq r$. Thus 2° holds. •

42.7 Theorem. Let A be a linearly compact ring with identity, let R be the radical of A, and let E be a unitary A-module. If A/R is artinian and if E is linearly compact for the R-topology, then E is finitely generated.

Proof. By hypothesis, E/RE is a discrete, linearly compact A-module, which we may regard as a module over A/R. By 26.16 and 36.19, E/RE is a finitely generated (A/R)-module and hence a finitely generated A-module. Thus there is a finitely generated submodule M of E such that $E = M+RE$. By 28.18, M is closed, so by 36.18, $E = M$. •

A noncommutative generalization of 42.6 is based on the following theorem:

42.8 Theorem. *Let A be a locally centrally linearly compact ring with identity 1 whose center C is a topological ring having no proper open ideals, and let B be an open, centrally linearly compact subring of A containing 1. The center of B is $B \cap C$, and the following statements are equivalent:*

1° *$B \cap C$ is semilocal.*
2° *C is semilocal.*
3° *A has only finitely many maximal ideals.*
4° *Every C-submodule of A is closed.*
5° *Every ideal of A is closed.*
6° *C is an artinian ring, and A is a finitely generated C-module.*
7° *A satisfies the Ascending Chain Condition on closed ideals.*
8° *A satisfies the Descending Chain condition on closed ideals.*
9° *C contains an invertible topological nilpotent.*

If these statements hold, then $B \cap C$ is a complete semilocal noetherian ring, and B is a noetherian $(B \cap C)$-module whose induced topology is its natural topology.

Proof. By 41.8, the center of B is $B \cap C$. Thus $B \cap C$ is a strictly linearly compact ring. By 34.6, $B \cap C$ is topologically isomorphic to the cartesian product of a family of linearly compact local rings and thus possesses an orthogonal family $(e_\lambda)_{\lambda \in L}$ of idempotents such that $(B \cap C)e_\lambda$ is a linearly compact local ring for each $\lambda \in L$ and $\sum_{\lambda \in L} e_\lambda = 1$. We shall first show that each of 2°–9° implies 1°, that is, that L is finite.

For each $\lambda \in L$, let M_λ be a maximal ideal of Ce_λ, $[Ae_\lambda]$, and let $N_\lambda = M_\lambda + C(1-e_\lambda)$ $[N_\lambda = M_\lambda + A(1-e_\lambda)]$. Clearly C/N_λ $[A/N_\lambda]$ is isomorphic to Ce_λ/M_λ $[Ae_\lambda/M_\lambda]$ and, if $\lambda \neq \mu$, $N_\lambda e_\mu = Ce_\mu$ $[Ae_\mu] \neq M_\mu = N_\mu e_\mu$, whence $N_\lambda \neq N_\mu$. Thus each of 2° and 3° implies 1°. The ideal generated by $\{e_\lambda : \lambda \in L\}$ is dense in A since $\sum_{\lambda \in L} e_\lambda = 1$, and contains 1 if and only if L is finite. Thus 5° implies 1°, and clearly 4° implies 5°. Suppose that $(\lambda_i)_{i \geq 1}$ is a sequence of distinct members of L, and let $f_n = \sum_{i=1}^n e_{\lambda_i}$ for each $n \geq 1$. Then each f_n is an idempotent of $B \cap C$, and consequently $Bf_n = Af_n \cap B$. By 28.18, Bf_n is closed in B; hence by 4.11, Af_n is closed in A. Similarly $A(1 - f_n)$ is closed. Thus $(Af_n)_{n \geq 1}$ is a strictly increasing sequence of closed ideals of A, and $(A(1 - f_n))_{n \geq 1}$ is a strictly decreasing sequence of closed ideals. Consequently, each of 7° and 8° implies 1°, and 6° implies 7° by 27.8.

To show that 9° implies 1°, let b be an invertible topological nilpotent in C. Then $b^m \in B$ for some $m \geq 1$, and b^m is also an invertible topological nilpotent. Replacing b by b^m if necessary, we may thus assume that $b \in B \cap C$. For each $\lambda \in L$,

$$\lim_{n \to \infty} (be_\lambda)^n = \lim_{n \to \infty} b^n e_\lambda = 0,$$

so be_λ belongs to the maximal ideal of $(B \cap C)e_\lambda$, and hence $b^{-1}e_\lambda \notin (B \cap C)e_\lambda$. Thus $b^{-1}e_\lambda \notin B$ for all $\lambda \in L$. But $(b^{-1}e_\lambda)_{\lambda \in L}$ is summable by 10.16, and therefore $b^{-1}e_\lambda \in B$ for all but finitely many $\lambda \in L$ by 10.5. Thus L is finite.

Assume 1°. By 42.2, $B \cap C$ is a complete semilocal noetherian ring whose induced topology is its natural topology, C is an artinian ring, and 9° holds. By 27.17, 2° also holds. Let R be the radical of $B \cap C$. We shall show that the topology of the $(B \cap C)$-module B is its R-topology. As R^n is open in $B \cap C$ and as B is linearly topologized, B contains an open left ideal J such that $J \cap B \cap C \subseteq R^n$. As $J \cap B \cap C$ is open in C, $(J \cap B \cap C)C = C$ by hypothesis, and consequently there exist $a_1, \ldots, a_s \in J \cap B \cap C$ and $c_1, \ldots, c_s \in C$ such that $a_1 c_1 + \ldots + a_s c_s = 1$. Let $I = \{x \in B : c_i x \in B \text{ for all } i \in [1, s]\}$. then I is an open right ideal of B contained in $R^n B$, for if $x \in I$, then

$$x = a_1(c_1 x) + \ldots + a_s(c_s x) \in (J \cap B \cap C)B \subseteq R^n B.$$

Therefore $R^n B$ is open. Also, if L is an open left ideal of B, then $L \cap B \cap C \supseteq R^k$ for some $k \geq 1$, so $R^k B = BR^k \subseteq BL = L$. Therefore $(R^n B)_{n \geq 1}$ is a fundamental system of neighborhoods of zero.

By (1) of 24.16, $(B \cap C)/R$ is artinian. Hence by 42.7, B is a finitely generated $(B \cap C)$-module. Let $B = (B \cap C)y_1 + \ldots + (B \cap C)y_n$ where $y_1, \ldots, y_n \in B$. As noted earlier, C contains an invertible topological nilpotent b. Thus for any $z \in A$, there exists $m \geq 1$ such that $b^m z \in B$, so there exist $c_1, \ldots, c_n \in B \cap C$ such that $b^m z = c_1 y_1 + \ldots + c_n y_n$, whence

$$z = (b^{-m} c_1)y_1 + \ldots + (b^{-m} c_n)y_n \in Cy_1 + \ldots + Cy_n.$$

Therefore A is a finitely generated C-module, so 6°, 7°, and 8° hold. If J is a C-submodule of A, then $B \cap J$ is a $(B \cap C)$-submodule of B; $B \cap J$ is closed in B by 24.14, so B is closed in A by 4.11. Thus 4° and 5° hold.

Finally, a maximal ideal of A is a primitive ideal by 26.6 and hence contains the radical R of A by 26.8. Thus, to establish 3°, it suffices to show that A/R has only finitely many ideals. Since 6° holds, A is an artinian C-module and a fortiori an artinian ring. Therefore A/R is a semisimple artinian ring 26.16 and 27.4. By 27.14, A/R is isomorphic to the cartesian product of finitely many rings with identity, each having no proper nonzero ideal. Thus A/R has only finitely many ideals by 24.12. •

42.9 Theorem. *Let A be a topological ring with identity. The following statements are equivalent:*

1° *A is a locally centrally linearly compact ring whose characteristic is either zero or a prime, and the center of A is a local topological ring that has no proper open ideals.*

2° *A is a Hausdorff finite-dimensional algebra over a complete, discretely valued field, and the center of A is a Cohen algebra.*

3° *A is a Hausdorff, indecomposable, finite-dimensional algebra over a complete, discretely valued field.*

Proof. Assume 1°. By 41.4, A contains an open, centrally linearly compact subring B such that $1 \in B$, and by 41.8, the center of B is $B \cap C$, where C is the center of A. Thus $B \cap C$ is strictly linearly compact, so C is locally strictly linearly compact. By 42.6, C is a finite-dimensional Cohen algebra over a complete discretely valued field K. By 42.8, A is a finitely generated C-module and hence a finite-dimensional algebra over K. Thus 2° holds, and clearly 2° implies 3°.

Assume 3°. Then C is also indecomposable, and A is a locally centrally linearly compact ring by 41.7. As C is a finite-dimensional algebra with identity, C is an artinian ring and hence is semilocal by 27.17. Thus as C is indecomposable, C is local by 37.7. Therefore 1° holds. •

43 Locally Centrally Linearly Compact Rings

From our results in §42 we may easily characterize those topological rings with identity that are cartesian products of finitely many finite-dimensional algebras over complete, discretely valued fields. We shall say that a ring is *squarefree* if the additive order of each of its nonzero elements is either infinite or a squarefree integer, that is, one not divisible by the square of a prime.

Analogous to the remark preceding 34.1 is the following: If e is an idempotent in a topological ring A, the epimorphism f from the additive group A to the additive group Ae [eA], defined by $f(x) = xe$, [$f(x) = ex$], is a topological epimorphism, since if U is a neighborhood of zero in A, $U \cap Ae \subseteq Ue$ [$U \cap eA \subseteq eU$].

43.1 Theorem. *Let A be a topological ring with identity. The following statements are equivalent:*

1° *A is a locally centrally linearly compact ring whose center C is a topological ring that has no proper open ideals, and any one and hence all of the following equivalent conditions hold:*

(a) *C is semilocal.*

(b) *A has only finitely many maximal ideals.*
(c) *Every C-submodule of A is closed.*
(d) *Every ideal of A is closed.*
(e) *C is an artinian ring, and A is a finitely generated C-module.*
(f) *A satisfies the Ascending Chain Condition on closed ideals.*
(g) *A satisfies the Descending chain Condition on closed ideals.*

$2°$ *A is a locally centrally linearly compact ring whose center contains an invertible topological nilpotent.*

If A is squarefree, the following is equivalent to each of $1°$ and $2°$:

$3°$ *A is topologically isomorphic to the cartesian product of finitely many Hausdorff finite-dimensional algebras with identity over complete, discretely valued fields.*

Proof. If A is a locally centrally linearly compact ring whose center has no proper open ideals, then A contains an open, centrally linearly compact subring B such that $1 \in B$ by 41.4, so (a)–(g) of $1°$ are equivalent by 42.8, and also $1°$ implies $2°$ by that theorem. If b is an invertible topological nilpotent of C and if J is an open ideal of C, then $b^m \in J$ for some $m \geq 1$, so $J = C$ since b^m is invertible. Thus $2°$ implies $1°$ by 42.8.

Assume that $A = \prod_{i=1}^n A_i$, where each A_i is a finite-dimensional Hausdorff algebra with identity e_i over a complete, discretely valued field K_i, and let $|..|_i$ be an absolute value on K_i defining its topology. For each $i \in [1, n]$, let $\alpha_i \in K_i$ be such that $0 < |\alpha_i|_i < 1$. Then $(\alpha_1 e_1, \ldots, \alpha_m e_m)$ is an invertible topological nilpotent belonging to the center of A, and A is locally centrally linearly compact by 41.7 and 41.3. Thus $3°$ implies $2°$.

Finally, assume that A is squarefree and that $1°$ holds. As C is artinian, by 37.7 there exist orthogonal idempotents $e_1, \ldots, e_n$ in C such that $\sum_{i=1}^n e_i = 1$ and each Ce_i is a local artinian ring. By the preceding remark, $x \to xe_i$ is a topological epimorphism from A to Ae_i. Consequently, Ae_i is locally strictly linearly compact by 41.2. Moreover, as $x \to xe_i$ is a continuous epimorphism from C to Ce_i, Ce_i has no proper open ideals. Clearly A is the topological direct sum of $Ae_1, \ldots, Ae_n$, and Ce_i is the center of Ae_i for each $i \in [1, n]$. As Ce_i is local and A squarefree, the characteristic of Ae_i is either zero or a prime by 21.2. Consequently, $3°$ holds by 42.9. •

Every locally centrally linearly compact ring with identity whose center has no proper open ideals arises in a natural way from the rings described in the preceding theorem:

43.2 Theorem. *Let A be a topological ring with identity. The following statements are equivalent:*

1° A is a locally centrally linearly compact ring whose center C is a topological ring that has no proper open ideals.

2° A is topologically isomorphic to the local direct sum of topological rings $(A_\lambda)_{\lambda \in L}$ relative to open subrings $(B_\lambda)_{\lambda \in L}$, where each A_λ is a ring with identity, B_λ is a centrally linearly compact ring that contains the identity of A_λ, and the center C_λ of A_λ is a semilocal topological ring that has no proper open ideals.

Proof. Assume 1°. By 41.4 there is an open, centrally linearly compact subring B of A that contains 1, and by 41.8, the center of B is $B \cap C$. Consequently, $B \cap C$ is strictly linearly compact, and therefore by (2) of 34.6, there is an orthogonal family $(e_\lambda)_{\lambda \in L}$ of idempotents in $B \cap C$ such that $\sum_{\lambda \in L} e_\lambda = 1$ and $(B \cap C)e_\lambda$ is a linearly compact local ring.

As noted before 43.1, for each $\lambda \in L$, $p_\lambda : x \to xe_\lambda$ is a topological epimorphism from A to Ae_λ. Consequently, Be_λ is an open, centrally linearly compact subring of Ae_λ by 41.2. The center of Ae_λ is clearly Ce_λ, which therefore has no proper open ideals since C does not. Consequently, by 41.8, the center of Be_λ is $Be_\lambda \cap Ce_\lambda$, which is $(B \cap C)e_\lambda$ and therefore is a strictly linearly compact local ring. By 42.8, Ce_λ is a semilocal ring.

Let $\Phi : A \to \prod_{\lambda \in L}(Ae_\lambda)$ be defined by $\Phi(x) = (xe_\lambda)_{\lambda \in L}$ for each $x \in A$. By 34.3, the restriction Φ_B of Φ to B is a topological isomorphism from the B-module B to the B-module $\prod_{\lambda \in L}(Be_\lambda)$. Since each $e_\lambda \in C$, Φ_B is an isomorphism from the ring B to the ring $\prod_{\lambda \in L}(Be_\lambda)$. For each $\lambda \in L$, Be_λ itself is the largest subring of Ae_λ in which Be_λ is an ideal because Be_λ contains the identity e_λ of Ae_λ. Thus the local direct sum D of $(Ae_\lambda)_{\lambda \in L}$ relative to $(Be_\lambda)_{\lambda \in L}$ is the subring $\bigoplus_{\lambda \in L}(Ae_\lambda) + \prod_{\lambda \in L}(Be_\lambda)$ of $\prod_{\lambda \in L}(Ae_\lambda)$. If $\mu \in L$ and if $x \in Ae_\mu$, then $\Phi(x) = (x_\lambda)_{\lambda \in L}$ where $x_\mu = x$ and $x_\lambda = 0$ for all $\lambda \neq \mu$. Therefore $\bigoplus_{\lambda \in L}(Ae_\lambda) \subseteq \Phi(A)$, and hence as $\prod_{\lambda \in L}(Be_\lambda) \subseteq \Phi(A)$, $D \subseteq \Phi(A)$. Conversely, if $x \in A$, then $(xe_\lambda)_{\lambda \in L}$ is summable and $x = \sum_{\lambda \in L} xe_\lambda$ by 10.16, so $xe_\lambda \in Be_\lambda$ for all but finitely many $\lambda \in L$ by 10.4. Thus $D = \Phi(A)$. As Φ_B is a topological isomorphism from the open subring B of A to the open subring $\prod_{\lambda \in L}(Be_\lambda)$ of D, Φ is a topological isomorphism from A to D.

Assume 2°. Then $\prod_{\lambda \in L} B_\lambda$ is open in the local direct sum D of $(A_\lambda)_{\lambda \in L}$, so D is locally centrally linearly compact by 41.3. The center C of D is clearly $(\prod_{\lambda \in L} C_\lambda) \cap D$. Let J be an open ideal of C. For each $\lambda \in L$, the projection of J on C_λ is open in C_λ and hence is C_λ. Thus $\bigoplus_{\lambda \in L} C_\lambda \subseteq J$ by 24.12, so $J = C$ as J is closed and $\bigoplus_{\lambda \in L} C_\lambda$ is dense in C. Therefore 1° holds. •

In applying these theorems to locally compact rings, we wish to allow for a connected factor. For this, we need the following three lemmas:

43.3 Lemma. *If E is a locally compact, connected, unitary left [right] topological module over a squarefree, locally centrally linearly compact ring A with identity whose center is a topological ring that contains no proper open ideals, then $E = \{0\}$.*

Proof. We assume that E is a left A-module. By 43.2 and 43.1 there is an orthogonal family $(e_\lambda)_{\lambda \in L}$ of idempotents in the center of A whose sum is 1 such that each Ae_λ is the direct sum of finitely many Hausdorff finite-dimensional algebras with identity over complete, discretely valued fields. Suppose that there is a nonzero element $x \in E$. Then

$$x = 1.x = (\sum_{\lambda \in L} e_\lambda).x = \sum_{\lambda \in L}(e_\lambda.x)$$

by 10.16, so there exists $\alpha \in L$ such that $e_\alpha.E \neq \{0\}$. Thus $e_\alpha.E$ is a nonzero topological unitary module over Ae_α. In view of the nature of Ae_α, there are complete, discretely valued fields $K_1, \ldots, K_n$ such that, if e_j is the identity element of K_j for each $j \in [1, n]$, then

$$(\sum_{j=1}^n e_j).e_\alpha = 1.e_\alpha = e_\alpha,$$

so for some $i \in [1, n]$, $e_i.E = e_i.(e_\alpha E) \neq \{0\}$. Thus $e_i.E$ is a nonzero Hausdorff vector space over K_i. Now $e_i.E$ is connected since it is a continuous image of E. Moreover, $e_i.E$ is clearly closed in E and hence is locally compact. Thus $e_i.E$ is a connected, locally compact vector space over K_i. By 16.2 and 13.8, $e_i.E$ is finite-dimensional and hence, by 15.9, is topologically isomorphic to K_i^m for some $m \geq 1$. But K_i^m is totally disconnected since K_i is a discretely valued field. Thus $E = \{0\}$. •

43.4 Lemma. *Let A be a Hausdorff ring, C the connected component of zero. If A has a left [right] identity, if C is locally compact and not the zero ideal, and if A/C is a square-free, locally centrally linearly compact ring with identity whose center is a topological ring that has no proper open ideals, then C is a finite-dimensional Hausdorff algebra over $\mathbb{R}$, the right [left] annihilator of C in C is $\{0\}$, and C contains a nonzero idempotent.*

Proof. We shall assume that A has a left identity element e. Let ϕ be the canonical epimorphism from A to A/C. By 35.4, C contains a connected, compact ideal K such that $CK = (0)$ and C/K is a finite-dimensional topological $\mathbb{R}$-algebra. Since A/C is a ring with identity, its identity element is $\phi(e)$. Therefore as $CK = (0)$, K is a unitary topological module over A/C. By 43.3, $K = (0)$; therefore C is a finite-dimensional Hausdorff algebra over

R. Let $J = \{c \in C : Cc = \{0\}\}$. Then J is a closed ideal of C and also a closed subspace of the $\mathbb{R}$-algebra C and hence is connected. Therefore as $CJ = \{0\}$, J is a unitary topological (left) module over A/C, so $J = \{0\}$ by 43.3.

For each $c \in C$, let R_c be the linear operator on the $\mathbb{R}$-vector space C defined by $R_c(x) = xc$ for all $x \in C$. Since $J = \{0\}$, $R : c \to R_c$ is an antimonomorphism from C into the $\mathbb{R}$-algebra $\text{End}_{\mathbb{R}}(C)$ of all linear operators on C (that is, R is an additive monomorphism satisfying $R_{cd} = R_d \circ R_c$ for all $c, d \in C$). As C is a finite-dimensional algebra, its radical is nilpotent by 27.15, so C is not a radical ring as $J = (0)$, and consequently C contains a nonnilpotent element u. The sequence $(Cu^n)_{n \geq 1}$ of subspaces of C is decreasing, so for some $m \geq 1$, $Cu^s = Cu^m$ for all $s \geq m$. Let $v = u^m$. Then $Cv \neq (0)$ as v is not nilpotent, and $Cv^2 = Cv$. Consequently, the restriction S of R_v to Cv is an automorphism of the R-vector space Cv. Therefore the characteristic polynomial $X^n + \ldots + \alpha_1 X + \alpha_0$ of S has a nonzero constant term α_0. Let

$$p = -\alpha_0^{-1}(v^n + \ldots + \alpha_1 v).$$

Then $p \in C$, $R_p(C) \subseteq Cv$, and by the Cayley-Hamilton theorem, $R_p(x) = x$ for all $x \in Cv$. Hence R_p is a projection on Cv, so as R is an antimonomorphism, p is a nonzero idempotent in C. •

43.5 Lemma. *Let A and A' be topological rings with identity elements 1 and $1'$ respectively, and let ϕ be a continuous homomorphism from A to A' such that $\phi(1) = 1'$. If C is a subring of A containing 1 such that the topological ring C contains no proper open ideals and if C' is a subring of A' containing $\phi(C)$, then the topological ring C' has no proper open ideals.*

Proof. Let J' be an open ideal of C'. Then there is an open set U' in A' such that $U' \cap C' = J'$. Moreover, $\phi^{-1}(U') \cap C$ is an (open) ideal of C. Indeed, if $x \in C$ and $y \in \phi^{-1}(U') \cap C$, then $\phi(x) \in \phi(C) \subseteq C'$ and

$$\phi(y) \in \phi(\phi^{-1}(U') \cap C) \subseteq U' \cap C' = J',$$

so

$$\phi(xy) = \phi(x)\phi(y) \in C'J' = J',$$

and therefore

$$xy \in \phi^{-1}(U' \cap C') \cap C \subseteq \phi^{-1}(U') \cap C.$$

Similarly, if $x, y \in \phi^{-1}(U') \cap C$, then $x - y \in \phi^{-1}(U') \cap C$. Consequently, $\phi^{-1}(U') \cap C = C$ by hypothesis. Therefore $1 \in C \subseteq \phi^{-1}(U')$, so $1' \in U' \cap C' = J'$, and therefore $J' = C'$. •

43.6 Theorem. *Let A be a Hausdorff ring with identity such that the connected component C of zero is locally compact and either $A = C$ or A/C is squarefree and locally centrally linearly compact. If the center of A has no proper open ideals, then C is a finite-dimensional Hausdorff algebra with identity over $\mathbb{R}$, the center of A/C has no proper open ideals, and A is the topological direct sum of C and a ring topologically isomorphic to A/C.*

Proof. By 35.6, we may assume that $A \neq C$ and $C \neq \{0\}$. By 43.5, applied to the centers of A and of A/C, the center of A/C is a topological ring having no proper open ideals. Consequently, it suffices to prove that C has an identity element that is in the center of A.

For each $c \in C$, let $L_c(x) = cx$ for all $x \in C$. By 43.4, C is a finite-dimensional topological algebra over $\mathbb{R}$, and the left annihilator of C in C is (0). Therefore $L : c \to L_c$ is a monomorphism from C to the $\mathbb{R}$-algebra $\text{End}_{\mathbb{R}}(C)$ of all linear operators on C. Now 0 is an idempotent of C; let e be an idempotent of C such that $L_e(C)$ is maximal in the set of all the subspaces $L_p(C)$, ordered by inclusion, where p is an idempotent of C.

Suppose that $(1 - e)C \neq (0)$. Let ϕ be the canonical epimorphism from A to A/C and let ϕ' be the restriction of ϕ to $(1 - e)A$. As noted before 43.1, the function π from the additive topological group A to the additive topological group $(1 - e)A$, defined by $\pi(x) = (1 - e)x$ for all $x \in A$, is a topological epimorphism. Since $e \in C$, $\phi = \phi' \circ \pi$; therefore as both ϕ and π are topological epimorphisms, ϕ' is a topological epimorphism from $(1 - e)A$ to A/C by 5.3. The kernel of ϕ' is $(1 - e)A \cap C$, which is the ideal $(1 - e)C$ of C. Therefore $(1 - e)A/(1 - e)C$ is topologically isomorphic to A/C. Clearly $(1 - e)C$ is closed in C and hence is locally compact; $(1 - e)C$ is the continuous image of a connected set and hence is connected. As the connected component of zero in $(1 - e)A$ is clearly contained in the subset $(1 - e)A \cap C$, therefore, $(1 - e)C$ is the connected component of zero in $(1 - e)A$. By 43.4 applied to $(1 - e)A$, which has the left identity $1 - e$, $(1 - e)C$ has a nonzero idempotent f. As $f = (1 - e)f$, $ef = 0$. Let $M = L_e(C)$, $N = L_f(C)$, and let $g = e + f$. Then $L_g(C) \subseteq M + N$. As $ef = 0$, $L_g(m - L_f(m)) = m$ for each $m \in M$, and $L_g(n) = n$ for each $n \in N$, so $L_g(M + N) \supseteq M + N$. The restriction of L_g to $M + N$ is therefore an automorphism of the $\mathbb{R}$-vector space $M + N$. Applying the Cayley-Hamilton theorem as in the proof of 43.4, we conclude that there is an idempotent $h \in C$ such that L_h is a projection on $M + N$. Since $M \cap N = (0)$ (as $ef = 0$) and since $N \neq (0)$, we thus obtain a contradiction.

Therefore $(1 - e)C = (0)$, so L_e is the identity linear operator on C, and consequently e is the identity element of C. Moreover, e is a central idempotent, for if $x \in A$, then ex and xe belong to C, so $ex = (ex)e = e(xe) = xe$. Thus A is the topological direct sum of eA, which is C, and

$(1-e)A$, which is topologically isomorphic to A/C. •

43.7 Theorem. *Let A be a squarefree locally compact ring with identity, and let C be its center. The following statements are equivalent:*

$1°$ *C is a topological ring having no proper open ideals, and any one and hence all of the following equivalent conditions hold:*

(a) *C is semilocal.*
(b) *A has only finitely many maximal ideals.*
(c) *Every C-module of A is closed.*
(d) *Every ideal of A is closed.*
(e) *C is an artinian ring, and A is a finitely generated C-module.*
(f) *A satisfies the Ascending Chain Condition on closed ideals.*
(g) *A satisfies the Descending Chain Condition on closed ideals.*

$2°$ *C contains an invertible topological nilpotent.*

$3°$ *A is topologically isomorphic to the cartesian product of finitely many Hausdorff finite-dimensional algebras with identity over nondiscrete locally compact fields.*

Proof. Assume that C has no proper open ideals, and let A_0 be the connected component of zero. Then A/A_0 is a totally disconnected locally compact ring by 5.16 and hence is locally centrally linearly compact by 4.21 and 4.20. By 43.6, A is the topological direct sum of A_0 and a subring A_1 topologically isomorphic to A/A_0, and A_0 is a finite-dimensional Hausdorff algebra with identity over $\mathbb{R}$. Let C_0 and C_1 be the centers of A_0 and A_1 respectively. Then C is the direct sum of C_0 and C_1 and statements (a)–(g) all hold if C and A are replaced respectively by C_0 and A_0, because they are finite-dimensional $\mathbb{R}$-algebras with identity. (The concluding part of the proof of 42.8 shows that every finite-dimensional algebra with identity over a field has only finitely many maximal ideals.) It readily follows that each of (a)–(g) is valid for C_1 and A_1 if and only if it is valid for C and A. Thus the equivalence of (a)–(g) for C and A follows from the equivalence of (a)–(g) for C_1 and A_1, established in 42.8.

As shown in the proof of 43.1, any finite-dimensional algebra with identity over a field topologized by a proper absolute value contains an invertible topological nilpotent. Consequently, $2°$ holds for C if and only if it holds for C_1. The equivalence of $1°$–$3°$ therefore follows from 43.1 and 16.2. •

Exercises

In these exercises, it is understood that a finite-dimensional vector space or algebra over a field, furnished with a complete absolute value, is topologized with the unique Hausdorff topology that makes it a topological vector space or algebra over that field.

EXERCISES 421

43.1 The following statements are equivalent for a topological ring A with identity:

1° A is a locally centrally linearly compact semisimple ring all of whose ideals are closed and whose center is a topological ring having no proper open ideals.

2° A is a locally centrally linearly compact semisimple ring whose center is a semilocal topological ring having no proper open ideals.

3° A is a locally centrally linearly compact semisimple ring whose center contains an invertible topological nilpotent.

4° A is the topological direct sum of finitely many rings, each the ring of all linear operators on a finite-dimensional vector space over a complete, discretely valued field.

43.2 The following statements are equivalent for a topological ring A with identity:

1° A is a locally centrally linearly compact semisimple ring whose center is a topological ring having no proper open ideals.

2° A is topologically isomorphic to the local direct sum of topological rings $(A_\lambda)_{\lambda \in L}$ relative to open subrings $(B_\lambda)_{\lambda \in L}$, where for each $\lambda \in L$, A_λ is the toopological direct sum of finitely many rings, each the ring of all linear operators on a finite-dimensional vector space over a complete, discretely valued field, and B_λ is centrally linearly compact.

43.3 The following statements are equivalent for a topological ring A with identity:

1° A is a locally compact semisimple ring all of whose ideals are closed and whose center is a topological ring that has no proper open ideals.

2° A is a locally compact semisimple ring whose center is a semilocal topological ring that has no proper open ideals.

3° A is a locally compact semisimple ring whose center contains an invertible topological nilpotent.

4° A is the topological direct sum of finitely many rings, each the ring of all linear operators on a finite-dimensional vector space over a nondiscrete locally compact field.

43.4 The following statements are equivalent for a topological ring A with identity:

1° A is a locally compact semisimple ring whose center is a topological ring that has no proper open ideals.

2° A is topologically isomorphic to $A_0 \times A_1$, where A_0 is the topological direct sum of finitely many rings, each the ring of all linear operators on a finite-dimensional vector space over $\mathbb{R}$ or $\mathbb{C}$, and A_1 is the local direct sum of topological rings $(A_\lambda)_{\lambda \in L}$ relative to open, compact subrings $(B_\lambda)_{\lambda \in L}$, where for each $\lambda \in L$, A_λ is the topological direct sum of finitely many rings, each the ring of all linear operators on a finite-dimensional vector space over a discretely valued, locally compact field.

43.5 The following statements are equivalent for a commutative topological ring A with identity:

1° A is a locally strictly linearly compact semisimple ring that has no proper open ideals, and every ideal of A is closed.

2° A is a locally strictly linearly compact, semilocal, semisimple ring that has no proper open ideals.

3° A is a locally strictly linearly compact semisimple ring that contains an invertible topological nilpotent.

4° A is topological direct sum of finitely many rings, each a complete, discretely valued field.

43.6 The following statements are equivalent for a commutative topological ring A with identity:

1° A is a locally strictly linearly compact semisimple ring that has no proper open ideals.

2° A is topologically isomorphic to the local direct sum of topological ring $(A_\lambda)_{\lambda \in L}$ relative to open subrings $(B_\lambda)_{\lambda \in L}$, where for each $\lambda \in L$, A_λ is the topological direct sum of finitely many rings, each a complete, discretely valued field, and B_λ is a strictly linearly compact subring.

43.7 The following statements are equivalent for a commutative topological ring A with identity:

1° A is a locally compact semisimple ring that has no proper open ideals, and every ideal of A is closed.

2° A is a locally compact, semilocal, semisimple ring that has no proper open ideals.

3° A is a locally compact semisimple ring that has an invertible topological nilpotent.

4° A is the topological direct sum of finitely many rings, each a nondiscrete locally compact field.

43.8 (Goldman and Sah [1965]) The following statements are equivalent for a commutative topological ring A with identity:

1° A is a locally compact semisimple ring that has no proper open ideals.

2° A is topologically isomorphic to $A_0 \times A_1$, where A_0 is topologically isomorphic to the cartesian product of finitely many topological fields, each either $\mathbb{R}$ or $\mathbb{C}$, and A_1 is the local direct sum of topological rings $(A_\lambda)_{\lambda \in L}$ relative to compact open subrings $(B_\lambda)_{\lambda \in L}$, where for each $\lambda \in L$, A_λ is the topological direct sum of finitely many subrings, each a discretely valued, locally compact field.

43.9 Let V be the valuation ring of a locally compact field F whose topology is defined by a discrete valuation, let M be the maximal ideal of V, and let $B = \{(x,y) \in V \times V : x - y \in M\}$. Let L be an infinite set, and for each $\lambda \in L$, let $A_\lambda = F \times F$, $B_\lambda = B$. (a) Each B_λ is a compact, open, subring containing the identity element of A_λ. The local direct sum A of $(A_\lambda)_{\lambda \in L}$ relative to $(B_\lambda)_{\lambda \in L}$ is thus a locally compact ring. (b) A is not isomorphic to the local direct sum A' of a family of rings $(A'_\mu)_{\mu \in M}$ with identity relative to subrings $(B'_\mu)_{\mu \in M}$, where each B'_μ contains the identity of A'_μ and is the direct sum of finitely many local rings. [Observe that there is a neighborhood of zero in A that contains no idempotent serving as the identity of a local subring.] (c) Which of the properties listed in 1° and 2° of Theorem 43.2 are unsatisfied by A and the A_λ's?

CHAPTER XI

HISTORICAL NOTES

We conclude with some historical remarks concerning the material presented in the preceding chapters, with the exception of those topics covered in *Topological Fields*. The use of topologies to facilitate the discussion of certain topics in commutative algebra is the subject of §44. A brief history of the development of the theory of locally and linearly compact rings is given in §45, and we conclude with a discussion of some topics not covered in the book: duality theory, embedding theory, and theorems concerning the existence of topologies on rings. This chapter is thus a continuation of the historical remarks constituting the final chapter of *Topological Fields*.

44 Topologies on Commutative Rings

Traditionally, a complete, discretely valued field of characteristic zero, the maximal ideal of whose valuation ring is generated by the prime number p, has been called a p-adic field. In our terminology, the valuation ring of a p-adic field is a Cohen ring of characteristic zero whose residue field has characteristic p, and consequently a p-adic field is simply the quotient field of such a Cohen ring.

The structure theory of complete, discretely valued fields was first presented by Hasse and Schmidt [1932]: (a) a complete, discretely valued field whose valuation ring is equicharacteristic is isomorphic to the field of power series in one variable over its residue field (22.1); (b) for any field k of prime characteristic p, there is a p-adic field whose residue field is isomorphic to k (22.8), and any isomorphism from the residue field k_1 of a p-adic field K_1 to the residue field k_2 of a p-adic field K_2 is induced by an isomorphism from K_1 to K_2 (or equivalently, by an isomorphism from the valuation ring of K_1 to that of K_2) (22.11); (c) a complete, discretely valued field of characteristic zero whose residue field has prime characteristic p is an Eisenstein extension of a p-adic field (22.7).

Hasse and Schmidt's proof of (b) depended, however, on an unproved statement, subsequently shown to be incorrect by MacLane [1938a], affirming that an extension k of a field k_0 of prime characteristic is the union of an increasing sequence of subfields each of which has a separating transcendence basis over k_0. This statement mattered in Hasse and Schmidt's proof of (b) only when k was imperfect, however, and thus their proof was valid whenever k was perfect. Later, Schmidt and MacLane [1941] refined the erroneous statement into two theorems, one of which affirmed that if k preserves p-independence over k_0, that is, if each p-independent subset of k_0 is also p-independent in k (the notion of p-independence arises naturally from that of a p-basis, introduced by Teichmüller [1936a]), then k is the union of a transfinite sequence of subfields, each countably generated and preserving p-independence over its predecessor, if it has one. Suitably modified by these theorems, Hasse and Schmidt's original proof of (b) is then valid in general.

Meanwhile, however, arguments of Teichmüller [1936b,c] amplified, generalized, and simplified at certain points by MacLane [1938b], established the structure theorems in complete generality. Both Teichmüller and Witt [1936] observed that the ring of Witt vectors with coefficients in a perfect field k of prime characteristic is a Cohen ring whose residue field is isomorphic to k, and using Teichmüller's theorem that established the existence of multiplicative representatives, both completed the proof of (b) for perfect k. Teichmüller then essentially reduced the problem of proving (b) in general to this already established special case by showing that if L is the p-adic field whose residue field is the smallest perfect extension of k, L contains a unique p-adic subfield K with residue field k that contains the multiplicative representatives of a given p-basis of k. Eliciting a generalization of this theorem from Teichmüller's proof, MacLane used it to prove a theorem yielding explicitly the uniqueness part of (b) in complete generality.

MacLane [1938b] also observed that a simple extension theorem established the existence of a p-adic field with prescribed residue field (22.8). Ostrowski [1932] had shown that a valuation v of a field K with residue field k could be extended to a valuation of a finite-dimensional extension field that has the same value group as v and a residue field k-isomorphic to a given finite-dimensional extension k' of k (his proof, though stated only for real valuations, is valid in general). MacLane [1937] rediscovered this theorem and supplemented it with the analogue for a simple transcendental extension k' of k to conclude that if k' is any extension of k, there is an extension of v to an extension field of K that has the same value group as v and a residue field k-isomorphic to k' (*Topological Fields*, Exercise 32.23).

The proof given here of the uniqueness part of (b) (22.9-22.11), which uses the existence of Cohen subrings in a (not necessarily noetherian) Hausdorff, complete local ring (21.20), was derived by Wehrfritz [1979] from a proof given by Rees.

Krull [1938] initiated the study of local noetherian rings and, in particular, developed the dimension theory of such rings. Krull [1928] had already given a description of the intersection of the powers of an ideal in a noetherian ring. This description implied that the intersection of the powers of the maximal ideal of a local noetherian ring A was the zero ideal (20.16) and thus enabled Krull to introduce a Hausdorff ring topology on A, called here the natural topology of A, for which the powers of the maximal ideal form a fundamental system of neighborhoods of zero. Krull [1938] showed that every ideal of A is closed (20.16), that a finite set generating an ideal of A also generated its closure in $\widehat{A}$ (24.6), and that $\widehat{A}$ is again a local noetherian ring furnished with its natural topology (24.7).

Krull [1938] also inferred from Hensel's Lemma (see *Topological Fields*, 32.11) that if the residue field k of a complete regular local ring A had characteristic zero, then A contained a subfield mapped isomorphically onto k by the canonical mapping, and consequently that A was isomorphic to the ring of formal power series in finitely many variables over k. This led him to conjecture that the Hasse-Schmidt theorems for complete, discrete valuation rings were simply the one-dimensional special cases of theorems describing complete, regular local rings. Specifically, he conjectured that if the residue field k of a complete, regular local ring A of dimension n had prime characteristic p, then A was isomorphic to the power series ring in n variables over k if A had characteristic p, whereas if A had characteristic zero and p did not belong to the square of its maximal ideal, then A was isomorphic to the power series ring in $n - 1$ variables over a Cohen subring. Krull also conjectured that every complete, local noetherian ring is an epimorphic image of a complete regular local ring (23.6).

I. S. Cohen [1945] verified these conjectures in a fundamental paper by applying his theorem (21.20) that a complete, local noetherian ring contains what has historically been called a coefficient subfield or subring but, as in in Samuel [1953] or Godement [1956], is here called a Cohen subfield or subring. Cohen's proof required consideration of local rings Hausdorff for their natural topologies whose maximal ideals are finitely generated, and in passing he proved that such a ring, if complete, is noetherian (23.6). Nagata [1949] simplified Cohen's proof in certain respects, but his proof of the existence of Cohen subrings in a local ring that is Hausdorff and complete for its natural topology (21.20) was seriously flawed. Correct proofs of this were given independently by Narita [1955a], who used theorems concerning p-bases, and Geddes [1954, 1955], whose conceptually simpler proof is presented here (21.11–21.13, 21.17–21.10).

Chevalley [1943] introduced semilocal noetherian rings and proved that the natural topology of such a ring is Hausdorff ((4) of 24.16). He showed that a complete, semilocal noetherian ring A is the direct sum of finitely many complete, local noetherian rings (24.17), and that if E is an A-module such that E/RE is finitely generated, where R is the radical of A, then E is finitely generated (cf. 42.7). He also established that the natural topology of a complete, semilocal noetherian ring A is the weakest metrizable ideal topology on A (36.35 and 36.33). In a circuitous way, Chevalley showed that a semilocal noetherian ring A is a dense subring of a complete, semilocal noetherian ring $\hat{A}$ whose natural topology induces on A its natural topology (24.17). En route, he proved that the ring $B[[X]]$ of power series over a noetherian ring B is noetherian (23.2); the proof given here, which was presented by Kaplansky [1970], uses I. S. Cohen's theorem [1949] that a commutative ring with identity is noetherian if each of its prime ideals is finitely generated (20.9). Chevalley [1943] also showed that the completion of a noetherian integral domain for the topology defined by a maximal ideal M is a local ring ((2) of 24.7), and that if B is a commutative ring finitely generated over a semilocal noetherian subdomain A, then B is a semilocal noetherian ring whose natural topology induces on A its natural topology (39.13). The discussion here of the completion of a semilocal noetherian ring (24.7) is similar to that given by Yoshida and Sakuma [1953].

Zariski [1945] broadened in a natural way Krull's investigation of the topology determined by the maximal ideal of a local noetherian ring by introducing the J-topology on a commutative noetherian ring with identity A, where J is any ideal of A (24.1). His proof that the completion $\hat{A}$ of A for a Hausdorff J-topology is noetherian is given here ((1) of 24.7), and he also showed that the topology of $\hat{A}$ is its $\hat{J}$-topology, where $\hat{J}$ is the closure of J in $\hat{A}$ (24.5). Using the primary decomposition of an ideal in a noetherian ring, Zariski verified that every ideal in $\hat{A}$ is closed (Exercise 24.2) and concluded that the closure of an ideal F of A in $\hat{A}$ is $\hat{A}F$ (24.6).

Zariski [1945] also showed that every ideal of A is closed for the J-topology if and only if J is contained in the radical of A (24.14); such J-topologies have been called Zariski topologies (Samuel [1953], Zariski and Samuel [1960]). The term has been broadened here to mean any Hausdorff ideal topology for which all ideals are closed, since that is an appropriate context for much of Chevalley's work [1943] on semilocal noetherian rings and Zariski's principal theorem for Zariski J-topologies (Exercise 24.2). With only minor modifications, Zariski's proof of that theorem establishes the very general result that the continuous extension of the identity map of a group G to a homomorphism from the completion of G for a Hausdorff group topology to the completion of G for a weaker Hausdorff group topology is a monomorphism, provided that there is a fundamental system of neighborhoods of the identity for the stronger topology that are closed for the weaker (7.20), and consequently that if G is complete for the weaker, it is complete for the stronger (7.21).

Using I. S. Cohen's theorem that a complete local ring whose maximal ideal is finitely generated is noetherian, Nagata [1950a] showed that a semilocal ring is noetherian if and only if its maximal ideal is finitely generated and its natural topology is a Zariski topology (24.18).

The Artin-Rees Lemma, so-named by H. Cartan in lectures given in January 1954, asserts that if M is an ideal of a noetherian ring A and if F is a submodule of a finitely generated A-module E, then there exists k such that $M^n E \cap F = M^{n-k}(M^k E \cap F)$ for all $n \geq k$. The Lemma, which became generally known in 1953, was obtained independently by Artin, who did not publish a proof, and Rees, who did [1955b]. The proof of the "non-uniform" case given here (20.11), presented by Kaplansky [1970], is due to Herstein.

The dimension theory of local noetherian rings is the creation of Krull [1938]; the presentation here (§§37-38) largely follows that of Kaplansky [1970]. The structure theory of complete local noetherian rings (39.16, 39.18, 39.20, 39.29) is due to I. S. Cohen [1945]. Nagata [1953b] and Mori [1955] independently proved that a complete local noetherian domain is Japanese (40.9). The proof given here is a special case of a proof of Tate [1962] contained in personal notes entitled "Rigid analytic spaces"; the proof was published by Grothendieck [1964], and the notes were published, first in Russian translation in 1969, then in English in 1971.

45 Locally and Linearly Compact Rings

Just as Hensel initiated the study of topological fields by constructing the p-adic number fields without, however, the use of topological concepts, so also Prüfer [1924], also without their use, initiated the study of topological rings a quarter of a century later by identifying, in modern terminology, the completion $\widehat{K}$ of an algebraic number field K for the supremum of all the valuation topologies defined by the prime ideals of its ring of integers D with $\prod_{P \in \mathcal{P}} \widehat{K}_P$, where $\mathcal{P}$ is the set of all nonzero prime ideals of D and $\widehat{K}_P$ is the completion of K for the valuation defined by P (see Topological Fields, 28.16). Von Neumann [1925] recast Prüfer's theory in topological language (but not in the language of valuation theory). At the time of Prüfer's and von Neumann's investigation, the theory of real valuations was barely under way; their theorem, however, directly implied a significant special case of Ostrowski's approximation theorem for real valuations [1932], which did not appear until 1935.

Van Dantzig [1931a,b] first formally defined a topological ring and proved that, under certain conditions later shown always to hold, a Hausdorff topological ring admitted a completion. Subsequently, he investigated the completion of a ring for a Hausdorff ideal topology [1934a] and demonstrated certain decomposition theorems (subsumed in Exercise 28.8 of *Topological Fields*).

Mining the structure theory of locally compact abelian groups for information about the connected component of zero in a locally compact ring marked the next significant advance. Jacobson and Taussky [1935] applied the Pontriagin-van Kampen theorem to show that a separable, locally compact, connected ring lacking nonzero bilateral annihilators is a finite-dimensional topological algebra over the real numbers (35.5). In the same manner, Anzai [1943] obtained Jacobson and Taussky's theorem and explicitly demonstrated that a compact ring lacking nonzero left or right annihilators is totally disconnected. Otobe [1944a,b] refined both results and, in particular, showed that the Jacobson-Taussky hypothesis of separability was unneeded. Two theorems presented here subsume these results: the first (32.2), due to Kaplansky [1947c] and based on the existence of sufficiently many characters on a locally compact abelian group (32.1), asserts that the connected component of zero in a locally compact ring annihilates on the right [left] any left [right] bounded additive subgroup; the second (35.2), due to Braconnier [1946] and based on the Pontriagin-van Kampen structure theorem for locally compact abelian groups (35.1), elicits the relation between the connected component of zero in a locally compact ring and the ideal that is the union of all compact additive subgroups.

Major progress in the theory of compact and locally compact rings was made by Kaplansky in 1946-52. He determined [1946], for example, the structure of compact semisimple rings (32.6). Using it and a lifting theorem for idempotents (34.22), he showed [1946] that a compact commutative ring A is the cartesian product of compact local rings and a compact radical ring (34.23). Kaplansky also showed [1946] that if A is a compact local ring such that R^2 is open, then A is a local noetherian ring whose topology is its natural topology (a noncommutative generalization is given in 36.39, in view of 36.35).

Throughout, Kaplansky's work exhibits the utility of advertibly open rings in the general theory, as, for example, in his proof [1949b] that the radical of a locally compact ring is not only closed, but the intersection of the closed regular maximal right ideals (35.6-35.10). Kaplansky also described, first [1947c], bounded, locally compact, semisimple rings (32.10), and later [1952a], right bounded, locally compact, semisimple rings (cf. 32.9).

The ring of all linear operators on a finite-dimensional vector space over a nondiscrete, locally compact field K, furnished with its unique Hausdorff topology as a finite-dimensional algebra over the center of K, is a locally compact simple ring, and a natural problem is to determine conditions insuring that a nondiscrete, locally compact primitive or simple ring A has this description. Using a theorem of Jacobson [1935] (32.11), Kaplansky [1952a] proved that a torsionfree, locally compact, primitive ring had this description (32.13), but exhibited [1947c] a locally compact primitive ring of prime characteristic that did not (Exercise 32.4). Kaplansky [1952a] showed that a simple locally compact ring A having a minimal left ideal admitted this description (32.16), and Skorniakov [1964] showed that if A had no proper open left ideals, then A had a minimal left ideal and hence was such a ring (35.16), but that not all nondiscrete, locally compact simple rings admitted this description (pp. 275-280). Kaplansky's proof was based on a theorem, ascribed to Litoff, asserting that if A is a simple ring with a minimal left ideal, there is a division ring D such that each finite subset of A is contained in a subring of A isomorphic to the ring of all linear operators on a finite-dimensional vector space over D. The first available proof of this theorem (Jacobson and Rickart [1950], Jacobson [1956]) depended on the duality theory of simple rings with minimal left ideals; Faith and Utumi [1962] have given a more elementary proof; Phạm Ngọc Ánh [1982] has generalized the theorem. The proof of Kaplansky's theorem given here (32.14-16), due to the author [1985], is considerably longer than Kaplansky's, but is more elementary as it is based on standard topological considerations rather than Litoff's theorem. Kaplansky's structure theorem for locally compact strongly regular rings [1949b] (Exercise 35.18) was generalized by Skorniakov's description [1977] of all locally compact biregular rings (Exercises 35.7-12).

Baire's theorem, asserting that locally compact and complete metric spaces are Baire spaces (9.4), was applied by Mazur and Orlicz [1948] to establish that scalar multiplication of a complete, metrizable vector space over $\mathbb{R}$ was jointly continuous if it was separately continuous in each variable. Their proof establishes, more generally, that a separately continuous $\mathbb{Z}$-bilinear function from the product of two Hausdorff commutative groups, one metrizable, the other a Baire space, to a third topological commutative group is jointly continuous (9.5). This subsumes an earlier theorem of Arens [1946] that if multiplication is separately continuous in each variable for a complete, metrizable additive group topology on a ring A, then A is a topological ring (9.6).

Otobe [1944c] first proved the continuity of inversion for a locally compact ring topology on a division ring (that is, that a locally compact ring topology on a division ring is a division ring topology). Kaplansky [1946, 1947c] generalized Otobe's theorem in several ways, but Ellis's decisive result [1956a,b] that a locally compact topology on a group for which the group composition is separately continuous in each variable is a group topology (6.13) immediately implies the continuity of adversion in a locally compact, advertibly open ring (11.11). Similarly, the theorem that a complete, metrizable topology on a group for which translations are continuous is a group topology (6.13), together with the theorem that the topology of an open subset of a complete metric space is defined by a complete metric, implies the continuity of adversion in a complete, metrizable, advertibly open ring (11.9).

Linear compactness had its origins in the work of Prüfer, who, in purely algebraic language, introduced [1923a] what in modern terminology is the Leptin topology associated to the discrete topology on an abelian group (*i.e.*, a $\mathbb{Z}$-module) G, defined algebraically [1923b] the completion $\widehat{G}$ of G for this topology, showed that $\widehat{G}$ is linearly compact (Exercises 33.1-2), and investigated its structure. Pietrkowski [1930] reformulated Prüfer's results in modern terminology. Krull [1940a,b] discussed linearly compact modules over the ring of p-adic integers.

Lefschetz [1942] introduced linear compactness in the context of vector spaces. He showed that a linearly compact vector space is the cartesian product of discrete, one-dimensional spaces (Exercise 28.4), and established a duality theorem between linearly compact vector spaces and discrete vector spaces that extended the ordinary duality for finite-dimensional vector spaces (an alternative presentation was provided by Dieudonné [1949a]).

Results of further investigations into linear compactness were not published, however, until the 1950s. Lefschetz [1942], Dieudonné [1949a], and Zelinsky [1952] derived the basic elementary properties of linearly compact vector spaces and modules (28.3-7); in particular, Zelinsky extended Lefschetz's observation that a discrete linearly compact vector space is finite-dimensional by showing that a discrete linearly compact module cannot contain a direct sum of infinitely many nonzero submodules (cf. 28.21).

Zelinsky and Leptin are primarily responsible for the progress made in the theory of linearly compact rings and modules during the 1950s. Extending theorems of Kaplansky [1946] (32.6 and 34.23), Zelinsky [1952] showed that an ideally topologized (or bounded), linearly compact, semisimple ring is the cartesian product of matric rings over division rings (29.9), and [1949] that a strictly linearly compact commutative ring is the direct sum of a strictly linearly compact radical ring and the cartesian product of strictly linearly compact local rings (34.6) (Dikranjan and Orsatti [1984a] gave the algebraic argument needed to extend the result to arbitrary linearly compact rings). Zelinsky [1952] also showed that a complete local noetherian ring is linearly compact for the discrete topology (cf. 36.35) and that a valuation is maximal if and only if its valuation ring is linearly compact (*Topological*

Fields, 31.21).

Leptin's earliest work on linear compactness [1954a] concerned linearly compact abelian groups ($\mathbb{Z}$-modules). He determined the structure of discret, linearly compact abelian groups (30.10) and characterized the largest divisible subgroup of a linearly compact topological p-primary group (cf. 36.24). His fundamental contributions to the general theory of linearly compact modules and rings begin with his demonstration [1954e] that of all the Hausdorff linear topologies on a module weaker than a linear compact topology $\mathcal{T}$, there is a weakest $\mathcal{T}_*$, called here the Leptin topology associated to $\mathcal{T}$ (33.8), characterized it (38.5-8), and established the preservation, under the replacement of $\mathcal{T}$ by $\mathcal{T}_*$, of important properties (35.9-11). He also demonstrated the useful fact that, under a continuous homomorphism from one linearly compact module to another, the image of the adherence of a filter base of cosets of submodules is the adherence of the image of the filter base (28.20). Leptin also introduced strict linear compactness, gave criteria for a linear topology to be strictly linear compact (28.15, 33.18), established that a strictly linearly compact topology is a minimal Hausdorff linear topology (28.13), and proved basic properties of permanence (28.11, 28.16-17).

Leptin also [1954e] established general properties of linearly compact rings (29.3-5, 29.15), proved that the radical of a linearly compact ring is closed (29.12), and generalized Zelinsky's theorem for ideally topologized, linearly compact semisimple rings by showing that a linearly compact semisimple ring is the cartesian product of rings, each the ring of all linear operators on a discrete vector space, furnished with the topology of pointwise convergence (29.7). The Wedderburn-Artin theorem for semisimple artinian rings (27.14) is simply the discrete special case of this theorem.

Leptin [1956] also showed that a linearly compact module over a strictly linearly compact ring is strictly linearly compact (33.19), a fact he needed in his demonstration that the Leptin topology associated to a linearly compact ring topology is strictly linearly compact if and only if the radical of the ring is transfinitely nilpotent (33.24). For ideally topologized (or bounded) linearly compact rings, Leptin showed that the radical R is transfinitely nilpotent if and only if $\bigcap_{n=1}^{\infty} R^n = \{0\}$ (33.25).

If A is a ring with radical R and if A/R is topologically isomorphic to the cartesian product of a family $(A'_\lambda)_{\lambda \in L}$ of (semisimple) topological rings, any attempt to identify A with the cartesian product of a family $(A_\lambda)_{\lambda \in L}$ of rings such that for all $\lambda \in L$, A_λ/R_λ is topologically isomorphic to A'_λ where R_λ is the radical of A_λ, leads to the problem of lifting an orthogonal family of idempotents from A/R to A. For linearly compact rings, lifting a single idempotent is always possible (34.1), as shown by Zelinsky [1952] in the commutative case and Leptin [1956] in general (the proof given here, due to Widiger [1987], incorporates ideas from both proofs). In the commutative case, lifting an idempotent can be done in only one way (34.2). Leptin also showed that any family $(e_\lambda)_{\lambda \in L}$ of idempotents of A determining an orthogonal family of idempotents in A/R with sum 1 effects a decomposition of the A-module A into the cartesian product of the left ideals $(Ae_\lambda)_{\lambda \in L}$ (34.3). Leptin erred in stating that if A is strictly linearly compact, any family of idempotents in A lifting an orthogonal family of idempotents in A/R is summable (see 34.7), but Widiger [1972, 1987] did establish that a summable, orthogonal family of idempotents in A/R may be lifted to a summable, orthogonal family of idempotents in A (34.4-5). One consequence of these lifting theorems is that a strictly linearly compact ring with identity which modulo its radical is the ring of all n by n matrices over a division ring K is the ring of all n by n matrices over a strictly linear compact ring that modulo its radical is isomorphic to K (34.9). This was established by Kaplansky [1946] for compact rings (34.24) and by Leptin [1956] in general.

A significant extension of Leptin's structure theorem for linearly compact semisimple rings is Ryan's structure theorem [1980] for semisimple linearly topologized rings possessing an open left ideal that is linearly compact for its induced topology (31.8). Crucial to her proof is Wiegandt's theorem [1966a] that such a ring possesses a minimal left ideal (31.2); the proof given here is Ryan's amplification of Wiegandt's original argument. Among the consequences of Ryan's theorem is Kaplansky's earlier [1952a] description of right bounded, locally compact semisimple rings (cf. 32.9).

Extending Kaplansky's earlier theorem [1946] on compact local rings, Numakura [1955b, 1981], Jans [1957] and the author [1971b] have all obtained special cases of the theorem that the topology of a strictly linearly compact ring with identity and radical R is the radical topology if that topology is Hausdorff and the closure of R^2 is open (36.27).

A natural inquiry is to determine the nature of linearly compact rings whose closed ideals are also linearly compact rings. Widiger [1972] showed that if A is a bounded, linearly compact ring whose radical R is a strictly linearly compact ring, then necessaily every closed ideal of A is a strictly linearly compact ring, and he gave a structure theorem for such rings (34.18). In particular, in such a ring, the radical is strongly linearly compact, that is, any filter base of closed additive subgroups has a nonzero intersection (34.13). This led Phạm Ngọc Ánh [1976c, 1977b] to investigate linearly compact rings whose radical is strongly linearly compact; he obtained [1977b] a structure theorem (modified in 34.17 by the addition of a hypothesis) extending Widiger's original result. A more restrictive condition is that every closed subring of the ring be linearly compact. Ursul, in a series of papers (the earliest in collaboration with Andrunakevich and Arnautov) has investigated these rings.

Strictly linearly compact rings form a natural domain for extending theorems about artinian rings, simply because strictly linearly compact rings are projective limits of discrete artinian rings, and the discrete strictly linearly compact rings are precisely the artinian rings. Widiger's theorem mentioned above is an example of a theorem that was first obtained in the discrete case for artinian rings. The author [1976] undertook an investigation of a more restrictive class of rings, those linearly compact for the radical topology (such rings are necessarily strictly linearly compact (36.17)). In such rings the topological torsion ideal has a unique topological supplement (36.23) S, which is divisible and topologically torsionfree and has a left identity, a theorem yielding in the discrete case the classical decomposition of an artinian ring into the direct sum of its torsion ideal and a unique ideal S, which is divisible and torsionfree and has a left identity (36.4). The consequences of the lack of divisible primary torsion groups in an artinian ring (Exercise 36.6) are mirrored in rings linearly compact for the radical topology that lack both divisible primary torsion groups and copies of the topological additive groups of the p-adic number fields (such groups are called pathological here) (36.26).

The radical topology arises naturally in investigations into the nature of rings linearly compact for the discrete topology, for if the radical topology of such a ring A is Hausdorff, then it is the Leptin topology associated to the discrete topology and hence is the weakest Hausdorff linear topology on A (36.33). Zelinsky [1952] identified complete, semilocal noetherian rings (36.29) and valuation rings of maximal valuations (*Topological Rings*, 31.21 and 31.12) as rings linearly compact for the discrete topology, and the theorem just mentioned extends Chevalley's theorem [1943] that the natural topology of a complete semilocal ring is its weakest metrizable ring topology. A description of rings with identity whose radical topology is Hausdorff for which the discrete topology is linearly compact was undertaken by the author [1971b], and the results have been extended here by replacing the hypothesis that an identity exists with the weaker hypothesis that the additive group lacks pathological subgroups (36.33). Such rings are always (left) noetherian, and one consequence (Exercise 36.6) of their description is Hopkins' classical theorem [1939] that an artinian ring with a left identity element is noetherian. Of the equivalent conditions of 36.33, Hinohara [1960] established, for rings with identity whose radical topology is Hausdorff, that 8° implies 4° and 1°, Kurke [1967] established, for commutative rings with identity, the equivalence of 1°, 2°, 7°, and 9°, and Phạm Ngọc Ánh [1977a] established, for rings with identity, the equivalence of 2°

and 7°. The characterizations of linearly compact, commutative noetherian rings (36.37-38) and of compact, commutative noetherian rings (36.41) were given by the author [1967c, 1960a] for rings with identity.

Using Leptin's theorem that the radical of a strictly linearly compact ring is transfinitely nilpotent, the author [1972] characterized topological fields whose topologies are given by a complete discrete valuation as nondiscrete locally strictly linearly compact fields (41.6). The proof of the extension of this result to division rings (41.10), which generalizes Jacobson's earlier theorem [1935] that the topology of a totally disconnected locally compact division ring is given by a discrete valuation, depends on a theorem of Artin and Whaples [1942] (18.15), as did Kaplansky's earlier proof [1947a] that the topology of a locally compact division ring is given by an absolute value.

The author [1967a, 1968b] characterized finite products of finite-dimensional algebras with identity over nondiscrete locally compact fields (cf. 43.7). Earlier, Goldman and Sah [1965] had obtained a structure theorem for commutative locally compact semisimple rings with identity having no proper open ideals (Exercise 43.8), and later [1968] completed a thorough invesitagion of locally compact semisimple rings with identity having no proper open left ideals. The structure theory presented here (§43) for squarefree locally centrally linearly compact rings whose centers have no proper open ideals and their extensions by a connected locally compact ideal is due to Lucke and the author [1972].

46 Category, Duality, and Existence Theorems

The possibility of constructing duality theories for certain classes of topological modules, especially linearly topologized modules, to mirror the classical Pontriagin-van Kampen duality theory of commutative locally compact groups has attracted attention ever since topological rings and modules began to be investigated. Duality theory is not presented here, since it became clear at least by 1958 that categorical concepts beyond the scope of this book were essential to it. From the beginning, linear compactness has played an essential role. The first theory presented was Lefschetz's duality theory [1942] of linearly compact and discrete vector spaces, already mentioned. Kaplansky generalized this theory to a duality theory of linearly compact and discrete modules over a complete discrete valuation ring [1952b] and demonstrated its utility by deriving from it certain theorems of Krull [1940b] and Vilenkin [1946a,b,c] concerning topological modules over the p-adic integers. Schöneborn [1953b, 1956] and Leptin [1954b,c,d, 1956] also developed and extended somewhat Kaplansky's duality theory. Their results were placed in a common framework by Fleischer [1959] and later subsumed, together with Matlis's duality theorem [1958] concerning noetherian and artinian modules over a complete local noetherian ring, by Macdonald [1962] in his duality theory of Hausdorff linearly topologized modules over a complete local noetherian ring A, furnished with its natural topology. If the injective envelope A^* of the residue field of A (regarded as an A-module) is given the discrete topology and if, for every Hausdorff linearly topologized A-module M, M^* is the A-module of all continuous homomorphisms from M to A^*, furnished with the linear topology having as a fundamental system of neighborhoods of zero the annihilators of the submodules of M that are linearly compact for the induced topology and Hausdorff for the natural topology, then the evaluation homomorphism e from M to M^{**} (defined by $e(x)(y^*) = y^*(x)$ for all $x \in M$, $y^* \in M^*$) is an open isomorphism and is, furthermore, a topological isomorphism if and only if a submodule U of M is open whenever every submodule of M/U is closed for the quotient topology and the natural topology of every finitely generated submodule of M/U is discrete. In particular, this establishes a duality between linearly compact A-modules and linearly topologized Hausdorff A-modules M for which a submodule U is open whenever the natural topology of every finitely generated submodule of M/U is discrete.

In clarifying work, Müller [1970] observed that all duality theories for linearly topologized modules thus far constructed, when restricted to the class of discrete modules, were Morita dualities, and he showed that every Morita duality for rings (with identity) A and B (defined on full subcategories of (unitary) left A-modules and right B-modules that are closed under the formation of finite direct sums, submodules, and quotient modules and contain all finitely generated modules, and induced by a bimodule $_AU_B$ having certain properties) may be extended to a duality for the categories A-top and top-B of all Hausdorff, linearly topologized left A- and right B-modules, Specifically, let $_AU_B$ be given the discrete topology, and for each Hausdorff linearly topologized left A-module [right B-module] M, let M^* be the right B-module [left A-module] of all continuous homomorphisms from M to $_AU_B$, furnished with the topology of pointwise convergence. Then the canonical evaluation homomorphism e from M to M^{**} is an isomorphism carrying the topology of M to a topology stronger than but having the same closed submodules as that of M^{**}. Conversely, Müller [1971] showed that if there is a duality between full subcategories of A-top and top-B, each containing all discrete modules, then A and B possess a Morita duality induced by a bimodule $_AU_B$ such that for any M in either category, the dual module assigned to M by the given duality is the module M^* defined above.

Earlier, Müller [1969] had shown that if A has a Morita duality induced by a bimodule $_AU_B$, then A and $_AU_B$ are necessarily A-linearly compact modules for the discrete topology, and the reflexive A-modules are precisely those that are linearly compact for the discrete topology. Müller's criterion for the existence of a Morita duality, that A be linearly compact for the discrete topology, strengthens Osofsky's earlier condition [1965] that $A/\mathrm{Rad}(A)$ be artinian and that each idempotent in $A/\mathrm{Rad}(A)$ arise from an idempotent in A, in view of Leptin's theorems (29.14, 34.1). Rings (with identity) linearly compact for the discrete topology and Hausdorff for the radical topology are necessarily noetherian (36.33), a fact yielding Müller's earlier [1968] theorem that a ring Hausdorff for the radical topology and having a Morita duality is necessarily noetherian.

A commutative ring (with identity) linearly compact for the discrete topology is necessarily the direct sum of finitely many local rings (34.6 and 28.21). The commutative rings (with identity) linearly compact for the discrete topology and Hausdorff for the radical topology are precisely the complete semilocal noetherian rings (36.35). As previously noted, Zelinsky [1952] showed that complete local noetherian rings and valuations rings of maximal valuations were linearly compact for the discrete topology. Other examples exist: Vámos [1976] has given an example of a non-noetherian local ring having proper zero-divisors that is linearly compact for the discrete topology, and Wiseman [1982] has constructed an integral domain D linearly compact for the discrete topology whose quotient field, furnished with the discrete topology, is not a linearly compact D-module. A culminating result is Phạm Ngọc Ánh's theorem [1989] that every commutative ring (with identity) that is linearly compact for the discrete topology has a Morita duality.

Further developments in duality theory build largely on the papers just discussed. For example, Menini and Orsatti [1981], Phạm Ngọc Ánh [1981b], and Dikranjan and Orsatti [1984a], generalized Müller's construction by allowing A and B to be linearly compact rings that are not necessarily discrete. Other contributors to the theory since 1971 include Abrams, Aragona, Baccella, Bazzoni, Gregorio, Hutchinson, Jansen, Lorenzini, Mader, Márki, Mazan, Onodera, Peters, Roselli, Sandomierski, Stöhr, Stoyanov, Vámos, Woodcock, and Zelmanowitz.

The use of categorical concepts in the investigation of linearly topologized rings and modules has proved fruitful not only in duality theory but elsewhere as well. New proofs of old theorems (such as the structure theorem for semisimple linearly compact rings with identity), new insights about old objects (such as the nature of a linearly compact module furnished with its Leptin topology), and new theorems not involving categorical concepts (see, for example, Menini [1983b] and Dikranjan and Orsatti [1984a]) have already been obtained.

Categorical concepts have also been decisive in generalizing the notion of the quotient field of an integral domain. A central construction in the modern theory of rings of quotients depends on a certain type of filter of left (or right) ideals, called a Gabriel filter, which actually is a fundamental system of neighborhoods of zero for a linear ring topology, called a Gabriel topology (cf. Stenström [1975]) (Exercise 28.5). Thus topological rings play an accessory role in the algebraic theory of rings of quotients. A natural problem is that of determining conditions under which a subring of a generalized ring of quotients of a Hausdorff ring admits a Hausdorff ring topology inducing on A a topology weaker than or identical with its given topology. The earliest theorem, that if A is a Hausdorff integral domain for which multiplication by any nonzero element is an open mapping, then its quotient field admits a Hausdorff field topology inducing on A a topology weaker than its given topology (cf. 11.2), is due to Gelbaum, Kalish, and Olmsted [1950]. In particular, for every Hausdorff ring topology on a field, there is a weaker Hausdorff field topology (11.3). Correl [1958] showed that if K is a topological field whose completion is a locally compact ring, there is a weaker Hausdorff field topology on K the completion of which is a locally compact field (Exercise 26.7). Using the same argument, Mutylin [1967] showed that if the invertible elements in the completion of a field K for a metrizable ring topology form an open set, there is a weaker metrizable field topology on K whose completion is a field (*Topological Fields*, 14.14). Other contributors to this problem include Anthony [1970], Arnautov [1978], Carini [1976, 1978], Davison [1969], Eckstein [1973], Endo [1963, 1964], Facchini [1979], Gomez Pardo [1982], Halter-Koch [1971], Jebli [1974], R. L. Johnson [1967], Koh [1967b], Luedeman [1969a,b, 1978], A. O. Nazarov [1982], Schiffels and Stenzel [1983], Vizitei [1978], and the author [1961, 1971a].

Hinrichs' construction [1963] of Hausdorff, additively generated ring topologies (that is, ring topologies for which no proper additive subgroup is open) on $\mathbb{Z}$ inspired research in two directions: First, Mutylin [1965] exhibited the plenitude of ring topologies on $\mathbb{Q}$ by constructing an additively generated ring topology that is not stronger than the usual archimedean topology, and in particular is not locally bounded.

Second, motivated by Hinrichs' work, Kiltinen answered affirmatively [1967] the question: Does every infinite field admit a nondiscrete Hausdorff field topology? Only algebraic extensions of finite fields need be considered, for as noted by Nagata, Nakayama and Tuzuku [1953] and later by Kiltinen [1967] and Mutylin [1968], every other infinite field admits a proper valuation. Kiltinen proved that every countable integral domain admits a nondiscrete Hausdorff ring topology; hence by the theorem of Gelbaum, Kalish, and Olmsted, every countable field and hence every infinite field admits a nondiscrete Hausdorff field topology. In a parallel development,

Arnautov [1968a] constructed Hausdorff additively generated ring topologies on $\mathbb{Z}$, then [1969e] proved that every countable ring admits a nondiscrete Hausdorff ring topology. From this, Mutylin [1968] concluded that every infinite field admits a nondiscrete Hausdorff field topology.

Extending further the techniques of Hinrichs and Kiltinen, Zobel [1972] defined a class of ring topologies on $\mathbb{Q}$ he called "direct", some of which had unusual completions, such as an integral domain that is not a field (*Topological Fields*, Exercises 13.2-13.14) and contains no nonzero topological nilpotents. Recently, Heckmanns [1989] constructed a much simpler example of Hausdorff topological field whose completion is an integral domain that is not a field.

The results of Kiltinen and Arnautov made possible the determination of the number of field topologies on an infinite field K. The maximum number of topologies on K is $2^{2^{card(K)}}$. Podewski [1972a] showed that that is, in fact, the number of field topologies if K is denumerable, and Kiltinen [1972] showed that in general there are $2^{2^{card(K)}}$ field topologies on K, no two of which are topologically isomorphic, and none of which is the supremum of a family of locally bounded ring topologies. Independently, Heine [1971, 1972] established the analogous result for ring topologies on K. In another direction, Mutylin [1967] established the existence of at least a continuum of metrizable, non-locally bounded field topologies on $\mathbb{Q}$ for which the completion of $\mathbb{Q}$ is a topological field.

A natural sequel to Kiltinen's and Arnautov's work is the question: Does every infinite commutative ring admit a nondiscrete Hausdorff ring topology? Hochster [1968] gave criteria for a commutative ring with identity to admit a nondiscrete Hausdorff ideal topology and showed, in particular, that a commutative noetherian ring with identity admits a nondiscrete, Hausdorff ideal topology if and only if it is not artinian (Exercises 33.9-11). Hochster and Kiltinen [1969] showed that any infinite commutative ring with identity admitted a nondiscrete Hausdorff ring topology, and Arnautov [1969d] obtained the same answer for arbitrary infinite commutative rings, a result also independently obtained by Hagglund [1972]. All these solutions built on Kiltinen's earlier theorem establishing the existence of nondiscrete Hausdorff field topologies on infinite fields.

BIBLIOGRAPHY

Topological rings, exclusive of topological algebras over the real or complex numbers and of topological rings that algebraically are fields or division rings, are the principal subject of this bibliography. Papers on related topics mentioned in the text (especially in the historical notes or in the exercises) and in the Errata are also included.

For historical reasons, the year in a citation of a research article refers to the year in which the paper was accepted for publication, or for earlier papers, the year in which it was completed, if that information is available in the paper (otherwise, it refers to the year of publication). Thus the year in the citation indicates more accurately when the paper was completed than the year it was published.

Abellanas, Pablo. [1982] Topology of Krull homomorphisms. Revista Matematica Hispano-Americana (4) 42 (1982), 105-112.

Abrams, Gene D. [1990] (with Phạm Ngọc Ánh and L. Márki) A topological approach to Morita equivalence for rings with local units. Rocky Mountain Journal of Math. 22 (1992), 405-416.

Ahsan, J. [1980] (with M. Ismail) On semisimple linearly compact rings. Mathematica Japonica 26 (1981), 21-23.

Albu, Toma. [1984] On composition series of a module with respect to a set of Gabriel topologies. *Abelian groups and modules (Udine, 1984)*, pp. 467-476. CISM Courses and Lectures 287, Springer, Vienna, 1984.

Alekseĭ, Stepanyu Fedorovich. [1972] (with V. I. Arnautov) Free topological modules. Matematicheskie Issled. 7 (1972), No. 3(25), 3-18, 267. [1973] The connectedness of free topological modules. *Ibid.* 8 (1973) No. 3(29), 136-139, 182. [1974] The boundedness of a free topological module. *Ibid.* 9 (1974), No. 3(33), 3-14, 210-211. [1975] Free topological modules in certain classes. *Ibid.* 10 (1975), No. 2(36), 3-15, 280. [1975] (with V. I. Arnautov and M. I. Vodinchar) The boundedness of free topological modules. *Ibid.* 10 (1975), No. 2(36), 16-27, 280. [1977] (with R. K. Kalistru The completeness of free topological modules. *Ibid.* No. 44 (1977), 164-173, 181-2.

[1982] (with R. K. Kalistru) Necessary conditions for the completeness of a free topological module. *Ibid.* No. 65 (1982), 3-8, 152. [1986] The existence of free topological modules in a class of modules that are bounded. *Ibid.* No. 105 (1988), 3-7, 194.

Al-Ezeh, H. [1989] Equality of two topologies defined on a commutative ring. Arabian Journal for Science and Engineering 14 (1989), 477-479.

Ameziane, Hassani Rachid. [1988] Espaces de suites sur un module β et γ-dualité. Atti della Accademia Peloritana dei Periolanti Classe Sci. Fis. Mat. Natur. 66 (1988), 301-316 (1989).

Andrunakievich, Vladimir Aleksandrovich. [1953] The radical in generalized Q-rings. Izvestiiā Akademiĭ Nauk SSSR. Seriiā Matematicheskaiā 18 (1954), 419-426. [1966] (with V. I. Arnautov) Invertibility in topological rings. Doklady Akademiĭ Nauk SSSR 170 (1966), 755-758. [1973] (with V. I. Arnautov and M. I. Ursul) Wedderburn decomposition of hereditarily linearly compact rings. *Ibid.* 211 (1973), 15-18. [1974] (with V. I. Arnautov and M. I. Ursul) Wedderburn decomposition of hereditarily linearly compact rings. Matematicheskie Issled. 9 (1974) No. 1(31), 3-16. [1976] (with V. I. Arnautov) Weakly Boolean topological rings. Doklady Akademiĭ Nauk SSSR 228 (1976), 1265-1268. [1977] (with V. I. Arnautov) Weakly Boolean topological rings. Matematicheskie Issled. No. 44 (1977), 3-23, 177.

Ang, C. H. [1971] (with K. Chew) Rings with ω-nilpotent radicals. Nanta Mathematica 5 (1971), 1-11.

Ánh, Phạm Ngọc. See Phạm Ngọc Ánh.

Anthony, Joby Milo. [1970] Topologies for quotient fields of commutative integral domains. Pacific Journal of Math. 36 (1971), 585-601.

Anzai, Hirotada. [1943] On compact topological rings. Nihon Gakushiin. Proceedings of the Japan Academy 19 (1943), 613-615.

Aragona, Jorge. [1971] Un resultat d'extension d'applications linéaires continues dans les modules localement compacts. Comptes Rendus Hebdomadaires des Séances de l'Acad. des Sci. Sér. A-B 274 (1972), A444-A446. [1973] Dualité dans les modules topologiques sur an anneau unitaire commutatif localement compact. *Ibid.* 277 (1973), A979-A981.

Arens, Richard Friederich. [1946] Linear topological division algebras. Bulletin of the American Mathematical Soc. 53 (1947), 623-630.

Arezzo, Domenico. [1989] m-adic topologies and completions. Atti della Accademia Peloritana dei Pericolante Classe Sci. Fis. Mat. Natur. 67 (1989), 119-140.

Armstrong, Kenneth William. [1969] A Cauchy completion for function rings. Mathematische Zeitschrift 113 (1970), 145-153.

Arnautov, Vladimir Ivanovich. [1963a] The topological Baer radical and the decomposition of a ring. Buletinul Akademieĭ de Ştiinţe a R.S.S. Moldovenşti. Matematica 1963, No. 11, 79-81 (1964). [1963b] The topological Baer radical and the decomposition of a ring. Sibirskiĭ Matematicheskiĭ Zhurnal 5 (1964), 1209-1227. [1964a] On the theory of topological rings. Doklady Akademiĭ Nauk SSSR 157 (1964), 12-15. [1964b] On the theory of topological rings. Sibirskiĭ Matematicheskiĭ Zhurnal 6 (1965), 249-261. [1965a] Pseudonormability of topological rings. Proc. Fourth Conf. Young Moldavian Scientists, Sept. 1964 (Phys. Math.), pp. 51-54. Akademiĭa Nauk Moldavskoĭ SSR, Kishinev, 1965. [1965b] A criterion of pseudonormability of topological rings. Algebra i Logika 4, no. 4 (1965), 3-24. [1965c] Topologically weakly regular rings. Studies in Algebra and Math. Analy., pp. 3-10. Izdat. "Karta Moldoveniǎski", Kishinev, 1965. [1965d] Topological rings of a given local weight. Studies in General Algebra (Sem.), pp. 25-36. Akademiĭa Nauk Moldavskoĭ SSR, Kishinev, 1965. [1966] (with V. A. Andrunakievich) Invertibility in topological rings. Doklady Akademiĭ Nauk SSSR 170 (1966), 755-758. [1967] Topologizations of countable rings. Sibirskiĭ Matematicheskiĭ Zhurnal 9 (1968), 1251-1261. [1968a] Topologizations of the ring of integers. Buletinul Akademiei de Ştiinţe a R.R.S. Moldoveneşti, 3, no. 1 (1968), 3-15. [1968b] Complementary radicals in topological rings. I, II. Matematicheskie Issled. 3 (1968), No. 2(8), 16-30; ibid. 4 (1969), No. 1(11), 3-15. [1968c] Radicals in rings with a basis of group neighborhoods of zero. Ibid. 3 (1968), No. 4(10), 3-17 (1969). [1968] (with M. I. Vodinchar) Radicals of topological rings. Ibid. 3 (1968), No. 2(8), 31-61. [1969a] Semitopological isomorphism of topological rings. Ibid. 4 (1969), no. 2(12), 3-16. [1969b] Algebraic radicals in topological rings. Ibid. 4 (1969), No. 2(12), 116-122. [1969c] Imbedding of radicals. Ibid. 4 (1969), No. 3, 3-20. [1969d] Nondiscrete topologizability of infinite commutative rings. Doklady Akademiĭ Nauk SSSR 194 (1970), 991-994. [1969e] Nondiscrete topologizability of countable rings. Ibid. 191 (1970), 747-750. [1969f] Radicals in restricted rings. Buletinul Akademiei de Ştiinţe a R.R.S. Moldoveneşti, 1969, no. 2, 3-8. [1970a] Nondiscrete topologizability of infinite commutative rings. Matematicheskie Issled. 5 (1970), No. 4(18), 3-15. [1970b] An example of an infinite ring that permits only discrete topologization. Ibid. 5 (1970), No. 3(17), 182-185. [1972a] The topologizability of infinite rings. Ibid. 7 (1972), No. 1(23), 3-15. [1972b] The topologizability of infinite modules. Ibid. 7 (1972), No. 4(26), 241-243, 259. [1972] (with S. F. Alekseĭ) Free topological modules. Ibid. 7 (1972) No. 3(25), 3-18, 267. [1973] (with V. N. Vizitiei) Extending a locally bounded field topology to an algebraic extension of the field. Doklady Akademiĭ Nauk SSSR 216 (1974), 477-480. English translation, including corrections: Soviet Mathematics Doklady 15 (1974), 808-812. [1973] (with V. Andrunakievich and M. I. Ursul) Wedderburn decomposition of hereditarily linearly compact rings. Ibid. 211 (1973), 15-18. [1974] (with V. Andrunakievich and M. I. Ursul) Wedderburn decomposition of hereditarily linearly compact rings. Matematicheskie Issled. 9 (1974) No. 1(31), 3-16. [1974] (with V. N. Vizitei) Continuation of a locally bounded topology of a field to its algebraic extension. Buletinul Akademiei de Ştiinţe a R.R.S. Moldovenşti, 1974, no. 2, 29-43, 94. [1975] (with S. F. Alekseĭ and M. I. Vodinchar) The boundedness of free topological modules. Matematicheskie Issled. 10 (1975) No. 2(36), 16-27, 280. [1975] (with M. I. Vodinchar) The local compactness of a free topological module. Ibid. 10 (1975) No. 3(37), 3-14, 238. [1976] (with V. A. Andrunakievich) Weakly Boolean topological rings. Doklady Akademiĭ Nauk SSSR 228 (1976), 1265-1268. [1977] Algebraic radicals in topological rings (supplement). Matematicheskie Issled. No. 44 (1977), 36-41, 178.]1977] (with V. A. Andrunakievich) Weakly Boolean topological rings. Ibid. No. 44 (1977), 3-23, 177. [1978] Extension of the topology of a commutative ring to some of its rings of quotients. Ibid. No. 48 (1978), 3-13, 167. [1979] Compact rings with a finite number of topological generators. Ibid. No. 53 (1979), 3-5, 221. [1979a] (with M. I. Ursul) Imbedding of topological rings into connected ones. Ibid. No. 49 (1979), 11-15, 159. [1979b] (with M. I. Ursul) Uniqueness of a linearly compact topology in rings. Ibid. No. 53 (1979), 6-14, 221. [1981] Extension of a locally bounded field topology to a prime transcendental extension of the field. Algebra i Logika 20 (1981), 511-521, 599. [1981a] (with A. V. Mikhalëv) An example of a commutative completely regular topological semigroup without embeddings in multiplicative semigroups of topological rings. Vestnik Moskovskogo Universiteta Seriiā I, Matematika, Mekhanika 36, no. 6 (1981), 68-70, 117. [1981b] (with A. V. Mikhalëv) Continuation of topologies onto semigroup rings. Sixteenth All-Union Algebra Conference, part 1. Leningrad, 1981. [1981] (with M. I. Vodinchar and A. V. Mikhalëv) *Introduction to the Theory of Topological Rings and Modules*, Kishinev, 1981. [1982] (with A. V. Mikhalëv) Sufficient conditions for extension of topologies of a group and a ring to their group ring. Vestnik Moskovskogo Universiteta Seriiā I, Matematika, Mekhanika 38, no. 5 (1983), 25-33. [1982] (with M. I. Ursul) Quasicomponents of topological rings and modules. Izvestiiā Akademiĭ Nauk Moldavskoĭ SSR: Ser. Fiz.-Tekh. Mat. Nauk 1984, no. 1, 9-13. [1983] (with E. I. Marin and A. V. Mikhalëv) Necessary conditions for the extension of a group and a field topology to their group algebra. Vestnik Moskovskosgo Universiteta Seriiā I: Matematika Mekhanika 1984, no. 6, 58-61, 112. [1985a] (with A. V.

Mikhalëv) On the extension of a group and a ring topology to their group ring. Uspekhi Matematicheskikh Nauk 40 (1985), No. 4 (244), 135-136. [1985b] (with A. V. Mikhalëv) Extension of a group and a ring topology to their group ring. Matematicheskie Issled. No. 85 (1985), 8-20, 152. [1987] (with A. Mikhalëv) Extension of topologies of a ring and a discrete monoid to a semigroup ring. Matematicheskie Issled. No. 111 (1989), 9-23, 140. [1988] (with S. Glavatskiĭ, M. Vodinchar, and A. Mikhalëv) *Constructions of Topological Rings and Modules*. "Shtiintsa", Kishinev, 1988. [1989] (with K. Beidar, S. Glavatskiĭ, and A. Mikhalëv) Intersection property in the radical theory of topological algebras. *Proceedings of the International Conference on Algebra Dedicated to the Memory of A. I. Mal'cev Part 2* (Novosibirsk, 1989), 227-235, Contemporary Mathematics 131, pp. 205-225, American Mathematical Soc., Providence, RI, 1992.

Artin, Emil [1942] (with G. Whaples) The theory of simple rings. American Journal of Math. 65 (1943), 87-107.

Aurora, Silvio. [1956] Multiplicative norms for metric rings. Pacific Journal of Math. 7 (1957), 1279-1304. [1958] On power multiplicative norms. American Journal of Math. 80 (1958), 879-894. [1969] A representation theorem for certain connected rings. Pacific Journal of Math. 31 (1969), 563-567.

Baccella, Giuseppe. [1989] (with A. Orsatti) On generalized Morita bimodules and their dualities. Rendiconti della Accademia Nazionale delle Scienze detta dei XL. Memorie di Matematica (5) 13 (1989), no. 1, 323-340.

Baer, Reinhold. [1928] Zur Topologie der Gruppen. Journal für die Reine und Angewandte Mathematik 160 (1929), 208-226. [1931] (with H. Hasse) Zusammenhang und Dimension topologischer Körperräume. Journal für die Reine und Angewandte Mathematik 167 (1932), 40-45.

Baker, Ann C. [1963] Systems of linear equations over a topological module. Quarterly Journal of Math. Ser. (2) 15 (1964), 327-336.

Ballet, Bernard. [1967] Structure des modules artiniens (cas commutatif). Comptes Rendus Hebdomadaires des Séances de l'Acad. des Sci. Sér. A-B 266 (1968), A1-A4. [1968] Structure des anneaux strictement linéairement compacts commutatifs. *Ibid.* 266 (1968), A1113-A1116. [1970] Topologies linéaires sur un anneau cohérent. *Ibid.* 270 (1970), A1209-A1211. [1971] Topologies linéaires et modules artiniens. Journal of Algebra 41 (1976), 365-397. [1972a] Sur les modules linéairement compacts. Bulletin de la Société Mathématique de France 100 (1972), 345-351. [1972b] Topologies linéaires et modules absolutment purs. Publications des Séminaires de

Mathématiques de l'Université de Rennes. *Colloque d'algèbre commutative* *(1972)*. Exp. No. 7. Université de Rennes, Rennes 1972.

Ballier, Friedhorst. [1954] Über lineartopologische Algebren. Journal für die Reine und Angewandte Mathematik 195 (1956), 42-75 (1955).

Barot, Jiří. [1955] Remark on inverse elements in topological rings. Casopis Pro Pestovani Matematiky 80 (1955), 241-243.

Batho, Edward Hubert. [1957] Non-commutative semi-local and local rings. Duke Mathematical Journal 24 (1957), 163-172. [1959] A note on a theorem of I. S. Cohen. Portugaliae Mathematica 18 (1959), 187-192.

Baxter, Willard Ellis. [1968] Topological rings with property (Y). Pacific Journal of Math. 30 (1969), 563-571.

Bazzoni, Silvana. [1975] Dualita sul completamento naturale di un anello noetheriano. Rendiconti del Seminario Matematico della Universita di Padova 55 (1976), 63-80. [1978] Pontryagin type dualities over commutative rings. Annali di Matematica Pura ed Applicata (4) 121 (1979), 373-385. Correction *ibid.* 123 (1980), 403-404.

Beattie, Margaret. [1989] (with M. Orzech) Prime ideals and finiteness conditions for Gabriel topologies over commutative rings. Rocky Mountain Journal of Math. 22 (1992), 423-439.

Beckenstein, Edward. [1977] (with L. Narici and C. Suffel) *Topological algebras*. North-Holland Mathematics Studies, Vol. 24. Notas de Matemática, No. 60. North-Holland, Amsterdam-New York-Oxford, 1977.

Beidar, K. I. [1989] (with V. Arnautov, S. Glavatskiĭ, and A. Mikhalëv) Intersection property in the radical theory of topological algebras. *Proceedings of the International Conference on Algebra Dedicated to the Memory of A. I. Mal'cev Part 2* (Novosibirsk, 1989), 227-235, Contemporary Mathematics 131, pp. 205-225, American Mathematical Soc., Providence, RI, 1992.

Bel'nov, V. K. [1972] The metrization of polynomial rings. Bulletin de l'Académie Polonaise des Sciences Sér. sci. math. astronom. phy. 22 (1974), 1227-1233.

Bendixson, I. [1884] Sur la puissance des ensemble parfaits de points. Svenska Vetenskapsakademien, Stockholm, Handlingar, Bihan, vol. 9, no. 6.

Blair, Robert Louie. [1976] On the commutativity of certain topological rings. *Ring theory conference, Ohio University, Athens, Ohio 1976*, pp. 231-242. Lecture Notes in Pure and Applied Mathematics, Vol. 25, Marcel Dekker, Inc., New York, 1977.

Bokalo, B. M. [1985] (with I. Guran) Boundedness in topological rings. Vīsnik L'vīvs'kogo Ordena Lenīna Derzhavogo Unīversitetu imeni Ivana Franka. Serīya Mekahnīko-Matematichna No. 24 (1985), 60-64, 116.

Bourbaki, Nicolas. [1942] *Topologie générale*. Ch. III-IV. Act. Sci. Ind., no. 916. Hermann & Cie., Paris, 1942. [1961] *Algèbre Commutative*. Ch. 3: Graduations, filtrations, et topologies. Ch. 4: Idéaux premiers associés et décomposition primaire. Act. Sci. Ind. no. 1293. Hermann, Paris, 1961. [1964] *Algèbre Commutative*. Ch. 5: Entiers. Ch. 6: Valuations. Act. Sci. Ind. no. 1308. Hermann, Paris, 1964. [1983] *Algèbre Commutative*. Ch. 8: Dimension. Ch. 9: Anneaux locaux noethériens complets. Masson, Paris, 1983.

Braconnier, Jean. [1946] Sur les modules localement compacts. Comptes Rendus Hebdomadaires des Séances de l'Acad. des Sci. 222 (1946), 527-529.

Brandal, Willy. [1991] Completions of commutative topological rings. Communications in Algebra 20 (1992), 3381-3391.

Brungs, Hans Heinrich. [1990] (with G. Törner) I-compactness and prime ideals. Publicationes Mathematicae 40 (1992), 291-295.

Budach, Lothar. [1960a] Aufbau der gangen Abschliessung einartiger Noetherscher Integritätsbereiche. I. Mathematische Nachrichten 25 (1963), 5-17. [1960b] Verallgemeinerte lokale Ringe. *Ibid.* 25 (1963) 65-81. [1961] Aufbau der ganzen Abschliessung einartiger Noetherscher Integritätsbereiche. II. *Ibid.* 25 (1963), 129-149. [1966] Beziehungen zwischen gewissen Topologien in Noetherschen Ringen. *Symposium on General Topology and Its Relation to Modern Analysis and Algebra, 2d, Prague, 1966*, pp. 77-82. Academia, Prague, 1967. [1971] Über den Verband aller Lineartopologien eines Noetherschen Ringes. Beiträge zur Algebra und Geometrie 2 (1973), 123-126 (1974).

Cahen, Paul-Jean. [1982] (with F. Grazzini and Y. Haouat) Intégrité du complété et théoreme de Stone-Weierstrass. Annales scientifiques de l'Université de Clermont-Ferrand II. Mathématiques 21 (1982), 47-58.

Cantor, Georg. [1883] Ueber unendliche, lineare Punktmannichfaltigkeiten. Mathematische Annalen 23 (1884), 453-488.

Carini, Luisa. [1976] On the extension of the m-adic topology of a factorial domain to its field of fractions. Rendiconti del Circolo Mathematico di Palermo (2) 25 (1976), 136-144. [1978] On the connection between a metric of a factorial domain and that of its field of fractions. *Ibid.* 28 (1979), 220-228.

Caruth, A. J. [1982] Two theorems on complete commutative rings. Bulletin of the London Mathematical Soc. 15 (1983), 601-603.

Chase, Stephen Urban. [1972] Function topologies on abelian groups. Illinois Journal of Math. 7 (1963), 593-608.

Chevalley, Claude. [1943] On the theory of local rings. Annals of Mathematics (2) 44 (1943), 690-708.

Chew, Kim Lin. [1971] On complete rings. Nanta Mathematica 5 (1971), 88-91. [1971] (with C. Ang) Rings with ω-nilpotent radicals. Ibid. 5 (1971), 1-11.

Chin Dang Khoï. [1977] Algebraic radicals in the class of topological rings with an $\aleph_1$-topology. Matematicheskie Issled. No. 44 (1977), 109-116, 181. [1978] Lower radicals of topological rings. Ibid. No. 48 (1978), 161-166, 171. [1979] The relation between radicals of topological rings and the radicals of rings without topology. Ibid. No. 53 (1979), 174-194, 226-227.

Ciumaşu, Dorin. [1972] Sur des certaines structures topologiques dans un anneau de séries formelles. I, II. Buletinul Institutului Politehnic din Iasi. Sect. I. Matematica, Mecanica Teoretica, Fizica 18 (22) (1972), fasc. 3-4, secţ. I, 33-41; ibid. 19 (23) (1973), fasc. 3-4, secţ. I, 31-39.

Civin, Paul. [1960] Involutions on locally compact rings. Pacific Journal of Math. 10 (1960), 1199-1202.

Cohen, Irving S. [1945] On the structure and ideal theory of complete local rings. Transactions of the American Mathematical Soc. 59 (1946), 54-106. [1949] Commutative rings with restricted minimum condition. Duke Mathematical Journal 17 (1950), 27-42.

Cohen, Jo-Ann Deborah. [1979] Locally bounded topologies on the rings of integers of a global field. Canadian Journal of Math. 33 (1981), 571-584. [1980] Topologies on the ring of integers of a global field. Pacific Journal of Math. 93 (1981), 269-276. [1984] Topologies on the quotient field of a Dedekind domain. Ibid. 117 (1985), 51-67. [1986] (with K. Koh) On the group actions of the units on non-units in a compact ring. Journal of Pure and Applied Algebra 51 (1988), 231-239. [1987] (with K. Koh) The group of units in a compact ring. Ibid. 54 (1988), 167-179. [1988a] (with K. Koh) Half-transitive group actions in a compact ring, Ibid. 60 (1989), 139-153. [1988b] (with K. Koh) Involutions in a compact ring. Ibid. 59 (1989), 151-168. [1988c] (with K. Koh) A characterization of the p-adic integers. Communications in Algebra 17 (1989), 631-636. [1989] (with K. Koh) The subring generated by the units in a compact ring. Ibid. 18 (1990), 1617-1620. [1990] (with K. Koh) The subgroup generated by the involutions in a compact ring. *Ibid.* 19 (1991), 2923-2954. [1991] (with K. Koh) The structure of compact rings. Journal of Pure and Applied Algebra 17 (1992), 117-129.

Corner, Anthony Leonard Southern. [1965] Endomorphism rings of torsion-free abelian groups. *International Conference on the Theory of Groups, Proceedings. Australian National University, Canberra, August 1965*, pp. 59-65. Gordon and Breech, New York-London-Paris, 1967.

Correl, Ellen. [1958] On topologies for fields. Ph.D. dissertation, Purdue University, Lafayette, Indiana. [1967] Topologies on quotient fiels. Duke Mathematical Journal 35 (1968), 175-178.

Costinescu, Olga. [1977] Quelques problems concernant les intégrals relativement invariantes sur un anneau topologique. Analele Ştiintifica Ale Universitatii "Al. I. Cuza" din Iasi. Serie noua. Sec. I.a. Matematica 23 (1977), 257-265.

Cude, Joe E. [1969] Compact integral domains. Pacific Journal of Math. 32 (1970), 615-619.

Curtis, Charles Whittlesey. [1953] The structure of nonsemisimple algebras. Duke Mathematical Journal 21 (1954), 79-85.

van Dantzig, David. [1931a] Studien over topologische algebra. Amsterdam, H. J. Paris, 1931. [1931b] Zur topologische Algebra I. Komplettierungstheorie. Mathematische Annalen 107 (1933), 587-626. [1934a] Zur topologischen Algebra, II. Abstrakte b_ν-adische Ringe. Compositio Mathematica 2 (1935), 201-233. [1934b] Zur topologischen Algebra, III. Brouwersche und Cantorsche Gruppen. *Ibid.* 3 (1936), 408-426. [1935] Neuere Ergebnisse der topologischen Algebra. Matematicheskiĭ Sbornik 1 (43) (1936), 665-675. [1936] Nombres universels ou ν!-adiques avec une introduction sur l'algèbre topologique. Annales Scientifiques de l'École Normale Superieure (3) 53 (1936), 275-307. [1950] Topologico-algebraic reconnoitering. (Seven lectures on topology) Centrumreeks, no. 1. Math. Centrum Amsterdam, pp. 56-79. J. Noorduijm en Zoon, Gorichem, 1950.

Datuashvili, Tamara Iraklievna. [1983] The global homological dimension of trivial linear topological extension of rings. Trudy Tbilisskogo Matematicheskogo Instituta 74 (1983), 25-38.

Dauns, John. [1968] (with K. Hoffman) *Representations of Rings by Sections*. Memoirs of the American Mathematical Society, No. 83. American Mathematical Soc., Providence, R.I., 1968.

Davison, Thomas Matthew Kerr. [1969] On rings of fractions. Canadian Mathematical Bulletin 13 (1970), 425-430.

Dawkins, Brian P. [1966] (with I. Halperin) The isomorphisms of certain continuous rings. Canadian Journal of Math. 18 (1966), 1333-1344.

Day, B. J. [1982] A note on powers of Hausdorff fields. Journal of Pure and Applied Algebra 28 (1983), 71-73.

Dazord, Jean. [1966a] Sur les anneaux filtrés de Gelfand. Comptes Rendus Hebdomadaires des Séances de l'Acad. des Sci. Sér. A-B 262 (1966), A326-A328. [1966b] Anneau filtré de Gelfand. Publications du Département de Mathématiques (Lyon) 3 (1966), fasc. 1, 41-53. Correction, ibid. fasc. 2, 71-72.

Deshpandi, V. K. [1979] Completions of Noetherian hereditary prime rings. Pacific Journal of Math. 90 (1980), 285-297. [1980] A decomposition theorem for the completions of noncommutative Noetherian rings. Journal of the Indian Mathematical Soc. (N.S.) 45 (1981), no. 1-4, 109-115 (1984).

D'Este, Gabriella. [1980] On topological rings which are endomorphism rings of reduced torsion-free groups. Quarterly Journal of Mathematics Ser. (2) 32 (1981), 303-311.

Dieudonné, Jean Alexandre. [1945] Sur les corps topologiques connexes. Comptes Rendus Hebdomadaires des Séances de l'Acad. des Sciences 221 (1945), 396-398. [1949a] Linearly compact spaces and double vector spaces over sfields. American Journal of Math. 73 (1951), 13-19. [1949b] Matrices semi-finies et espaces localement linéairement compacts. Journal für die Reine und Angewandte Mathematik 188 (1950), 162-166. [1951] Sur les sous-espaces linéairement compacts. Boletim da Sociedade de Matemática de São Paulo 6 (1951), 53-60 (1952).

Dikranjan, Dikran. [1973] Minimal precompact topologies in rings. Godishnik na Sofiiskiĭa Universitet, Matematicheski Fakultet 67 (1972/3), 391-397 (1976). [1976] Precompact minimal topologies on Noetherian domains. Doklady Bŭlgarskoĭ Akademii Nauk 29 (1976), 1561-1562. [1978] Minimal ring topologies and Krull dimension. Proceedings of the fourth Colloquium on Topology, pp. 357-366. Colloquia Mathematica Sociétatis János Bolyai 23, North Holland, Amsterdam, 1980. [1980] (with W. Więsław) Rings with only ideal topologies. Commentarii Mathematici Universitatis Sancti Pauli 29 (1980), 157-167. [1981] Extension of minimal ring topologies. General topology and its relation to modern analysis and algebra V (Prague, 1981), pp. 98-103. Sigma Ser. Pure Math. 3, Helderman, Berlin, 1983. [1982] Minimal topological rings. Serdica 8 (1982), 149-165. [1984a] (with A. Orsatti) On the structure of linearly compact rings and their dualities. Rendiconti della Accademia nazionale delle scienze detta dei XL. Parte I, Memorie di matematica (5) 8 (1984), no. 1, 143-184. [1984b] (with A. Orsatti) Linearly compact rings. Supplemento ai Rendiconti del Circolo Matematico di Palermo (2), No. 4 (1984), 59-74. [1984c] (with A. Orsatti) On the structure of linearly compact rings and their dualities. *Abelian Groups and Modules (Udine, 1984)*, 415-439. CISM Courses and Lectures, 187. Springer, Vienna-New York, 1984. [1988] Radicals and Duality. I. *General Algebra 1988*, 45-64. North-Holland, Amsterdam, 1990. [1988] (with A. Orsatti) On an unpublished manuscript of Ivan Prodanov concerning locally compact modules and their dualities. Communications in Algebra 17 (1989), 2739-2771. [1990] (with E. Gregorio and A. Orsatti) Kasch bimodules. Rendiconti del Seminario Matematico della Universita di Padova 85 (1991), 147-160. [1992] (with A. Tonolo) On the lattice of linear module topologies. Communications in Algebra 21 (1993), 275-298.

Dinh Van Huynh. [1973] Über eine Klasse von linear kompakten Ringen. Publicationes Mathematicae 22 (1975), 231-233. [1977] Über linear kompakte Ringe. Acta Mathematica Academiae Scientiarum Hungaricae 36 (1980), 1-5. [1979] On the maximal regular ideal of a linearly compact ring. Archiv der Mathematik 33 (1979/80), 232-234. [1984a] Some results on linearly compact rings. Ibid. 44 (1985), 39-47. [1984b] Some results on rings with chain conditions. Mathematische Zeitschrift 191 (1986), 43-52.

Dosta, S. I. [1977] Galois extensions of linear topological rings. Vestnik Moskovskogo Universiteta. Seriiâ I. Matematika, Mekhanika 1979, no. 4, 43-46, 102.

Eckstein, Frank [1967a] Complete semi-simple rings with ideal neighborhoods of zero. Archiv der Mathematik 18 (1967), 587-590. [1967b] Semi-direct splitting of the radical. Abhandlungen aus dem Mathematischen Seminar der Univerität Hamburg 32 (1968), 61-72. [1971] (with G. Michler) On compact open subrings of semi-simple locally compact rings. Archiv der Mathematik 23 (1972), 10-18. [1973] Topological rings of quotients and rings without open left ideals. Communications in Algebra 1 (1974), 365-376.

Edelman, Jorge. [1970] (with A. Larontonda) Quasi-norms in rings. Revista de la Union Matematica Argentina 25 (1970/71), 357-361.

Elkins, Bryce L. [1972] A Galois theory for linear topological rings. Pacific Journal of Math. 51 (1974), 89-107.

Ellis, Robert. [1956a] A note on the continuity of the inverse. Proceedings of the American Mathematical Soc. 8 (1957), 372-373. [1956b] Locally compact transformation groups. Duke Mathematical Journal 24 (1957), 119-125.

El Miloudi, Mahrani. [1989] (with A. Mohamed) Les S-anneaux et propriétés de tonnelage dans les modules topologiques. Atti della Accademia Peloritana dei Pericolanti Classe Sci. Fis. Mat. Natur. 67 (1989), 275-292.

Endo, Mikihiko. [1963] On the embedding of topological rings into quotient rings. Tokyo Daigaku. Rigakubu. Journal of the Faculty of Science, the University of Tokyo. Section I. 10 (1964), 196-214. [1964] On the embedding of topological rings into quotient rings, II. Tokyo Daigaku. Kyoyo Gakubu. Scientific Papers of the College of General Education of the University of Tokyo 14 (1964), 51-54.

Facchini, Alberto. [1979] On the ring of quotients of a Noetherian commutative ring with respect to the Dickson topology. Rendiconti del Seminario Matematico della Universita di Padova 62 (1980), 233-243.

Faith, Carl Clifton. [1962] (with Y. Utumi) On a new proof of Litoff's Theorem. Acta Mathematica Academiae Scientarium Hungaricae 14 (1963), 369-371.

Fakruddin, Syed M. [1971] Linearly compact modules over noetherian rings. Journal of Algebra 24 (1973), 544-550. [1983] On topologies over rings. International Journal of Mathematics and Mathematical Sciences 8 (1985), 197-199.

Fischer, Hans R. [1964] (with H. Gross) Quadratische Formen und lineare Topologien. III. Tensorprodukte linearer Topologien. Mathematische Annalen 160 (1965), 1-40.

Fleischer, Isidore Bernard. [1955] Modules of finite rank over Prüfer rings. Annals of Mathematics (2) 65 (1957), 250-254. [1959] Über Dualität in linear-topologischen Moduln. Mathematische Zeitschrift 72 (1960), 439-445.

Freudenthal, Hans. [1935] Einige Sätze über topologische Gruppen. Annals of Math. (2) 37 (1936), 46-56.

Fuchs, Laszlo. [1955] (with T. Szele) On Artinian rings. Acta Scientiarum Mathematicarum Hungaricae 17 (1956), 30-40. [1967] Note on linearly compact abelian groups. Journal of the Australian Mathematical Society 9 (1969), 433-440. [1985] (with L. Salce) *Modules over Valuation Domains*. Lecture Notes in Pure and Appl. Math. 97. Marcel Dekker, New York, 1985.

Fuster, Robert. [1984] (with A. Marquina) Geometric series in incomplete normed algebras. American Mathematical Monthly 91 (1984), 49-51.

Gabriel, Pierre. [1961] Des catégories abéliennes. Bulletin de la Société Mathématique de France 90 (1962), 323-448.

Galuzzi, Massimo. [1971] Proprieta' della *a*-topologie su un anello commutativo. Rendiconti dell'Istituto Lombardo di Scienze e Lettere, Milan. Sci. math. fis. chim. geol. A 105 (1971), 579-583.

Garcia Hernandez, J. L. [1984] (with J. Gomez Pardo) V-rings relative to Gabriel topologies. Communications in Algebra 13 (1985), 59-83.

Geddes, Archibald. [1954] A short proof of the existence of coefficient fields for complete, equicharacteristic local rings. Journal of the London Mathematical Society 29 (1954), 334-341. [1955] On the embedding theorems for complete local rings. Proceedings of the London Mathematical Society (3) 6 (1956), 343-354.

Gelbaum, Bernard Russell. [1950] (with G. Kalisch and J. Olmsted) On the embedding of topological groups and integral domains. Proceedings of the American Mathematical Soc. 2 (1951), 807-821.

Ghika, Alexandru. [1952] Les topologies définies sur un A-module par une A-seminorme. Academia Republicii Populare Romîne. Buletin Ştiinţific. Seria Matematica, Fizică, Chimie 4 (1952), 563-583.

Gil de Lamadrid, Jesus. [1956] (with J. Jans) Note on connectedness in topological rings. Proceedings of the American Mathematical Soc. 8 (1957), 441-442.

Gilmer, Robert. [1972] On ideal-adic topologies for a commutative ring. Enseignement Mathématique (2) 18 (1972), 201-204 (1973).

Glavatskiĭ, Sergeĭ Timofeevich. [1980] Topological quasi-injective and quasi-projective modules. *Abelian Groups and Modules*, pp. 31-44, 135-136. Tomsk. Universitet. 1980. [1988] (with V. Arnautov, M. Vodinchar, and A. Mikhalëv) *Constructions of Topological Rings and Modules*. "Shtiintsa", Kishinev, 1988. [1989] (with V. Arnautov, K Beidar, and A. Mikhalëv) Intersection property in the radical theory of topological algebras. *Proceedings of the International Conference on Algebra Dedicated to the Memory of A. I. Mal'cev* Part 2 (Novosibirsk, 1989), 227-235, Contemporary Mathematics 131, pp. 205-225, American Mathematical Soc., Providence, RI, 1992.

Goblot, Rémi. [1970] Sur les anneaux linéairement compacts. Comptes Rendus Hebdomadaires des Séances de l'Acad. des Sci., Paris, Sér. A-B 270 (1970), A1212-A1215.

Godement, Roger. [1956] Topologies m-adiques. Séminaire H. Cartan et C. Chevalley 1955/56. Géometrie algébrique. Exp. 18. Secrétariat mathématique, Paris, 1956.

Golan, Jonathan S. [1987] *Linear Topologies on a Ring: an Overview*, Pitman Research Notes in Mathematics 159, Longman Scientific and Technical, Harlow, Essex, UK, 1987.

Goldman, Oscar. [1965] (with C. Sah) On a special class of locally compact rings. Journal of Algebra 4 (1966), 71-95. [1968] (with C. Sah) Locally compact rings of special type. *Ibid.* 11 (1969), 363-454. [1968] Rings and modules of quotients. *Ibid.* 13 (1969), 10-47. [1973] A Wedderburn-Artin-Jacobson structure theorem. *Ibid.* 34 (1975), 64-73.

Gomez Pardo, José Louis. [1982] (with N. Rodrigues Gonzalez) Quotient rings of coherent rings with respect to a Gabriel topology. *Proceedings of the ninth conference of Portugese and Spanish mathematicians*, 1 (Salamanca 1982), 119-122.

Goodearl, Kenneth Ralph. [1974] Simple regular rings and rank functions. Mathematische Annalen 214 (1975), 267-287. [1976] Completions of regular rings. *Ibid.* 220 (1976), 229-252.

Gould, Gerald C. [1961] Locally unbounded topological fields and box topologies on products of vector spaces. Journal of the London Mathematical Soc. 36 (1961), 273-281.

Gouyon, Luce. [1966] Sur le choix de certains "P-compacités". Comptes Rendus Hebdomadaires des Séances de l'Acad. des Sci. Sér. A-B 263 (1966), A345-A347.

Gravett, Kenneth Albert Henry. [1955] Note on a result of Krull. Proceedings of the Cambridge Philosophical Society, Mathematical and Physical Sciences 52 (1956), 379.

Grazzini, Fulvio. [1982] (with P. Cahen and Y. Haouat) Intégrité du complété et théorème de Stone-Weierstrass. Annales scientifiques de l'Université de Clermont-Ferrand II. Mathématiques 21 (1982), 47-58.

Greco, Silvio. [1971] (with P. Salmon) *Topics in m-adic Topologies*. Ergebnisse der Mathematik und ihrer Grenzgebiete, Vol. 58, Springer-Verlag, New York-Berlin, 1971.

Green, Edward Lewis. [1979] Rings with Noetherian completions. *Commutative Algebra* (Fairfax, Virginia, 1979), pp. 233-255. Lecture Notes in Pure and Applied Math., 68, Dekker, New York, 1982.

Gregorio, Enrico. [1987a] Generalized Morita equivalence for linearly topologized rings. Rendiconti del Seminario Matematico della Universita di Padova 79 (1988), 221-246. [1987b] Dualities over compact rings. Ibid. 80 (1988), 151-174 (1989). [1988a] Classical Morita equivalence and linear topologies; applications to quasi-dualities. Communications in Algebra 18 (1990), 1137-1146. [1988b] Tori and continuous dualities. Rendiconti della Accademia nazionale delle scienze detta dei XL. Parte I. Memorie di matematica (5) 13 (1989), 211-221. [1990a] On a class of linearly compact rings. Ibid. 19 (1991), 1313-1325. [1990b] Topologically semiperfect rings. Rendiconti del Seminario Matematico della Universita di Padova 85 (1991), 265-290. [1990] (with D. Dikranjan and A. Orsatti) Kasch bimodules. Ibid. 85 (1991), 147-160.

Gross, Herbert. [1964] (with H. Fischer) Quadratische Formen und lineare Topologien. III. Tensorprodukte linearer Topologien. Mathematische Annalen 160 (1965), 1-40.

Grothendieck, Alexandre. [1960] Éléments de géometrie algébrique. I. Le langage des schémas. Institut des Hautes Études Scientifiques Publ. Math. No. 4 (1960). [1964] Éléments de géometrie algébrique. IV. Étude locale des schémas et des morphismes de schémas. I. Ibid. No. 20 (1964).

Grover, V. K. [1990] (with N. Sankaran) Projective modules and approximation couples. Compositio Mathematica 74 (1990), 165-168.

Guérindon, Jean. [1956] Sur une famille d'équivalences en théorie des idéaux. Comptes Rendus Hebdomadaires des Séances de l'Acad. des Sci. 242 (1956), 2693-2695. [1965] Une classe d'homomorphismes naturels. Ibid. 260 (1965), 383-386. [1970] Topologies linéaires sur les anneaux de polynômes. Bulletin des Sciences Mathématiques (2) 95 (1971), 241-259. [1975] Séries restreints et compacité linéaire. Séminaire P. Dubreil, F. Aribaud et M.-P. Malliavin (28e année: 1974/75), Algèbre, Exp. 7. Secrétariat Mathématique, Paris, 1975. [1976] Sur les modules linéairement compacts. Groupe d'Étude d'Algèbre. (Marie-Paule Malliavin, 2e année; 1976/77), Exp. 1. Secrétariat Mathématique, Paris, 1978.

Guran, Igor' Iosifovich. [1985] (with B. Bokalo) Boundedness in topological rings. Vīsnik L'vīvs'kogo Ordena Lenīna Derzhavogo Unīversitetu imeni Ivana Franka. Serīya Mekahnīko-Matematichna No. 24 (1985), 60-64, 116.

Gwynne, W. D. [1970] (with J. C. Robson) Completions of non-commutative Dedekind prime rings. Journal of the London Mathematical Society (2) 4 (1971), 346-352.

Hacque, Michel. [1983] Complétions intrinseques et co-intrinseques. Communications in Algebra 12 (1984), 1931-1988.

Hadzien, Dzh. [1975] (with T. Sarymsakov and S. Nasirov) A description of the ideals of a certain class of rings. Doklady Akademiĭ Nauk SSSR 225 (1975), 1018-1019.

Hagglund, Lee O. [1972] Existence of proper topologies in commutative rings. Dissertation, Duke University, Durham, North Carolina, 1972.

Haghany, Ahmad. [1984] The canonical topology of certain rings. Bulletin of the Iranian Mathematical Society 11 (1984), 39-43.

Haley, David K. [1969a] Equationally compact noetherian rings. *Conference on Universal Algebra, Queen's University, 1969, Proceedings* (Queen's papers in pure and applied mathematics, no. 25), pp. 259-267. Queen's University, Kingston, Ont., 1970. [1969b] On compact commutative Noetherian rings. Mathematische Annalen 189 (1970), 272-274. [1972] Equationally compact Artinian rings. Canadian Journal of Math. 25 (1973), 273-283. [1973] A note on compactifying Artinian rings. Ibid. 26 (1974), 580-582. [1979] Equational compactness in rings. With applications to the theory of topological rings. Lecture Notes in Mathematics, 745. Springer, Berlin-Heidelberg-New York, 1979.

Halperin, Israel. [1961] Von Neumann's arithmetics of continuous rings. Acta Litterarum ac Scientiarum Regiae Universitatis Hungaricae Francisco-Jospheniae. Sectio Scientiarum Mathematicarum 23 (1962), 1-17. [1964] Regular rank rings. Canadian Journal of Math. 17 (1965), 709-719. [1966] (with B. Dawkins) The isomorphisms of certain continuous rings. Ibid. 18 (1966), 1333-1344.

Halter-Koch, Franz. [1971] Idealtheorie der offen einbettbaren kommutativen topologischen Ringe. Journal für die Reine und Angewandte Mathematik 256 (1972), 168-172.

Haouat, Youssef. [1982] (with P. Cahen and F. Grazzini) Intégrité du complété et théorème de Stone-Weierstrass. Annales scientifiques de l'Université de Clermont-Ferrand II. Mathématiques 21 (1982), 47-58.

Harada, Manabu. [1959] Note on raising idempotents. Journal of the Institute of Polytechnics, Osaka City University Series A 10 (1959), 63-65.

Hartmann, Peter. [1987] Stellen und Topologien von Schiefkörpern und Alternativkörpern. Archiv der Mathematik 51 (1988), 274-282.

Hasan, Ali Hussain. [1976] On the completion of a Noetherian ring in the natural topology. Revue Roumaine de Mathématiques Pures et Appliquées 22 (1977), 483-485.

Hasse, Helmut. [1931] (with R. Baer) Zusammenhang und Dimension topologischer Körperräume. Journal für die Reine und Angewandte Mathematik 167 (1932), 40-45. [1932] (with F. Schmidt) Die Struktur diskret bewerteter Körper. Ibid. 170 (1933), 4-63.

Hasse, Klaus. [1966] Homogene Topologien in Polynomringen. *Symposium on General Topology and Its Relation to Modern Analysis and Algebra, 2d, Prague, 1966*, pp. 77-82. Academia, Prague, 1067.

Heckmanns, Ulrich. [1989] Beispiel eines topologischen Körpers, dessen Vervollständigung ein Integritätsring, aber kein Körper ist. Archiv der Mathematik 57 (1991), 144-148.

Heine, J. [1971] Ringtopologien auf nichtalgebraischen Körpern. Mathematische Annalen 199 (1972), 205-211. [1972] Ring topologies of type N of fields. General Topology and Its Applications 3 (1973), 135-148.

Heuer, Gerald Arthur. [1972] Continuous multiplications in R^2. Mathematics Magazine 45 (1972), 72-77.

Hinohara, Yukitoshi. [1960] Note on noncommutative semilocal rings. Nagoya Mathematical Journal 17 (1960), 161-166.

Hinrichs, Lowell A. [1962] The existence of topologies on field extensions. Transactions of the American Mathematical Soc. 113 (1964), 397-405. [1963] Integer topologies. Proceedings of the American Mathematical Soc. 15 (1964), 991-995.

Hochster, Melvin. [1968] Rings with nondiscrete ideal topologies. Proceedings of the American Mathematical Soc. 21 (1969), 357-362. [1969] Existence of topologies for commutative rings with identity. Duke Mathematical Journal 38 (1971), 551-554. [1969] (with J. Kiltinen) Commutative rings with identity have ring topologies. Bulletin of the American Mathematical Soc. 76 (1970), 419-1420.

Hoffman, Karl Heinrich. [1968] (with J. Dauns) *Representations of Rings by Sections*. Memoirs of the American Mathematical Society, No. 83. Amer. Math. Soc., Providence, R.I., 1968.

Hopkins, Charles [1938] Rings with minimal condition for left ideals. Annals of Math. 40 (1939), 712-730.

Hutchinson, John J. [1979] The completion of a Morita context. Communications in Algebra 8 (1980), 712-742.

Ikushima, Isaku. [1950] G-radical of topological rings. Osaka Institute of Science and Technology. Journal. Part I: Mathematics and Physics 2 (1950), 81-84.

Isac, George. [1968] Modules topologiques ouvert-injectifs. Revue Roumaine de Mathématiques Pures et Appliquées 14 (1969), 1017-1024.

Iseki, Kiyoshi. [1952] Sur le G-radical d'un anneau topologique. Comptes Rendus Hebdomadaires des Séances de l'Acad. des Sci. 234 (1952), 1938-1939. [1953a] On 0-dimensional compact ring. Mathematica Japonicae 3 (1953), 37-40. [1953b] On the Brown-McCoy radical in topological rings. Academia Brasileira de Ciencias, Rio de Janeiro. Anais 25 (1953), 79-86.

Ismail, M. [1980] (with J. Ahsan) On semisimple linearly compact rings. Mathematica Japonica 26 (1981), 21-23.

Iwasaki, Koziro. [1960] Integral transforms and self-dual topological rings. Nihon Gakushiin. Proceedings 36 (1960), 529-532.

Iwasawa, Kenkichi. [1951] On the rings of valuation vectors. Annals of Math. (2) 57 (1953), 331-356.

Jacobson, Nathan. [1935] Totally disconnected locally compact rings. American Journal of Math. 58 (1936), 433-449. [1935] (with O. Taussky) Locally compact rings. Proceedings of the National Academy of Sciences U.S.A. 21 (1935), 106-108. [1937] A note on topological fields. American Journal of Math. 59 (1937), 889-894. [1950] (with Charles E. Rickart) Jordan homomorphisms of rings. Transactions of the American Mathematical Soc. 69 (1950), 479-502. [1956] Structure of Rings. *American Mathematical Society Colloquium Publications* vol. 37, Providence, R. I., 1956.

Janiszewski, Zygmunt. [1915] On cuts of the plane made by continua. Prace Matematyczno-fizyczne 26 (1915), 11-63.

Jans, James Patrick. [1957] Compact rings with open radical. Duke Mathematical Journal 24 (1957), 573-577.

Jansen, Willen G. [1984] (with J. Zelmanowitz) Duality modules and Morita duality. Journal of Algebra 125 (1989), 257-277. [1988] (with J. Zelmanowitz) *Duality for Module Categories*, Algebra Berichte 59, Verlag Reinhard Fischer, München, 1988.

Jebli, Ahmed. [1971a] Sur certaines topologies linéaires. Comptes Rendus Hebdomadaire des Séances de l'Acad. des Sci. Sér. A-B 274 (1972), A444-A446. [1971b] Topologies et valuations essentielles. Ibid. 272 (1971), A1173-A1174. [1972a] Sur certains topologies linéaires. Séminaire de Théorie des Nombres, 1971-1972 (Univ. Bordeaux I, Talence), Exp. 7. Lab. de Théorie des Nombres, Centre Nat. Recherche Sci., Talence, 1972. [1972b] Sur certaines topologies linéaires. Publications des Séminaires de Mathématiques de l'Université de Rennes. Colloque d'algèbre commutative (1972).

Exp. 7. Univ. Rennes, Rennes, 1972. [1974a] Corps des fractions et topologies artiniennes. Comptes Rendus Hebdomadaires des Séances de l'Acad. des Sci. Sér. A 278 (1974), 973-976. [1974b] Topologies linéaires du corps des fractions d'un anneau noethérien. Publications des Séminaires de Mathématiques de l'Université de Rennes 1974, fasc. 2, 1-44. [1975] Topologies linéaires sur le corps des fractions d'un anneau de polynômes. Comptes Rendus Hebdomadaires des Séances de l'Acad. des Sci. Sér. A-B 280 (1975), A701-A703.

Jensen, Christian U. [1969] On the vanishing of $\varprojlim^{(i)}$. Journal of Algebra 15 (1970), 151-166.

Johnson, Eugene W. [1968] A note on quasi-complete local rings. Colloquium Mathematicum 21 (1970), 197-198.

Johnson, Johnny Albert. [1976] Quasi-completeness in local rings. Mathematica Japonica 22 (1977), 183-184.

Johnson, Robert Leroy. [1967] Rings of quotients of topological rings. Mathematische Annalen 179 (1969), 203-211.

Kalisch, Gerhard Karl. [1950] (with B. Gelbaum and J. Olmsted) On the embedding of topological groups and integral domains. Proceedings of the American Mathematical Soc. 2 (1951), 807-821.

Kalistru, Rel' Konstantinovich. [1977] (with S. Alekseĭ) The completeness of free topological modules. Matematicheskie Issled. No. 44 (1977), 164-173, 181-2. [1979] Completeness of a free topological module generated by a discrete space. Ibid. No. 53 (1979), 98-102, 223. [1981] Completeness of a free topological module generated by a zero-dimensional space. Ibid. No. 62 (1981), 57-64, 149. [1985] Completeness of a topological polynomial ring. Ibid. No. 85 (1985), 71-82, 154.

Karpelevich, Fridrikh Izrailevich. [1948] Pseudonorms in the rings of integers. Uspekhi Matematicheskikh Nauk 3, No. 5 (17) (1948), 174-177 (1949).

Kaplan, Samuel. [1952] Cartesian products of reals. American Journal of Math. 74 (1952), 936-954.

Kaplansky, Irving. [1946] Topological rings. American Journal of Math. 69 (1947), 153-183. [1947a] Topological methods in valuation theory. Duke Mathematical Journal 14 (1947), 527-541. [1947b] Dual rings. Annals of Math. (2) 49 (1948), 689-701. [1947c] Locally compact rings. American Journal of Math. 70 (1948), 447-459. [1948a] Polynomials in topological fields. Bulletin of the American Mathematical Soc. 54 (1948), 909-916.

[1948b] Topological rings. *Ibid.* 54 (1948), 809-286. [1949a] The Weierstrass theorem in fields with valuations. Proceedings of the American Mathematical Soc. 1 (1950), 356-357. [1949b] Locally compact rings. II. American Journal of Math. 73 (1951), 20-24. [1951] Modules over Dedekind rings and valuation rings. Transactions of the American Mathematical Soc. 72 (1952), 327-340. [1952a] Locally compact rings. III. American Journal of Math. 74 (1952), 929-935. [1952b] Dual modules over a valuation ring. I. Proceedings of the American Mathematical Soc. 4 (1953), 213-219. [1970] *Commutative Rings*. Allyn and Bacon, Boston, 1970. [1974] *Topics in Commutative Ring Theory*. Lecture Notes. Department of Mathematics, University of Chicago, 1974.

Katayama, Hisao. [1985] Maximal linear topologies and the complement of linear topologies. Mathematical Journal of Okayama University 27 (1985), 97-105.

Keller, Hans Arwed. [1971] Stetigkeitsfragen bei linear-topologischen Clifford-algebren. Dissertation, Universität Zurich, 1971.

Kertész, Andor. [1969] (with Alfred Widiger) Artinsche Ringe mit artinschem Radical. Journal für die Reine und Angewandte Mathematik 242 (1970), 8-15. [1987] *Lectures on Artinian Rings*. Akadémiai Kiadó, Budapest, 1987.

Khromulyak, O. M. [1991] (with E. G. Zelenyuk and I. V. Protasov) Topologies on countable groups and rings. Doklady Akademiĭā Nauk Ukraïns'koĭ RSR 1991, no. 8, 8-11, 1992.

Khuri, Soumaya Makdissi. [1983] Endomorphism rings and Gabriel topologies. Canadian Journal of Math. 36 (1984), 193-205.

Kiltinen, John Oscar. [1967] Inductive ring topologies. Transactions of the American Mathematical Soc. 134 (1968), 149-169. [1968] Embedding a topological domain in a countably generated algebraic ring extension. Duke Mathematical Journal 37 (1970), 647-654. [1969] (with M. Hochster) Commutative rings with identity have ring topologies. Bulletin of the American Mathematical Soc. 76 (1970), 419-420. [1972] On the number of field topologies on an infinite field. Proceedings of the American Mathematical Soc. 40 (1973), 30-36.

Kiriyak, L. L. [1990] Topology of free topological algebras with the Malt'- tsev condition and a k-space. Buletinul Academiei de Ştiinţe a R.S.S. Moldovensti. Matematica 1990, no. 3, 7-13, 77.

Kirku, Pavel Ivanovich. [1982a] Precompact topologizability of modules. Matematicheskie Issled. No. 65 (1982), 88-90, 154. [1982b] Locally compact topologizability of modules over finite rings. Izvestiĭā Akademiĭ nauk

Moldavskoĭ SSR: Seriiā fiziko-tekhnicheskikh i Matematicheskikh nauk no. 1, 3-7, 1983. [1984] Locally compact topologizable modules over Dedekind rings. Matematicheskie Issled. No. 76 (1984), 47-72.

Klingen, Jutta. [1977] Über eine topologische Verallgemeinerung von Krullringen. Mathematische Nachrichten 91 (1979), 59-76.

Klyushin, A. V. [1977a] Openly Noetherian rings. Vestnik Moskovskogo Universiteta. Seriiā I. Matematika, Mekhanika 1979, no. 5, 48-51, 88. [1977b] The connected component of a locally compact open Artinian ring. Matematicheskie Zametki 26 (1979), 263-267, 318. [1981] Compact monothetic rings. Ibid. 33 (1983) 893-899.

Koh, Kwangil. [1967a] On the set of zero divisors of a topological ring. Canadian Mathematical Bulletin 10 (1967), 595-596. [1967b] On compact prime rings and their rings of quotients. Ibid. 11 (1968), 563-568. [1986] (with J. Cohen) On the group actions of the units on non-units in a compact ring. Journal of Pure and Applied Algebra 51 (1988), 231-239. [1987] (with J. Cohen) The group of units in a compact ring. Ibid. 54 (1988), 167-179. [1988a] (with J. Cohen) Half-transitive group actions in a compact ring, Ibid. 60 (1989), 139-153. [1988b] (with J. Cohen) Involutions in a compact ring. Ibid. 59 (1989), 151-168. [1988c] (with J. Cohen) A characterization of the p-adic integers. Communications in Algebra 17 (1989), 631-636. [1989] (with J. Cohen) The subring generated by the units in a compact ring. Ibid. 18 (1990), 1617-1620. [1990] (with J. Cohen) The subgroup generated by the involutions in a compact ring. Ibid. 19 (1991), 2923-2954. [1991] (with J. Cohen) The structure of compact rings. Journal of Pure and Applied Algebra 17 (1992), 117-129.

Kohn, Samuel. [1970] (with D. Newman) Multiplication from other operations. Proceedings of the American Mathematical Soc. 27 (1971), 244-246.

Kolman, Bernard. [1967] On a theorem in complete $\mathfrak{A}$-adic rings. Proceedings of the American Mathematical Soc. 19 (1968), 681-684.

Kovács, István. [1954] Infinite rings without infinite proper subrings. Publicationes Mathematicae 4 (1955), 104-108.

Kowalsky, Hans-Joachim. [1953] Beiträge zur topologischen algebra. Mathematische Nachrichten 11 (1954), 143-185.

Krull, Wolfgang. [1928] Primidealketten in allgemeinen Ringbereichen. Heidelberger Akademie des Wissenschaften, Math.-Natur. Kl., Sitzungberichte, 1928, No. 7. [1938] Dimensionstheorie in Stellenringen. Journal für die Reine und Angewandte Mathematik 179 (1938), 204-226. [1940a]

Über separable, abgeschlossene Abelsche Gruppen. *Ibid.* 182 (1940), 235-241. [1940b] Über separable, insbesondere kompakte separable Gruppen. *ibid.* 184 (1942), 19-48.

Kurata, Yoshiki. [1991] (with T. Shigeyuki) Linearly compact dual bimodules. Mathematical Journal of Okayama University 33 (1991), 149-154.

Kurke, Herbert. [1967] Topologische Methoden in der Theorie der kommutativen Ringe. Mathematische Nachrichten 39 (1969), 33-85.

Lafon, Jean-Pierre. [1955] Anneaux noethériens. Séminaire H. Cartan et C. Chevalley 1955/56. Géometrie algébrique. Exp. 2. Secrétariat, mathématique, Paris, 1956. [1961] Complétions d'un anneau local dans certaines topologies. Comptes Rendus Hebdomadaires des Séances de l'Acad. des Sci. 254 (1962), 53-55. [1962] Anneaux henséliens. Bulletin de la Société Mathématique de France 91 (1963), 77-107.

Lambek, Joachim. [1971] Localization and completion. Journal of Pure and Applied Algebra 2 (1972), 343-370. [1972] Noncommutative localization. Bulletin of the American Mathematical Soc. 79 (1973), 857-872. [1975] (with G. Michler) On products of full linear rings. Publicationes mathematicae 24 (1977), 123-127.

Langmann, Klaus. [1990] m-adische Topologie, Fréchettopologie und der Cartansche Abgeschlossenheitssatz. Archiv der Matematik 57 (1991), 456-459.

Larotonda, Angel. [1970] (with J. Edelman) Quasi-norms in rings. Revista de la Union Matematica Argentina 25 (1970/71), 357-361.

Le Donne, Attilio. [1976] Sull'algebra degli endomorfismi di una classe di moduli ridotti e senza torsione sui domini di Dedekind. Rendiconti del Seminario Matematico della Universita di Padova 56 (1976), 215-226.

Lefschetz, Solomon. [1942] *Algebraic Topology.* American Mathematical Society Colloquium Publ. vol. 27. American Math. Soc., New York, 1942.

Lemaire, Claude. [1966] Sur le complété d'un anneau-module topologique. Académie Royale des Sciences, des Lettres et des Beaux-Arts de Belgique (5) 52 (1966), 390-394.

Leptin, Horst. [1954a] Über eine Klasse linear kompakter abelscher Gruppen. I. Abhandlungen aus dem Mathematischen Seminar der Universität Hamburg 19 (1954), 23-40 (1954/55). [1954b] Über eine Klasse linear kompakter abelscher Gruppen. II. *Ibid.* 19 (1955), 221-243. [1954c] Abelsche Gruppen mit kompakten Charaktergruppen und Dualitätstheorie gewisser linear topologischer abelscher Gruppen. *Ibid.* 19 (1955), 244-263.

[1954d] Zur Dualitätstheorie projektiver Limites abelscher Gruppen. *Ibid.* 19 (1955), 264-268. [1954e] Linear kompakte Moduln und Ringe. Mathematische Zeitschrift 62 (1955), 241-267. [1956] Linear kompakte Moduln und Ringe. II. *Ibid.* 66 (1957), 289-327.

Le Tien Tam. [1948] (with Nguyen Van Khuê) On topological properties of Noetherian topological modules. Acta Mathematica Vietnamica 9 (1984), 201-212.

Levin, Martin D. [1971] Locally compact modules. Journal of Algebra 24 (1973), 25-55.

Liepold, Frank. [1990] Zur Vervollständigung bewerteter Schiefkörper. Resultate der Mathematik 19 (1991), 122-142.

Lipkina, Zoĩa Semenovna. [1964a] On the pseudonormability of topological rings. Sibirskiĭ Matematicheskiĭ Zhurnal 6 (1965), 1046-1052. [1964b] Locally bicompact rings without zero divisors. Doklady Akademiĭ Nauk SSSR 161 (1965), 523-525. [1966a] Locally bicompact rings without zero divisors. Izvestiĩa Akademiĭ Nauk SSSR. Seriĩa Mat. 31 (1967), 1239-1262. [1966b] Bicompact rings with finite images. Sibirskiĭ Matematichcskiĭ Zhurnal 9 (1968), 720-722. [1973] Structure of compact rings. *Ibid.* 17 (1973), 1346-1348, 1368. Letter to the editor, *ibid.* 20 (1979). 1162-1163.

Livenson, E. [1936] An example of a non-closed connected subgroup of the two-dimensional vector space. Annals of Mathematics (2) 38 (1937), 920-922.

Lluis Riera, Emilio. [1954] On the open ideals in Zariski rings. Boletin de la Sociedad Matematica Mexicana 11 (1954), 33-34.

Lorenzini, Anna. [1979] Compact and linearly topologized modules and duality. Annali dell'Universita di Ferrara. Sezione 7, Scienze Matematiche 25 (1979), 169-182.

Lorimer, Joseph Wilson. [1982] (with Y. Park) On linearly compact commutative regular rings. Comptes Rendus Mathématiques de l'Académie des Sciences 4 (1982), 159-164. [1990] The classification of compact right chain rings. Forum Mathematicum 4 (1992), 335-347.

Lu, Chin Pi. [1969] Local rings with noetherian filtrations. Pacific Journal of Math. 36 (1971), 209-218. [1971] A generalization of Mori's theorem. Proceedings of the American Mathematical Soc. 31 (1972), 373-375. [1975] Factorial modules. Rocky Mountain Journal of Math. 7 (1977), 125-139. [1978] Purity of linearly topological rings in overrings. Journal of Algebra 59

(1979), 290-301. [1979] Quasicomplete modules. Indiana University Mathematics Journal 29 (1980), 277-286. [1980] Quasi-complete rings. Tamkang Journal of Mathematics 12 (1981), 245-256.

Lucas, Dean. [1977] A multiplication in N-space. Proceedings of the American Mathematical Soc. 74 (1979), 1-8.

Lucke, James Bennett. [1968] Commutativity in locally compact rings. Pacific Journal of Math. 32 (1970), 187-196. [1972] (with S. Warner) Structure theorems for certain topological rings. Transactions of the American Mathematical Soc. 186 (1973), 293-298.

Luedeman, John Keith. [1969a] On the embedding of topological domains into quotient fields. Manuscripta Mathematica 3 (1970), 213-226. [1969b] On the embedding of compact domains. Mathematische Zeitschrift 115 (1970), 113-116. [1978] On the embedding of topological rings into rings of quotients. Bulletin de la Société Mathématique de Belgique. Sér. A 30 (1978), 91-98.

Macdonald, Ian Grant. [1962] Duality over complete local rings. Topology 1 (1962), 213-235.

MacLane, Saunders. [1937] The uniqueness of the power series representation of certain fields with valuations. Annals of Mathematics (2) 39 (1938), 370-382. [1938a] Steinitz field towers for modular fields. Transactions of the American Mathematical Soc. 46 (1939), 23-45. [1938b] Subfields and automorphism groups of p-adic fields. Annals of Mathematics (2) 40 (1939), 423-442. [1941] (with F. Schmidt) The generation of inseparable fields. Proceedings of the National Academy of Science U.S.A. 27 (1941), 583-587.

Mader, Adolph. [1980] (with R. Mines) Completions of linearly topologized vectorspaces. Journal of Algebra 74 (1982), 317-327. [1981] Duality and completions of linearly topologized modules. Mathematische Zeitschrift 179 (1982), 325-335. [1983] Completions via duality. *Abelian Group Theory* (Honolulu, Hawaii 1983), 562-568. Lecture Notes in Math. 1006, Springer, Berlin-New York, 1983.

Makharadze, Daredzhan Mikhailovna. [1960] Locally nilpotent radical in locally bounded topological rings. Trudy Akademiiā Nauk Gruzinskoĭ SSR, Tiflis, Vychislitel'nyi Tsentr. 2 (1961), 21-28.

Makharadze, L. M. [1956a] Locally nilpotent ideals in topological rings. Matematicheskiĭ Sbornik (N.S.) 41 (83) (1957), 395-414. [1956b] Topological nilpotent rings with minimal condition. Uspekhi Matematicheskikh Nauk (N.S.) 12 (1957), no. 4 (76), 181-186. [1964] On the nilpotency of topological rings. Akademiĩa Nauk Gruzinskoĭ SSR, Tiflis. Soobshcheniĩa 40 (1965), 257-261. [1970] A generalized locally nilpotent radical in topological rings. Sakharthvelos SSSR Mecnierebatha Akademiĩa Moambe (Tbilisi) (former Akademiĩa Nauk Gruzinskoĭ, Soobshcheniĩa, Tiflis) 59 (1970), 25-28. [1973] The generalized radical in topological rings. Trudy Akademiĩa Nauk Gruzinskoĭ, Tiflis. Vychislitel'nyi tsentr 12 (1973), no. 1, 132-139.

Manchanda, Pammy. [1990] (with S. Singh) Locally bounded topologies on $F(X)$. Indian Journal of Pure and Applied Math. 22 (1991), 313-321. [1991] (with S. Singh) Seminormability of certain ring topologies on Noetherian Krull domains. Indian Journal of Math. 33 (1991), no. 1, 19-23.

Marin, Elena Ivanovna. [1983] (with V. Arnautov and A. Mikhalëv) Necessary conditions for the extension of a group and a field topology to their group algebra. Vestnik Moskovskosgo Univ. Ser. I: Matematika Mekhanika 1984, no. 6, 58-61, 112.

Márki, László. [1986] (with R. Mlitz and R. Wiegandt) A note on radical and semisimple classes of topological rings. Acta Scientiarum Mathematicarum 51 (1987), 145-151. [1990] (with G. Abrams and Phạm Ngọc Ánh) A topological approach to Morita equivalence for rings with local units. Rocky Mountain Journal of Math. 22 (1992), 405-416.

Marot, Jean. [1979] Topologies d'anneau principal. Comptes Rendus Hebdomadaires des Séances de l'Acad. des Sci. Sér. A-B 288 (1979), A713-A715.

Marquina, Antonio. [1984] (with R. Fuster) Geometric series in incomplete normed algebras. American Mathematical Monthly 91 (1984), 49-51.

Marubayashi, Hidetoshi. [1984] Linearly compact modules over HNP rings. Osaka Journal of Math. 22 (1985), 835-844.

Mascart, Henri. [1964a] Sur quelques propriétés élémentaires des modules topologiques. Comptes Rendus Hebdomadaires des Séances de l'Acad. des Sci. 258 (1964), 1683-1685. [1964b] Sur l'invariance par homothétie du filtre des voisinages de l'origine dans un module topologique. Ibid. 258 (1964), 3148-3150. [1965] Sur l'utilisation de la polarité dans la théorie des modules topologiques. Ibid. Sér. A-B 262 (1966), A16-A19. [1967] Les modules topologiques: Resultats récents. Revista de la Academia de Ciencias Exactas, Fisico-Quimicas Naturales de Zaragoza (2) 22 (1967), 189-203. [1972] La convergence locale dans un module topologique. Revue Roumaine de Mathématiques Pures et Appliquées 17 (1972), 1381-1384.

Massagli, Robert A. [1972] On a new radical in a topological ring. Pacific Journal of Math. 45 (1973), 577-584.

Massaza, Carla. [1972] Sugli anelli che soddisfano alla condizione topologica di Artin-Rees. Rendiconti del Seminario Matematico della Universita di Padova 49 (1973), 205-215. [1973] Anelli topologicamente bezoutiani e condizione topologica di Artin-Rees. Annali dell'Universita di Ferrara. Sezione 7, Scienze Matematiche 19 (1974), 37-50.

Mathiak, Karl. [1989] Completions of valued skew fields. Journal of Algebra 150 (1992), 257-270.

Matlis, Eben. [1958] Injective modules over Noetherian rings. Pacific Journal of Math. 8 (1958), 511-528.

Maurer, I. [1956] Sur la topologisation des anneaux. Studii şi Cercetări de Matematică 8 (1957), 177-180.

Mazan, Michel. [1976a] Dual topologique d'un module quelconque. Bollettino della Unione Matematica Italiana Ser. A. Fascicoli Ordinari 13 (1976), 586-591. [1976b] Espaces de suites sur un anneau topologique. Rendiconti del Circolo Matematico di Palermo 25 (1976), 145-157. [1981] Étude de la dualité dans les modules. Rendiconti del Seminario Matematico (Turin, Italy) 40 (1982), no. 2, 63-87 (1983). [1982] Dualité dans les modules topologiques. Revista Matematica Hispano-Americana (4) 42 (1982), 56-62.

Mazur, Stanisław. [1948] (with W. Orlicz) Sur les espaces métriques linéaires. I. Studia Mathematica 10 (1948), 184-208.

Meijer, A. R. [1982] (with P. Smith) The complement of Gabriel topologies. Journal of Pure and Applied Algebra 31 (1984), 119-137.

Menini, Claudia. [1976] Topologie di caratteri per moduli su anelli Noetheriani. Annali dell'Universita di Ferrara. Sezione 7, Scienze Matematiche 23 (1977), 45-58. [1977] On E-compact modules over SISI rings. Ibid. 23 (1977), 195-207 (1978). [1977] (with A. Orsatti) Duality over a quasi-injective module and commutative $\mathcal{F}$-reflexive rings. Symposia Mathematica, Vol. 23 (Conference on abelian groups and their relationship to the theory of modules, INDAM, Rome, 1977), pp. 145-179. Academic Press, London, 1979. [1980] Linearly compact rings and strongly quasi-injective modules. Rendiconti del Seminario Matematico della Universita di Padova 65 (1981), 251-262; Errata, ibid. 69 (1983), 305-306. [1980] (with A. Orsatti) (Good dualities and strongly quasi-injective modules. Annali di Matematica Pura ed Applicata (4) 127 (1981), 187-230. [1981] (with A. Orsatti) Dualities between categories of topological modules. Communications in Algebra 11 (1983), 21-66. [1983a] Linearly compact rings and selfcogenerators. Rendiconti del Seminario Matematico della Universita di

Padova 72 (1984), 99-116. [1983b] A characterization of linearly compact modules. Mathematische Annalen 271 (1985), 1-11. [1983] (with A. Orsatti) Topologically left Artinian rings. Journal of Algebra 93 (1985), 475-508.

Michler, Gerhard O. [1971] (with F. Eckstein) On compact open subrings of semi-simple locally compact rings. Archiv der Mathematik 23 (1972), 10-18. [1975] (with J. Lambek) On products of full linear rings. Publicationes Mathematicae 24 (1977), 123-127.

Mikhalëv, A. V. [1961] An isomorphism of rings of continuous endomorphisms. Sibirskiĭ Matematicheskiĭ Zhurnal 4 (1963), 177-186. [1981a] (with V. Arnautov) An example of a commutative completely regular topological semigroup without embeddings in multiplicative semigroups of topological rings. Vestnik Moskovskogo Univ. Ser. I, Matematika, Mekhanika 36, no. 6 (1981), 68-70, 117. [1981b] (with V. Arnautov) Continuation of topologies onto semigroup rings. Sixteenth All-Union Algebra Conference, part 1. Leningrad, 1981. [1981] (with V. Arnautov and M. I. Vodinchar) *Introduction to the Theory of Topological Rings and Modules*. Kishinev, 1981. [1982] (with V. Arnautov) Sufficient conditions for extension of topologies of a group and a ring to their group ring. Vestnik Moskovskogo Univ. Ser. I, Matematika, Mekhanika 38, no. 5 (1983), 25-33. [1983] (with E. I. Marin and V. Arnautov) Necessary conditions for the extension of a group and a field topology to their group algebra. Vestnik Moskovskosgo Univ. Ser. I: Matematika Mekhanika 1984, no. 6, 58-61, 112. [1985a] (with V. Arnautov) On the extension of a group and a ring topology to their group ring. Uspekhi Matematicheskikh Nauk 40 (1985), No. 4 (244), 135-136. [1985b] (with V. Arnautov) Extension of a group and a ring topology to their group ring. Matematicheskie Issled. No. 85 (1985), 8-20, 152. [1987] (with V. Arnautov) Extension of topologies of a ring and a discrete monoid to a semigroup ring. Matematicheskie Issled. No. 111 (1989), 9-23, 140. [1988] (with V. Arnautov, S. Glavatskiĭ, and M. Vodinchar) *Constructions of Topological Rings and Modules*. "Shtiintsa", Kishinev, 1988. [1989] (with V. Arnautov, K. Beidar, and S. Glavatskiĭ) Intersection property in the radical theory of topological algebras. *Proceedings of the International Conference on Algebra Dedicated to the Memory of A. I. Mal'cev* Part 2 (Novosibirsk, 1989), 227-235, *Contemporary Mathematics 131*, pp. 205-225, American Mathematical Soc., Providence, RI, 1992.

Miller, Robert W. [1972] (with D. Turnidge) Some examples from infinite matrix rings. Proceedings of the American Mathematical Soc. 38 (1973), 65-67.

Mines, Ray. [1980] (with A. Mader) Completions of linearly topologized vectorspaces. Journal of Algebra 74 (1982), 317-327.

Mlitz, Rainer. [1986] (with L. Márki and R. Wiegandt) A note on radical and semisimple classes of topological rings. Acta Scientiarum Mathematicarum 51 (1987), 145-151.

Močkoř, Jiří. [1977] The completion of Prüfer domains. Proceedings of the American Mathematical Soc. 67 (1977), 1-10.

Mohamed, Aamri. [1989] (with M. El Miloudi) Les S-anneaux et propriétés de tonnelage dans les modules topologiques. Atti della Accademia Peloritana dei Pericolanti Classe Sci. Fis. Mat. Natur. 67 (1989), 275-292.

Molinelli, Silvia. [1974] Sui prodotti di anelli e certi completamenti non M-adici. Atti della Accademia delle Scienze de Torino 108 (1974), 745-756.

Mori, Yoshiro. [1955] On the integral closure of an integral domain. II. Kyoto Kyokiu Daigaku kiyo B., No. 7 (1955), 19-30.

Müller, Bruno J. [1968] On Morita duality. Canadian Journal of Math. 21 (1969), 1338-1347. [1969] Linear compactness and Morita duality. Journal of Algebra 16 (1970), 60-66. [1970] Duality theory over linearly topologized modules. Mathematische Zeitschrift 119 (1971), 63-74. [1971] All duality theories for linearly topologized modules come from Morita dualities. *Rings, Modules and Radicals*. Bolyai János Matematikai Társulat. Colloquia Mathematica, No. 6, pp. 357-360. North-Holland, Amsterdam, 1973. [1984] Morita duality—a survey. *Abelian Groups and Modules (Udine, 1984)*, 395-414. CISM Courses and Lectures, 287. Springer, Vienna, 1984.

Müller, Vladimír. [1990] On topologizable algebras. Studia Mathematica 99 (1991), 149-153.

Mulvey, Christopher J. [1978] On a condition for the representability of a topological ring. Communications in Algebra 7 (1959), 995-998.

Murato, Kentaro. [1986] A ring topology based on submodules over an Asano order of a ring. Mathematical Journal of Okayama University 28 (1986), 61-64 (1987).

Mutylin, A. F. [1965] An example of a nontrivial topologization of the field of rational numbers. Complete locally bounded fields. Izvestiiā Akademiĭ Nauk SSSR Seriiā Matematicheskaiā 30 (1966), 873-890. Corrections, *ibid.* 32 (1968), 245-246. Notes added to English translation: American Mathematical Society Translations (2) 73 (1968), 159-179. [1967] Connected complete locally bounded fields. Complete not locally bounded fields. Matematicheskiĭ Sbornik 76 (118) (1968), 454-472. [1968] Completely simple commutative topological rings. Matematicheskie Zametki 5 (1969), 161-171.

Nagata, Masayoshi. [1949] On the structure of complete local rings. Nagoya Mathematical Journal 1 (1950), 63-70. Corrections, ibid. 5 (1953), 145-147. [1950a] On the theory of semi-local rings. Nihon Gakushiin, Proceedings 26 (1950), 131-140. [1950b] Note on the subdirect sums of rings. Nagoya Mathematical Journal 2 (1951), 49-53. [1951a] Some studies on semi-local rings. Ibid. 3 (1951), 23-30. [1951b] On the theory of Henselian rings. Ibid. 5 (1953), 45-57. [1953a] Some remarks on local rings. Ibid. 6 (1953), 53-58. [1953b] Some remarks on local rings. II. Kyoto Daigaku. Rigakubu. Memoirs Series A. Mathematics 28 (1954), 109-120. [1953] (with T. Nakayama and T. Tuzuku) On an existence lemma in valuation theory. Nagoya Mathematical Journal 6 (1953), 59-61. [1958] Note on coefficient fields of complete local rings. Ibid. 32 (1959/60), 91-92. [1962] Local Rings. Interscience Tracts in Pure and Applied Mathematics, No. 13. Interscience, New York-London, 1962.

Nakayama, Tadasi. [1953] (with M. Nagata and T. Tuzuku) On an existence lemma in valuation theory. Nagoya Mathematical Journal 6 (1953), 59-61.

Narici, Lawrence Robert. [1977] (with E. Beckenstein and C. Suffel) Topological Algebras. North-Holland Mathematics Studies, Vol. 24. Notas de Matemática, No. 60. North-Holland, Amsterdam-New York-Oxford, 1977.

Narita, Masao. [1955a] On the structure of complete local rings. Nihon Sūgakkai. Journal (Journal of the Mathematical Soc. of Japan) 7 (1955), 435-443. [1955b] On the structure of complete local rings. Proceedings of the International Symposium on Algebraic Number Theory, Tokyo & Nikko, 1955, pp. 251-253. Science Council of Japan, Tokyo, 1956.

Nasirov, S. N. [1975] (with T. Sarymsakov and Dzh. Hadzien) A description of the ideals of a certain class of rings. Doklady Akademiĭ Nauk SSSR 225 (1975), 1018-1019.

Năstăsecu, Constantin. [1968] (with N. Popescu) On the localization ring of a ring. Journal of Algebra 15 (1970), 41-56.

Nazarov, A. O. [1982] Open imbeddings of topological rings. Vestnik Moskovskogo Universiteta. Seriĭa I, Matematika, Mekhanika 1983, no. 1, 51-54, 101. [1986] Localization in the catgory of topological modules. Ibid. 1988, no. 3, 5-9, 113.

Nazarov, O. A. [1985] Localizations of compact rings with respect to a filter of open ideals. Algebra, Logic and Number Theory, pp. 65-68, Moskovskiĭ Gosudarstvennyĭ Universitet, Moscow, 1986.

von Neumann, Johan [John]. [1925] Zur Prüferschen Theorie der idealen Zahlen. Acta Litterarum ac Scientiarum Regiae Universitatis Hungaricae Francisco-Jospheniae. Section Scientiarum Mathematicarum 2 (1924/26), 193-227.

Newman, Donald J. [1970] (with S. Kohn) Multiplication from other operations. Proceedings of the American Math. Soc. 27 (1971), 244-246.

Ng, Shu-bun. [1972] (with S. Warner) Continuity of positive and multiplicative functionals. Duke Mathematical Journal 39 (1972), 281-284.

Nguyễn Van Khuê. [1984] (with Le Tien Tam) On topological properties of Noetherian topological modules. Acta Mathematica Vietnamica 9 (1984), 201-212.

Nguyễn Việt Dũng. [1987] On linearly compact rings. Archiv der Mathematik 51 (1988), 327-331.

Nicholson, W. Keith. [1984] (with B. Sarath) Rings with a largest linear topology. Communications in Algebra 13 (1985), 769-780.

Nishimura, Jun-ichi. [1987] Ideal-adic completions of Noetherian rings. Revue Romaine de Mathématique Pures et Appliquées 33 (1988), 369-373.

Novoselov, E. V. [1961] Integration on a bicompact ring and its applications to number theory. Izvestiiā Vysshikh Uchebnykh Zavedniĭ Matematika 1961 (22), 66-79.

Numakura, Katsumi. [1955a] Theory of compact rings. Mathematical Journal of Okayama University 5 (1955), 79-93. [1955b] Theory of compact rings II. Ibid. 5 (1956), 103-111. [1959] A note on Wedderburn decompositions of compact rings. Nihon Gakushiin. Proceedings 35 (1959), 313-315. [1961] Theory of compact rings. III. Compact dual rings. Duke Mathematical Journal 29 (1962), 107-123. [1981] Notes on compact rings with open radical. Czechoslovak Mathematical Journal 33 (108) (1983), 101-106.

Oberst, Ulrich. [1969] Duality theory for Grothendieck categories. Bulletin of the American Mathematical Soc. 75 (1969), 1401-1407. [1971] (with H.-J. Schneider) Kommutative, F-linear kompakte Ringe. Journal of Algebra 25 (1973), 316-363.

Øfsti, Audun. [1965] On the structure of a certain class of locally compact rings. Mathematica Scandinavia 18 (1966), 134-142.

Olmsted, John Meigs Hubbell. [1950] (with B. Gelbaum and G. Kalisch) On the embedding of topological groups and integral domains. Proceedings of the American Mathematical Soc. 2 (1951), 807-821.

O'Neil, John D. [1990] Slender modules over various rings. Indian Journal of Pure and Applied Mathematics 22 (1991), 287-273.

Onodera, Takeshi. [1971] Linearly compact modules and cogenerators. Hokkaido Daigaku. Rigakubu. Journal of the Faculty of Science, Hokkaido University. Series I 22 (1972), 116-125. [1973] Linearly compact modules and cogenerators. II. Hokkaido Mathematical Journal 2 (1973), 243-251. [1975] On balanced projectives and injectives over linearly compact rings. Ibid. 5 (1976), 249-256. [1978] A note on linearly compact modules. Ibid. 8 (1979), 121-125.

Orlicz, Władysław. [1948] (with S. Mazur) Sur les espaces métriques linéaires. I. Studia Mathematica 10 (1948), 184-208.

Orsatti, Adalberto. [1976] Dualita per alcune classi di module E-compatti. Annali di Matematica Pura ed Applicata (4) 113 (1977), 211-235. [1977] (with C. Menini) Duality over a quasi-injective module and commutative $\mathcal{F}$-reflexive rings. Symposia Mathematica, Vol. 23 (Conference on abelian groups and their relationship to the theory of modules, INDAM, Rome, 1977), pp. 145-179. Academic Press, London, 1979. [1980] (with V. Roselli) A characterization of discrete linearly compact rings by means of a duality. Rendiconti del Seminario Matematico della Universita di Padova 64 (1981), 219-234. [1980] (with C. Menini) Good dualities and strongly quasi-injective modules. Annali di Matematica Pura ed Applicata (4) 127 (1981), 187-230. [1981] (with C. Menini) Dualities between categories of topological modules. Communications in Algebra 11 (1983), 21-66. [1982] Linearly compact rings and the theorems of Leptin. Bollettino della Unione Matematica Italiana (6) 1-A (1982), 331-357. [1983] (with C. Menini) Topologically left Artinian rings. Journal of Algebra 93 (1985), 475-508. [1984a] (with D. Dikranjan) On the structure of linearly compact rings and their dualities. Rendiconti della Accademia nazionale delle scienze detta dei XL. Parte I, Memorie di matematica (5) 8 (1984), no. 1, 143-184. [1984b] (with D. Dikranjan) Linearly compact rings. Supplemento ai Rendiconti del Circolo Matematico di Palermo (2), No. 4 (1984), 59-74. [1984c] (with D. Dikranjan) On the structure of linearly compact rings and their dualities. Abelian Groups and Modules (Udine, 1984), 415-439. CISM Courses and Lectures, 287. Springer, Vienna-New York, 1984. [1988] (with D. Dikranjan) On an unpublished manuscript of Ivan Prodanov concerning locally compact modules and their dualities. Communications in Algebra 17 (1989), 2739-2771. [1989] (with G. Baccella) On generalized Morita bimodules and their dualities. Rendiconti della Accademia Nazionale delle Scienze detta dei XL. Memorie di Matematica (5) 13 (1989), no. 1, 323-340. [1990] (with D. Dikranjan and E. Gregorio) Kasch bimodules. Rendiconti del Seminario

Matematico della Universita di Padova 85 (1991), 147-160.

Orzech, Morris. [1989] (with M. Beattie) Prime ideals and finiteness conditions for Gabriel topologies on commutative rings. Rocky Mountain Journal of Mathematics 22 (1992), 423-439.

Osofsky, Barbara L. [1965] A generalization of quasi-Frobenius rings. Journal of Algebra 4 (1966), 373-387.

Ostrowski, Alexander. [1915] Über einige Lösungen der Funktionalgleichung $\phi(x) \cdot \phi(y) = \phi(xy)$. Acta Mathematica 41 (1918), 271-284. [1932] Untersuchungen zur arithmetischen Theorie der Körper. Mathematische Zeitschrift 39 (1935), 269-404.

Otobe, Yosikazu. [1944a] On quasi-evaluations of compact rings. Nihon Gakushiin. Proceedings 20 (1944), 278-282. [1944b] Note on locally compact simple rings. Ibid. 20 (1944), 283.

Park, Young Lim. [1982] (with J. Lorimer) On linearly compact commutative regular rings. Comptes Rendus Mathématiques de l'Académie des Sciences 4 (1982), 159-164. [1986] Linearly compact left duo rings and their structure spaces. Kyungpook Mathematical Journal 26 (1986), 37-41.

Peters, Fritz Eduard. [1970] Natürliche Pontrjagin-Dualität topologischer Moduln. BMBW-GMD-26. Gesellschaft für Mathematik und Datenverarbeitung, Bonn, 1970.

Petrich, Mario. [1974] *Rings and Semigroups*. Lecture Notes in Mathematics, Vol. 380. Springer-Verlag, Berlin-New York, 1974.

Petrucci, Silvana. [1972] Sui lemmi di preparazione per gli anelli di serie ristrette. Rendiconti del Seminario Matematico della Universita di Padova 49 (1973), 125-135.

Phạm Ngọc Ánh. [1976a] On semisimple classes of topological rings. Annales Universitatis Scientiarum Budapestimensis de Rolando Eötvös Nominatae. Sectio Mathematica 20 (1977), 59-70. [1976b] Die additive Struktur der in engeren Sinne linear kompakten Ringe. Studia Scientiarum Mathematicarum Hungarica 11 (1976), 205-210 (1978). [1976c] Über die Struktur linear kompakter Ringe. Acta Mathematica Academiae Scientiarum Hungaricae 31 (1978), 61-73. [1977a] On the theory of linearly compact rings. I. Beiträge zur Algebra und Geometrie No. 9 (1980), 13-27. [1977b] Über die Struktur linear kompakter Ringe. II. Acta Mathematica Academiae Scientiarum Hungaricae 34 (1979), 245-251 (1980). [1977] (with R. Wiegandt) Linearly compact semisimple rings and regular modules. Mathematica Japonicae 23 (1978/9), 335-338. [1980] A note on linearly compact rings. Acta Mathematica Academiae Scientiarum Hungaricae 39 (1982), 55-58. [1981a] Duality over Noetherian rings with a Morita duality. Journal of Algebra 75 (1982), 275-285. [1981b] Duality of modules over topological rings. Ibid. 75 (1982), 395-425. [1981c] On a problem of B. J. Müller. Archiv der Mathematik 39 (1982), 303-305. [1982] On Litoff's theorem. Studia Scientiarum Mathematicarum Hungarica 16 (1981), 255-259. [1984a] A representation of locally Artinian categories. Journal of Pure and Applied Algebra 36 (1985), 221-224. [1984b] Direct products of linearly compact rings. Rendiconti del Seminario Matematico della Universita di Padova 76 (1986), 45-58. [1989] Morita duality for commutative rings. Communications in Algebra 18 (1990), 1781-1788. [1990] (with G. Abrams and L. Márki) A topological approach to Morita equivalence for rings with local units. Rocky Mountain Journal of Math. 22 (1992), 405-416.

Pietrkowski, Stephan. [1930] Theorie der unendlichen Abelschen Gruppen. Mathematische Annalen 104 (1931), 535-569.

Podewski, Klaus-Peter. [1972a] The number of field topologies on countable fields. Proceedings of the American Mathematical Soc. 40 (1973), 33-38. [1972b] Transcendental extensions of field topologies on countable fields. Ibid. 39 (1973), 39-41. [1975] Topologisierung algebraischer Strukturen. Revue Roumaine de Mathématiques Pures et Appliquées 22 (1977), 1283-1290.

Poneleit, Volker. [1976] Lineare Kompaktheit und die Zerlegung endlich erzeugter Moduln bei einreihigen Duoringen. Mitteilungen aus dem Mathematischen Seminar Giessen 121 (1976), 85-92.

Pontriâgin, Lev Semenovich. [1931] Über stetige algebraische Körper. Annals of Mathematics (2) 33 (1932), 163-174. [1934] Theory of topological commutative groups. Ibid. 35 (1934), 361-388. [1939] *Topological Groups*. Princeton University Press, Princeton, 1939. [1954] *Topological Groups*, 2nd edition. Gosurdarstv. Inzdat. Tekhn.-Theor. Lit., Moscow, 1954.

Pop, Horia Călin. [1983] On the structure of noncomutative complete local rings. Revue Roumaine de Mathématiques Pures et Appliquées 29 (1984), 495-498.

Popescu, Nicolae. [1968] (with C. Năstăsecu) On the localization ring of a ring. Journal of Algebra 15 (1970), 41-56. [1973] *Abelian Categories with Applications to Rings and Modules*. London Mathematical Society Monographs, No. 3. Academic Press, London-New York, 1973.

Probert, G. A. [1968] Local rings whose maximal ideal is principal as a right ideal. Proceedings of the London Mathematical Society (3) 19 (1969), 403-420.

Prodanov, Ivan. [1973] Minimal compact representations of algebras. Godishnik na Sofiiskiĩa Universitet, Fakultet Po Matematika I Mekhanika 67 (1972/73), 507-542 (1976). [1976] Precompact minimal topologies on some torsion free modules. Ibid. 69 (1974/75), 157-163.

Protasov, I. V. [1991] (with E. G. Zelenyuk and O. M. Khromulyak) Topologies on countable groups and rings. Doklady Akademiĩa Nauk U-kraïns'koï RSR 1991, no. 8, 8-11, 1992.

Prüfer, Heinz. [1923a] Theorie der Abelschen Gruppen. I. Grundeigenschaften. Mathematische Zeitschrift 20 (1924), 165-187. [1923b] Theorie der Abelschen Gruppen. II. Ideale Gruppen. Ibid. 22 (1925), 222-249. [1924] Neue Begrundung der algebraischen Zahlentheorie. Mathematische Annalen 94 (1925), 198-243.

Rees, David. [1954a] Valuations associated with a local ring (I). Proceedings of the London Mathematical Society (3) 5 (1955), 107-128. [1954b] Valuations associated with ideals. Ibid. 6 (1956), 161-174. [1955a] Valuations asociated with ideals (II). Journal of the London Mathematical Society 31 (1956), 221-228. [1955b] Two classical theorems of ideal theory. Proceedings of the Cambridge Philosophical Society, Mathematical and Physical Sciences 52 (1956), 155-157. [1955c] Valuations associated with a local ring (II). Journal of the London Mathematical Society 31 (1956), 228-235. [1956] Filtrations as limits of valuations. Ibid. 32 (1957), 97-102.

Reisel, Robert Benedict. [1955] A generalization of the Wedderburn-Malcev theorem to infinite dimensional algebras. Proceedings of the American Mathematical Soc. 7 (1956), 493-499.

Rhodes, C. P. L. [1972] A note on primary decompositions of a pseudovaluation. Pacific Journal of Math. 47 (1973), 507-513.

Rickart, Charles Earl. [1950] (with N. Jacobson) Jordan homomorphisms of rings. Transactions of the American Mathematical Soc. 69 (1950), 479-502.

Rigo, Thomas. [1977] (with S. Warner) Topologies extending valuations. Canadian Journal of Math. 30 (1978), 164-169.

Rios Montes, Jose. [1982] Some functors related to completion of modules with respect to a Gabriel filter. Revista Matematica Hispano-Americana (4) 42 (1982).

Roba, P. [1986] Propriété d'approximation pour les éléments algébriques. Compositio Mathematica 63 (1987), 3-14.

Robson, J. C. [1970] (with W. Gwynne) Completions of noncommutati Dedekind prime rings. Journal of the London Mathematical Soc. 33 (1958 181-185.

Rodinò, Nicola. [1984] Locally compact modules over compact rings. Rendiconti delle sedute della Accademia Nazionale dei Lincei. Cl. Sci. Fis. Mat. e Nat. 77 (1984), 61-63.

Rodrigues Gonzalez, Nieves. [1982] (with J. Gomez Pardo) Quotient rings of coherent rings with respect to a Gabriel topology. *Proceedings of the Ninth Conference of Portugese and Spanish mathematicians*, 1 (Salamanca 1982), 119-122.

Roos, Jan-Erik. [1968a] Sur la structure des catégories abéliennes localement noethériennes. Comptes Rendus Hebdomadaires des Séances de l'Acad. des Sci. Sér. A-B 266 (1968), A701-A704. [1968b] Locally noetherian categories and generalized strictly linearly compact rings. Applications. *Category Theory, Homology Theory and their Applications*, II (Battelle Institute Conference, Seattle, Washington, 1968, Vol. Two), 197-277. Lecture Notes in Mathematics, Vol. 92. Springer, Berlin, 1969.

Roquette, Peter. [1958] Abspaltung des Radikals in vollständigen lokalen Ringen. Abhandlungen aus dem Mathematischen Seminar der Universität Hamburg 23 (1959), 75-113.

Roselli, Valter. [1980] (with A. Orsatti) A characterization of discrete linearly compact rings by means of a duality. Rendiconti del Seminario Matematico della Universita di Padova 64 (1981), 219-234.

Rossignol, Antoine. [1972] Plongement d'un anneau topologique dans un anneau topologique unitaire. Application aux modules topologiques. Comptes Rendus Hebdomadaires des Séances de l'Acad. des Sci. Sér. A-B 274 (1972), A543.

Rothman, Neal Jules. [1965] A note on compact rings. Mathematische Zeitschrift 91 (1966), 179-184.

Rotman, Joseph Jonah. [1959] A note on completion of modules. Proceedings of the American Mathematical Soc. 11 (1960), 356-360.

Rutsch, Martin. [1960] Coeffizientenringe lokaler Ringe. Annales Universitatis Saraviensis 9 (1960/61), 163-196.

Ryan, Merilyn. [1980] Locally linearly compact semisimple rings. Journal für die Reine und Angewandte Mathematik 320 (1980), 86-96.

Sah, Chih-han. [1965] (with O. Goldman) On a special class of locally compact rings. Journal of Algebra 4 (1966), 71-95. [1968] (with O. Goldman) Locally compact rings of special type. Ibid. 11 (1969), 363-454.

Sakuma, Motoyoshi. [1953] (with M. Yoshida) A note on semilocal rings. Hiroshima Daigaku. Journal of Science of the Hiroshima University. Ser. A 17 (1953), 181-184.

Salce, Luigi. [1985] (with L. Fuchs) *Modules over Valuation Domains*. Lecture Notes in Pure and Applied Mathematics, 97. Dekker, New York, 1985.

Salmon, Paolo. [1971] (with S. Greco) *Topics in m-adic Topologies*. Ergebnisse der Mathematik und ihrer Grenzgebiete, vol. 58. Springer-Verlag, New York-Berlin, 1971.

Samuel, Pierre. [1953] *Algèbre locale*. Mémorial des Sciences Mathématiques, no. 123. Gauthier-Villars, Paris, 1953. [1954] Généralités sur l'algèbre locale. Séminaire Krasner 1953/54, Exp. 13. Faculté des Sciences, Paris, Paris, 1954. [1956] Progrès récents de l'algèbre locale. Colloque d'algèbre superieure tenu a Bruxelles du 19 au 22 décembre 1956, pp. 231-243. Centre Belge de Recherches Mathématiques. Gautier-Villars, Paris, 1957. [1958] (with O. Zariski) *Commutative Algebra*, vol. I. Van Nostrand, Princeton, N.J., 1958. [1959] Progrès récents d'algèbre locale. Notas de Matematica, No. 19. Instituto de Matematica Pura e Aplicada de Conselho Nacional de Pesquisas, Rio de Janeiro, 1959. [1960] (with O. Zariski) *Commutative Algebra*, vol. II. Van Nostrand, Princeton, N.J.-Toronto-London-New York, 1960.

Sandomierski, Francis Louis. [1971] Linearly compact modules and local Morita duality. Conference on Ring Theory, Park City, Utah, 1971. *Ring Theory*, pp. 333-346. Academic Press, New York, 1972.

Sankaran, N. [1990] (with V. Grover) Projective modules and approximation couples. Compositio Mathematica 74 (1990), 165-168.

Sarath, B. [1984] (with W. Nicholson) Rings with a largest linear topology. Communications in Algebra 13 (1985), 769-780.

Sarymsakov, Tashmuhamed Alievich. [1975] (with Dzh. Hadzien and S. Nasirov) A description of the ideals of a certain class of rings. Doklady Akademiĭ Nauk SSSR 225 (1975), 1018-1019.

Schiffels, Gerhard. [1983] (with M. Stenzel) Einbettung lokal-beschränkter topologischer Ringe in Quotientenringe. Resultate der Mathematik 7 (1984), 234-248.

Schmidt, Friederich Karl. [1932] (with H. Hasse) Die Struktur diskret bewerteter Körper. Journal für die Reine und Angewandte Mathematik 170 (1933), 4-63. [1941] (with S. MacLane) The generation of inseparable fields. Proceedings of the National Academy of Science U.S.A. 27 (1941), 583-587.

Schneider, Hans-Jürgen. [1971] (with U. Oberst) Kommutative, F-linear kompakte Ringe. Journal of Algebra 25 (1973), 316-363.

Schöneborn, Heinz. [1950a] Über Linearformenmoduln unendlichen Ranges. I. Primäre, kompakte Linearformenmoduln. Journal für die Reine und Angewandte Mathematik 189 (1951), 168-185. [1950b] Über Linearformenmoduln unendlichen Ranges. II. Nichtarchimedisch perfekt bewertete, operatorreduzierte Linearformenmoduln. Ibid. 189 (1952), 193-203. [1953a] Über gewisse Topologien in Abelschen Gruppen. I. Mathematische Zeitschrift 59 (1954), 455-473. [1953b] Über gewisse Topologien in Abelschen Gruppen. II. Ibid. 60 (1954), 17-30. [1956] Über den Zusammenhang zwischen Dualitäts- und Vollständigkeitseigenschaften bei gewisse topologischen Abelschen Gruppen. Ibid. 65 (1956), 429-441.

Schröder, Martin. [1987] Bewertungsringe von Schiefkörpern, Resultate und offene Probleme. Resultate der Mathematik 12 (1987), 191-206.

Schrot, Mary Delores. [1967a] Pseudo-valuations sur les domaines de Dedekind. Séminaire Delange-Pisot-Poitou: 1966/67, Théorie des Nombres, Fasc. 2, Exp. 10. Paris, 1968. [1967b] Pseudo-valuations sur les corps globaux. Ibid. Exp. 14.

Segal, Daniel. [1978] Congruence topologies in commutative rings. Bulletin of the London Mathematical Society 11 (1979), 186-190.

Seydi, Hamet. [1970] Sur la théorie des anneaux japonais. Comptes Rendus des Séances Hebdomadaires de l'Académie des Sci. Sér. A-B 271 (1970), A73-A75.

Shakenko, N. I. [1981] Topological rings of continuous real-valued functions. Uspekhi Matematicheskikh Nauk 37 (1982), no. 5 (227), 207-208.

Sharpe, David William. [1972] (with P. Vámos) *Injective Modules*. Cambridge Tracts in Mathematics and Mathematical Physics, No. 62. London-New York 1972.

Shell, Niel. [1977] Maximal and minimal ring topologies. Proceedings of the American Mathematical Soc. 68 (1978), 23-26. [1990] *Topological Fields and Near Valuations*. Monographs and Textbooks in Pure and Applied Mathematics, vol. 135. Dekker, New York, 1990.

Shigebuki, Tsuboi. [1991] (with T. Kurata) Linearly compact dual bimodules. Mathematical Journal of Okayama University 33 (1991), 149-154.

Simon, Anne-Marie. [1989] Some homological properties of complete modules. *Algebra and Geometry (Santiago de Compostela, 1989)*, 203-235. Álxebra, 54, Univ. Santiago de Compostela, Santiago de Compotela, 1990. [1990] Some homological properties of complete modules. Mathematical Proceedings of the Cambridge Philosophical Society 108 (1990), 231-246.

Singh, Surjit. [1973] Pseudo-valuation and pseudonorm. Rendiconti del Seminario Matematico della Universita di Padova 51 (1974), 83-96. [1985] Seminormability of certain ring topologies on Dedekind domains. Indian Journal of Pure and Applied Math. 16 (1985), 1465-1471. [1990] (with P. Manchanda) Locally bounded topologies on $F(X)$. Ibid. 22 (1991), 313-321. [1991] (with P. Manchanda) Seminormability of certain ring topologies on Noetherian Krull domains. Indian Journal of Math. 33 (1991), no. 1, 19-23.

Skorniakov, Lev Anatol'evich. [1962] Locally bicompact biregular rings. Matematischeskiĭ Sbornik (N.S.) 62 (104) (1963), 3-13. [1964] Einfache lokal bikompakte Ringe. Mathematische Zeitschrift 87 (1965), 241-251. [1965] Locally bicompact biregular rings. Matematicheskiĭ Sbornik (N.S.) 69 (111) (1966), 663. [1977] On the structure of locally bicompact biregular rings. Ibid. 104 (146) (1977), 652-664.

Smith, James Thomas. [1972] Haar integrals on topological rings. American Mathematical Monthly 79 (1972), 267-270.

Stenström, Bo. [1969] On the completion of modules in an additive topology. Journal of Algebra 16 (1970), 523-540. [1975] *Rings of Quotients. An Introduction to the Methods of Ring Theory.* Die Grundlehren der mathematischen Wissenschaften, Vol. 217. Springer-Verlag, Berlin-Heidelberg-New York, 1975.

Stenzel, Michael. [1978] Inversenbildung und Komplettierung bei topologischen Ringen und Körpern. Dissertation, Universität Bielefeld, 1978. [1983] (with G. Schiffels) Einbettung lokalbeschränkter topologischer Ringe in Quotientenringe. Resultate der Mathematik 7 (1984), 234-248.

Stoev, P. V. [1987] (with Tshan Binh Tram) Polynormed modules. Doklady Bolgarskoĭ Akademii Nauk 41, no. 5 (1988), 15-17.

Stöhr, Karl Otto. [1969a] Dualitäten in der Kategorie der lokal kompakten Moduln über einem Dedekindschen Ring. Journal für die Reine und Angewandte Mathematik 239/240 (1969), 239-255. [1969b] Funktoren in der Kategorie der lokal kompakten Moduln. Ibid. 246 (1971), 180-188.

Stout, Lawrence Neff. [1976] A topological structure on the structure sheaf of a topological ring. Communications in Algebra 5 (1977), 695-705.

Stoyanov, Luchezar. [1983] Dualities over compact commutative rings. Rendiconti della Accademia nazionale delle scienze detta dei XL. Parte I, Memorie di matematica (5) 7 (1983), 155-176.

Subramanian, Hariharaier. [1968] Ideal neighbourhoods in a ring. Pacific Journal of Math. 24 (1968), 173-176.

Suffel, Charles. [1977] (with L. Narici and E. Beckenstein) *Topological algebras*. North-Holland Mathematics Studies, Vol. 24. Notas de Matemática, No. 60. North-Holland, Amsterdam-New York-Oxford, 1977.

Suzuki, Satoshi. [1970] Modules of high order differentials of topological rings. Journal of Mathematics of Kyoto University 10 (1970), 337-348.

Szász, Ferenc. [1962a] On topological algebras and rings I. Matematikai Lapok 13 (1962), 256-278. [1962b] On topological algebras and rings II. *Ibid.* 14 (1963), 74-87. [1966] Hinreichendene Bedingung für die Existenz eines Rechtseinelements in einem Ring. Publicationes Mathematicae 14 (1967), 151-152.

Szele, Tibor. [1955] (with L. Fuchs) On Artinian rings. Acta Scientiarum Mathematicarum 17 (1956), 30-40.

Tabaâ, Mohamed. [1980] Corps linéairement compacts. Comptes Rendus hebdomadaires des Séances de l'Acad. des Sci. Sér. A-B 290 (1980), A531-A532. [1989] Sur un question de compacité linéaire. Journal of Mathematics of Kyoto University 31 (1991), 807-812. [1993] Quelques remarques sur la compacité linéaire. Journal of Algebra.

Tanaka, Susumu. [1962] Torsion-free abelian groups over the ring of p-adic integers. Sugaku 14 (1962/63), 33-35.

Tate, John T. [1962] Rigid analytic spaces. Inventiones Mathematicae 12 (1971), 257-289.

Taussky, Olga. [1935] (with N. Jacobson) Locally compact rings. Proceedings of the National Academy of Science U.S.A. 21 (1935), 106-108. [1936a] Analytical methods in hypercomplex systems. Compositio Mathematica 3 (1936), 399-407. [1936b] Some problems of topological algebra. *International Congress of Mathematicians, Oslo, 1936. Comptes Rendus.* Vol. 2, pp. 31-32. Olso, 1937. [1936c] Zur topologischen Algebra. Ergebnisse eines mathematischen Kolloquiums 7 (1936), 60-61.

Teichmüller, Oswald. [1936a] p-Algebren. Deutsche Matematik 1 (1936), 362-388. [1936b] Über die struktur diskret bewerteter perfekter Körper. Akademie der Wissenschaften, Göttingen. Math.-Phy. Kl. Nachrichten 1 (1936), 151-161. [1936c] Diskret bewertete perfekte Körper mit unvollkommenen Restklassenkörper. Journal für die Reine und Angewandte Mathematik 176 (1936), 141-152.

Tewari, K. [1975] (with S. Verma) Compact rings and the natural topology. Bulletin of the Calcutta Mathematical Society 71 (1979), no. 2, 67-76.

Tonolo, Alberto. [1991] On the existence of a finest equivalent linear topology. Communications in Algebra 20 (1992), 437-455. [1992] (with D. Dikranjan) On the lattice of linear module topologies. Communications in Algebra 21 (1993), 275-298.

Törner, Günter. [1990] (with H. Brungs) I-compactness and prime ideals. Publicationes Mathematicae 40 (1992), 291-295.

Trinh Dăng Khôi. [1983] Strictly hereditary radicals in the class of all topological rings. Studia Scientiarum Mathematicarum Hungarica 20 (1985), 37-41.

Tsarelunga, B. I. [1988] Uniqueness of complete separable metrizable topologies on rings. Matematicheskie Issled. No. 118 (1990), 121-125, 141. [1990] Strengthening of topologies of compact rings. Buletinul Akademiei de Ştiinţie a R.S.S. Moldoveneşti 1990, no. 2, 49-52, 79.

Tshan Binh Tram. [1987] (with P. V. Stoev) Polynormed modules. Doklady Bolgarskoĭ Akademii Nauk 41, no. 5 (1988), 15-17.

Turnidge, Darrel R. [1972] (with R. Miller) Some examples from infinite matrix rings. Proceedings of the American Mathematical Soc. 38 (1973), 65-67.

Tuzuku, Tosiro. [1953] (with M. Nagata and T. Nakayama) On an existence lemma in valuation theory. Nagoya Mathematical Journal 6 (1953), 59-61.

Ursul [Ursu], Mikhail Ivanovich. [1973] (with V. Andrunakievich and V. Arnautov] Wedderburn decomposition of hereditarily linearly compact rings. Doklady Akademiĭ Nauk SSSR 211 (1973), 15-18. [1974] Product of hereditarily linearly compact rings. Matematicheskie Issled. 9 (1974), No. 4(34), 137-149, 177. [1974] (with V. Andrunakievich and V. Arnautov) Wedderburn decomposition of hereditarily linearly compact rings. Ibid. 9 (1974) No. 1(31), 3-16. [1975] Wedderburn decomposition and hereditarily linearly compact rings. Ibid. 10 (1975) No. 2(36), 238-246, 247. [1978a]

Wedderburn decomposition of monocompact rings. *Ibid.* No. 48 (1978), 134-145, 170-171. [1978b] Locally hereditarily linearly compact biregular rings without nilpotent elements. *Ibid.* No. 48 (1978), 146-160, 171. [1979] The relations between the concepts of compactness, linear compactness, and hereditary linear compactness in topological rings. *Ibid.* No. 53 (1979), 136-147, 226. [1979a] (with V. Arnautov) Imbedding of topological rings into connected ones. *Ibid.* No. 49 (1979), 11-15, 159. [1979b] (with V. Arnautov) Uniqueness of a linearly compact topology in rings. *Ibid.* No. 53 (1979), 6-14, 221. [1980] The product of hereditarily linear-compact rings. Uspekhi Matematicheskikh Nauk 35 (1980), no. 3 (213), 230-233. [1981] The relations between the concepts of compactness, linear compactness and hereditarily linear compactness in topological rings. II. Buletinul Akademieĭ de Shtiintsie a RSS Moldovenesht' 1982, no. 1, 54-55, 80. [1982] Nil-ideals in some topological rings. Matematicheskie Issled. No. 65 (1982), 140-151, 156. [1982] (with V. Arnautov) Quasicomponents of topological rings and modules. Buletinul Akademieĭ de Shtiintsie a RSS Moldovenesht' 1984, no. 1, 9-13. [1983a] Locally finite and locally projective nilpotent ideals of topological rings. Matematicheskiĭ Sbornik (N.S.) 125 (167) (1984), 291-305. [1983b] Compact nilrings. Matematicheskie Zametki 36 (1984), 839-845, 941. [1984] Retracts of locally compact rings. Matematicheskie Issled. No. 76 (1984), 133-147. [1986] Uniqueness of compact topologies on rings. *Ibid.* No. 105 (1988), 142-152, 197. [1987] Connectivity in weakly Boolean topological rings. Buletinul Akademiei de Ştiinţie a R.S.S. Moldoveneşti 1989, no. 1, 17-21, 79. [1988] Strengthening of the topologies of countably compact rings. Matematicheskie Issled. No. 118 (1990), 126-136, 142. [1991] *Compact Rings and Their Generalizations.* Ştiinţa, Kishinev, 1991.

Utumi, Yuzo. [1962] (with C. Faith) On a new proof of Litoff's Theorem. Acta Mathematica Academiae Scientiarum Hungaricae 14 (1963), 369-371.

Valabrega, Paolo. [1971a] Anelli Henseliani topologici. I, II. Annali di Matematica Pura ed Applicata (4) 91 (1972), 283-303, 305-316. [1971b] Proprieta dell'henselizzazione di anelli topologici e di valutazione. *Symposia Mathematica*, vol. 8, pp. 393-403. Academic Press, London, 1972.

Vámos, Peter. [1972] (with D. Sharpe) *Injective Modules.* Cambridge Tracts in Mathematics and Mathematical Physics, No. 62. London-New York 1972. [1973] Classical rings. Journal of Algebra 34 (1975), 114-129. [1976] Rings with duality. Proceedings of the London Mathematical Society (3) 35 (1977), 275-289.

Vasilach, Serge. [1968] Ensembles bornés dans les modules topologiques. Comptes Rendus Hebdomadaires des Séances de l'Acad. des Sci. Sér. A-B 267 (1968), A681-A683. [1970] Sur les ensembles bornés dans les modules topologiques. Rendiconti della Accademia nazionale dei Lincei. Classe di Scienze Fisiche, Matematiche e Naturali (8) 49 (1970), 258-260 (1971).

Verma, S. [1975] (with K. Tewari) Compact rings and the natural topology. Bulletin of the Calcutta Mathematical Soc. 71 (1979), no. 2, 67-76.

Vicknair, J. Paul. [1985] On valuation rings. Rocky Mountain Journal of Math. 17 (1987), 55-58.

Vidal, Robert. [1971] Sur les complétés d'anneaux pour la filtration du radical de Jacobson. Comptes Rendus Hebdomadaires des Séances de l'Acad. des Sci. Sér. A-B 272 (1971), A1638-A1641. [1975] Modules de type quasi fini sur un anneau non commutatif. Séminaire P. Dubreil, F. Aribaud et M.-P. Malliavin (28e année; 1974/75), Algèbre, Exp. No. 8. Secrétariat Mathématique, Paris, 1975. [1976a] Anneau principal non nécessairement commutatif. Comptes Rendus Hebdomadaires des Séances de l'Acad. des Sci. Sér. A-B 282 (1976), A87-A89. [1965b] Anneaux principaux, non nécessairement commutatif, séparés et complets pour la topologie de Krull. Ibid. 283 (1976), A143-A145. [1976c] Anneaux de valuation discrète, complets, non nécessairement commutatifs. Ibid. 283 (1976), A281-A284. [1976d] Représentations d'anneaux principaux, non nécessairement commutatifs, séparés et complets pour la topologie de Krull. Deuxième Colloque d'Algèbre Commutative (Rennes, 1976), Exp. No. 8. Univ. Rennes, Rennes, 1976. [1977a] Contre-exemple non-commutatif dans la théorie des anneaux de Cohen. Comptes Rendus Hebdomadaires des Séances de l'Acad. des Sci. Sér. A-B 284 (1977), A791-A794. [1977b] Un exemple d'anneau de valuation discrète complet, non commutatif, qui n'est pas an anneau de Cohen. Ibid. 284 (1977), A1489-A1492. [1980] Anneaux de valuation discrète complets non commutatifs. Transactions of the American Mathematical Soc. 267 (1981), 65-81.

Vilenkin, Naum Iakovlevich. [1946a] On direct decompositions of topological groups. Matematicheskiĭ Sbornik 19 (61) (1946), 85-154. [1946b] On direct decompositions of topological groups. II. Ibid. 19 (61) (1946), 311-340. [1946c] On the theory of weakly separable groups. Ibid. 22 (64) (1948), 135-177.

Vizitei [Vizitiu], Viktor Nikolaevich. [1973] (with V. Arnautov) Extending a locally bounded field topology to an algebraic extension of the field. Doklady Akademiĭ Nauk SSSR 216 (1974), 477-480. English translation, including corrections: Soviet Mathematics Doklady 15 (1974), 808-812. [1974] (with V. Arnautov) Continuation of a locally bounded topology of a field to its algebraic extension. Buletinul Akademiei de Ştiinţe a R.R.S. Moldoveneşti, 1974, no. 2, 29-43, 94. [1975] Extension of m-adic topologies of a field. Matematicheskie Issled. 10 (1975), no. 2(36), 64-78, 281-282. [1976] Topological rings with an m-unrefinable topology. Ibid. No. 38 (1976), 97-115, 206. [1978] Extensions of m-topologies of commutative rings. Ibid. No. 48 (1978), 24-39, 168. [1979] Uniqueness of the continuation of m-topologies of a field onto its algebraic extensions. Ibid. No. 53 (1979), 27-42, 222. [1982] Continuation of some topological rings onto their extensions. Ibid. No. 56 (1982), 17-26, 153. [1983] Continuous extension of differentiations of rings. Ibid. No. 74 (1983), 3-17.

Vodinchar, Mikhail Ivanovich. [1967] A criterion for pseudonormability of topological algebras. Matematicheskie Issled. 2 (1967) No. 1, 133-140. [1968] The minimal hypernilpotent radical in topological rings. Ibid. 3 (1968), No. 4(10), 29-50 (1969). [1968] (with V. Arnautov) Radicals of topological rings. Ibid. 3 (1968) No. 2(8), 31-61. [1969] Hereditary and special radicals in topological rings. Ibid. 4 (1969), No. 2(12), 17-31. [1975] (with S. Alekseĭ and V. Arnautov) The boundedness of free topological modules. Ibid. 10 (1975), No. 2(36), 16-27, 280. [1975] (with V. Arnautov) The local compactness of a free topological module. Ibid. 10 (1975), No. 3(37), 3-14, 238. [1979] Algebraic radicals and radical-semisimple classes of topological modules. Ibid. No. 53 (1070), 43-53, 222. [1081] (with V. Arnautov and A. Mikhalëv) *Introduction to the Theory of Topological Rings and Modules.* Kishinev, 1981. [1988] (with V. Arnautov, S. Glavatskiĭ, and S. Mikhalëv) *Constructions of Topological Rings and Modules.* "Shtiintsa", Kishinev, 1988.

Vogel, Annie. [1983] Completion and injective hull of a von Neumann regular ring. Communications in Algebra 12 (1984), 1787-1794.

Warner, Seth L. [1955] Polynomial completeness in locally multiplicatively-convex algebras. Duke Mathematical Journal 23 (1956), 1-11. [1956] Weakly topologized algebras. Proceedings of the American Mathematical Soc. 8 (1957), 314-316. [1958] Characters of cartesian products of algebras. Canadian Journal of Math. 11 (1959), 70-79. [1960a] Compact noetherian rings. Mathematische Annalen 141 (1960), 161-170. [1960b] Compact rings and Stone-Čech compactifications. Archiv der Mathematik 11 (1960), 327-332. [1961] Compact rings. Mathematische Annalen 145 (1961/62), 52-63. [1965] Locally compact simple rings having minimal left ideals. Transactions of the American Mathematical Soc. 125 (1966), 395-405. [1966a] Compactly generated algebras over discrete fields. Bulletin of the American Mathematical Soc. 73 (1967), 227-230. [1966b] Locally compact vector spaces and algebras over discrete fields. Transactions of the American Mathematical Soc. 130 (1968), 463-493. [1967a] Locally compact equicharacteristic semilocal rings. Duke Mathematical Journal 35 (1968),

179-190. [1967b] Locally compact semilocal rings. Bulletin of the American Mathematical Soc. 73 (1967), 906-908. [1967c] Linearly compact noetherian rings. Mathematische Annalen 178 (1968), 53-61. [1967d] Compact and finite-dimensional locally compact vector spaces. Illinois Journal of Math. 13 (1969), 383-393. [1968a] Two types of locally compact rings. Bulletin of the American Mathematical Soc. 74 (1968), 926-930. [1968b] Locally compact rings having a topologically nilpotent unit. Transactions of the American Mathematical Soc. 139 (1969), 145-154. [1969] Locally compact commutative artinian rings. Illinois Journal of Math. 16 (1972), 102-115. [1970a] Metrizability of locally compact vector spaces. Proceedings of the American Mathematical Soc. 27 (1971), 511-513. [1970b] Locally compact principal ideal domains. Mathematische Annalen 188 (1970), 317-334. [1970c] Sheltered modules and rings. Proceedings of the American Mathematical Soc. 30 (1971), 8-14. [1971a] Openly embedding local noetherian domains. Journal für die Reine und Angewandte Mathematik 253 (1972), 146-151. [1971b] Linearly compact rings and modules. Mathematische Annalen 197 (1972), 29-43. [1971c] Normability of certain topological rings. Proceedings of the American Mathematical Soc. 33 (1972), 423-427. [1972] A topological characterization of complete, discretely valued fields. Pacific Journal of Math. 48 (1973), 293-298. [1972] (with S. Ng) Continuity of positive and multiplicative functionals. Duke Mathematical Journal 39 (1972), 281-284. [1972] (with J. Lucke) Structure theorems for certain topological rings. Transactions of the American Mathematical Soc. 186 (1973), 65-90. [1976] Topological rings that generalize Artinian rings. Journal für die Reine und Angewandte Mathematik 293/294 (1977), 99-108. [1977] (with T. Rigo) Topologies extending valuations. Canadian Journal of Math. 30 (1978), 164-169. [1978] Generalizations of a theorem of Mutylin. Proceedings of the American Mathematical Soc. 78 (1980), 327-330. [1989] *Topological Fields*. North-Holland Mathematics Studies 157, Notas de Matémactics (126). North-Holland, Amsterdam-New York-Oxford-Tokyo, 1989.

Weber, Hans. [1976] Ringtopologien auf $\mathbb{Z}$ und $\mathbb{Q}$. Mathematische Zeitschrift 155 (1977), 287-298. [1981] Unabhängige Topologien, Zerlegung von Ringtopologien. *Ibid.* 180 (1982), 379-393. [1982] Bestimmung aller Integritätsbereiche, deren einzige lokalbeschränkte Ringtopologien Idealtopologien sind. Archiv der Mathematik 41 (1983), 328-336.

Wehrfritz, Bertram A. F. [1979] *Lectures around Complete Local Rings*. Queen Mary College Mathematical Notes. Queen Mary College, Department of Pure Mathematics, London, 1979.

Wei, Jing Dong. [1990] Density of primitive rings relative to a Gabriel topology. Acta Mathematica Sinica 33 (1990), 456-461.

Weiss, Edwin. [1954] Boundedness in topological rings. Pacific Journal of Math. 6 (1956), 149-158.

Whaples, George. [1942] (with E. Artin) The theory of simple rings. American Journal of Math. 65 (1943), 87-107.

Widiger, Alfred. [1971] Die Struktur einer Klasse linear kompakter Ringe. Rings, Modules and Radicals. Colloquia Mathematica Sociétatis János Bolyai 6, pp. 501-505. North-Holland, Amsterdam, 1973. [1972] Die Struktur einer Klasse linear kompakter Ringe. Beiträge zur Algebra und Geometrie 3 (1974), 139-159. [1974] Zur Zerlegung in engeren Sinne linear kompakter Ringe. Ibid. 4 (1975), 85-88. [1979] Zur Existenz eines (Rechts-, Links-) Einselementes in im engeren Sinne linear kompakten Ringen. Ibid. 12 (1982), 45-49. [1987] Linearly compact rings, Chapter XV of Lectures on Artinian Rings by Andor Kertész, Akadémiai Kiadó, Budapest, 1987, pp. 365-393.

Wiegandt, Richard. [1964] Über halbeinfache linearkompakte Ringe. Studia Scientiarum Mathematicarum Hungarica 1 (1966), 31-38. [1965a] Über linear kompakte reguläre Ringe. Bulletin de l'Académie Polonaise des Sciences. Sér. des sciences math. astronom. phys. 13 (1965), 445-446. [1965b] Über transfinit nilpotente Ringe. Acta Mathematica Academiae Scientiarum Hungaricae 17 (1966), 101-114. [1966a] Über lokal linear kompakte Ringe. Acta Litterarum ac Scientiarum Regiae Universitatis Hungaricae Francisco-Jospheniae. Sectio Scientiarum Mathematicarum 28 (1967), 255-260. [1966b] Investigations in the theory of linear compact rings. I, II. A Magyar Tudományos Akadémia Mathematikai és Fizikai Tudomanyok Ostályának Közleményei 16 (1966) 239-267, 333-363. [1971] Radicals coinciding with the Jacobson radical in linearly compact rings. Wissenschaftliche Beiträge der Martin-Luther-Universität. Halle-Wittenberg M 3 (1971), 195-199. [1974] On linearly compact primitive and semisimple rings. Appendix in: Rings and Semigroups by Mario Petrich, Lecture Notes in Mathematics, vol. 38, pp. 152-166. Springer-Verlag, Berlin-New York, 1974. [1977] (with Phạm Ngọc Ánh) Linearly compact semisimple rings and regular modules. Mathematica Japonica 23 (1978/79), 335-338. [1986] (with L. Márki and R. Mlitz) A note on radical and semisimple classes of topological rings. Acta Scientiarum Mathematicarum 51 (1987), 145-151.

Więsław, Witold. [1973] A remark on complete and connected rings. Bulletin de l'Académie Polonaise des Sciences. Série des sciences math. astronom. phys. 22 (1974), 15-17. [1978a] Locally bounded topologies on fields and rings. Mitteilungen der Mathematischen Gesellschaft in Hamburg. Band X, Heft 6 (1978), 481-508. [1978b] Locally bounded topologies on Euclidean rings. Archiv der Mathematik 31 (1978/79), 33-37. [1979] Locally

bounded topologies on some Dedekind rings. *Ibid.* 33 (1979/80), 41-44.
[1980] (with D. Dikranjan) Rings with only ideal topologies. Commentarii Mathematici Universitatis Sancti Pauli 29 (1980), 157-167. [1982] *Topological Fields.* Acta Universitatis Wratislaviensis No. 675, Wydawnictwa Uniwersytetu Wrocławskiego Seria Matematyka, Fizyka, Astronomia XLIII. Wrocław, 1982. [1988] *Topological Fields.* Monographs and Textbooks in Pure and Applied Mathematics, vol. 119. Dekker, New York and Basel, 1988.

Wiseman, Andrew N. [1982] Integral extensions of linearly compact domains. Communications in Algebra 11 (1983), 1099-1121.

Witkowski, Lech. [1975] On linearly topological modules. Demonstratio Mathematica 10 (1977), 317-327. [1976] On coalgebras and linearly topological rings. Colloquium Mathematicum 40 (1978/79), 207-218.

Witt, Ernst. [1936] Zyklische Körper und Algebren der Charakteristik p vom Grad p^n. Struktur diskret bewerteter perfekter Körper mit vollkommenem Restklassenkörper der Charakteristik p. Journal für die Reine und Angewandte Mathematik 176 (1936), 126-140.

Wolfson, Kenneth Graham. [1953] Some remarks on ν-transitive rings and linear compactness. Proceedings of the American Mathematical Soc. 5 (1954), 617-619. [1954] Annihilator rings. Journal of the London Mathematical Soc. 31 (1956), 84-104.

Woodcock, C. F. [1972] A note on duality over complete local rings. Journal of the London Mathematical Soc. (2) 6 (1973), 614-616.

Wu, Ling-erl Eileen Ting. [1966] Bicontinuous isomorphisms between two closed left ideals of a compact dual ring. Canadian Journal of Math. 18 (1966), 1148-1151.

Xu, Bang Teng. [1991] Stability of linear topologies over Noetherian rings. Journal of Mathematics (Wuhan) 11 (1991), 23-28.

Xue, Wei Min. [1991] Linearly compact modules over perfect rings. Advances in Mathematics (China) 20 (1991), 75-76. [1992] *Rings with a Morita Duality.* Lecture Notes in Mathematics 1523. Springer-Verlag, Berlin, 1992.

Yao, Mu Sheng. [1988] Linearly compact modules over perfect rings. K'o Hsueh T'ung Pao. Kexue Tongbao 33 (1988), 1048-1049.

Yen, Ti. [1955] Notes on linearly compact algebras. Proceedings of the American Mathematical Soc. 8 (1957), 698-701.

Yood, Bertram. [1962] Ideals in topological rings. Canadian Journal of Math. 16 (1964), 28-45. [1970] Closed prime ideals in topological rings. Proceedings of the London Mathematical Soc. (3) 24 (1972), 307-323. [1971] Incomplete normed algebras. Bulletin of the American Mathematical Soc. 78 (1971), 50-52.

Yoshida, Michio. [1953] (with M. Sakuma) A note on semi-local rings. Hiroshima Daigaku. Journal of Science of the Hiroshima University. Ser. A 17 (1953), 181-184.

Zariski, Oscar. [1945] Generalized semi-local rings. Summa Brasiliensis Mathematicae 1 (1946), 169-195. [1958] (with P. Samuel) *Commutative Algebra*, Vol. I. Van Nostrand, Princeton, N. J., 1958. [1960] (with P. Samuel) *Commutative Algebra*, Vol. II. Van Nostrand, Princeton, N. J., 1960.

Zelenyuk, E. G. [1991] (with I. V. Protasov and O. M. Khromulyak) Topologies on countable groups and rings. Doklady Akademiĭa Nauk Ukraïns'koĭ RSR 1991, no. 8, 8-11, 1992.

Zelinsky, Daniel. [1949] Rings with ideal nuclei. Duke Mathematical Journal 18 (1951), 431-442. [1951] Complete fields from local rings. Proceedings of the National Academy of Science U.S.A. 37 (1951), 379-381. [1952] Linearly compact modules and rings. American Journal of Mathematics 75 (1953), 79-90. [1953] Raising idempotents. Duke Mathematical Journal 21 (1954), 315-322.

Zelmanowitz, Julius Martin. [1978] Semisimple rings of quotients. Bulletin of the Australian Mathematical Soc. 19 (1978), 97-115. [1983] Duality theory for quasi-injective modules. NATO Advanced Study Institute on methods in ring theory (1983: Antwerp, Belgium). *Methods in Ring Theory*, pp. 551-566, Reidel, Dordrecht-Boston, Mass., 1984. [1984] *Duality Theory for Quasi-injective Modules*. Algebra Berichte 46. Verlag Reinhard Fischer, München, 1984. [1984] (with W. Jansen) Duality modules and Morita duality. Journal of Algebra 125 (1989), 257-277. [1988] (with W. Jansen) *Duality for Module Categories*, Algebra Berichte 59, Verlag Reinhard Fischer, München, 1988.

Zobel, Robert. [1972] Direkte Gruppen- und Ringtopologien. Dissertation, Technische Universität Carolo-Wilhelmina zu Braunschweig, 1972.

Zöschinger, Helmut. [1982] Linear-kompatke Moduln über noetherschen Ringen. Archiv der Mathematik 41 (1983), 121-130.

ERRATA

In addition to occasional typographical errors, which readers will readily recognize, four significant errors in *Topological Fields* have come to light since its publication in 1989.

First, the proof of (3) of Theorem 24.13 on pages 229-230 of *Topological Fields* is incorrect and should be replaced by the proof of Theorem 15.14 on pages 118-119 of this book.

Second, Ulrich Heckmanns has persuaded me that the proof of Theorem 31.10 is incorrect. It should be replaced by the proof of Theorem 28.7 on page 234 of this book.

Professor Sibylla Priess-Crampe of the University of Munich has kindly offered a correction of the statement "No examples of straight or even minimally topologized division rings are currently known other than locally retrobounded division rings" on page 225, lines 2-3. In the statement, both occurrences of "division rings" should be replaced by "fields". Indeed, let V be a valuation ring of a division ring K (that is, a subring of K such that for all $x \in K^*$, either $x \in K$ or $x^{-1} \in K$) such that $\{Va : A \in V^*\}$ is not a fundamental system of neighborhoods of zero for a ring topology. Schröder [1987] and Hartmann [1987] independently showed that $\{aVb : a, b \in V^*\}$ (or equivalently, $\{cVc : c \in V^*\}$, since $(ab)V(ab) \subseteq aVb$) is a fundamental system of neighborhoods of zero for a Hausdorff division ring topology on K, and Hartmann showed that it was a minimal Hausdorff ring topology on K. Mathiak [1989] and Liepold [1990] independently gave examples showing that the completion of such a topological division ring could have exactly two maximal ideals and thus not be a division ring. Such examples are therefore neither locally retrobounded by Theorem 13.9 nor straight by Theorem 13.4. They also show that the statement of Kowalsky's theorem, the completion of a field for a minimal Hausdorff ring topology is a field (Exercise 24.13, page 235 of *Topological Fields*), no longer holds if "field" is replaced by "division ring".

Lastly, in the final preparation of the text of *Topological Fields*, the following entries intended for the bibliography were lost:

Cohn, Paul Moritz. [1953] An invariant characterization of pseudo-valuations on a field. Cambridge Philos. Soc. Proc. 50 (1954), 159-177. [1953] (with K. Mahler) On the composition of pseudo-valuations. Nieuw Arch. Wisk. (3) 1 (1953), 161-198. [1980] (with M. Mahdavi-Hezavehi) Extension of valuations on skew fields. *Ring Theory Antwerp 1980*, pp. 28-41. Lecture Notes in Math. 825, Springer, Berline-Heidelberg New York, 1980. [1981] On extending valuations in division algebras. Magyar Tudomanyos Akademia. Studia Sci. Math. Hungar. 16 (1981), 65-70.

Freudenthal, Hans [1935] Einige Sätze über topologische Gruppen. Ann. of Math. (2) 37 (1936), 46-56.

Lorenzen, Paul [1953] Über die Komplettierung in der Bewertungstheorie. Math. Z. 59 (1953), 84-87.

Index of Names

This index of names does not include occurrences of names in the Bibliography, nor occurrences of names in such expressions as "Hausdorff space".

Abrams, G., 437
Andrunakevich, V., 68, 434
Ánh, Phạm Ngọc, see Phạm Ngọc Ánh
Anthony, J., 438
Anzai, H., 429
Aragona, J., 437
Arens, R., 430
Arnautov, V., 68, 111, 250, 434, 438–39
Artin, E., 150, 428, 435
Aurora, S., 109

Baccella, G., 437
Baer, R., 101, 132, 140–41
Baire, R., 430
Bazzoni, S., 437
Bendixson, I., 132
Blair, R., 314
Braconnier, J., 429

Cantor, G., 132
Carini, L., 438
Cartan, H., 428
Chevalley, C., 205, 426–27, 434
Cohen, I. S., 166, 186, 391, 426–28
Correl, E., 32, 226, 438
Cude, J., 315

Dantzig, D. van, 428
Davison, T., 438
Dieudonné, J., 133, 431
Dikranjan, D., 431, 437
Dinh Van Huynh, 316

Eckstein, F., 438
Ellis, R., 430
Endo, M., 438

Facchini, A., 438
Faith, C., 430
Fleischer, I., 435
Freudenthal, H., 54
Frobenius, F., 123
Fuchs, L., 358

Geddes, A., 426
Gelbaum, B., 438
Gel'fand, I., 88, 131, 226
Godement, R., 426
Goldman, O., 422, 435
Gomez Pardo, J., 438
Gould, G., 86
Gregorio, E., 437
Grothendieck, A., 428

Hagglund, L., 439
Halter-Koch, F., 438
Hartmann, P., 487
Hasse, H., 101, 132, 424–26
Heckmanns, U., 439, 487
Heine, J., 439
Hensel, K., 428
Herstein, I., 428
Hinohara, Y., 434
Hinrichs, L., 438–39
Hochster, M., 293–94, 439
Hopkins, C., 358, 434
Huchinson, J., 437

Jacobson, N., 314, 404, 429–30, 435
Janiszewski, Z., 132
Jans, J., 433
Jansen, W., 437
Jebli, A., 438
Johnson, R., 438

Kalish, G., 438
Kampen, E. van, 317
Kaplansky, I., 231, 250, 281, 316,
    327, 330, 427–431, 433, 435
Kertész, A., 316
Kiltinen, J., 438–39
Koh, K., 438
Kovács, I., 358
Kowalsky, H.-J., 111, 487
Krull, W., 426–8, 431, 435
Kurke, H., 294–95, 434

Lafon, J.-P., 205
Lefschetz, S., 240, 431, 435
Leptin, H., 431–33, 435, 437
Liepold, F., 487
Lipkina, Z., 315
Litoff, 430
Livenson, E., 133
Lorenzini, A., 437
Lucke, J., 313–14, 435
Luedeman, J., 438

Macdonald, I., 436
MacLane, S., 424–25
Mader, A., 437
Márki, L., 437
Mathiak, K., 487
Matlis, E., 435
Mazan, M., 437
Mazur, S., 88, 131, 430
Menini, C., 437
Mori, Y., 428

Müller, B., 436–37
Mutylin, A., 438–39

Nagata, M., 197, 397, 426, 428, 438
Nakayama, T., 438
Narita, M., 426
Nazarov, A., 438
Neumann, J. von, 428
Ng, S., 54
Numakura, K., 433

Øfsti, A., 315
Olmsted, J., 438
Onodera, T., 437
Orlicz, W., 430
Orsatti, A., 431, 437
Osofsky, B., 437
Ostrowski, A., 88, 129, 131, 165, 425, 428
Otobe, Y., 429–30

Peters, F., 437
Phạm Ngọc Ánh, 251, 294, 405, 430, 433–34, 437
Pietrkowski, S., 431
Podewski, K.-P., 439
Pontriãgin, L., 88, 124, 317
Prieß-Crampe, S., 487
Prüfer, H., 292, 428, 431

Rees, D., 425, 428
Rickart, C., 430
Roselli, V., 437
Ryan, M., 433

Sah, C., 422, 435
Sakuma, M., 427
Samuel, P., 426–27
Sandomierski, F., 437

Schiffels, G., 438
Schmidt, F., 424-26
Schöneborn, H., 435
Schröder, M., 487
Skorniakov, L., 328-30, 430
Stenström, B., 438
Stenzel, M., 438
Stöhr, K., 437
Stoyanov, L., 437
Szele, T., 358

Tate, J., 428
Taussky, O., 429
Teichmüller, O., 425
Tuzuku, T., 438

Ursul, M., 250, 433
Utumi, Y., 430

Vámos, P., 437
Vilenkin, N., 435
Vizitei, V., 438

Warner, S., 54, 86-7, 119, 250, 294,
    357, 405, 430, 433-35, 438
Wehrfritz, B., 425
Whaples, G., 150, 435
Widiger, A., 315-16, 433- 34
Wiegandt, R., 251, 433
Wiseman, A., 437
Witt, E., 425
Woodcock, C., 437

Yood, B., 226
Yoshida, M., 427

Zariski, O., 205, 240, 427-28
Zelinsky, D., 431-32, 434, 437
Zelmanowitz, J., 437
Zobel, R., 439

Index of Symbols and Definitions

A^*, 1
$A^\times$, 1
A^a, x^a, 82
A_P, 363
$A[[X_1, X_2, \ldots, X_m]]$, 195
$|\cdot|_{p,c}$, 8
$|\cdot|_p$, 8, 146
$|\cdot|_\infty$, 9
Ann, 206, 210, 241, 292
$\mathbb{C}, \mathbb{H}, \mathbb{N}, \mathbb{Q}, \mathbb{R}, \mathbb{R}_{>0}, \mathbb{Z}$, 1
$\mathfrak{C}(\mathcal{T})$, 285
card(X),
$D(H)$, 346
diam, 69
dim(A), 371
$D(J)$, 217
$\mathrm{End}_K(E)$, 207
End(E), 206
$e(v'/v)$, 156
$f(v'/v)$, 156
$\widehat{G}$, 50, 56, 62
ht(P), 368, 371
IJ, 170
J-topology, 197
J_*, 289
$K((X)), K[[X]]$, 148
$[L:K]$, 154
$\varinjlim_{\lambda \in L} E_\lambda$, 39
$\varprojlim_{J \in \mathcal{J}}(A/J)$, 41
$N_{D/K}$, 127
(N 1)–(N 5), 2
$|n|$, 196
$n.x$, 11
ord, 147, 152
(PF), 39
$P(J)$, 217

$Q(A)$, 81
$Q_{top}(A)$, 376
$\mathbb{R}_\infty$, 134
rad(J), 364
$S[A, \mathbb{N}^m]$, 194
$S(K, G)$, 151
$S(K, \mathbb{Z})$, 147
$S^{-1}A$, 81
$\sigma_K(E, E')$, 119
$\bigoplus_{\lambda \in L} M_\lambda$, 112-13
$\sum_{\alpha \in A} x_\alpha$, 72
$\mathfrak{S}_{\lambda \in L} A_\lambda$, 334
$\mathrm{Tr}_{L/K}(a)$, 392
$\mathcal{T}^*$, 294
$\mathcal{T}_*$, 284
(TG 1)–(TG 2), 13
(TGB 1)–(TGB 2), 20
(TGN 1)–(TGN 2), 19, 22
(TGN 3), 22
(TM 1)–(TM 3), 11
(TM 4)–(TM 6), 17
(TMN 1)–(TMN 3), 22
(TR 1)–(TR 3), 1
(TR 4)–(TR 5), 17
(TRN 1)–(TRN 2), 21
vdim(A), 373
$\mathbb{Z}(p^\infty)$, 254

Absolute value, 5
Absolute value of a real valuation, 136
Absolute value to base c, 136
Adherence of a filter base, 55
Adherent point, 55
Adherent point (net), 71
Adverse, 82

Advertible (left, right), 82
Advertible (left, right) ideal, 222
Advertibly complete, 86
Advertibly open, 83
Algebra, 12
Algebra norm, 12
Algebra of formal power series, 147
Algebra topology, 12
Algebraic supplement, 114
Algebraically dense subring, 263
Almost integral element, 388
Annihilator, 206, 210, 241, 292, 332
Approximation theorem, 98
Archimedean absolute value, 7
Archimedean-ordered group, 140
Artin–Wedderburn Theorem, 229
Artinian algebra, 226
Artinian module, 226
Artinian ring, 226
Ascending Chain Condition, 167

Baire space, 69
Basic divisible p-primary group, 254
Bijection, 32
Bijective function, 32
Bilinear function, 16
Biregular ring, 327
Boolean ring, 314
Bounded (left, right) ring, 91
Bounded (left, right) subset of ring, 89
Bounded module, 91
Bounded subset of module, 88
Bounded (left, right) topology, 91

Canonical embedding in an inductive limit, 276
Canonical embedding of residue field, 156

Canonical epimorphism to a quotient group, 33
Canonical homomorphism to a projective limit, 41
Canonical surjection to a quotient space, 32
Cauchy filter (base), 55
Cauchy net, 71
Cauchy's Condition, 72
Central idempotent, 327
Centrally linearly compact ring, 398
Character of a locally compact group, 267
Circle composition, 82
Circulation, 82
Closed linear filter base, 232
Cluster point, 55
Codimension, 189
Cohen algebra, 408
Cohen ring, 174
Cohen subring (subfield), 174
Compact space, 3
Complete local ring, 174
Complete metrizable commutative group, 50
Complete norm, 2
Complete real valuation, 138
Complete semilocal ring, 202
Complete subset of a Hausdorff commutative group, 56
Complete topological module, 64
Complete topological ring, 64
Completely integrally closed integral domain, 388
Completion of a Hausdorff commutative group, 56
Completion of a metrizable commutative group, 50
Completion of a topological ring,

64
Condensation point, 132
Conditionally simple topological ring, 329
Constant formal power series, 192
Constant term of formal power series, 192
Converges (filter base), 55
Converges (net), 71
Core, 102

Dense ring of linear operators, 206
Density Theorem, 208
Descending Chain Condition, 230
Development, 144
Diameter, 69
Dimension of a local noetherian ring, 371
Direct sum of submodules, 112
Direct sum of subrings, 112–13
Directed set, 38
Direction, 38
Discrete central idempotent, 328
Discrete valuation, 141
Divisible group, ring, ideal, 251. 331
Division ring topology, 12

Eisenstein exension relative to a valuation, 189
Eisenstein extension, 386
Eisenstein polynomial, 186
Eisenstein polynomial relative to a valuation, 187
Epimorphism, 32
Equicharacteristic, 174
Equivalent absolute values, 6
Equivalent real valuation, 137
Extension Theorem, 126

Faithful module, 211
Family of projections, 112
Field topology, 12
Filter, 18
Filter base, 18
Filter base associated to a sequence, 55
Filter base generated by a net, 71
Finite length, 359
First Baire category subset, 69
Formal power series, 147, 192
Fundamental sequence of bounded sets, 93
Fundamental system of neighborhoods, 18

Gabriel topology, 241
Gel'fand ring, 226
Group topology, 13

Hausdorff group, 24
Hausdorff module, 24
Hausdorff ring, 24
Height of a prime ideal, 368
Height of an ideal, 371
Hyperplane, 115

Ideal supplement of torsion ideal, 333
Ideal topological supplement, 346
Ideal topology, 22
Improper absolute value, 6
Improper valuation, 137
Indecomposable ring (algebra), 408
Inductive limit, 276
Injection, 32
Injective function, 32
Integral closure, 379
Integral element, 377
Integral over, 379

Integrally closed integral domain, 387
Integrally closed subring, 379
Invariant metric, 48
Isomorphism, 32

Japanese integral domain, 391
Jordan-Hölder sequence, 359

Krull dimension, 371

Largest divisible ideal, 332
Left invariant metric, 48
Length of a decreasing sequence of submodules, 359
Length of a module, 360
Leptin topology, 284
Linear filter (base), 232
Linear form, 114
Linear operator, 112
Linear topology, 232
Linear transformation, 112
Linearly compact module, 232
Linearly compact ring, 234
Linearly compact topology, 232
Linearly topologized module, 232
Local direct sum, 262
Local direct sum topology, 262-3
Local domain, 172
Local ring, 172
Localization, 364
Locally (left, right) bounded ring, 91
Locally bounded module, 91
Locally bounded topology, 91
Locally centrally linearly compact ring, 398
Locally compact space, 3
Locally metacompact ring, 313

Locally retrobounded division ring (field), 95
Locally retrobounded topology, 95
Locally strictly linearly compact ring, 398
Locally without central idempotents, 328

Maximal ideal of valuation subring, 140
Maximal linear filter, 233
Maximal linearly compact topology, 294
Maximal regular left (right) ideal, 216
Meager subset, 69
Metacompact ring, 313
Minimal central idempotent, 327
Minimal Hausdorff linear topology, 235
Minimal left (right) ideal, 212
Minimal prime ideal, 362
Minimal prime ideal over an ideal, 362
Minimum Condition, 230
Module, 11
Module topology, 11
Monomorphism, 32
Multilinear function, 118
Multiplicative set, 81

n-fold transitive, 206
Natural topology of a semilocal ring, 202
Natural topology of local ring, 172
Neighborhood, 15
Net, 71
Nil (left, right) ideal, 220
Nil ideal, 177

Nil ideal of bounded index, 177
Nil ideal of index r, 177
Nilpotent ring of index n, 174
Nilpotent ring or ideal, 174
Noetherian module, 166
Noetherian ring, 166
Nonarchimedean absolute value, 7
Nondiscrete central idempotent, 328
Norm on a ring, 2
Norm on a vector space, 12
Norm on an algebra, 12
Norm on commutative group, 49
Norm-bounded subset, 102
Normable ring, 3
Normed algebra, 12
Normed ring, 3
Normed space, 12
Nowhere dense subset, 69

Open function, 33
Opposite ring, 213
Order, 147, 152
Order topology, 148, 152
Order valuation, 148
Orthogonal family of idempotents, 297
Outer direct sum of modules, 112

p-adic absolute value, 8, 146
p-adic development, 150
p-adic number field, 146
p-adic topology, 8
p-adic valuation, 8
Pairwise relative prime ideals, 200
Pathological subgroup, 338
Perfect subset, 132
Pointwise convergence, 241
Primary component, 252
Primary group, ring, ideal, 252, 331

Prime ideal, 169, 405
Primitive (left, right) ideal, 216
Primitive (left, right) ring, 211
Primitive ring of endomorphisms, 206
Principal Ideal Theorem, 367
Principal prime, 368
Projection, 114
Projective family of sets, 39
Projective limit, 39
Proper absolute value, 6
Proper subset, 25
Proper valuation, 137
Proper zero-divisor, 5

Quotient ring, 81
Quotient topology, 33

Radical ideal, 364
Radical of a ring, 219
Radical of an ideal, 364
Radical ring, 219
Radical topology, 330
Ramification index, 156
Ramified nonequicharacteristic local ring, 384
Rankfree, 111
Rare subset, 69
Real valuation, 134
Regular (left, right) ideal, 216
Regular local ring, 373
Regular maximal left (right) ideal, 216
Regular topological space, 21
Relatively compact subset, 93
Relatively prime ideals, 200
Representative set, 142
Residue class degree, 156
Residue division ring (field), 136

Residue field of local ring, 172
Retrobounded set, 95
Right invariant metric, 48
Ring, 1
Ring of p-adic numbers, 146
Ring of formal power series, 147, 194
Ring topology, 1
Ring with continuous adversion, 82
Ring with continuous inversion, 80
Ring with identity, 1

Semilocal ring, 202
Seminorm, 111
Semisimple module, 286
Semisimple ring, 219
Sequentially retrobounded, 101
Shelter of a module, 283
Shelter of an ideal, 292
Sheltered ideal, 292
Sheltered submodule, 283
Simple convergence, 241
Simple module, 211, 286
Simple ring (algebra), 328
Small set, 55
Spectral norm, 105
Squarefree ring, 414
Standard basis, 115
Stem field, 186
Straight division ring or field, 94
Straight topological vector space, 94
Strictly biregular subring, 327
Strictly linearly compact, 235
Strongly linearly compact module, 251
Strongly regular ring, 329
Sum, 71
Summable, 71

Supplement, 114
Support, 151
Surjection, 32
Surjective function, 32
Symmetric subset of a group, 14

Topological algebra, 12
Topological direct sum, 113
Topological division ring, 12
Topological epimorphism, 35
Topological field, 12
Topological group, 13
Topological homomorphism, 35
Topological isomorphism, 35
Topological module, 11
Topological nilpotent, 6
Topological monomorphism, 35
Topological p-primary component, 334
Topological p-primary group, ring, 334
Topological quotient ring, 376
Topological ring, 1
Topological supplement, 114
Topological torsion group, subgroup ring, ideal, 334
Topological vector space, 12
Topologically minimal central idempotent, 329
Topology defined by a norm, 49
Topology defined by a valuation, 137
Topology of pointwise convergence, 241
Topology of simple convergence, 241
Torsion group, ring, ideal, 252, 331
Torsionfree group, ring, ideal, 267, 331
Total quotient ring, 80

Total subspace, 119
Totally disconnected space, 25
Trace, 392
Transfinite index, 289
Transfinitely nilpotent ideal, 289
Trivial module, 11
Trivial ring, 1

Ultrametric, 48
Ultranorm on a ring, 49
Ultranorm on a commutative group, 49
Uniform topology, 334
Uniformizer, 141
Uniformly continuous function, 59
Unitary module, 11
Unramified nonequicharacteristic local ring, 384

Valuation ideal, 136
Valuation ring, 136
Valuation subring, 140
Value group, 136
Vector dimension of a local noetherian ring, 373
Vector topology, 11
von Neuman regular ring, 251

Zariski module topology, 240
Zariski ring topology, 205
Zero ring, 1

www.ingramcontent.com/pod-product-compliance
Ingram Content Group UK Ltd.
Pitfield, Milton Keynes, MK11 3LW, UK
UKHW020657050526
12271UKWH00003B/8